Lecture Notes on Clinical Chemistry

Lecture Notes on Clinical Chemistry

L. G. WHITBY
MD, PhD, FRCP, FRCPE, FRCPath
Professor of Clinical Chemistry
University of Edinburgh

I. W. PERCY-ROBB
MB, PhD, FRCPE, FRCPath
Professor in Pathological Biochemistry
University of Glasgow

Formerly Reader in Clinical Chemistry
University of Edinburgh

A. F. SMITH
MD, FRCPE, FRCPath
Senior Lecturer in Clinical Chemistry
University of Edinburgh

THIRD EDITION

BLACKWELL SCIENTIFIC PUBLICATIONS
OXFORD LONDON EDINBURGH
BOSTON PALO ALTO MELBOURNE

Editorial Offices:
Osney Mead, Oxford, OX2 0EL
8 John Street, London, WC1N 2ES
9 Forrest Road, Edinburgh, EH1 2QH
52 Beacon Street, Boston
Massachusetts 02108, USA
706 Cowper Street, Palo Alto
California 94301, USA
99 Barry Street, Carlton
Victoria 3053, Australia

First published 1975
Revised Reprint 1977
Reprinted 1978
Second Edition 1980
Third Edition 1984

Typeset by Scottish Studios & Engravers Ltd., Glasgow
Printed and bound by Adlard & Son Ltd., Dorking, Surrey.

DISTRIBUTORS

USA
Blackwell Mosby Book Distributors
11830 Westline Industrial Drive
St. Louis, Missouri, 63141

Canada
Blackwell Mosby Book Distributors
120 Melford Drive, Scarborough
Ontario, M1B 2X4

Australia
Blackwell Scientific Book Distributors
31 Advantage Road, Highett
Victoria 3190

British Library
Cataloguing in Publication Data

Whitby, L. G.
Lecture notes on clinical chemistry.
3rd ed.
1. Chemistry, Clinical
I. Title II. Percy-Robb, I.W.
III. Smith, A.F. (Alistair Fairley)
616.07'56 RB40

ISBN 0-632-01232-3

Contents

	Preface to Third Edition	vii
	Abbreviations	viii
	Introduction	xiii
1	Requesting Laboratory Investigations	1
2	Reports of Laboratory Work	15
3	Chemical Tests Performed Outside the Laboratory	28
4	Water, Sodium and Potassium Disturbances	39
5	Acid–Base Balance and Oxygen Transport	64
6	Renal Disease	83
7	Plasma Protein Abnormalities	111
8	Enzyme Tests in Diagnosis	138
9	Liver Disease	169
10	Gastrointestinal Tract Disease	192
11	Disorders of Carbohydrate Metabolism	213
12	Disorders of Plasma Lipids	242
13	Deficiencies of Vitamins and Trace Elements	262
14	Disorders of Calcium and Magnesium Metabolism	277
15	Iron, Porphyrin and Haemoglobin Metabolism	305

16 Disorders of Purine Metabolism 327

17 Abnormalities of Thyroid Function 336

18 Steroid Hormones 354

19 Hypothalamic and Pituitary Hormones 387

20 Clinical Chemistry in Obstetrics and Gynaecology 397

21 Inherited Metabolic Disorders 418

22 Clinical Chemistry in Paediatrics 431

23 Clinical Chemistry in Geriatrics 443

24 Neoplastic Disease 449

25 The CNS and CSF 465

26 Therapeutic Drug Monitoring and Chemical Toxicology 472

27 Computers in Clinical Chemistry 490

28 Screening for Disease Using Chemical Tests 500

Appendix 1 Units Used in Reporting 517

Appendix 2 Acid–Base Terminology 522

Index 527

Preface to Third Edition

This book has been written primarily for medical students in the clinical years of their course, and for doctors in clinical practice seeking guidance in the selection and interpretation of chemical investigations. However, it should appeal also to clinical biochemists and chemical pathologists preparing for postgraduate professional qualifications and to medical laboratory scientific officers and technologists. Indeed, we believe the book has something to offer to all interested in the contributions of clinical chemistry to the detection, diagnosis and management of illness.

We assume that readers have access to textbooks of biochemistry, physiology and clinical medicine, but some overlap with these has been included to ensure necessary background and continuity.

The objectives of this book are stated in the Introduction, and we shall be very glad to receive suggestions from readers or criticisms that might help us attain these objectives more nearly in the future. Drawing on our experience of the use made of laboratory facilities, and on published statements that describe the experience of others, it would seem that these objectives have still not been attained by teachers of clinical chemistry or chemical pathology.

In preparing this edition, every chapter has been carefully reviewed and several have been extensively revised to take account of recent advances, to eliminate obsolete material, and to improve their presentation. The chapter numbering remains unchanged from the second edition, so as to enable this edition to be used in association with our related book of multiple choice questions.

Miss Doreen Fisher has again borne the brunt of the work involved in typing the various drafts and we thank her most sincerely; she has been ably assisted by Mrs. Cathy Veitch. We also wish to thank Mr. Nigel Palmer for his continued interest and encouragement.

1984

L.G. Whitby
I.W. Percy-Robb
A.F. Smith

Abbreviations

Symbols and abbreviations used to a very limited extent, or having a special meaning in a particular context, are defined in the relevant part of the text and have been omitted from this list.

Concentrations. Square brackets, e.g. [X], are used to indicate 'the concentration of substance X'.

Units of measurement. Abbreviations, and prefixes to indicate multiples and submultiples of these units, are described in Appendix 1 (p. 517).

ACTH	adrenocorticotrophic hormone
ADH	antidiuretic hormone (vasopressin)
ADP	adenosine 5′-pyrophosphate
AFP	α-fetoprotein
ALA	5-aminolaevulinic acid
ALT	alanine aminotransferase (alanine transaminase)
AMP	adenosine 5′-phosphate
Cyclic AMP	adenosine 3′, 5′-phosphate
Apo A, etc.	apolipoprotein A, etc.
APRT	adenine phosphoribosyl transferase
AST	aspartate aminotransferase (aspartate transaminase)
ATP	adenosine 5′-triphosphate
BT PABA	benzoyl-tyrosyl-*p*-aminobenzoic acid
°C	degrees Celsius (or Centigrade)
CAH	congenital adrenal hyperplasia
Cal	kilocalorie (1000 calories)
CBG	cortisol-binding globulin (transcortin)
CCK	cholecystokinin
CCK-PZ	cholecystokinin-pancreozymin
CEA	carcinoembryonic antigen
CF	cystic fibrosis
CHE	cholinesterase
CK	creatine kinase

CNS	central nervous system
CoA	coenzyme A
C-peptide	connecting peptide (in pro-insulin)
CRF	corticotrophin-releasing factor
CSF	cerebrospinal fluid
Da	dalton
DDAVP	desaminocys-1-8-D-arginine vasopressin
DHA	dehydro*epi*androsterone
DHCC	dihydroxycholecalciferol (e.g. 1:25–DHCC)
DHT	dihydrotestosterone
DNA	deoxyribonucleic acid
DPG	diphosphoglycerate
EC	Enzyme Commission
ECF	extracellular fluid
ECG	electrocardiogram
EDTA	ethylenediaminetetra-acetate
FFA	free fatty acids
FSH	follicle-stimulating hormone
GC-MS	gas chromatography-mass spectrometry
GFR	glomerular filtration rate
GGT	gamma-glutamyl transferase
GH	growth hormone
GH-RF	growth hormone-releasing factor
GIP	glucose-dependent insulinotrophic peptide
GMP	guanosine 5′-phosphate
Cyclic GMP	guanosine 3′, 5′-phosphate
Gn-RH	gonadotrophin-releasing hormone
GOT	see AST
GPT	see ALT
G6PD	glucose-6-phosphate dehydrogenase
h	hour
Hb	haemoglobin
Hb-F	fetal haemoglobin
HbO_2	oxyhaemoglobin
Hb-S	sickle-cell haemoglobin
HCC	hydroxycholecalciferol (e.g. 25-HCC)
HCG	(human) chorionic gonadotrophin
HDL	high density lipoprotein

HGPRT	hypoxanthine-guanine phosphoribosyl transferase
5-HIAA	5-hydroxyindoleacetic acid
HLA	human leucocyte antigens
HMMA	4-hydroxy, 3-methoxy mandelic acid
HPA	hypothalamic-pituitary-adrenal (axis)
HPG	hypothalamic-pituitary-gonadal (axis)
HPL	(human) placental lactogen
HPLC	high-performance liquid chromatography
5-HT	5-hydroxytryptamine
5-HTP	5-hydroxytryptophan
ICF	intracellular fluid
IDL	intermediate density lipoprotein
IgA, etc.	immunoglobulin, class A (similarly IgG, IgM, etc.)
IMP	inosine 5′-phosphate
ISE	ion-selective electrode
I.U.	International Unit (also i.u.)
L	litre
LCAT	lecithin cholesterol acyl transferase
LD	lactate dehydrogenase (isoenzymes, LD_1 etc.)
LDL	low density lipoprotein
LE	lupus erythematosus
LH	luteinizing hormone
LPH	lipotrophin
M	mean
MCV	mean cell volume
min	minute
mmHg	millimetres of mercury
mol. mass	molecular mass
MEN	multiple endocrine neoplasia
MSAFP	maternal serum α-fetoprotein
MSH	melanocyte-stimulating hormone
NAD	nicotinamide-adenine dinucleotide
NADH	nicotinamide-adenine dinucleotide (reduced form)
NADP	nicotinamide-adenine dinucleotide phosphate
NIH	National Institutes of Health
OGTT	oral glucose tolerance test
17-OHCS	17-hydroxycorticosteroids
PABA	*p*-aminobenzoic acid
PBG	porphobilinogen

P_{CO_2}	Appendix 2 (p. 522)
PCV	packed cell volume
pH	Appendix 2 (p. 522)
PKU	phenylketonuria
P_{O_2}	Appendix 2 (p.522)
PRA	plasma renin activity
PRIH	prolactin release-inhibiting hormone
PRPP	5-phosphoribosyl-1-pyrophosphate
$PS\beta_1G$	pregnancy-specific β_1-glycoprotein
PTH	parathyroid hormone
RIA	radioimmunoassay
SD	standard deviation
SHBG	sex hormone-binding globulin
SI	Système International
sp. gr.	specific gravity
T3	tri-iodothyronine
T4	thyroxine
TBG	thyroxine-binding globulin
TBP	thyroxine-binding proteins
TBPA	thyroxine-binding pre-albumin
TDM	therapeutic drug monitoring
TIBC	total iron-binding capacity
TRH	thyrotrophin-releasing hormone
TSH	thyroid-stimulating hormone
U (mU etc.)	unit (milliunit, etc.)
UDP	uridine 5′-pyrophosphate
U-TBP	unsaturated component of TBP
UV	ultraviolet (light)
VIP	vasoactive intestinal peptide
VLDL	very low density lipoprotein
VMA	see HMMA
WHO	World Health Organization

Atomic symbols. The following are used in this book:

Br	Bromine	Hg	Mercury	O	Oxygen
C	Carbon	I	Iodine	P	Phosphorus
Ca	Calcium	K	Potassium	S	Sulphur
Cl	Chlorine	Mg	Magnesium	Se	Selenium
Co	Cobalt	Mn	Manganese	Si	Silicon
Cr	Chromium	Mo	Molybdenum	Sn	Tin
Cu	Copper	N	Nitrogen	Tc	Technetium
F	Fluorine	Na	Sodium	V	Vanadium
Fe	Iron	Ni	Nickel	Zn	Zinc
H	Hydrogen				

Introduction

Clinical chemistry brings together knowledge and skills gained from the scientific disciplines of biochemistry, chemistry, physiology and related subjects, and applies them to the problems of detection, diagnosis, treatment and prevention of disease in man. Teaching of clinical chemistry has to fulfil two sets of functions:

(1) *To provide general information* about the biochemical basis of disease, and about the principles of laboratory diagnosis for (A) conditions with a biochemical pathogenesis and (B) disease states in which biochemical derangements occur as secondary manifestations.

(2) *To supply specific guidance* on the clinical value of chemical investigations, (A) indicating their range of applications and limitations as well as (B) relating results of laboratory tests to the processes of clinical diagnosis and management as these might apply in the case of individual patients.

This book places greater emphasis on the first set of functions. It would be impracticable in the space available to present detailed individual case histories and to use these to illustrate points of principle. However, the clinical value of chemical investigations is discussed, if only in general terms, and the book can be extended in its application by readers, if they relate our necessarily general treatment to their specific questions or problems affecting individual patients.

Our educational objectives can be classified in terms of knowledge, skills and attitudes. **Readers should acquire knowledge of:**

(1) the commonly available laboratory examinations that may be requested for the investigation and management of patients suffering from the commoner diseases that affect one or more of the various systems;

(2) the commoner, and especially the avoidable, reasons for sending unsatisfactory specimens to the clinical chemistry laboratory;

(3) the discomfort and danger to patients associated with the performance of certain investigations;

(4) the expensive and time-consuming nature of some investigations.

Medical students and doctors should develop the ability to:

(1) perform chemical tests that are suitable for side-room use;
(2) use deductive reasoning in selecting laboratory investigations appropriate to the diagnosis and management of individual patients;
(3) obtain appropriate specimens for the satisfactory performance by the laboratory of various tests requested;
(4) interpret the results of chemical investigations in terms of their understanding of each patient's illness, and then act appropriately;
(5) interpret clinical features in relation to the underlying pathological process and, where possible, describe the biochemical changes as a result of which particular symptoms or signs have appeared.

As far as attitudes are concerned, readers should acquire:

(1) a critical approach to the selection and interpretation of laboratory investigations;
(2) concern for patients, including a desire not to subject them to unnecessary investigations;
(3) a responsible approach to the use of laboratory resources;
(4) an awareness that laboratory staff form part of the health team, motivated by concern for the welfare of patients;
(5) a preparedness to discuss with clinical and laboratory colleagues results that are difficult to explain, or which they do not understand.

This is not a laboratory textbook of practical clinical chemistry, so descriptions of practical details relevant to the performance of tests have only been included where needed for their understanding, or to draw attention to pitfalls associated with improper performance. There are many variations in the ways of performing even apparently standard tests. Users of clinical chemistry services should, therefore, seek guidance locally on the detailed practical requirements of laboratories.

The investigative aspects of clinical chemistry can perhaps be best described as the application of chemical techniques to the study of disease in man. However, such investigations are not all carried out in clinical chemistry laboratories. For tests where there might be discussion about the most appropriate laboratory specialty to perform the test, we have based our decisions whether or not to describe such tests in this book on our assessment of what seems to be the most widespread current practice.

Chapter 1

Requesting Laboratory Investigations

Doctors expect clinical chemistry laboratories to provide reliable data in response to requests for investigations. They also expect laboratories to issue reports sufficiently quickly for the data to be of real help in coming to decisions in respect of the clinical problems that gave rise to the requests for chemical investigations. In addition, doctors expect laboratories to offer a range of tests appropriate to changing clinical circumstances. This requires laboratory staff to be active in research and in development work, much of this undertaken in collaboration with clinical staff and leading to the introduction of new procedures. As a corollary, obsolete tests need to be discontinued, and clinical staff kept informed of the changing spectrum of available tests.

Clinical chemistry laboratories in most parts of the world have for years been experiencing rapidly and continuously increasing workloads, due partly to greater use of tests of established value, and partly to the introduction of new tests. The ability to handle growing workloads, and at the same time improve analytical reliability, has owed much to improvements in technical methods, particularly to the introduction of automatic and work-simplified methods of analysis and, more recently, the use of computers. It is worth describing the main patterns of test request procedures used by doctors, i.e. requests for (1) discretionary tests, and (2) screening investigations.

Discretionary tests

Most chemical investigations requested for diagnostic purposes have in the past been selected on the basis of the sum of knowledge gained from (1) the history of the patient's illness, (2) observations made during the clinical examination, (3) results of simple chemical 'side-room' tests and (4), where available, findings from other special examinations such as X-ray or haematological tests. The results of these chemical investigations need to be interpreted in relation to all the available data concerning a patient if they are to contribute fully to the diagnostic process; it is only occasionally possible to make a definitive diagnosis solely on the basis of chemical findings.

There should, in general, be a definite clinical reason for the performance of each and every discretionary test, and laboratories would be helped if senior clinicians were to encourage their junior staff to exercise judgement when initiating requests for discretionary tests. Attempts should be made to eliminate requests for unnecessarily repetitive work. Some aspects of the diagnostic process and of problems relating to the management of illness will now be considered.

ESTABLISHING A DIAGNOSIS

Examples of chemical tests which are diagnostic by themselves are measurements which specifically identify some of the inborn errors of metabolism and determinations which characterize some of the syndromes resulting from endocrine dysfunction. Even with these examples, however, firm diagnoses already made on clinical and other evidence may only be being refined, amplified or confirmed by chemical investigations, and a quantitative measure of the severity of the illness obtained as a baseline for subsequent monitoring of progress (see below).

Much more often, doctors make a provisional diagnosis and then use chemical investigations to seek confirmation of this. The measurement of plasma [urate] to confirm a clinical diagnosis of gout, or the measurement of plasma creatine kinase activity in patients with suspected myocardial infarction, provide two common examples. In those cases where several possible diagnoses have to be considered, chemical tests may help to clarify the differential diagnosis, for instance when investigating the cause of jaundice or of an acute attack of abdominal pain.

MANAGEMENT OF ILLNESS AND MONITORING PROGRESS

Many chemical measurements can assist in the management of patients without necessarily contributing further to the diagnostic process. In the initial assessment of a patient, for example, the appropriate chemical investigations indicate the severity of the illness (e.g. the degree of dehydration, the depth of jaundice) in more precise terms than can be deduced from clinical examination. Similarly, during the course of an illness, chemical tests are often used to monitor the effects of treatment. Examples include tests, the results of which influence (1) the choice of intravenous fluid therapy before and after major surgical operations, or (2) the composition of the gas mixture being administered to patients requiring assisted ventilation.

DETECTION OF COMPLICATIONS

Some requests for discretionary investigations include features of both a diagnostic and a management or monitoring function. For instance, an important source of requests for chemical tests comprises measurements carried out to detect possible adverse side-effects of therapy; if the results for these tests become abnormal, they may indicate the onset of complications and have thereby served both a diagnostic and a management function. Examples include the regular performance of plasma enzyme activity measurements (e.g. ALT or AST) on patients receiving potentially hepatotoxic drugs such as the monoamine oxidase inhibitors, or the monitoring of plasma thyroxine and TSH concentrations to detect the hypothyroidism which may develop in patients who have been treated for hyperthyroidism.

In summary, discretionary tests are requested (1) as part of the diagnostic process, or (2) for assisting in the management of illness already diagnosed and under treatment, or (3) for both these reasons. A doctor should always be able to give a clinical reason for requesting each discretionary test.

Screening investigations

Clinical chemistry departments are making increasing use of mechanized equipment, often containing microprocessor technology. One consequence of these developments in instrumentation has been a greater capacity for analytical work. Another consequence has been a growing tendency to organize the performance of laboratory investigations into groups, sometimes for analytical convenience, but in some instances because the routine performance of all the tests in the group has been found to be more informative than carrying out solely those determinations specifically requested on a discretionary basis.

Technical developments have facilitated the introduction of groups of investigations into hospital laboratory work, and these same test groupings have been incorporated into programmes of screening investigations carried out on (1) patients and (2) the apparently healthy ambulant population.

To distinguish these screening investigations on patients from screening the healthy population, some have advocated the expression 'profiles' (e.g. admission profiles) when referring to groups of non-discretionary tests performed on patients. This semantic distinction has not been widely adopted, and it is probably least confusing to adopt 'screening' as the general term, and to qualify the use of the word in respect of its application, to patients or to the healthy population.

SCREENING INVESTIGATIONS ON PATIENTS

Each group of tests usually includes some discretionary tests and, in addition, other tests that the laboratory finds it convenient to perform at the same time because of the way the analytical work has been organized into particular combinations of tests. These groups of investigations do not always include discretionary tests, however, for instance when admission screening is automatically performed on all patients admitted to hospital, regardless of diagnosis.

Examples of screening investigations on patients, as frequently performed, include an 'electrolyte group' and a group of 'liver function tests'. Even though the clinician may only have requested plasma urea or K^+ determination, many laboratories now find it more convenient to perform a group of tests (e.g. urea, Na^+, K^+, Cl^-, total CO_2 and creatinine), and in many instances to report all the results; the clinical interpretation of individual results is, in this case, sometimes assisted by having all sets of data available. Similarly, with 'liver function tests', laboratories often carry out a group of measurements (e.g. plasma albumin, bilirubin, alkaline phosphatase, aminotransferase and gamma-glutamyl transferase) even though only one of these investigations has been specifically requested (e.g. plasma bilirubin).

Developments in automatic chemical equipment have led many laboratories to extend the concept of grouped analyses considerably. In several countries, instruments capable of performing 10—20 different analyses on each specimen of plasma or serum examined are used to carry out analyses whenever one or more of the tests in the group is requested. Furthermore, mainly in the U.S.A. but to an increasing extent elsewhere, doctors can now ask laboratories to perform as many as 25 different chemical determinations, on small (2 mL) specimens of blood obtained from any or all their patients when first admitted to hospital. Doctors may request these tests as a single group even though, on strictly clinical grounds, only 1 or 2 of the analyses might have been indicated from among all the tests included in the group, if being requested on a discretionary basis. Since, with some patients, there might have been no indication for the performance of any chemical investigations on a discretionary basis, this practice is sometimes described as admission screening. So far, in Britain, few studies of admission screening and its cost-effectiveness have been reported.

Developments in multi-channel analytical equipment have in part been influenced by assessments of previous patterns of requests for chemical investigations. By their very nature, however, these technical developments have tended to focus interest on, and even restrict interest to, those assays that can be performed on the automatic equipment already available. Apart from simplifying laboratory organization, and in some instances reducing staff and operating costs, screening patients by means of multi-channel chemical investigations has had little impact on clinical practice. This is because the range of tests so far included in the screening programmes has never done more than reveal, in only a *small* percentage of the patients investigated, unexpected abnormalities that were later shown to be the first step towards a change in diagnosis or an important alteration in treatment.

In many instances, unexpected abnormalities revealed by the performance of unrequested (non-discretionary) investigations have remained unexplained, despite much extra investigation. In others, the unrequested abnormal results have been overlooked or ignored by the clinician responsible for the patient, since these unsolicited findings did not appear to be relevant to the principal or even the differential diagnosis. For instance, an abnormally high plasma [calcium] might be overlooked in a patient admitted to a gynaecology ward for a hysterectomy. Unexplained abnormal findings can constitute a serious disadvantage of screening investigations, and potentially important abnormalities that are ignored represent a waste of screening facilities, whether in hospital or general practice.

SCREENING INVESTIGATIONS ON THE WELL POPULATION

These investigations are carried out on *apparently healthy individuals,* occasionally as single investigations but much more often in groups as part of a multi-phasic screening programme, for detecting hitherto unsuspected illness. Multi-phasic screening programmes may include some clinical observations (e.g. blood pressure, height, weight) as well as radiological, haematological and chemical investigations.

Prevention is generally accepted to be better than cure, and this belief underlies most screening programmes. Some diseases arise as a result of disordered biochemical mechanisms, and a progressive development of such a disease process can be visualized (Fig. 1.1). The objective of these screening programmes is to detect disease at an earlier stage than might otherwise be the case, so as to be able to institute treatment that will arrest or even reverse the pathological process.

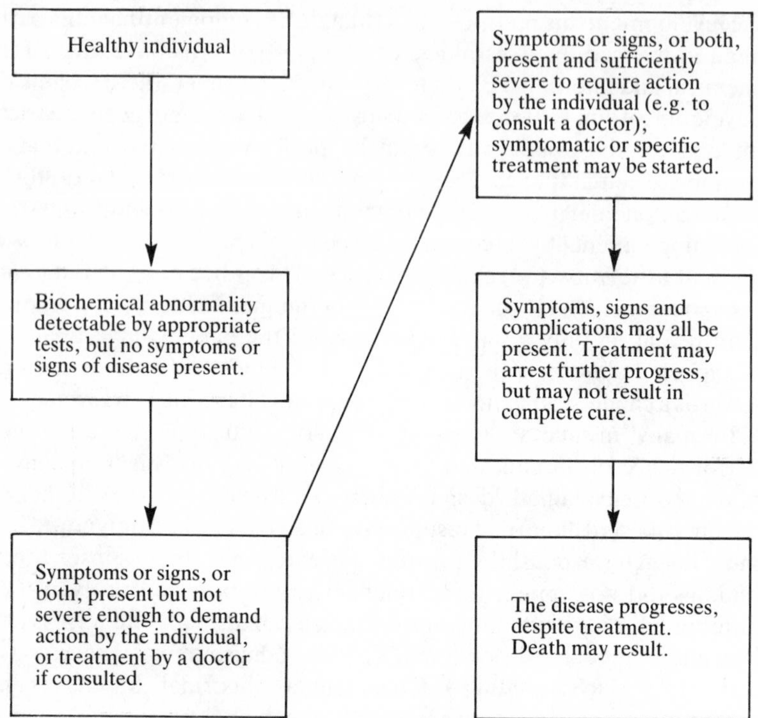

FIG 1.1. *Theoretical progression of a disease process.* Screening programmes aim to detect the abnormality early, preferably before symptoms or signs appear, but otherwise before they demand action by the individual.

For diseases having a biochemical basis, the aim is to correct the underlying biochemical abnormality.

It must be emphasized that multi-phasic screening based on a large spectrum of non-specific chemical tests has proved to be of very limited value, when applied to the apparently healthy population. Examples of limited screening programmes, based on restricted ranges of chemical tests and designed to detect specific disease entities, are considered in several chapters.

Discretionary or screening investigations?

The foregoing discussion has emphasized that screening by means of chemical investigations has not yet been shown to have much diagnostic value, even in patients in whom there may be both clinical and other evidence on which to base medical decisions. In view of this, the place of clinical chemistry in the diagnosis and management of disease presented in this book will mainly be discussed in terms of the selection and interpretation of tests requested on a discretionary basis or as part of limited, mainly organ-specific groups of investigations that include some discretionary tests. This is an important statement. By emphasizing the continuing role of discretionary methods of requesting tests, we are stressing the need for doctors to be able to explain the basis of each request for a chemical investigation, at the time each request is made. Asher's catechism is particularly relevant in this context:

(1) Why do I request this test?
(2) What will I look for in the result?
(3) If I find what I am looking for, will it affect my diagnosis?
(4) How will this investigation affect my management of the patient?
(5) Will this investigation ultimately benefit the patient?

It seems to be general experience that the ability to answer the questions in this catechism *at the time of initiating a laboratory request* greatly increases the likelihood that subsequently, when the results of tests become available, the doctor will be able to understand and interpret the data. The converse is also true. Only in Chapter 28 do we return to detailed consideration of screening programmes.

REQUESTING LABORATORY INVESTIGATIONS: PRACTICAL CONSIDERATIONS

Most quantitative chemical investigations are carried out on blood specimens. The next most frequently examined material is urine, and a small percentage of the laboratory's work is carried out on specimens of faeces. Other materials submitted for chemical analysis include cerebrospinal fluid, gastric, duodenal and jejunal secretions, amniotic fluid, saliva, sweat, pathological fluids obtained by paracentesis, calculi and samples of dietary intake. Recently, interest has begun to focus on the chemical analysis of cellular constituents and biopsy material. It is also important to remember that many chemical tests can now be performed in the side-room or at the bedside (Chapter 3).

There are no 'best' procedures for carrying out chemical analyses. It is, therefore, impracticable to state a comprehensive series of rules or guidelines relating to the collection of specimens. However, it is important to stress the need to follow the directions provided by each laboratory about the procedures to be adopted when collecting specimens and submitting them for analysis, if reliable results are to be obtained. Specimens collected or preserved under unsuitable conditions may *appear* to be suitable for analysis since the laboratory may be unable to detect that the specimen was collected under unsatisfactory conditions. However, it must be appreciated by those who submit unsatisfactory specimens to laboratories that the results of analyses carried out on these specimens will often be inaccurate, and may not properly relate to the condition of the patient at the time the specimen was collected.

There are many factors to be taken into account if reliable results are to be reported by clinical chemistry laboratories. Factors that are principally under the control of doctors initiating requests for laboratory investigations will now be discussed.

Identification of patients and of specimens

It is essential to collect each specimen from the correct patient and to identify the specimen in such a way that the report issued by the laboratory can be unequivocally matched up with the correct set of patient's records. Reliable laboratory work amongst other things depends, therefore, upon:

(1) The method used for identifying each patient uniquely and reproducibly.
(2) The completion of a laboratory request form bearing the data necessary for identifying the patient unequivocally.
(3) The collection of specimens appropriate to each request for analysis.
(4) The unequivocal identification of specimens and the association of each with its corresponding request form.

There is no perfect method for identifying patients, but the simplest way of identifying individuals satisfactorily is by means of an unique registration number or alphanumeric coding; this assumes that each patient's unique identification data can be made available for quoting as soon as required. Most hospitals adopt their own system of unique registration numbers for linking records on individual patients, but often there is a considerable delay before these numbers are issued, especially in the case of urgent admissions, and many doctors fail to enter full identification data for patients on laboratory request forms even when these data are readily available to them.

The accurate completion of request forms can be a time-consuming process, and many forms are only partially completed. This may not prevent laboratories from carrying out analyses nor from issuing reports. However, when a request form is incomplete, this increases the likelihood that the corresponding report will carry insufficient identification data for it to be matched up with the patient's case notes.

It is just as important to label each specimen unequivocally as it is to complete the request form. Otherwise, there are several places where the link between the specimen and its request form can be broken (e.g. during transport to the laboratory or during the preliminary stages of preparing the specimen for analysis).

Collection and preservation of blood specimens

Many factors associated with specimen collection can influence the results of chemical investigations and thus affect their validity and interpretation. The grossest examples of carelessness in specimen collection include the collection of specimens from the wrong patient, or the incorrect identification of specimens after collection. Other factors of importance in the collection of specimens and in their subsequent handling prior to analysis are discussed below in general terms. Specific examples relating to particular tests are mentioned in the appropriate chapters.

BEFORE COLLECTING A BLOOD SPECIMEN

Factors to consider include the patient's previous diet and any necessary dietary preparation before the test is carried out, whether drug treatment might affect the results, whether the time of day for collection of specimens might be important, and many other factors discussed later in relation to the establishment of reference values (p. 15)

Dietary constituents are likely to interfere with some chemical measurements; detailed guidance about foods to be avoided, where this is necessary, are available from the laboratory. In other instances, the interpretation of chemical tests may be influenced by the patient's recent food intake. For example, the dietary intake of carbohydrate in the previous few days influences a patient's response to a glucose tolerance test. Also, the intake of calcium and of phosphate may greatly influence their output in the urine.

Drugs. Many exert adverse effects on the chemical composition of blood. Special mention is needed of the diverse effects of oral contraceptives (Chap. 20). Other drugs may disturb metabolic processes sufficiently to produce marked abnormalities, for instance effects of phenobarbitone on plasma GGT activity, or of methyldopa on urinary HMMA output; where possible, such drugs should be stopped before the investigations are carried out.

Some drugs influence one method of analysis but not an alternative procedure. It can be important, therefore, to give details of drug treatment on request forms, and it may be helpful to discuss with laboratory staff the treatment that a patient is receiving.

Diurnal variation. The concentration in plasma of several substances varies considerably at different times of the day (e.g. iron, cortisol). For determining the concentration of such substances, the directions provided by the laboratory as to the appropriate times for collecting specimens must be observed. If not, it may be impossible to interpret the results satisfactorily as reference values may not be available relating to the time of collection adopted.

Other factors. The results of several investigations are influenced by whether the patient has been ambulant or in bed, or has recently been taking exercise. Examples are given later.

AT THE TIME OF COLLECTING THE SPECIMEN

Factors to consider include the posture of the patient, the choice of skin-cleansing agent, the selection of a suitable vein, the amount of venous stasis, and the steps to be taken so as to avoid haemolysis.

A standard technique should be adopted as far as possible when collecting blood specimens. The ideal would be for most non-urgent specimens to be collected from patients who have fasted overnight, and who have been lying down for at least 20 min prior to collection. All specimens should be obtained with a minimum of venous stasis. These conditions should be attainable for most non-urgent investigations on in-patients in hospital, but are only partially applicable to ambulant patients and to emergency admissions.

Standardization of technique is worth the effort as it makes possible the closer comparison of successive sets of results obtained on individual patients. When the posture is changed from lying to standing, within 15 min there may be an increase of as much as 13% in the concentration of all the non-filterable components of the blood (e.g. plasma proteins and compounds bound or partly bound to protein), due to redistribution of fluid in the extracellular space.

The skin must be clean over the area selected for collecting the blood specimen. In choosing the skin-cleansing agent, however, alcohol and methylated spirits are unsuitable if blood [ethanol] is to be determined. Alcohol can also cause haemolysis.

A limb into which an intravenous infusion is being given must not be selected as the site of venepuncture unless particular care is taken. If specimens are to be collected from an infusion site, the needle or cannula must first be thoroughly flushed out with blood.

Temporary venous stasis is frequently necessary to reveal a vein prior to venepuncture. It is important, however, to avoid prolonged venous stasis as this can markedly raise the concentration of plasma proteins and other non-diffusible substances. Whenever possible, it is desirable to release the tourniquet, if used, after entering the vein and before collecting the sample of blood.

Haemolysis can occur for many different reasons, and careful attention to technique is essential if this common cause of unsatisfactory specimens is to be avoided. Haemolysis renders blood specimens unsuitable for plasma K^+, magnesium and some enzyme activity measurements. To reduce the likelihood of haemolysis, the specimen should be collected with only moderate suction and the needle should be removed from the syringe before transferring blood to the specimen container. The transfer should then be performed slowly, leaving behind any froth that may be in the syringe. If there is anticoagulant or preservative in the specimen tube, only the appropriate amount of blood should be added and the anticoagulant or preservative then dissolved in the sample by repeated gentle inversion of the tube. The tube must not be shaken vigorously and it should be transported carefully to the laboratory. When collecting capillary specimens it is essential that the selected site be warm and there should be a free flow of blood; squeezing not only dilutes the blood sample with tissue fluid but increases the risk of causing haemolysis.

STORAGE OF BLOOD SPECIMENS AND THEIR TRANSPORT TO THE LABORATORY

Before sending the specimen to the laboratory, the container must be correctly and sufficiently labelled, and clearly identifiable in relation to its associated request form, on which full details about the patient and the tests to be performed should be entered.

Blood specimens should be delivered to the laboratory as soon as possible. In general, this means within two hours of their collection, but special arrangements are needed for some specimens (e.g. for acid-base measurements, p.68) because of their lack of stability. Some specimens may be stable for considerably longer, especially if plasma or serum is first separated from the cells. Directions for storage of plasma or serum specimens (e.g. in a refrigerator overnight in a general practitioner's surgery) can be obtained from the laboratory; with few exceptions, whole blood specimens should not be stored in this way.

Several changes may occur in blood samples following collection unless the precautions already mentioned are observed. The following are examples of the commoner and more important changes that occur in blood prior to the separation of plasma or serum from the cells:

(1) Glucose is converted to lactate, as a result of glycolysis occurring in the erythrocytes. This process is inhibited by fluoride.
(2) Several substances pass through the erythrocyte membrane, or may be added in significant amounts to plasma as a result of red cell destruction insufficient to cause detectable haemolysis. Important examples are K^+, lactate dehydrogenase and the two aminotransferases (ALT and AST).
(3) Loss of CO_2 occurs since the $P\text{CO}_2$ of blood is much higher than in air. This leads to passage of HCO_3^- into the erythrocytes, and movement of Cl^- from the erythrocytes into the plasma (the chloride shift or Donnan equilibrium).
(4) Plasma [phosphate] increases due to hydrolysis of organic ester phosphates in the erythrocytes.
(5) Some of the more labile plasma enzymes lose activity, for instance the prostatic component of acid phosphatase.

Collection and preservation of urine and faecal specimens

Alterations in the chemical composition of urine, on standing, are mentioned on p.29. These changes can be delayed by the use of the appropriate preservative (e.g. dilute hydrochloric acid), or by suitable conditions of refrigeration, and sometimes both a preservative and refrigeration are needed.

With some urine specimens to be submitted for chemical examination, directions for obtaining a clean or mid-stream specimen need to be followed. For timed collections, where *all* the urine passed within a specified period is to be collected (e.g. 24-hour specimens), the time-limits need to be clearly stated and the patient must understand what is required if co-operation is to be ensured in the collection of the complete specimen; written instructions for the patient to follow, and written directions for nursing staff, can be of great assistance. If a preservative needs to be added to the urine specimen, to ensure stability of the material to be analyzed, with long (e.g. 24-hour) collections the container must be mixed after each addition of urine to the specimen container.

The performance of quantitative measurements on faeces is aesthetically unpleasant, and precise analytical procedures may be difficult because of the bacteriologically dangerous nature of the specimens. Preservation of faecal specimens prior to chemical analysis is usually effected by refrigeration.

Accurate quantitative work with faecal specimens can, as a rule, only be satisfactorily performed with patients who are fully co-operative, and who are being looked after by trained staff in a metabolic unit. Even then, it can be difficult to ensure the completeness of faecal collections and to define precisely the time relating to the collection. Some recommend that a coloured marker (e.g. carmine dye) be given at the start of the collection period, and a further dose of the marker at the end of the period. All the faeces passed between the time of appearance of the first marker (retaining this specimen) and the time of appearance of the second marker (discarding this specimen) are collected and stored for analysis. This technique makes no allowance for the loss of one of the intermediate specimens, and some investigators have instead advocated the use of inert, non-absorbable indicators such as chromium sesquioxide or polyethylene glycol that become uniformly distributed in the faecal material. The quantity of the substance to be measured (e.g. faecal fat) can then be related to the amount of the inert indicator.

Difficulties in obtaining satisfactory analytical data from faecal specimens, coupled with improvements in intubation techniques, together explain the growing interest in the collection and chemical analysis of gastric, duodenal, jejunal and ileal contents for the study of gastrointestinal function (Chapter 10.).

Urgent requests for investigations outside normal hours of laboratory work

Most hospital laboratories employ technical staff on a single-shift system. Outside these hours, requests for laboratory investigations are usually undertaken by these same staff taking part in a roster system and providing an on-call service for these emergency duties. The detailed arrangements vary considerably from one laboratory to another.

The number of fully trained staff available to a laboratory for performing analyses outside normal hours is limited, being restricted to one or occasionally two at any one time. Since these staff may have to carry out work deriving from a large number of hospital beds, it is important to restrict requests at these times to measurements required for the immediate investigation and management of each patient. Other investigations, which may be needed as baseline assessments but which do not have to be reported immediately, can usually be satisfactorily handled by collecting the relevant specimens and storing them under suitable conditions for later analysis, during normal working hours.

Many laboratories use different analytical techniques under emergency conditions from those routinely employed during normal working hours. Many emergency procedures are chosen for their rapidity and ease of performance, and provide data that suffice to answer the urgent problem. The fact that different methods may be used can, however, lead to difficulties in interpretation if the results from the emergency procedure are closely compared with data obtained later by the routine method; quite marked differences may be observed between the results for analyses carried out *on the same specimen,* by the rapid emergency method and by the slower daytime routine method, which is usually more specific. It is also worth noting that, although quality control checks are applied by many laboratories to out-of-hours emergency work, these checks are often less comprehensive than the daytime quality control programmes. These analytical considerations provide cogent reasons for limiting emergency requests to the conduct of truly urgent investigations.

Some laboratories have had difficulty in finding sufficient staff willing to man out-of-hours emergency rosters, and able to work quickly and reliably at these times. This has provided industrial firms with the stimulus to develop an increasing range of ward-based, side-room or bedside equipment for performing urgently required chemical investigations (Chapter 3).

FURTHER READING

ASHER, R. (1954). Straight and crooked thinking in medicine. *British Medical Journal,* **2,** 460–462.

DURBRIDGE, T.C., EDWARDS, F., EDWARDS, R.G., and ATKINSON, M. (1976). Evaluation of benefits of screening tests done immediately on admission to hospital. *Clinical Chemistry,* **22,** 968–971.

FLEMING, P.R. and ZILVA, J.F. (1981). Work-loads in chemical pathology: too many tests? *Health Trends,* **13,** 46–49.

Leading Article (1979). The value of diagnostic tests. *Lancet,* **1,** 809–810.

WHITEHEAD, T.P. and WOOTTON, I.D.P. (1974). Biochemical profiles for hospital patients. *ibid*, **2,** 1439–1443.

WILDING, P., ZILVA, J.F. and WILDE, C.E. (1977). Transport of specimens for clinical chemistry analysis. *Annals of Clinical Biochemistry,* **14,** 301–306.

YOUNG, R.M. and PAYNE, R.B. (1981). Effectiveness of out-of-hours biochemistry investigations. *British Medical Journal,* **283,** 289–291.

Chapter 2

Reports of Laboratory Work

Most reports issued by clinical chemistry laboratories contain sets of numerical data, sometimes accompanied by comments on the results. Few reports are purely descriptive. In order to interpret and make the best use of laboratory reports, doctors need to be able to answer two important questions:

(1) Are these results normal or abnormal?
(2) Has a significant change occurred since the previous set of results?

The answer to the first question is based on comparisons with reference values. The second question only applies to patients who have been previously investigated; the approach to answering it requires consideration of criteria on which to base a decision whether numerical changes in results for individual tests are significant or not significant. Both questions inevitably impinge on the subject of reliability of laboratory work, on topics such as accuracy and precision of methods, and on sources of variation.

REFERENCE VALUES

The proper interpretation of chemical results on specimens from individual patients ideally requires a set of baseline data obtained for each individual before becoming ill. Using patients as their own controls or sources of baseline values is a practice widely adopted before elective major surgery, e.g. pre-operative determination of plasma [urea]. For most patients, however, baseline data obtained when they were healthy will not be available unless regular medical examinations, perhaps performed annually and which include a range of chemical investigations, become a practical proposition and are accepted as being worthwhile. In the absence of such baseline sets of data for individuals, data for patients have to be compared with reference values established by studying appropriate population groups.

If a group of healthy people is selected and chemical analyses are performed on blood samples obtained from them, provided the conditions of collecting blood have been standardized, the results for each of the analyses are found to cluster around one value. It is possible to depict this distribution pattern graphically by plotting the frequency of each result. With many constituents, the distribution pattern is symmetrical and has a characteristic shape, called the Normal or Gaussian distribution. For other blood constituents, for example plasma [bilirubin], [iron], [urea] and alkaline phosphatase activity, the distribution is asymmetrical or skewed.

The results of many chemical analyses performed on healthy individuals can be expressed in simple mathematical terms by calculating the mean (M) and standard deviation (SD); the SD is a measure of the scatter of the results. For variables with a Gaussian distribution, the results are symmetrically distributed about the arithmetic mean, and 95% of the population is included in the range M ± 2 SD; other data relating to this pattern of distribution are given in Table 2.1.

Arbitrarily, it has been conventional to describe the 'normal range' for a chemical constituent as the spread of values which includes 95% of the healthy population (i.e. M ± 2 SD for variables with a Gaussian distribution). However, since about 5% of healthy individuals will by definition fall outside this range, the term 'normal range' is now held to be a misnomer. Instead, it is recommended that 'normal range' be replaced by *reference values*. This latter expression avoids the danger inherent in the use of 'normal range' whereby inferences might be drawn that all results inside the 'normal range' are normal, and all results outside the 'normal range' are abnormal.

TABLE 2.1. Data for a measurement that has a Normal or Gaussian distribution in the healthy population.

Observed value falling within stated limits of mean (M)	Number of observations (%)
M ± 1 SD*	68
M ± 2 SD	95
M ± 3 SD	99.7
Difference between observed result (R) and mean value (M)	**Probability that a normal healthy individual would have this result**
Over 2 SD, but less than 3 SD	Less than 5%
Over 3 SD	Less than 1%

*The SD in this case is the value established when determining the reference values for the measurement. The value for this SD takes into account both analytical and biological sources of variation.

Data for substances that are found to be asymmetrically distributed in the healthy population may be able to be treated in the same way as symmetrically distributed data, after employing a suitable mathematical conversion or transformation to bring them into a symmetrically distributed form. This often takes the form of a log-transformation, in which each result is converted to its corresponding logarithm and the logarithmic data plotted. If the frequency distribution of the logarithmic data is symmetrical, calculation of M and SD is then performed using these logarithmic values, and the results for M and for M $\pm$ 2 SD converted back into their corresponding non-logarithmic values to obtain the mean and the upper and lower reference values.

Two points must be emphasized about the above discussion. Firstly, the statistical calculations (which are based on the assumption that the data are distributed in a certain way) are convenient because they are easy to process numerically. It may sometimes be more valid, however, to analyse the data in a way which makes no assumptions about distribution; details of these special techniques may be obtained from the original articles. Secondly, the upper and lower limits of the reference values to be used in the interpretation of data obtained from patients have to be selected arbitrarily. This process of selection aims to minimize the degree of overlap between the results from healthy and diseased populations, but some overlap is inevitable. This subject is further discussed (p. 509) when considering the sensitivity and specificity of investigations which form part of screening programmes.

In practice, it is the responsibility of each laboratory, in collaboration with clinical colleagues, to establish sets of reference values. Largely because of difficulties in obtaining adequate numbers of healthy individuals, statistical techniques have been developed for deriving reference values from studies on patients with relatively minor illnesses which, for the most part, would not be expected to affect the results of the chemical investigations. One example of such groups is provided by ambulant patients attending general practitioners' surgeries (Fig. 2.1).

Differences in methodology, as well as differences between populations, mean that there may be considerable variations between the reference values quoted by different laboratories. Some laboratories print reference values on their report forms and others provide this information in pocket handbooks. The appropriate sets of values must be used by doctors when interpreting results reported for their patients. It is unwise to attempt to memorize these values, and unfair to expect students to learn them, while there is still diversity

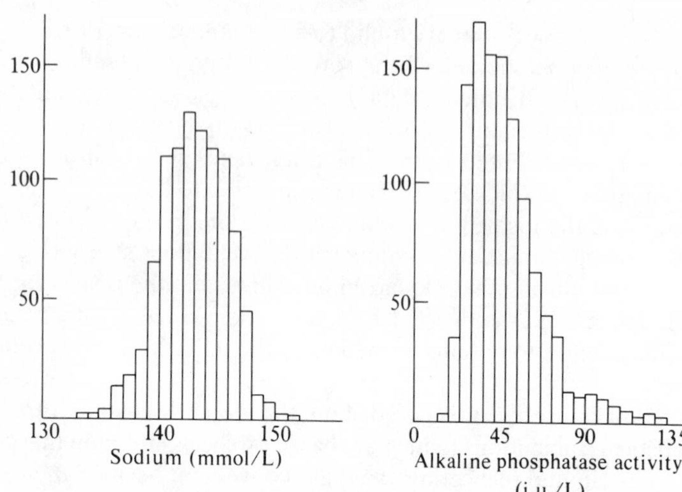

FIG. 2.1. These histograms show the frequency with which values for plasma [sodium] and alkaline phosphatase activity were observed in a group of 1048 ambulant patients attending their general practioners' surgeries. By calculating the values for M and SD for the whole group, and then recalculating these values after the exclusion of observations that were more than 3 SD from the M value, reference values can be calculated (with the alkaline phosphatase data, these calculations were performed after logarithmic transformation). Data from Percy-Robb *et al.* (1971), *Brit. Med. J.*, **1**, 596-599.

among the units of measurement, and so much variation between laboratories.

Examples of reference values are given in Table 2.2. Where we give reference values in this book, these in general relate to the healthy adult population except where otherwise stated. Where we use the word 'abnormal' when discussing quantitative data, this means that the results are outside the range of reference values. For many qualitative or semi-quantitative tests, the terms normal and abnormal clearly continue to be appropriate.

To conclude this discussion of reference values, it might seem that the recommendation that 'normal ranges' be replaced by 'reference values' is unimportant. The change, however, implies an alteration in the approach to quantitative data. It is now common practice to discuss the *probability* that a particular set of results does or does not indicate the presence of disease, and this concept applies to all data, whether or not they fall within the reference values for the measurements.

Are these results abnormal?

Whenever a result falls outside the laboratory's range of reference

TABLE 2.2. Reference values

Examples of commonly requested clinical chemistry determinations, with results expressed in units that accord with the recommendations of the Système International d'Unités (SI recommended units) and in units which the SI system was intended to replace. Units are discussed in Appendix I (p. 517).

Constituent	SI recommended units		Other units	
PLASMA OR SERUM				
(for venous blood, except as stated)				
No numerical change				
Sodium	132–144	mmol/L	132–144	meq/L
Potassium	3.3–4.7	mmol/L	3.3–4.7	meq/L
Chloride	95–107	mmol/L	95–107	meq/L
Total CO_2 (bicarbonate)	24–30	mmol/L	24–30	meq/L
Anion gap (AG_1, p. 76)	10–20	mmol/L	10–20	meq/L
Change in the decimal point				
Albumin	36–47	g/L	3.6–4.7	g/100 mL
Protein, total (serum)	60–80	g/L	6.0–8.0	g/100 mL
Marked numerical differences				
Bilirubin (total)	2–20	µmol/L	0.1–1.2	mg/100 mL
Calcium	2.12–2.62	mmol/L	8.5–10.5	mg/100 mL
			4.25–5.25	meq/L
Cholesterol (total)	3.6–6.7	mmol/L	140–260	mg/100 mL
Creatinine	55–150	µmol/L	0.6–1.6	mg/100 mL
Glucose (fasting specimen)	3.6–5.8	mmol/L	65–105	mg/100 mL
Iron (males)	14–32	µmol/L	80–180	µg/100 mL
Iron (females)	10–28	µmol/L	60–170	µg/100 mL
Iron-binding capacity	47–72	µmol/L	250–400	µg/100 mL
Magnesium	0.7–1.0	mmol/L	1.6–2.4	mg/100 mL
			1.4–2.0	meq/L
$P\text{CO}_2$ (arterial blood)	4.5–6.1	kPa	34–46	mmHg
Phosphate, as P (fasting specimen)	0.8–1.4	mmol/L	2.5–4.5	mg/100 mL
$P\text{O}_2$ (arterial blood)	12–15	kPa	90–112	mmHg
Urate (males)	0.12–0.42	mmol/L	2.0–7.0	mg/100 mL
Urate (females)	0.12–0.36	mmol/L	2.0–6.0	mg/100 mL
Urea (males, below 50)	2.5–6.6	mmol/L	15–40	mg/100 mL
URINE				
Examples with marked numerical differences				
Calcium (low calcium diet)	1.2–3.8	mmol/24 h	50–150	mg/24 h
Creatinine	10–20	mmol/24 h	1.0–2.0	g/24 h
5-HIAA	10–45	µmol/24 h	2–9	mg/24 h
HMMA	10–30	µmol/24 h	2–6	mg/24 h
Urate	1.2–3.0	mmol/24 h	200–500	mg/24 h
Urea	170–600	mmol/24 h	10–36	g/24 h

values, non-pathological reasons that could account for an abnormal result of the magnitude observed, even in healthy individuals, should be briefly reviewed. In addition, there is the known probability that the reference range does not, by definition, include every healthy individual (Table 2.1). Failure to take proper account of these reasons sometimes leads to mistakes in interpreting the significance of 'abnormal' results.

Results which, because they fall outside the reference range, might be taken to indicate the presence of disease cannot always be interpreted as having this meaning. We shall mention several non-pathological factors which may be sufficient to make results on patients fall outside the laboratory's reference values. Indeed, an awareness of these same factors is essential if a laboratory is to establish its reference values properly. Some of the factors to be considered cause variations in individuals, others are due to differences between individuals. In addition, factors relating to the method of analysis adopted by the laboratory may be relevant, although it is not necessary for doctors to know the technical details. Common examples in each of these categories are mentioned below:

Factors affecting the individual

Several factors need to be considered both in establishing reference values and in interpreting results of analyses carried out on specimens collected from individual patients:

(1) *Diet.* Variations in diet can affect plasma [cholesterol], the response to glucose tolerance tests, urinary calcium excretion, and the results of many other measurements. Before carrying out these tests on patients, the possible influence of diet on the result should be considered.

(2) *Menstrual cycle.* Several substances show variation with the phase of the cycle. Examples include plasma [iron], and the plasma concentrations and urinary excretion of the pituitary gonadotrophins, ovarian steroids and their metabolites.

(3) *Muscular exercise.* Measurements affected include plasma creatine kinase activity and blood [pyruvate] and [lactate].

(4) *Posture.* Proteins and all protein-bound constituents show significant differences in concentration between blood samples collected from upright and from recumbent individuals. Examples include plasma albumin, calcium, cholesterol, cortisol and thyroxine concentrations.

(2) *Time of day.* Several plasma constituents show diurnal variation (variation with the time of day), or a nychthemeral rhythm (variation with the sleeping-waking cycle). Examples include plasma [iron] and plasma [cortisol].

(6) *Drugs.* These can have marked effects on chemical results, and the occurrence of interference from drugs emphasizes the importance of laboratories knowing what their methods actually measure, and of their being provided with relevant information about drug treatment that patients may be receiving. Attention should be drawn particularly to the effects of oral contraceptives (p.397).

Even when all these conditions (diet, posture, etc.) have been standardized prior to the collection of samples, drug effects eliminated, and factors attributable to the imprecision of laboratory measurements (see later) have been taken into account, there remains the possible effect of residual individual variation. This can, at times, be enough to account for 'abnormal' results being observed in individuals for whom, at other times, the concentration of the same constituent for that individual falls within the laboratory's reference range.

The magnitude of *residual individual variation* depends on the substance under consideration; examples are given in Table 2.3. Clearly, large changes in the plasma concentrations or activities of some substances may occur before some 'abnormal' results can necessarily be attributed to pathological as distinct from physiological causes.

Factors responsible for variations between groups of individuals

These variables have to be taken into account both when establishing

TABLE 2.3. Residual individual variation of plasma constituents.

Substance	Day-to-day, within-individual coefficient of variation (approximate %)
Sodium	1
Calcium	1–2
Potassium	5
Urea	10
Iron	25
Aspartate aminotransferase	25
Alanine aminotransferase	25

reference values and in selecting the appropriate set of reference values for interpreting results on individual patients:

(1) *Age*. Several measurements show marked physiological variation with age, including plasma [phosphate], [urea] and alkaline phosphatase activity.
(2) *Race*. Racial differences have been described for plasma [cholesterol] and [protein]. It can, however, be difficult to distinguish the supposed effects of race from effects of dietary and other factors.
(3) *Sex*. Many substances show differences in concentration between the sexes. Examples include plasma creatinine, iron, urate and urea concentrations, and the various sex hormones.

Factors attributable to the laboratory

Without going into technical details of analytical performance, it is nevertheless important to mention this group of factors:

(1) *Method of analysis*. There are many different analytical methods available for determining most chemical constituents. Some differ in only minor respects from alternative techniques, but some may show important differences in their specificity (e.g. measurements of plasma glucose as distinct from whole blood reducing substances), and others differ in the principles of measurement (e.g. plasma sodium may be measured by flame photometry or by an ion-selective electrode). These factors can have an important bearing on the results reported for the same determination when carried out in different laboratories.
(2) *Standards of performance*. The analytical reliability with which measurements are made varies considerably. Poor standards of precision widen the reference values for the substance being determined, because the SD is greater, while poor accuracy affects the level at which the mean value is set (Fig. 2.2).

Has a significant change occurred?

All analytical measurements, however carefully made, are subject to errors in performance. For example, if the same sample of plasma is analysed for [K^+] five times by one analyst, using the same technique each time but in ignorance of the fact that the samples all came from the identical specimen originally, the results obtained are most unlikely to be all the same. More probably, a set of figures like 3.6, 3.4, 3.6, 3.5 and 3.5 mmol/L would be recorded.

Clinical chemistry laboratories maintain systems of quality control which allow them to determine the magnitude of these errors in analytical performance, under circumstances which do not introduce observer bias into the handling of the quality control specimens. These data provide indices of analytical precision (depicted diagrammatically in Fig. 2.2), usually expressed in the form of standard deviations (SD). It should be noted that the SD values being considered here relate to standards of analytical performance. They are *not* the same as the SD values previously discussed in relation to the variability of chemical results obtained on specimens from different individuals, and used in the establishment of reference values. The SD in the reference value studies contains both a contribution derived from the biological variability of the population being studied and a contribution due to lack of analytical precision.

If the value for the analytical SD for a measurement is known, from the results of properly conducted quality control procedures, it is possible to calculate the probability with which two results on specimens obtained on two different occasions from the same patient, and for which the findings are *not* the same, can have the numerical difference explained solely on the basis of lack of analytical precision. It is, of course, important to adopt standard conditions for specimen collection, as otherwise additional factors might be introduced which could contribute to the differences observed between two specimens collected from the same patient on the two occasions.

If we consider the case where there has been no change in the *true* value of the result (i.e. the condition of the patient has not changed and the specimens were obtained under identical conditions, so that the

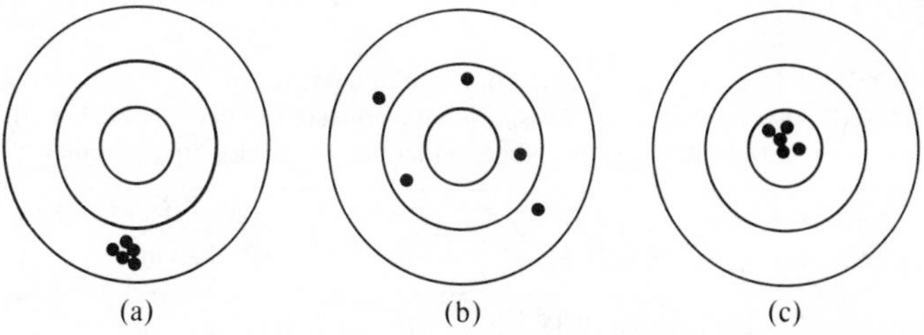

FIG. 2.2. Diagrammatic representation, using shooting targets, of the concepts of accuracy and precision. In (a), the holes are closely grouped but are all below and well away from the centre of the target, illustrating good precision but poor accuracy. In (b) the holes are widely dispersed, but considered as a group they centre around the middle of the target, illustrating poor precision but (on average) an accurate result. In (c) there is both good accuracy and good precision.

two sets of results should have been identical), then in 95% of instances the two results reported will differ from each other by less than 3 SD, the SD value in this context being the measure of the laboratory's analytical precision. While recognizing that this is an over-simplification, we recommend that biological significance should *not* be attributed to changes in analytical results where the extent of the change observed is less than three times the laboratory's stated analytical precision. Table 2.4 expresses this same guidance, but in a different way; its data can be applied to comparisons between two single observations, again on the assumption that the specimens have been collected under identical conditions.

The data in Table 2.4 have important clinical and analytical implications. This can be seen from Table 2.5, which provides information based on the analytical precision (SD) obtained in the author's laboratory at the time of writing.

Clinically, from the interpretative aspect, it is necessary to recognize that the *lack of precision* of quantitative laboratory measurements imposes limitations on the medical significance to be attached to some of the changes observed in the results of investigations. Failure to recognize that analytical limitations exist, and to understand the implications, can mean that too much attention will be paid to small changes in results reported on individual patients. One possible consequence is that treatment might be altered on the basis of laboratory data when, in fact, these data provide insufficient justification for making any change in treatment.

The wide spread shown by some sets of reference values (Table 2.2) reflects the presence of (1) biological variation between individuals and (2) methodological short-comings due to lack of analytical precision, as exemplified by the data in Table 2.5. Biological variation

TABLE 2.4. The probability that a difference of a particular magnitude will be observed between results of two specimens, collected under identical conditions but on different occasions, when in fact the two values are the same.

Difference between results (Number of analytical SD)	Probability
More than 1 SD, but less than 2 SD	Less than 1 in 2 (<0.5)
More than 2 SD, but less than 2.8 SD	Less than 1 in 5 (<0.2)
More than 2.8 ($2\sqrt{2}$) SD	1 in 20 (0.05)
More than 3 SD, but less than 4 SD	Less than 1 in 20 (<0.05)
More than 4 SD	Less than 1 in 200 (<0.005)

Note. Biological sources of variation have been exluded from consideration in determining the probabilities.

TABLE 2.5. Differences that need to be observed between two sets of results on specimens collected from one patient under identical conditions for differences to be significant at the 5% level.

Analysis	Concentration range over which precision data obtained	Analytical precision (SD)	Change needed for two results to differ at the 5% level
Albumin (g/L)	20–55	0.6	1.78
Bilirubin (μmol/L)	(a) up to 50	0.8	2.2
	(b) over 50	2.0	5.6
Calcium (mmol/L)	1.30–3.50	0.03	0.08
Cholesterol (mmol/L)	3.4–11.5	0.20	0.6
Potassium (mmol/L)	2.0–7.3	0.08	0.2
Sodium (mmol/L)	110–155	1.10	3.0
Urea (mmol/L)	(a) up to 16	0.15	0.4
	(b) over 16	0.35	1.0

cannot be eliminated, but some narrowing of reference values can be obtained (1) by improvements in analytical technique, and (2) by standardizing the conditions under which specimens are collected. Narrow reference ranges represent a desirable goal, as their availability should mean that investigations could be used in future to detect smaller but nevertheless still significant departures from normality in individuals.

It is worth noting that results obtained from patients frequently fall within the range of reference values for healthy people. Furthermore, it is sometimes found, by serial observations on these patients, that changes which are significant, according to the criteria shown in Table 2.4, may occur without the results for the patients necessarily moving outside the range of reference values. In other words, the chemical state of the patient may have changed significantly, possibly for a pathological reason in that individual, without the results becoming 'abnormal' as judged in relation to reference values. This again illustrates the potentially misleading concept inherent in the expression 'normal range', and the advantages which should accrue from obtaining reference data for individuals rather than having to continue to rely on reference values obtained from studies on apparently healthy groups of individuals.

THE CLINICAL USE OF DATA

When provided with a set of results from the laboratory, the doctor needs to review each result and its relationship to the distribution of values in the healthy population. This is where reference values with all their present shortcomings have to be considered — few laboratories, for instance, report data on individual patients in relation to sets of reference values that take into account the effects of age, sex or other factors. The doctor also needs to take account of the distribution of results for a particular measurement in defined disease groupings, and of the approximate prevalence of the various disease entities in the population with which he is concerned. There is, however, very little definitive information available to assist with these last two considerations, which become very important when planning case-finding and screening programmes (p. 508).

The increasing use of screening methods of investigation has shown that doctors often experience difficulty in assimilating sets of data and appreciating their full significance when faced with a mass of figures. To obviate these undesirable effects, some computer-assisted systems of reporting incorporate ways of drawing attention to results classified as abnormal on the basis of stipulated reference values (e.g. by marking them with asterisks). Others have used criteria for 'exception reporting', in which only results classified as abnormal are reported to the doctor as a routine; the rest of the results in the group of tests may then be assumed by the doctor to be within the reference values when reviewing his assessment of the patient. Whichever method of reporting results is adopted, it is desirable for the doctor to ask himself a series of questions in respect of each item of laboratory data, as follows:

(1) Does this result fit in with my previous clinical assessment of the patient?
(2) If not, can I explain the discrepancy?
(3) Does the result alter my diagnosis and approach to the management of the patient's illness?
(4) If I am unable to explain a result, what do I propose to do next about the finding?

Special problems may arise with results that are unexpected and cannot immediately be explained in terms of the clinical findings. All too often such results are designated as laboratory errors, and either ignored or the test automatically repeated without further thought.

We would stress the desirability of discussing unexplained abnormalities with laboratory staff, particularly if the result is about to be written off as a 'laboratory error'. Such discussion may lead to checks which show that an error did indeed occur in the laboratory, for instance a simple mistake in calculation, or an obvious error in transcription, the detection of which allows the correct result to be reported. At other times, clinical feedback can prove to be the first evidence of a systematic fault in analytical performance, and such liaison then makes a positive contribution to the laboratory's programme of quality control. Discussion may also result in previously unexplained findings receiving an explanation that may still be of help in diagnosis or in the subsequent management of the patient, or it may draw the clinician's attention to hitherto unsuspected sources of interference (e.g. drug effects) to be avoided in future, for instance when repeating the investigation.

The main uses of the data contained in clinical chemistry reports can, in summary, be classified as follows:

(1) Assisting in the diagnostic process.
(2) Providing data that are of value in assessing the severity of disease, and so helping in the formulation of the prognosis.
(3) Monitoring the progress of patients and their response to treatment.
(4) Providing information that may in the long term lead to a better understanding of the disease process.
(5) Serving a precautionary function. This can apply to patients where failure to perform a certain investigation might result in a charge of professional negligence (at present, this is more relevant to diagnostic radiology and microbiology than to clinical chemistry).

These possible uses of clinical chemistry data clearly interrelate with the questions that should have been considered at the time of initiating laboratory requests (p. 7). Failure to take these questions properly into account in the first place is liable to lead to instances of misuse, overuse and even frank abuse of laboratory services.

FURTHER READING

BOLD, A.M. and WILDING, P. (1978). *Clinical Chemistry Companion,* Oxford: Blackwell Scientific Publications.

CARAWAY, W.T. and KAMMEYER, C.W. (1972). Chemical interference by drugs and other substances with clinical laboratory test procedures. *Clinica Chimica Acta,* **41,** 395-434.

COPELAND, B.E. (1975). The medical communication pathway. *New England Journal of Medicine,* **293,** 41-42.

HEALY, M.J.R. (1969). Normal values from a statistical viewpoint. *Annals of Clinical Biochemistry,* **6,** 12.

HOLLAND, W.W. and WHITEHEAD, T.P. (1974). Value of new laboratory tests in diagnosis and treatment. *Lancet,* **2,** 391-394.

MCPHERSON, K., HEALY, M.J.R., FLYNN, F.V., PIPER, K.A.J. and GARCIA-WEBB, P. (1978). The effect of age, sex and other factors on blood chemistry in health. *Clinica Chimica Acta,* **84,** 373-397.

Chapter 3

Chemical Tests Performed Outside the Laboratory

Several chemical investigations can nowadays be carried out satisfactorily in a doctor's consulting room, or in a ward side-room, or in some cases by patients themselves. All that is required is a working area consisting of a bench-top, sink and if possible a sluice (if specimens of urine or faeces are to be examined), and the requisite materials.

We shall subdivide the chemical investigations that can be performed by clinicians, medical students, nursing staff and other non-laboratory personnel into the following categories:

(1) Simple side-room tests, qualitative or semi-quantitative, on urine or faecal specimens.

(2) Simple side-room tests, semi-quantitative or quantitative, mostly performed on blood specimens.

(3) Quantitative tests, performed mostly on blood or plasma with more complex equipment that can, however, be quickly and reliably operated as 'black boxes' after very little instruction.

In the last few years, for various reasons, there has been an increasing tendency for laboratory services to be centralized, each laboratory serving several hospitals, some of which might be a long way from the laboratory. This has led to growing interest in ways of providing on the spot those investigations that are often required urgently, and for which delays in performance would be unacceptable for medical or analytical reasons.

Developments in reagent packaging and in equipment manufacture have reached the stage where several quantitative chemical investigations required for close support of seriously ill patients (e.g. blood glucose and acid-base measurements) can be performed reliably, after short courses of instruction, by non-laboratory personnel in a ward or operating theatre environment. Examples of these will be considered, to illustrate the fact that it is not necessary for all chemical investigations to be sent to a laboratory, as long as the staff on the spot have taken the time to acquire the necessary skills, *and are motivated to use these skills thereafter in practice.*

Side-room tests on urine

A wide range of simple tests and observations can be carried out on urine. The most suitable specimens for analysis are fresh specimens, passed early in the morning when the urine specific gravity is usually at its highest. Careful technique is required with precise adherence to simple instructions if reliable results are to be obtained. Cleanliness is particularly important. Fresh specimens are required because of changes which occur in urine on standing. These changes, all of which are slowed by refrigerating the urine, include:

(1) Destruction of glucose by bacteria.
(2) Conversion of urea to ammonia by bacteria, with consequent fall in $[H^+]$ and tendency for phosphates to precipitate.
(3) Oxidation of urobilinogen to urobilin, which occurs spontaneously.

Despite the growing importance of quantitative chemical determinations, and the consequent increasing dependence of medical practice on clinical chemistry departments, there is still a definite place in the practice of medicine for performing qualitative and semi-quantitative chemical side-room tests.

Appearance

Urine is normally a clear fluid, varying in colour from orange to colourless. The colour usually reflects the concentration of urochrome. It may appear abnormal for a number of reasons, and the list grows with the increasing number of abnormal appearances described due to drugs and their metabolites (Table 3.1). The patient may complain about this abnormal appearance, and the clinician should at least look at a specimen of urine and enquire further about diet and drug intake before proceeding to request laboratory investigations.

Specific gravity (sp. gr.)

Measurement of urinary sp. gr. may be used as an index of renal concentrating power, e.g. in a patient with thirst or frequency of micturition. However, the following points need to be emphasized:

(1) Urinary osmolality (p. 90), not urinary sp. gr., is the true index of renal concentrating power.
(2) Information gained from measuring sp. gr. of a random sample of urine is limited. However, it is worth measuring the sp. gr. of an

TABLE 3.1. Abnormal appearances of urine specimens.

Appearance	Examples of causes	Further action needed to explain appearance, preferably to be taken before initiating laboratory investigations
Cloudy	Precipitation of phosphates	Precipitate increases if specimen warmed, dissolves if dilute acetic acid added.
Brick-red deposit	Urate	Precipitate dissolves on heating specimen.
Abnormal colours		
Dark specimen, or darkens on standing	Oxidation of various compounds including porphobilinogen, urobilinogen, homogentisic acid and melanogen.	Examine fresh specimen for porphobilinogen and urobilinogen. Also, allow a fresh specimen to stand to confirm the colour change.
Greenish brown or mahogany colour	Conjugated bilirubin	Test for bilirubin and for urobilinogen.
Green or blue colour	Drugs, e.g. methylene blue	Check medication.
Grey, brown or black colour	Drugs, e.g. methyldopa or iron sorbitol.	Check medication.
Pink, red, orange or brown colours	Beetroot	Ask about intake.
	Drugs, e.g. phenindione, phenolphthalein and desferrioxamine.	Check medication, and ask about purgatives (some contain phenolphthalein).
	Haemoglobin, methaemoglobin and myoglobin	Examine fresh specimen microscopically and test for blood.
	Porphyrins.	Test for porphobilinogen.
Yellow colour	Drugs, e.g. the tetracyclines.	Check medication.

Note: Several conditions require more detailed examination of urine, blood, etc., in the chemical laboratory, and other examinations not listed in the comments column (e.g. X-ray, haematological or microbiological investigations). Some of the causes mentioned are rare.

early morning specimen before proceeding to tests of renal concentrating ability based on osmolality measurements.

(3) Measurements of sp. gr. using a hydrometer should be made after cooling urine specimens to 15°C, since the instruments used in side-rooms are calibrated for use at this temperature.

(4) Measurements of sp. gr. using commercially available test strips (e.g. Ames Co.) correlate reasonably well with measurements made with a hydrometer unless there is marked glycosuria.

Causes of increased urinary sp. gr. include dehydration, heavy proteinuria, glycosuria (e.g. glucose, mannitol), excretion of X-ray contrast media, and high doses of antibiotics (e.g. carbenicillin) that have been given parenterally and that are excreted unchanged in the urine.

CHEMICAL TESTS

Table 3.2 lists the principal tests that can be carried out on urine under side-room conditions, either with commercial test strips and reagents, or with classical 'wet chemistry' methods. The commercial firms supply clear and precise directions for the use of their products, and it is unnecessary to repeat these here except to stress that the instructions must be strictly followed. Otherwise, false positive and false negative results may be obtained and the tests fall into disrepute because of apparent lack of reliability. The commercially available tests have almost entirely replaced the wet chemistry methods, details of which can be found in the literature. The main roles of these tests are:

(1) Part of the complete clinical examination of any patient.

(2) Part of the follow-up of patients with jaundice, diabetes mellitus, renal tract disease, etc.

(3) One component of screening programmes (Chap. 28).

Side-room chemical tests can provide important data additional to the information obtained from the history of the patient's illness and the findings on initial physical examination. This is especially true of the tests for glucose and protein in urine; the results of these tests, in many patients, narrow down the initial differential diagnosis, thereby helping the doctor to choose the most appropriate further investigations (laboratory, radiological, etc).

Chemical side-room tests performed as part of follow-up clinical examinations may provide some of the earliest evidence of complications (e.g. proteinuria) developing in the course of an illness.

TABLE 3.2. Tests that can be rapidly performed on urine specimens under side-room conditions using simple, commercially available test materials.

Substance to be tested for	Comment
Bilirubin	Semi-quantitative test.
Blood	Semi-quantitative test for haem derivatives. The test does not differentiate between haematuria, haemoglobinuria and myoglobinuria.
Glucose	Qualitative test, based on glucose oxidase, that is practically specific for glucose.
$[H^+]$	Fresh specimen required.
Ketone bodies	Semi-quantitative test that detects acetoacetate and (but with much less sensitivity) acetone.
Protein	Semi-quantitative test, most sensitive for detection of albumin. Very insensitive for Bence-Jones protein.
Reducing substances	Semi-quantitative test which detects reducing sugars (e.g. glucose, galactose, lactose, fructose).
Urobilinogen	Fresh specimen required, cooled to room temperature before testing. Semi-quantitative test that may also react with porphobilinogen.

The principal suppliers of the test materials required for the substances listed in Table 3.2 are Ames Division of Miles Laboratories Ltd., and the Boehringer Corporation (London) Ltd. These companies also manufacture simple analysers for blood or plasma [glucose], suitable for use on wards or by patients. Details of the full and up-to-date range of their products can be obtained from these manufacturers or their representatives. Their addresses are:

Ames Division of Miles Laboratories Ltd., PO Box 37, Stoke Court, Stoke Poges, Slough SL2 4LY.

Boehringer Corporation (London) Ltd., Bell Lane, Lewes, East Sussex BN7 1LG.

It is, therefore, clearly worthwhile to have carried out the same tests when the patient was first examined, to provide a baseline.

The simplicity of many side-room tests and stability of the reagents means that they can be performed by patients themselves (e.g. in day-to-day management of diabetes mellitus), and they can all be carried out by nurses or physician assistants after a little instruction. However, the very simplicity of side-room chemical tests can be a danger, as they are often carelessly performed with consequent loss of reliability. Several studies have shown that the best results with these tests are obtained with the single-test or limited multi-test materials, partly because the recording of several sets of results obtained in a very short time tends to be inaccurate. Nevertheless, the multi-test (e.g. six or more tests) urine-testing materials offer many advantages, and are undoubtedly here to stay. All it is necessary to emphasize, therefore, is that the test materials must be used correctly.

Medical students and doctors tend to rely on nursing staff to carry out side-room tests. This may be acceptable for routine practice, but it is nevertheless important that students and doctors should gain and retain experience in the performance of these tests, since the results of side-room investigations are usually the responsibility of clinical staff. Doctors may need to instruct and supervise other staff in the performance of side-room tests, and those of their patients expected to carry the tests out as part of their therapeutic control. It is clearly desirable that doctors should be able to discuss side-room test results with patients, using data collected and recorded by the patients or their relatives, and to deal with any difficulties in interpreting results; these difficulties are sometimes due to faults in technique.

Side-room tests for blood in faeces

The detection of gastrointestinal blood loss may be very important for clinical diagnosis, e.g. it may be an early sign of carcinoma of the colon. Tests for occult blood in faeces are the simplest and most widely available means of detecting quite small losses of blood. The various chemical tests, most of them now available from commercial sources, all depend on the pseudoperoxidase activity of haem. Unfortunately, selection of the most appropriate test and the interpretation of results are not simple, for the following reasons:

(1) Normal individuals may lose up to 2 mL blood each day in the faeces.
(2) Meat and some vegetables in the diet, and certain iron preparations, may all give rise to false positive results for occult blood tests, since they also have pseudoperoxidase activity.
(3) Reducing agents, e.g. ascorbic acid, may cause false negative results if taken in sufficient quantity.
(4) Although all the chemical tests are designed to detect the pseudoperoxidase activity present in faeces, the various tests differ widely in their sensitivity.
(5) Blood lost into the lumen of the gastrointestinal tract is unlikely to be evenly dispersed throughout the faeces. Unless the faecal specimen is homogenized, sampling errors are likely to occur.

Because of these difficulties, in practice all chemical methods of testing for faecal occult blood give rise to both false positive and false negative results.

False positive results often occur with the more sensitive methods of detection. They are a nuisance as they can cause unnecessary further investigation of the patient. Their incidence can be reduced by repeating the test after the patient has been placed for 3 days on a diet that excludes meat and green vegetables.

False negative results occur with the less sensitive tests. They may be dangerous, not just a nuisance. They give the doctor and the patient a false sense of security, and can seriously delay diagnosis.

In view of the importance of detecting the presence of gastrointestinal bleeding, and the problems mentioned above, interest is likely to focus increasingly on the use of isotopic methods. An example of this approach is the intravenous injection of ^{51}Cr-labelled erythrocytes followed by measuring the radioactivity excreted in the faeces; this technique is relatively expensive and time-consuming, but it can provide an indication of the severity of blood loss. Improvements in endoscopic procedures are also changing the approach to diagnosis.

Side-room chemical tests on blood specimens

Simple tests have been developed for application to samples of blood, plasma or serum under side-room conditions. These have been carefully compared with laboratory-based procedures, and the glucose procedures especially can now be recommended for use, particularly when laboratory facilities are not readily available. The best correlations between laboratory results for blood [glucose] and results obtained by clinicians and patients on specimens examined under side-room conditions, and the best precision with the side-room observations, have been obtained when the test strip materials have been used with the meters that the manufacturers (Ames Co. and Boehringer Mannheim) have developed specially for use with their test materials.

Side-room measurements of blood [glucose] are quick to perform, and have found their main application (1) in the investigation of hyperglycaemia, and (2) in monitoring the control of diabetic patients on a regular (e.g. day-to-day) basis. They have also been used to follow the blood [glucose] response in insulin-induced hypoglycaemia tests, but should only be used for this purpose by experienced and careful operators because of the serious dangers resulting from inaccurate observations at low blood [glucose].

Similar techniques have been developed for measuring blood [urea]. Also test materials designed primarily for use with urine can sometimes be used for qualitative or semi-quantitative measurements on plasma and serum specimens, e.g. test strips for bilirubin and for acetoacetate.

Chemical equipment suitable for use by non-laboratory staff

In the last few years, two developments have tended to bring some chemical measurements out of the laboratory, and to have them performed instead close to the patient:

(1) Increasing numbers of patients are being cared for in specialized units, where there are recurring and predictable needs for certain types of chemical measurement, and where the results are required as quickly as possible.
(2) Several instrument manufacturers have progressively simplified the operational procedures needed to use some of their newer instruments, without detriment to the reliability of the results.

For several years many labour wards have had their own equipment for monitoring fetal blood $[H^+]$ on capillary samples obtained from the scalp, and this equipment has been used successfully by clinicians and other non-laboratory staff.

The best recognized use of complex analytical equipment outside the clinical chemistry laboratory lies in the increasing 'bedside' availability of automated systems for performing blood gas measurements. These depend on specific and reliable electrodes for pH, $P\text{CO}_2$ and $P\text{O}_2$ determinations. Such systems are simple to operate and much effort has been expended on building in checks to avoid or to detect misuse of the equipment. The various operations include automatic calibration, quality control checks, wash cycles and fault-recognition procedures, precisely controlled by a minicomputer or by a microprocessor. Results appear as a digital display, and in some cases as a printout.

The range of sophisticated chemical equipment that can be used safely and reliably, sometimes at considerable distances from the laboratory responsible for its maintenance, now extends far beyond just the blood gas analyser systems (Table 3.3). In some cases, the equipment can be used with whole blood, e.g. some ion-selective electrode systems can measure calcium, potassium and sodium. However, for analyses liable to be adversely affected by haemolysis, we would strongly recommend that they be performed on plasma or serum rather than on whole blood; seriously misleading data can sometimes be provided if a specimen is even slightly haemolysed and the haemolysis has gone undetected, as could be the case for potassium measurements.

TABLE 3.3. Examples of chemical equipment that can be operated satisfactorily by clinical staff, if regularly serviced and calibrated by laboratory staff.

Equipment	Measurements	Examples of location
Bilirubinometer	Plasma [bilirubin]	Obstetric units
Blood gas analyser	Blood $[H^+]$, Pco_2, Po_2 plasma $[HCO_3^-]$ etc.	Intensive care units Obstetric units
Glucose analyser	Blood or plasma [glucose]	Diabetic clinics
Osmometer	Plasma or urine osmolality	Intensive care units
Ion-selective electrode systems	Plasma $[K^+]$ and $[Na^+]$	Intensive care units
Systems	*Quantitative measurements on plasma or serum*	
Du Pont ACA*, various configurations of the equipment available	These systems can be used to perform over 40 measurements, e.g. the concentration of albumin, bilirubin, calcium, cholesterol, creatinine, glucose, iron, magnesium, K^+, protein (total), Na^+, urate, urea and some drugs, as well as several enzyme activities (amylase, ALT, AST, CK, GGT, etc.)	

*Du Pont (U.K.) Ltd., Wedgewood Way, Stevenage, Herts SG1 4QN.

The need to centrifuge blood specimens prior to analysis of plasma or serum on ward-based chemical equipment means that satisfactory conditions for specimen handling and separation of plasma or serum have to be provided; these include an adequate working area, properly ventilated, a suitable centrifuge and proper disposal facilities. The non-laboratory staff who are to use these facilities need to be instructed in the importance of clean and safe working practices, and a workable code of practice should be laid down. In the U.K., the work-places where ward-based laboratory equipment is to be used are liable to be inspected by the Health and Safety Executive.

There are two main obstacles to the widespread introduction of automatic blood gas analysers and other sophisticated chemical equipment into operating theatres, hospital wards and elsewhere outside the laboratory, and their subsequent satisfactory use there. The first obstacle is expense, but this may be justifiable in places where chemical investigations are likely to be needed urgently, especially out-of-hours, and where the laboratory is unable satisfactorily to meet this clinical requirement because of geographical or staffing considerations.

The second obstacle is that, in practice, non-laboratory staff are either not given adequate initial training in the use of the automated piece of equipment and, equally important, in general principles of good laboratory practice, or they fail to maintain these levels of performance in subsequent everyday practice.

Published data have often shown that non-laboratory staff can achieve results that compare favourably with data obtained in the laboratory. However, these reports cannot necessarily be taken as representative since they all tend to emanate from units that have a special interest. In our experience, unless there is constant vigilance, standards lapse and disasters are liable to occur due to failure to adhere to a simple set of rules, such as the following:

(1) Non-laboratory staff require to be properly trained in the operation of the equipment. They need to observe the guidelines for the proper collection and preparation (e.g. centrifuging) of specimens, and they should have readily available a simple set of instructions for operating the equipment.

(2) There should be a check-list on which staff should be required to record their compliance with directions before, during and after each set of observations.

(3) There should be a record book for entering results and any other essential data relating to the use of the equipment (e.g. calibration readings and results for quality control specimens).

(4) There should be a daily check of the equipment, preferably carried out by trained laboratory staff.

(5) There should be a back-up method available, probably dependent upon laboratory staff.

Clinicians and other non-laboratory staff need to take a pride in the operation of blood gas analysers and other equipment that has, until recently, been found mostly in chemical laboratories. They cannot expect laboratory staff to carry out uncomplainingly the essential daily maintenance checks, repairs and cleaning up of such equipment when this has been carelessly used and frequently left in a dirty state.

Repeated misuse of ward-based chemical equipment inevitably adds to the cost of its maintenance, and shortens its working life. Also, when breakdowns occur, the clinical unit has to send specimens to the laboratory, where the back-up method is normally available. The reasons for such intermittent heavy dependence of specialized units on the laboratory for the performance of urgent chemical investigations are usually all too evident to the laboratory staff, and do not help to foster good relations.

FURTHER READING

ANDERSON, J.R., LINSELL, W.D. and MITCHELL, F.M. (1981). Guidelines on the performance of chemical pathology assays outside the laboratory. *British Medical Journal,* **282,** 743.

BELL, P.M. and WALSHE, K. (1983). Benefits of self monitoring of blood glucose. *ibid,* **286,** 1230–1231.

EVANS, S.E. and BUCKLEY, B.M. (1983). Biochemists nearer the patient? *ibid,* **287,** 1399–1400.

HARDCASTLE, J.D., FARRANDS, P.A., BALFOUR, T.W., CHAMBERLAIN, J., AMAR, S.S. and SHELDON, M.G. (1983). Controlled trial of faecal occult blood testing in the detection of colorectal cancer. *Lancet,* **2,** 1–4.

Leading article (1982). The place of nurses in management of diabetes. *ibid,* **1,** 145–146.

MARKS, V. (1983). Clinical biochemistry nearer the patient. *British Medical Journal,* **286,** 1166–1167 (and letter from S. C. Frazer (1983). *ibid,* **286,** 1440).

MINTY, B.D. and BARRETT, A.M. (1978). Accuracy of an automated blood-gas analyser operated by untrained staff. *British Journal of Anaesthesia,* **50,** 1031–1039.

RAYMAN, G., ELLWOOD-RUSSELL, M., SPENCER, P., PRENTICE, M., ROUSE, S. and WISE, P. (1983) Comparative accuracy of portable blood-glucose monitors. *Journal of the Royal College of Physicians of London,* **17,** 183–186.

SMITH, B.C., PEAKE, M.J. and FRASER, C.G. (1977). Urinalysis by use of multi-test reagent strips: Two dipsticks compared. *Clinical Chemistry,* **23,** 2337–2340.

WOOTTON, I.D.P. and FREEMAN, H. (1982). *Microanalysis in Medical Biochemistry,* 6th Edition, pp. 254–267. Edinburgh: Churchill Livingstone. (This provides information about classical 'wet chemistry' methods, many of which were at one time used under side-room conditions).

ZILVA, J.F. (1982). Qualitative biochemical urine testing; the case for selectivity. *Annals of Clinical Biochemistry,* **19,** 8–11.

Chapter 4

Water, Sodium and Potassium Disturbances

This chapter considers water balance and the principal univalent cations in the extracellular fluid (Na^+) and intracellular fluid (K^+). Measurements of plasma [Na^+] and [K^+], often accompanied by plasma [urea] and [total CO_2] and sometimes by plasma [Cl^-], together make up the most frequently requested group of discretionary tests presently carried out by clinical chemistry laboratories. Similar measurements on urine and on other fluids, especially when these fluids are being lost in large amounts, can provide useful additional information for the management of patients with severe or prolonged disturbances of fluid and electrolyte metabolism. However, laboratory data cannot substitute for careful clinical assessment and re-assessment of these patients. Discussion of fluid replacement therapy leads on to consideration of parenteral nutrition.

The concept of balance

The total quantity of water in the body normally remains the same from day to day. There is an overall state of balance between the amount of water taken in, or derived from various sources, and the amount eliminated by various routes. Balance between input and output is a general concept, applicable to water, Na^+, K^+, Cl^-, calcium, nitrogenous materials, energy, etc. Physiological changes in the body's composition, as part of growth and development (Table 4.1), occur gradually over an extended time-scale. There are also short-term diurnal variations in many inorganic constituents.

WATER AND SODIUM METABOLISM

The principal sources of water intake and components of water output are listed in Table 4.2, together with a set of illustrative daily values for healthy adults living in a temperate climate. To a first approximation, the easily measurable component of fluid intake (the amount of fluid drunk) normally equals the easily measurable component of fluid output, the 24-hour urinary excretion

TABLE 4.1. Distribution of water in an adult male and in an infant.

	Adult (weight 70 kg)			Infant (weight 3.0 kg)		
	Litres		Percentage of body weight	Litres		Percentage of body weight
Extracellular fluid		14	20		1.2	40
Plasma	3			0.25		
Extravascular fluid	11			0.95		
Intracellular fluid		28	40		0.9	30
Total body water		42	60		2.1	70

TABLE 4.2. Average daily water intake and output of a healthy adult in a temperate climate.

Intake	mL	Output	mL
Water drunk	1500	Urine volume	1600
Water in food	750	Faecal water content	50
Water from metabolism of food	250	Loss in expired air; Insensible perspiration	850
Total intake	2500	Total output	2500

In an average 70 kg adult, the total body water is about 42 L, and the total body Na^+ about 4200 mmol. Approximately 50% of the Na^+ is in the ECF, 40% in bone and 10% in ICF.

The normal daily intake of Na^+, K^+, and Cl^- and the amounts eliminated by various routes are shown in Table 4.3. The urinary excretion of Na^+ and Cl^- normally adjust quickly to variations in dietary intake but changes in K^+ output are less precisely regulated. Acute and severe disturbances of water and electrolyte balance may develop rapidly in disease, as will be evident from the data shown in Table 4.4. Normally the various intestinal secretions are reabsorbed,

TABLE 4.3. Data for the daily intake and output of Na^+, K^+ and Cl^- for a healthy adult (all values in mmol/24 h).

		Output		
Ion	Intake	Urine	Faeces	Skin
Na^+	100–200	100–200	<5	<5
K^+	20–100	20–100	<5	trace
Cl^-	100–200	100–200	<5	<5

TABLE 4.4. Secretions entering the digestive tract each day.

	Volume (L/24 h)	Approximate composition (mmol/L)				
		H^+	Na^+	K^+	Cl^-	HCO_3^-
Saliva	1–2	0	30	20	35	15
Gastric juice	1–4	20–120	20–120	10	150	0
Pancreatic juice	0.5–1.5	0	140–180	5	75	70–110
Bile	0.5–1	0	140	5	75	70
Intestinal secretions	2–4	0	140	5	110	35

Notes

(1) Average figures for composition are shown, except in the case of constituents for which the composition can vary widely, in which case ranges are given.

(2) The $[H^+]$ and $[Na^+]$ in gastric juice vary inversely with one another, together totalling approximately 120 mmol/L.

(3) The $[Na^+]$ and $[HCO_3^-]$ in pancreatic juice vary directly with one another, but the $[Cl^-]$ remains fairly constant.

and their loss as a result of vomiting, diarrhoea, paralytic ileus, etc., can quickly have serious effects.

Depletion of water and sodium

Water and Na^+ depletion usually occur together, and the clinical state of the patient shows features attributable to both water depletion and Na^+ depletion. States of water depletion (no depletion of Na^+) are uncommon, and states of Na^+ depletion (no depletion of water) can only develop under special circumstances. Nevertheless, discussion of these separate states of depletion helps with the understanding of the commonly occurring mixed state of water and Na^+ depletion, and will be considered first.

From first principles, the causes of any state of depletion can be considered in terms of (1) reduced intake, or (2) excessive losses by one or more routes (urine, intestinal tract, skin, lungs), or (3) a combination of these.

WATER DEPLETION

Reduction in water intake usually causes thirst, and the conscious response is to drink. If restoration of water balance is delayed, plasma osmolality rises and physiological responses occur, as follows:

(1) *Release of ADH,* which thereby reduces the continuing loss of water as urine.

(2) *Transfer of water* from the ICF to the ECF.
(3) *Renal tubular reabsorption* of Na^+ and Cl^- is increased.

These mechanisms all help to sustain the ECF volume, but only the first two help to minimise the tendency for the plasma osmolality to rise further.

The release of ADH causes urinary osmolality and sp. gr. to rise, but continuing loss of water via the kidneys is obligatory, if excretory functions are to be maintained; there is a consequent inevitable tendency for plasma osmolality to go on rising.

The increase in plasma osmolality caused by the reduction in water intake leads to a *transfer of water* from the ICF to the ECF, to restore equality of osmolality; any tendency to dehydration thus has effects on both the ICF and ECF (Fig. 4.1). This transfer of fluid from the ICF helps to maintain the plasma volume, and it is the loss of fluid from the ICF that is responsible for the sensation of thirst.

If the degree of water depletion becomes more severe, the third compensatory mechanism becomes important in the maintenance of ECF volume. *Renal tubular reabsorption* of Na^+ and Cl^- is increased,

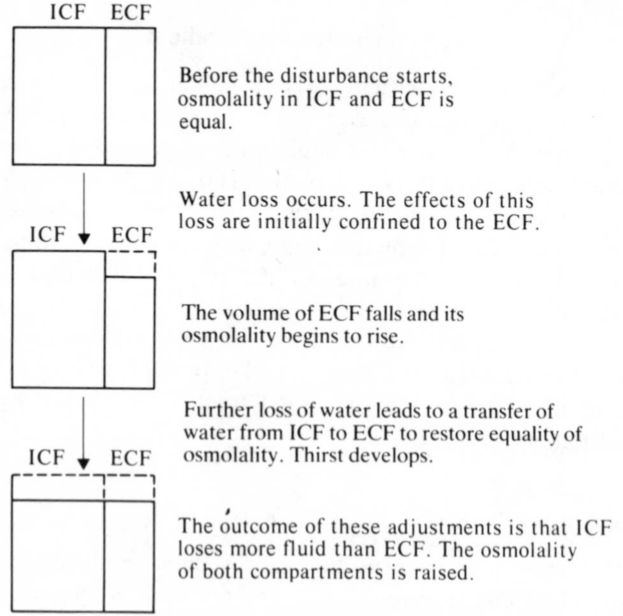

FIG. 4.1. *The development of a state of water depletion.* The dotted lines indicate the normal size of the ICF and the ECF compartments. If the water loss continues, there is transfer of both water and electrolytes between the ICF and ECF, but the osmolality of the two compartments remains equal.

accompanied by an obligatory but disproportionately small reabsorption of water. The effect on body fluid composition of this renal response to severe water depletion is a further rise in the osmolality of the ECF, and this leads to still further transfer of water from the ICF to the ECF. When dehydration reaches this degree of severity, the losses of water are shared by the ECF and the ICF in approximately equal proportions.

Even in health, when all three compensatory mechanisms are working normally, the body only has limited ability to adjust to severe reduction in water intake because of the continuing loss of water via the lungs, skin, faeces and urine. These obligatory losses total about 1300–1500 mL/24 h for a healthy adult in temperate climates. In febrile patients with temperatures above 39°C, or in the tropics, obligatory losses of water may be substantially greater and the effects of dehydration develop much more rapidly. Causes of pure or nearly pure water depletion are summarized in Table 4.5.

SODIUM DEPLETION

The volume and osmolality of the ECF both depend to a considerable extent upon the Na^+ content of the body. When this is reduced, due to the loss of Na^+-containing fluid, there is usually an isosmotic loss of fluid, initially almost entirely from the ECF. There is little or no alteration in plasma osmolality and, because there is therefore little or no transfer of fluid from the ICF, the thirst mechanism is not stimulated. Symptoms usually are vague, the patient complaining of weakness, apathy and loss of appetite. There is loss of skin elasticity

TABLE 4.5. Causes of water depletion.

	Examples
Reduced intake	
Unable to get water	Infancy, unconsciousness, extreme weakness
Water not available	People lost in a desert, shipwrecked
Inability to swallow	Any cause of dysphagia
Nausea	Various causes
Increased losses	
(1) Skin	Fever, hot and dry environment (tropics, stokers, etc.), thyrotoxicosis
(2) Lungs	Fever, hyperventilation
(3) Urine	Diabetes insipidus, nephrogenic diabetes insipidus

and hypotension. The reduction in ECF volume causes the following physiological responses:

(1) *Reduction in the glomerular filtration rate* (GFR).
(2) *Stimulation of aldosterone production.*
(3) *Secretion of antidiuretic hormone* (ADH).
(4) *Reduction in natriuretic hormone secretion.*

The first two of these responses are linked, because the reduction in GFR stimulates the renin-angiotensin system, and this leads to increased output of aldosterone (p. 373). Aldosterone exerts its effects on the distal convoluted tubule, promoting Na^+ reabsorption.

The secretion of ADH occurs in response to severe Na^+ depletion even if plasma [Na^+] is normal or decreased; ADH release occurs in response to both ECF volume depletion and hypertonicity of the ECF.

The fourth response, reduction in secretion of natriuretic hormone, is less certain. There is, however, growing evidence for the existence of a non-steroidal natriuretic hormone of hypothalamic origin, possibly similar in structure to the cardiac glycosides. Secretion of the natriuretic hormone is controlled by the left atrial pressure, and thus falls when ECF volume is depleted.

In health, these physiological responses all influence Na^+ conservation by the kidney and together prove to be so efficient in the face of Na^+ depletion that the urinary output of Na^+ can be reduced to as little as 1 mmol/day. Reduction in urinary Cl^- output closely parallels the fall in Na^+ excretion.

Patients may respond to the loss of Na^+-containing fluid by drinking water, or they may misguidedly be given water to drink as replacement or be treated inappropriately with parenteral fluids that do not contain Na^+. Under these latter circumstances it is possible for the initially isosmotic loss of Na^+-containing fluid, the common mixed deficiency state, to be changed into a pure deficiency of Na^+.

SIMULTANEOUS DEPLETION OF SODIUM AND WATER

This is by far the commonest form of Na^+ and water depletion. Physiological responses include reduction of the GFR, and release of both aldosterone and ADH. There is some redistribution of fluid between the ECF and ICF, but the ECF bears the fluid loss predominantly (Fig. 4.2).

Thirst and oliguria, symptoms of water depletion, occur and there may be non-specific symptoms attributable to Na^+ depletion. The physical signs, however, are predominantly those attributable to Na^+

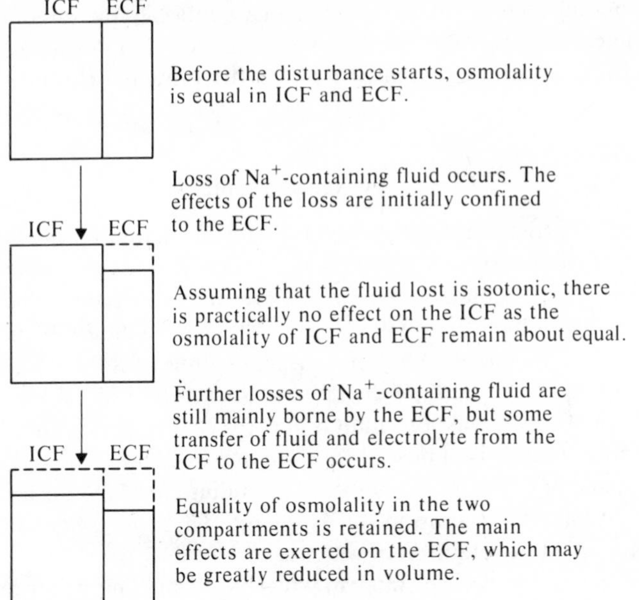

FIG. 4.2. The development of a mixed depletion of water and sodium.

depletion (loss of skin elasticity and hypotension). In addition, the tongue is dry and the intraocular tension low. The urine volume is usually low and there may be only traces of Na^+ in the urine.

The causes of mixed depletion of sodium and water, with examples of each, are summarized in Table 4.6.

CHLORIDE DEPLETION

Changes in Cl^- content of the body mostly follow passively the

TABLE 4.6. Causes of mixed depletion of water and sodium.

Route of loss	Examples
Alimentary tract	Vomiting; aspiration of gastrointestinal contents; watery diarrhoea (e.g. cholera); fistulous drainage; mucus-secreting tumours of colon; paralytic ileus
Urinary tract	Osmotic diuresis (e.g. diabetes mellitus); overdose of diuretics; renal tubular damage (e.g. diuretic phase of acute renal failure, relief of obstruction); chronic renal failure; mineralocorticoid insufficiency
Skin	Excessive sweating; extensive dermatitis; burns
Miscellaneous	Ascites; inferior vena caval thrombosis

corresponding changes in Na^+. The main exception to this general rule occurs in conditions involving loss of gastric secretion in which Na^+ is replaced, to a variable extent, by H^+ as the principal cation (Table 4.4).

Excess of water and sodium

Any tendency to overhydration normally results in inhibition of ADH release. If reduction in ADH output does not occur, or if the conditions responsible for the overhydrated state of the individual continue, a serious situation develops in which both the ECF and ICF are expanded, the increase in ICF causing swelling of cells. There is an upper limit to the capacity of the kidneys to excrete urine, even under physiological conditions; this limit is approximately 20 mL/min.

Retention of Na^+ is almost always accompanied by some retention of water, and of Cl^- in an amount corresponding closely to the amount of Na^+ retained. Retention of Na^+ and Cl^-, or other anions, unaccompanied by retention of water is very rare.

There are many disease states in which Na^+, Cl^- and water are all retained as components of an approximately isosmotic fluid. These patients are liable to become oedematous.

OEDEMA

The causes of generalized oedema can be classified aetiologically, the main categories being cardiovascular, hypoproteinaemic, nutritional and renal states. Their pathophysiological basis usually involves two or more of the following mechanisms:

(1) *Na^+ retention.* This is often due to secondary hyperaldosteronism (p. 375) in patients with cardiac failure, liver disease, etc. Aldosteronism, whether primary or secondary, does not by itself cause oedema, but needs some additional contributing factor (e.g. hypoproteinaemia, increased venous pressure). Reduction of GFR can sometimes, by itself, cause Na^+ retention.

(2) *Hypoproteinaemia.* This is usually due to defective synthesis or abnormal losses of proteins, especially albumin. More rarely, increased capillary permeability results in the leakage of albumin from the plasma compartment to the interstitial fluid, as in angioneurotic oedema. Whatever the cause of the hypoproteinaemia, reabsorption of interstitial fluid at the venous end of the capillaries is reduced.

(3) *Increased venous pressure*. The hydrostatic effect that this exerts interferes with the reabsorption of interstitial fluid at the venous end of the capillaries.

In all these states, there is an increase in the amount of Na^+-containing fluid in the body. However, as discussed later, these patients usually have a low plasma [Na^+], as the increase in total body Na^+ is not as great, proportionately, as the increase in total body water.

Investigation of water and sodium metabolism

By means of suitable isotope methods, it is possible to measure total body water (using ^3H_2O), ECF volume (using $^{82}Br^-$ or $^{35}SO_4^{2-}$), blood volume (by injecting ^{51}Cr-labelled erythrocytes) and plasma volume (by injecting ^{125}I-albumin). However, these measurements are not practicable in severely ill patients, nor are measurements of total body Na^+ (by *in vivo* neutron activation analysis to produce $^{24}Na^+$) or total exchangeable Na^+ (by isotope dilution using $^{24}Na^+$ or $^{22}Na^+$). There are, instead, several much simpler methods that can be used to provide sufficient information for most practical purposes.

The measurements to be considered here include plasma osmolality, [Na^+] and [Cl^-], and urinary osmolality or sp. gr., Na^+ and Cl^- output. The results require to be interpreted in relation to the history of the patient's illness, clinical findings, fluid balance charts and results of [haemoglobin] and PCV measurements, which are usually carried out in haematology departments. Plasma [K^+] and acid–base measurements are discussed later.

Plasma [urea] usually increases in patients with fluid depletion, and is widely used by clinicians as a crude index of the amount of fluid lost from the body. However, so many variables can affect plasma [urea], such as diet, renal function, hepatic function, tissue damage and blood loss, that the measurement is of limited value in assessing the state of hydration of a patient. Plasma [total protein] and [albumin] are also crude indices of the state of hydration for much the same reasons.

Osmolality measurements

The physiological control mechanisms discussed above help to maintain the volume of ECF and its osmolality constant. The osmolality of plasma and serum is normally between 285 and 295 mmol/kg. On the other hand, urine osmolality can vary markedly, depending on the state of hydration; it can fall as low as 80 mmol/kg or rise as high as 1200 mmol/kg, if renal function is normal. Urinary sp. gr. can similarly range from 1.000 to 1.040, relative to the sp. gr. of water at 15°C.

Plasma osmolality depends upon the number of particles of solute dissolved in a litre (or kg) of plasma water. As the principal extracellular ion, Na^+ is the largest single contributor to the plasma osmolality, the next largest contributors normally being its accompanying cations, Cl^- and HCO_3^-. As a general rule, plasma osmolality is approximately double the plasma $[Na^+]$. It is common practice, therefore, to use measurements of plasma $[Na^+]$ to provide an approximate index of plasma osmolality. The equipment for measuring plasma $[Na^+]$ is more widely available, and the techniques of measurement are more suited to automated analysis, than is the case with plasma osmolality.

Osmometers are now available in many hospital laboratories, and plasma osmolality should be measured if results of plasma $[Na^+]$ are proving difficult to interpret. The urine osmolality should be *appropriate,* when considered in relation to the plasma osmolality. If the urine does not reflect an appropriate osmolality, either there is inappropriate secretion of ADH (or an ADH-like substance) or there is impaired renal function.

Measurements of sodium

Two fundamentally different methods of measuring sodium are now routinely used in clinical chemistry laboratories. These are emission flame photometry and ion-selective electrode (ISE) systems. The results of both types of measurement are reported as concentrations and, under most circumstances, the two techniques give closely similar results.

Marked differences between these two techniques can occur with plasma (or serum) measurements in patients with severe hyperlipidaemia or hyperproteinaemia, in whom the lipids or the proteins, respectively, may occupy a significant proportion of the plasma volume. Unless the conditions are recognized and results expressed in terms of plasma water rather than in terms of plasma volume, as they normally are, the plasma $[Na^+]$ will be misleadingly low when measured by flame photometry or an ISE technique which uses *diluted* plasma. However, results for plasma $[Na^+]$ determined by an ISE technique that measures $[Na^+]$ in *undiluted* plasma will not be adversely affected. In these patients, results for plasma osmolality conform with estimates obtained by doubling the Na^+ results obtained by an ISE method that uses undiluted plasma, but may differ markedly from figures calculated from flame photometric data or from ISE measurements on diluted plasma.

It is important, for the proper interpretation of all plasma electrolyte measurements, to stress the limitations imposed by the fact that laboratory results are expressed in units of concentration, e.g. plasma $[Na^+]$ = 135 mmol/L. Considered in isolation, such data do not provide information about the *amount* of Na^+ in the body as a whole, nor in the ECF, nor even in the plasma. Instead, they merely indicate the *ratio* between the amount of Na^+ in the ECF (or plasma water) and the volume of the ECF.

The limitations inherent in plasma electrolyte measurements are illustrated in Fig. 4.3. It shows a substance (e.g. Na^+) entering the plasma space by various possible routes, at a net rate indicated by the arrow marked K_1; the same substance is shown leaving the plasma space at a net rate indicated by the arrow marked K_2. The total amount of any substance in the plasma compartment is influenced by the rate constants, K_1 and K_2. Input equals output when a steady state exists, i.e. when $K_1 = K_2$.

It is only possible to determine the total amount of a substance, such as the amount of Na^+ in the plasma compartment, if both plasma $[Na^+]$ and plasma volume are known. Although the plasma volume can be determined reliably, this measurement is at present only made in a very small percentage of the patients on whom plasma $[Na^+]$ and related chemical investigations are so frequently performed. Since the plasma volume is very often abnormal in patients with water and electrolyte disturbances, it follows that measurements of plasma $[Na^+]$ cannot by themselves be used to calculate the total content of Na^+ in the plasma compartment or in the ECF.

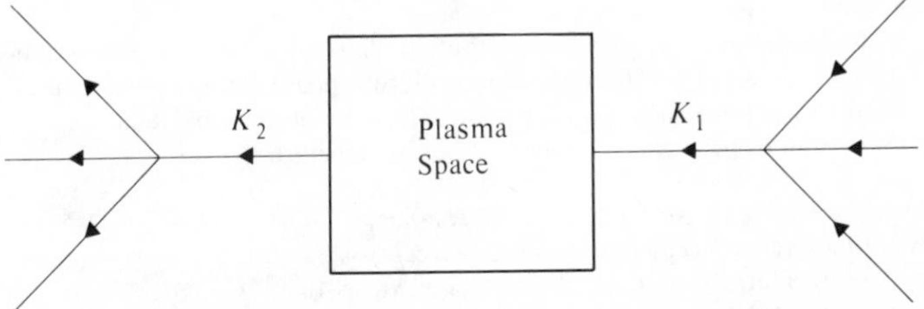

FIG. 4.3. *The movement of substances into and out of the plasma space*. Substances may enter from several sources (e.g. absorption from the gastrointestinal tract, transfer from ICF, renal tubular reabsorption, intravenous infusion), the net rate of entry being indicated by the arrow marked K_1. Substances may also leave by various routes (e.g. renal excretion, transfer to ICF, sweat, expired air, losses from gastrointestinal tract), the net rate of exit being indicated by the arrow marked K_2. At equilibrium, the rate constants are equal, i.e. $K_1 = K_2$.

Urinary Na^+ measurements, by flame photometry or ISE methods, can be used to monitor the daily excretion of Na^+ by this route. Similar measurements can be made, and are occasionally needed, on fluids lost from the gastrointestinal tract, but the losses of Na^+ in these can usually be gauged sufficiently closely from measurements of volume and the data for composition given in Table 4.4, and replacement therapy planned accordingly.

DECREASED PLASMA SODIUM CONCENTRATION (Table 4.7)

Hyponatraemia may be defined as a plasma $[Na^+]$ below 130 mmol/L (reference values, 132–144 mmol/L). Since Na^+ is the principal cation in the ECF, and by far the largest single contributor to plasma osmolality, hyponatraemia and hypo-osmolality always co-exist. A classification of decreased plasma $[Na^+]$, with examples of causes, is shown in Table 4.7. It will be evident that these causes are so diverse that it would be impossible to interpret a result for plasma $[Na^+]$ of, for example, 120 mmol/L without taking clinical features and other relevant data into account.

In patients with increased total body Na^+ and water retention giving rise to hyponatraemia, the urinary Na^+ is usually less than 20 mmol/L

TABLE 4.7. Hyponatraemia: classification and examples of causes.

With increased total body Na^+ and water retention
- Secondary hyperaldosteronism (e.g. cardiac failure, liver disease)
- Hypoproteinaemia (e.g. nephrotic syndrome, protein-losing gastroenteropathy)
- Reduced GFR (acute renal failure and severe chronic renal failure)

With decreased total body Na^+ and water deficit
- Abnormal fluid losses, inadequately replaced:
  (1) *Gastrointestinal* (e.g. vomiting, aspiration, diarrhoea, fistula, paralytic ileus)
  (2) *Renal* (e.g. excessive diuretic therapy, diuretic phase of acute renal failure, analgesic nephropathy, polycystic kidneys, mineralocorticoid deficiency)
  (3) *Skin* (e.g. tropical climate, febrile states, severe burns)

With normal total body Na^+ (excessive ADH effect)
- CNS disease (e.g. encephalitis, tumour, abscess)
- Drugs (e.g. chlorpropamide, carbamazepine, vincristine)
- Glucocorticoid deficiency
- Inappropriate ADH secretion (p. 395)

Artefact
- Dilution of blood specimen with intravenous fluid from a drip
- Marked hyperlipidaemia or hyperproteinaemia (if undetected and plasma $[Na^+]$ measured but not plasma osmolality)

unless renal failure is the underlying cause, in which case it is usually greater than 20 mmol/L and there is other chemical evidence of renal insufficiency (e.g. reduced creatinine clearance). In severe degrees of secondary hyperaldosteronism or severe hypoproteinaemia, urinary Na^+ may be less than 5 mmol/L.

In patients with decreased total body Na^+ and water deficit giving rise to hyponatraemia, urinary Na^+ measurements may help to determine the cause. If the hyponatraemia is due to a renal cause, urinary $[Na^+]$ may exceed 20 mmol/L, whereas hyponatraemia due to extrarenal Na^+ loss is usually associated with a urinary $[Na^+]$ of less than 10 mmol/L.

Patients in whom there is excessive ADH effect, giving rise to hyponatraemia, may be asymptomatic unless they develop water intoxication. Urinary Na^+ excretion changes in response to alterations in Na^+ intake.

It should be apparent, from the above discussion, that it can be very dangerous to treat patients with hyponatraemia by means of hypertonic sodium chloride infusions, under the mistaken impression that there is Na^+ depletion. Usually, these patients already have too much Na^+ in their bodies.

INCREASED PLASMA SODIUM CONCENTRATION (Table 4.8)

Hypernatraemia may be defined as a plasma $[Na^+]$ above 146 mmol/L (reference values 132–144 mmol/L). It occurs much less often than hyponatraemia and it is always accompanied by hyperosmolality. A classification of hypernatraemia, with examples of causes, is given in Table 4.8; it is nearly always due to water loss, due to one of the causes in the first two categories listed.

TABLE 4.8. Hypernatraemia: classification and examples of causes.

With low total body Na^+
- Extrarenal (e.g. severe sweating, diarrhoea)
- Renal (osmotic diuresis — e.g. glucose, mannitol, urea)

With normal total body Na^+
- High temperatures (tropics, febrile states)
- Diabetes insipidus (hypothalamic, nephrogenic)

With high total body Na^+
- Steroid excess (Cushing's syndrome, Conn's syndrome, steroid therapy)
- Iatrogenic (hypertonic Na^+ - containing infusions, dialysis against hypertonic Na^+ - containing fluids)
- Self-induced (ingestion of NaCl tablets)

Hypernatraemia with low total body Na^+ is caused by loss of hypotonic Na^+-containing fluids. If the loss is extrarenal, urinary osmolality is usually high (above 800 mmol/kg) and urinary [Na^+] less than 10 mmol/L. On the other hand, if the cause is renal, urinary osmolality is usually low (below 300 mmol/kg) and urinary [Na^+] greater than 20 mmol/L.

Hypernatraemia due solely to loss of water may occur in a hot environment or if the patient has a severe fever; in these patients, urinary osmolality is usually considerably above 300 mmol/kg. By contrast, patients with diabetes insipidus mostly pass urine with an osmolality less than 300 mmol/kg. In this group of patients, urinary Na^+ alters in response to changes in Na^+ intake.

Hypernatraemia with high total body Na^+ is much the least common of the three categories listed in Table 4.8. Plasma [Na^+] is only slightly increased, if indeed it is abnormal, in patients with Cushing's syndrome or primary hyperaldosteronism. The iatrogenic causes are usually self-evident. Urinary osmolality is above 300 mmol/kg and urinary [Na^+] usually greater than 20 mmol/L.

CHANGES IN PLASMA CHLORIDE CONCENTRATION

These in general accompany alterations in plasma [Na^+], a reduction in plasma [Na^+] being associated with a reduction in plasma [Cl^-], and *vice versa.* Because of this, in most instances no additional diagnostic information is provided by determining plasma [Cl^-], and many laboratories have stopped performing this test as a routine investigation. However, there are some conditions where this generalization does not hold:

Decreased plasma [Cl^-]. The electrolyte composition of gastric juice is markedly different from plasma (Table 4.4). Patients with substantial losses of gastric contents (e.g. due to pyloric stenosis) often show a decrease in plasma [Cl^-] that is disproportionately marked compared with any hyponatraemia that may develop. In addition, they show evidence of dehydration and metabolic alkalosis, with a markedly increased plasma [total CO_2].

Increased plasma [Cl^-], out of proportion to any accompanying increase in plasma [Na^+], may occur in (1) patients with metabolic acidosis due to chronic renal failure, ureteric transplants into the colon, renal tubular acidosis, or treatment with carbonic anhydrase inhibitors. It may also occur in (2) patients who develop respiratory alkalosis as a result of prolonged assisted ventilation, and (3) patients given excessive amounts of 'normal' or 'physiological' saline, which contains [NaCl] at 154 mmol/L, to be compared with the average normal plasma [Na^+] and [Cl^-] of 138 mmol/L and 101 mmol/L respectively.

POTASSIUM METABOLISM

The total body K^+ in an average 70 kg adult is about 3500 mmol, of which only about 50–60 mmol are in the ECF. Potassium is the main intracellular cation, and the [K^+] in ICF is about 25–30 times the [K^+] in ECF. There is a tendency for K^+ to diffuse out of cells, down this concentration gradient, just as there is a tendency for Na^+ to diffuse into cells. These ionic exchanges across the cell membrane are interlinked, and the maintenance of these gradients is dependent on the supply of energy, as ATP. In addition, the distribution of K^+ is influenced by a number of important factors, several of which may operate simultaneously in some patients.

Potassium is lost from the cells whenever dehydration occurs, and returns into the cells when dehydration is corrected. In addition, K^+ moves out of the cells in any acid-base disturbance resulting in an increase in ECF [H^+], the movement occurring mainly in exchange for H^+; conversely, a fall in ECF [H^+] causes K^+ to move into cells.

Insulin exerts a marked effect on the uptake of K^+ by cells, injections of insulin promoting the entry of K^+ and glucose into cells. Conversely, in conditions of relative lack of insulin (e.g. poorly controlled diabetes mellitus), there is a marked loss of K^+ from the ICF (as much as 500–1000 mmol). Aldosterone and cortisol influence the distribution of K^+ between the ECF and ICF, as do certain drugs (e.g. digitalis glycosides).

In catabolic states, whenever there is breakdown of intracellular protein (e.g. after a major operation, severe infection, starvation), K^+ is lost from the cells. Conversely, K^+ is taken up in anabolic states.

Depletion of potassium

There is a small obligatory daily loss of K^+ in the faeces and through the skin, but the main source of loss is in the urine (Table 4.3). The kidney responds less well to K^+ restriction than to Na^+ restriction and depletion of body K^+ can occur all too easily. Depletion of K^+ has adverse effects on renal function. Increased urinary loss of K^+ may be caused by:

(1) *Primary renal disease* (e.g. renal tubular acidosis).

(2) *Extrarenal factors* affecting the renal excretion of K^+ (e.g. alkalosis, diuretics, aldosteronism).
(3) *Dehydration.*
(4) *Catabolic states.*

Diuretics may cause K^+ depletion by a variety of mechanisms, but do so most commonly by decreasing proximal tubular reabsorption of Na^+, thereby causing the distal tubule to exchange more K^+ for Na^+ in order to retain Na^+; the urinary output of K^+ consequently rises.

Disturbances of acid-base balance and disturbances of K^+ metabolism often co-exist and deserve special consideration. In acute acidosis (e.g. acute renal failure) there is retention of K^+ by the kidney whereas some conditions giving rise to a chronic acidosis (e.g. distal renal tubular acidosis) are associated with loss of K^+ in the urine. Renal excretion of K^+ also tends to be increased in alkalosis, because K^+ and H^+ in the tubular cells compete with one another for exchange with Na^+ in the lumen of the distal tubules (p. 96); under these conditions the tubular cells retain H^+ preferentially and excrete K^+.

Excess of potassium

This is a potentially dangerous condition, caused by the ingestion or administration of K^+ in excess of the body's ability to eliminate it via the kidneys. Although invariably associated with an increased plasma $[K^+]$, potassium excess is only one of several causes of hyperkalaemia and the two terms must be distinguished from one another.

Investigation of potassium metabolism

Total body K^+ can be measured by a whole body counter, used to measure the naturally occurring isotope $^{40}K^+$. Total exchangeable K^+ can be measured by isotope dilution, using $^{42}K^+$. Neither of these measurements is readily practicable in severely ill patients, and both are expensive and time-consuming.

Plasma and serum $[K^+]$, measured by emission flame photometry or by an ISE system, are among the tests most frequently performed by clinical chemistry laboratories. As with the corresponding Na^+ measurements (p. 48), the two techniques usually give results that are closely similar; both are reported as concentrations. The reasons for discrepancies (when these occur) between the two methods are the same as for Na^+. Leucocyte $[K^+]$ can be measured, and provides an index of ICF $[K^+]$, but this test has not yet been widely accepted.

Urinary K^+ can be readily measured by flame photometry or ISE methods, as can K^+ output in other fluids, e.g. gastrointestinal secretions.

Alterations in plasma [K^+] give little information about the body content of K^+. The predominantly intracellular distribution of K^+ in the body, and the existence of so many factors that exert an influence on its distribution, make it difficult to relate changes in plasma [K^+] to changes in K^+ balance. Some patients may have developed a severe negative K^+ balance but still have a normal or even an increased plasma [K^+], due to a shift of K^+ from the ICF to the ECF.

Results for plasma [K^+] determinations need to be interpreted in relation to the history of the patient's illness, clinical state, fluid input and output charts, and assessment of the possible effects of complicating factors such as pyrexia and paralytic ileus. Other investigations that may prove helpful include plasma [urea], plasma [total CO_2] or full assessment of acid-base status, measurements of urinary K^+ output and of K^+ lost via abnormal routes (e.g. fistula). These investigations may all be required in a severely ill patient. In less acute circumstances, investigation of the kidney's ability to excrete H^+ may also be indicated.

DECREASED PLASMA POTASSIUM CONCENTRATION (Table 4.9)

Hypokalaemia may be defined as a plasma [K^+] below 3.0 mmol/L (reference values, 3.2–4.6 mmol/L). Symptoms attributable to hypokalaemia usually do not develop until plasma [K^+] is well below 3.0 mmol/L. Even then there may be no symptoms or, more often, only non-specific feelings of weakness and in some cases drowsiness and mental confusion. The onset of symptoms is related to the rapidity of onset of the hypokalaemia. There may be paralytic ileus, loss of tendon reflexes and cardiac arrhythmias.

The main groupings of causes of decreased plasma [K^+] are shown in Table 4.9. The subdivisions are not rigid, and alternative classifications are possible. More important, however, is the fact that these various causes may give rise to negative K^+ balance without necessarily producing hypokalaemia initially. Often, decreased plasma [K^+] is not observed until losses of water and Na^+ have been corrected, but without proper attention having been paid to K^+ requirements.

Inadequate intake of K^+ may occur after operations, especially major abdominal operations, or whenever intake is insufficient to keep pace with losses. Abnormal losses of K^+ may occur in gastrointestinal

TABLE 4.9. Hypokalaemia: classification of causes, with examples.

Inadequate oral intake (Very rarely a cause, by itself)

Abnormal losses
(1) *Gastrointestinal*
Chronic laxative abuse
Diarrhoea and vomiting
Gastric aspiration and fistulous drainage
Villous papilloma of the colon
(2) *Renal*
Diuretic therapy (e.g. thiazides, frusemide)
Hyperaldosteronism, primary and secondary
Glucocorticoid excess
Renal tubular acidosis
Bartter's syndrome

Abnormal losses plus inadequate replacement
(1) Post-operative states
(2) Diuretic phase of acute renal failure

Redistribution of K^+ between ECF and ICF
(1) Alkalosis, respiratory or metabolic
(2) Insulin overdose, insulinoma
(3) Familial periodic paralysis

Artefact
Specimen collected from infusion site

secretions (vomit, diarrhoea, fluid from a fistula) or in urine (e.g. thiazide diuretics, renal tubular acidosis, diuretic phase of acute renal failure, Bartter's syndrome).

Acute shifts of K^+ between ECF and ICF can be produced by increased plasma [insulin], and hypokalaemia may be observed during treatment of diabetic coma, or in patients with insulin overdosage or with insulinoma. Steroid-induced hypokalaemia is a feature of hyperaldosteronism and may be seen in patients with glucocorticoid excess (e.g. Cushing's syndrome, steroid therapy).

Decreased plasma [K^+] is frequently reported when blood specimens are collected from a limb into which an intravenous infusion is being administered. These results are spurious.

INCREASED PLASMA POTASSIUM CONCENTRATION (Table 4.10)

Hyperkalaemia can be defined as a plasma [K^+] above 5.0 mmol/L. However, symptoms attributable to hyperkalaemia usually do not develop until plasma [K^+] exceeds 6.0 mmol/L, and even then they are not always present. Symptoms, which tend to be vague, include a feeling of weakness and mental confusion. There may be loss of tendon reflexes and ECG abnormalities.

The causes of hyperkalaemia are listed in Table 4.10. In practice, in many laboratories, high plasma [K^+] is more often due to artefacts than to pathological causes. These artefacts are due to the deterioration of blood specimens that can occur between the time of specimen collection and the time of separation of plasma or serum from the cells. Separation should be effected within 2 h of collection, although changes in plasma [K^+] due to deterioration of the specimen may not be marked for as long as 4 hours.

Specimens of blood for electrolyte determinations frequently have to be collected in the evening and held overnight prior to analysis, for instance in a general practitioner's surgery, if a patient is not to be asked to return the following day to have the specimen collected. It is essential with these specimens to separate the plasma from the cells before storing the sample. Plasma should be aspirated gently (a clean syringe and needle are needed) and transferred to another container. Blood specimens collected for electrolyte analysis must not be refrigerated until plasma or serum has been separated from the cells. Refrigeration promotes the loss of K^+ from the erythrocytes, without

TABLE 4.10. Hyperkalaemia: classification of causes, with examples.

Excessive oral intake (Very rarely a cause, by itself)

Diminished renal excretion
- (1) Acute renal failure
- (2) Chronic renal failure
  - (a) If GFR falls below 20 mL/min
  - (b) K^+-sparing diuretics (e.g. amiloride, spironolactone)

Excessive intake in terms of excretory capacity
- (1) Post-operative states combined with excessive parenteral K^+
- (2) Renal insufficiency combined with excessive oral K^+

Redistribution of K^+ between ECF and ICF
- (1) Haemolysis (e.g. incompatible blood transfusion, disseminated intravascular coagulation, auto-immune states)
- (2) Acidosis, respiratory or metabolic
- (3) Tissue necrosis (e.g. major trauma, burns)

Artefact
- (1) Faulty technique when collecting specimens
- (2) Delay in separation of plasma/serum
- (3) Improper storage conditions

necessarily producing obvious evidence of haemolysis. Haemolysis, for instance due to freezing of blood samples, renders them unsuitable for many chemical analyses, especially determination of plasma [K^+].

Pathological causes of hyperkalaemia include acute and chronic renal failure, and K^+ overload is very liable to occur due to excess dietary intake in patients with renal insufficiency. Hyperkalaemia can be produced by excessive intake of K^+-containing drinks (e.g. fruit juices) or excessive intravenous administration of K^+-containing fluid, especially if this occurs during hypercatabolic states such as develop shortly after a major surgical operation.

Poorly controlled diabetes mellitus is associated with protein catabolism, dehydration, acidosis and insulin lack, all of which tend to promote the loss of K^+ from cells and hyperkalaemia is often observed. It is also a feature of severe adrenocortical insufficiency (Addison's disease).

The increased plasma [K^+] that is sometimes observed in patients with leukaemia has been shown to be due to abnormal fragility of the leucocytes. If plasma is separated soon after collection of the blood specimen, and with extra care when centrifuging, plasma [K^+] is usually found to be normal in these patients.

FLUID BALANCE: POSTOPERATIVE MANAGEMENT

Patients admitted for major elective surgery, and who might be liable to develop disturbances of water and electrolyte balance postoperatively, require preoperative clinical assessment and determination of a set of baseline values for plasma urea, Na^+, K^+ and total CO_2 concentrations. It is nowadays also common practice to assess liver function (p. 189) preoperatively.

Patients who present with disturbances of water and electrolyte metabolism require to have the severity of the disturbance assessed and corrective measures instituted. Assessment is usually on the basis of the history, physical examination and initial values for plasma urea and electrolyte measurements. The disturbance should, if possible, be corrected preoperatively in patients requiring surgical treatment.

Metabolic response to trauma

Accidental and operative trauma produce a number of metabolic effects. These include breakdown of protein, release of K^+ from cells and a consequential deficit due to urinary loss, temporary retention of water, utilization of glycogen reserves, gluconeogenesis, mobilization of fat reserves and a tendency to ketosis sometimes progressing to a metabolic acidosis. Several hormonal responses are involved, including increased secretion of adrenal corticosteroids, with temporary abolition of feedback control, aldosterone and ADH.

The metabolic responses to trauma are physiological and appropriate. These reponses are complex and explain why 'postoperative states' are such frequent causes of temporary disturbances in electrolyte metabolism. The complexity of the homeostatic mechanisms, and the various effects of their disturbance due to trauma, mean that there is no consistent pattern of abnormality although increased plasma [urea] is frequently observed. Injudicious fluid therapy, especially in the first 48 hours after operation, may 'correct' the chemical abnormalities (e.g. lower the plasma [urea]) but only by causing dangerous retention of fluid.

Postoperatively, any tendency for patients to develop disturbances of water and electrolyte balance can be minimised by regular follow-up clinical assessment, supplemented by careful maintenance of fluid balance and temperature records. Plasma urea and electrolyte concentrations may need to be measured daily, initially, and plasma [albumin] and other indices of liver function at weekly intervals.

Fluid balance charts

The value of maintaining accurate input and output charts cannot be overemphasized. Patients on whom these charts are required should have their fluid input and output regularly and accurately recorded with all volumes recorded in the same units of measurement. These observations should be extended to include the approximate volume of fluid lost via abnormal routes, as in drainage from a fistula. Fluid balance charts and records provide information that is of value in calculating the approximate volume and composition of fluid needed to replace continuing losses.

It is often helpful to determine the urinary excretion of Na^+ and K^+ in patients with postoperative disturbances of water and electrolyte metabolism, especially if the disturbance is severe and if it continues for 2 days or longer. In patients who are losing large volumes of gastrointestinal secretions, it may be sufficient to collect the secretions, measure their volume, and use data such as those contained in Table 4.4 for calculating the replacement fluid requirements. However, a more precise estimate of needs can be obtained when large amounts of fluid are being lost, by measuring the fluid $[Na^+]$ and $[K^+]$ as well as the volume.

The range of fluids available for use in replacement treatment varies from one hospital to another and local experience and preferences are important factors in everyday practice. Guidance as to the most appropriate fluid to be used in particular circumstances, as well as precautions to be observed in respect of quantity or rate of administration (e.g. when administering K^+-containing fluids) can be obtained from textbooks of medicine. Table 4.11 lists examples of some simple solutions widely used for parenteral administration.

Parenteral nutrition

It is now widely recognized that fluid replacement therapy, especially if this is to depend on prolonged administration of parenteral fluids, must provide more than simply water, Na^+, Cl^-, K^+ and the small amount of energy that is usually given in the form of glucose. Unless adequate amounts of energy are provided, extensive catabolism of the body's own energy stores will occur, and this only serves to delay the recovery and increase the risk of complications. Tissue healing is delayed unless the body's requirements for amino acids, vitamins and trace elements are met.

There are restrictions on the type of fluid that can be administered by a peripheral vein, and seriously ill patients can only be supported for limited periods in this way. Table 4.12 contains details of one parenteral regime that can be administered satisfactorily for 2–3 days to postoperative patients requiring parenteral feeding and in whom nutritional complications have not developed.

TABLE 4.11. Simple solutions for parenteral use.

	5% glucose*	Glucose-saline*	'Normal' saline	Hartmann's solution
Glucose (g/L)	50	43	—	—
Energy				
MJ/L	0.83	0.70	—	0.04
Cal/L	200	170	—	10
Ionic constituents (mmol/L)				
Na^+	—	31	154	130
K^+	—	—	—	5
Ca^{2+}	—	—	—	2
Mg^{2+}	—	—	—	2
Cl^-	—	31	154	112
Lactate	—	—	—	27

*Often called 5% dextrose and dextrose-saline.

Some patients may require prolonged administration of parenteral fluids. Nutritional deficiencies are very liable to develop, and can seriously affect prognosis, unless preventive measures are taken. Usually these patients require a central venous catheter, through which it is possible to administer hypertonic solutions and to provide adequate amounts of water, energy, amino acids, Na^+, K^+, Ca^{2+}, Mg^{2+}, Cl^-, phosphate, trace elements (Zn^{2+}, iron, Mn^{2+}, Cu^{2+}) and water-soluble vitamins without imposing an excess load of Na^+ or water.

Total parenteral nutrition for long periods and intravenous hyperalimentation are specialised and complex subjects that are beyond the scope of this book. Table 4.13 lists the unit content of six preparations that can be satisfactorily administered via a central venous catheter. It also indicates the number of units of each preparation to be combined in order to provide the six different types of intake described. Two of these regimes are designed to administer larger amounts of energy than are normally required by patients at rest in bed, these being examples of intravenous hyperalimentation.

Patients who are being maintained for long periods by total

TABLE 4.12. Example of a daily regimen for parenteral nutrition able to be given for 2–3 days via a peripheral vein (from Ellis *et al.*, 1976).

Solutions

Volumes (mL)	Description
1000	Vamin glucose *
1000	Dextrose (10% w/v)
500	Intralipid
500	Hartmann's solution (Table 4.11)

Additives

Potassium, multivitamin solution (added to dextrose or to Hartmann's solution)
Vitamin K as necessary, and folic acid (15 mg/week) i.m.
Trace elements (Addam solution*)

Amounts administered (approx.)

		Inorganic constituents		
Energy	6.7 MJ (1600 Cal)	Na^+	115	mmol
Water	2800 mL	K^+	25	mmol (plus additions)
Carbohydrate	200 g	Ca^{2+}	7.5	mmol
		Mg^{2+}	3	mmol
Protein	70 g	Phosphate	15	mmol
Fat	50 g	Zinc, iron, manganese and copper in trace amounts		

* KabiVitrum Ltd., Bilton House, Uxbridge Road, London W5 2TH.

TABLE 4.13. Examples of total parenteral nutrition regimes for severely ill patients (from Ellis *et al.*, 1976).

Solutions (Contents per 500 mL bottle)

	Solution	MJ	Carbohydrate (g)	Protein (g)	Fat (g)	Inorganic constituents
I	Vamin glucose*	1.36	50	35	—	Na, K, Cl, Mg, Ca
II	Aminosol*	0.69	—	48	—	Na, K, Cl
III	Electrolyte solution A with 20% glucose ∅	1.67	100	—	—	Mg, Ca, Cl (Zn, Mn)
IV	Electrolyte solution B with 20% glucose ∅	1.67	100	—	—	K, PO_4
V	Glucose 50%	4.18	250	—	—	—
VI	Intralipid 20%	4.18	—	—	100	—

Composition of regimen

	Combinations of solutions I–VI						Composition		
Description	I	II	III	IV	V	VI	MJ	Na (mmol)	K (mmol)
Low sodium	2	—	1	1	—	1	10.24	50	50
Standard	1	1	1	1	—	1	9.57	105	40
High sodium	—	2	1	1	—	1	8.90	160	30
Fat-free	1	1	1	1	1	—	9.57	105	40
Increased energy	1	1	1	1	1	1	13.75	105	40
Increased energy, low sodium	2	—	1	1	1	1	14.42	50	50

Additives

Multivitamin (B vitamin) preparation, folic acid and vitamin K. Also potassium, as required.

* KabiVitrum Ltd., Bilton House, Uxbridge Road, London W5 2TH.

∅ Travenol Laboratories Ltd., Caxton Way, Thetford, Norfolk IP24 3SE.

parenteral nutrition should have plasma [urea], [Na^+], [K^+] and [total CO_2], and urinary Na^+ and K^+ output determined daily. Liver function should be assessed weekly, as should plasma [calcium], [phosphate] and [magnesium]. Plasma [zinc] and [iron] need to be determined every 2–3 weeks if the illness is very prolonged. In addition to chemical measurements, these patients require regular haematological investigations (haemoglobin, full blood count, folate and vitamin B_{12}) and frequent blood cultures as the response to infection may be masked.

FURTHER READING

BARON, D.N. and LEVIN, G.E. (1978). Intracellular chemical pathology. In *Recent Advances in Clinical Biochemistry,* **1,** 153–174. Ed. K.G.M.M. Alberti. Edinburgh: Churchill Livingstone.

BERL, T. and HENRICH, W.L. (1980). Disorders of hydration. In *Renal Function Tests,* pp. 101–118. Ed. C.G. Duarte. Boston: Little, Brown & Company.

British National Formulary No. 7 (1984). Section on intravenous nutrition, pp. 283–286. London: British Medical Association.

ELLIS, B.W., STANBRIDGE, R. de L., FIELDING, L.P. and DUDLEY, H.A.F. (1976). A rational approach to parenteral nutrition. *British Medical Journal,* **1,** 1388–1391.

GENNARI, F.J. (1984). Serum osmolality: Uses and limitations. *New England Journal of Medicine,* **310,** 102–105.

HATFIELD, A.R.W. (1982). Hyperalimentation. *British Journal of Hospital Medicine,* **28,** 220–233.

JOHNSTON, I.D.A. (1972). The endocrine response to trauma. *Advances in Clinical Chemistry,* **15,** 255–285. New York: Academic Press.

JONES, N.F. (1970). Potassium deficiency: A reappraisal. In *Sixth Symposium on Advanced Medicine,* pp. 203–214. Ed. J.D.H. Slater. London: Pitman Medical.

LADEFOGED, K. and JARNUM, S. (1978). Long-term parenteral nutrition. *British Medical Journal,* **2,** 262–266.

WARDENER, H.E. de and MACGREGOR, G.A. (1982). The natriuretic hormone and essential hypertension. *Lancet,* **1,** 1450–1454.

WILLATTS, S.M. (1982). *Lecture Notes on Fluid and Electrolyte Balance.* Oxford: Blackwell Scientific Publications.

WOOLFSON, A.M.J. (1983). Artificial nutrition in hospital. *British Medical Journal,* **287,** 1004–1006.

Chapter 5

Acid–Base Balance and Oxygen Transport

The hydrogen ion concentration of extracellular fluid is normally regulated with great precision, despite the large amount of acid generated each day as an end-product of metabolism. Regulation depends on efficient buffer mechanisms and physiological responses. The lungs and kidneys are both involved in the elimination of the acids produced as a result of metabolism.

Carbonic acid (H_2CO_3), which is potentially volatile, is the acid generated in by far the largest amount as a result of metabolism, and the lungs (by excreting CO_2) eliminate about 25 000 mmol of H^+ each day. The role of the kidney might seem unimportant by comparison since renal excretion of H^+ only amounts to 40–80 mmol/24 h. However, the kidney excretes mainly non-volatile acids, which cannot be excreted by the lungs. It is for this reason that renal disease often gives rise to disturbances of acid–base balance. The maintenance of acid–base balance depends on efficient respiratory and renal excretion mechanisms.

The first part of this chapter is concerned with acid–base terminology, with the measurements that can be made, with ways of expressing results and with the interpretation of acid–base data. From the technical viewpoint, the bicarbonate/carbonic acid (HCO_3^-/H_2CO_3) buffer system is very important, and equipment is widely available for its detailed investigation. There are, however, other buffer systems that are far more important physiologically. For instance, erythrocytes contain the most efficient buffers in the blood and many proteins act as highly efficient buffers at physiological $[H^+]$ or pH values. The side-chains of proteins include groups with pKa values close to pH 7.40, or ionisation constants close to 40 nmol H^+/L, whereas the pKa of the HCO_3^-/H_2CO_3 system is 6.10.

Examples of physiological buffers are given in Table 5.1. For investigative purposes, it is perfectly satisfactory to use the HCO_3^-/H_2CO_3 system, since changes in $[H^+]$ that affect the composition of one buffer system simultaneously produce corresponding alterations in the other buffer systems in the body.

Carbon dioxide is produced in large quantities in the tissues as a result of the aerobic metabolism of organic compounds, and it diffuses

TABLE 5.1. Acid-base buffer systems in man

Acid		Conjugate base	Approximate pKa
H_2CO_3	$\rightleftharpoons$	$H^+ + HCO_3^-$	6.1
HPr	$\rightleftharpoons$	$H^+ + Pr^-$	6–8
$H_2PO_4^-$	$\rightleftharpoons$	$H^+ + HPO_4^{2-}$	7.2
NH_4^+	$\rightleftharpoons$	$H^+ + NH_3$	9.2

Pr = protein

into the ECF where the following reactions then occur:

$$H_2O + CO_2 \rightleftharpoons H_2CO_3 \qquad \text{Reaction (1)}$$
$$H_2CO_3 \rightleftharpoons H^+ + HCO_3^- \qquad \text{Reaction (2)}$$

These equations are both reversible. The rate of reaction (1) is greatly increased by carbonic anhydrase. The conditions resulting from the production of CO_2 tend to drive the equations towards the right, with the production of H^+. Generation of H^+ occurs mainly in the erythrocytes, because of the localization of carbonic anhydrase there. The H^+ thereby produced are mainly buffered by haemoglobin, and there is little alteration in $[H^+]$ as a result. In the pulmonary capillaries the reactions are drawn to the left, due to the gradient between the relatively high $P\text{CO}_2$ in plasma and the much lower $P\text{CO}_2$ in alveolar air. Again carbonic anhydrase plays a key role, catalysing the slow non-ionic reaction (1) in which carbonic acid dissociates to CO_2 and water.

The Henderson and Henderson–Hasselbalch equations

Two ways of expressing reaction (2) have been considered in discussions of acid–base balance. Historically, the first was the Henderson equation, which simply applies the Law of Mass Action:

$$[H^+] = K\frac{[H_2CO_3]}{[HCO_3^-]} \qquad \text{Equation (A)}$$

In this equation the parameter K is the first ionization constant of carbonic acid. The $[H_2CO_3]$ term can be replaced by the expression $S.P\text{CO}_2$, where S is the solubility coefficient of CO_2, since H_2CO_3 is in equilibrium with dissolved CO_2, as shown in reaction (1). At 37°C, S has the value 0.23 mmol/J if $P\text{CO}_2$ is expressed in kPa, and 0.03 mmol/J if $P\text{CO}_2$ is exressed in mmHg; K is 7.94×10^{-7}. Equation (A) thus becomes:

$$[H^+] = 7.94\frac{S.P\text{CO}_2}{[HCO_3^-]} \times 10^{-7} \qquad \text{Equation (B)}$$

A few years after Henderson described equation (A), Hasselbalch rearranged the Henderson equation into the corresponding logarithmic form, called the Henderson–Hasselbalch equation:

$$pH = pK + \log_{10} \frac{[HCO_3^-]}{[H_2CO_3]} \qquad \text{Equation (C)}$$

Again, the $[H_2CO_3]$ term can be replaced by $S.P\text{CO}_2$, the value of S depending on the units adopted for $P\text{CO}_2$ measurements; pK is 6.10. Equation (C) then becomes:

$$pH = 6.10 + \log_{10} \frac{[HCO_3^-]}{S.P\text{CO}_2} \qquad \text{Equation (D)}$$

In equations (B) and (D), $P\text{CO}_2$ is sometimes referred to as the respiratory component. This is because it is directly related to alveolar $P\text{CO}_2$ which, in turn, depends on respiratory function. The HCO_3^- term is similarly sometimes referred to as the metabolic or non-respiratory component since, in the absence of changes in $P\text{CO}_2$, changes in $[HCO_3^-]$ reflect alterations in $[H^+]$ or pH due, in the first place, to metabolic or non-respiratory causes. However, the expression 'metabolic (or non-respiratory) component' is potentially misleading, as discussed below, and it should be avoided.

The changes discussed in the previous paragraph are caused by changes in the equilibria of *chemical* reactions. These must be clearly distinguished from the acid–base changes which may occur as a result of *physiological* mechanisms operating to return plasma $[H^+]$ towards normal. Thus, if there is a rise in $P\text{CO}_2$, this will be immediately reflected by a rise in both plasma $[H^+]$ and $[HCO_3^-]$ due to a shift to the right in the chemical reactions (1) and (2). Only after a delay of several hours, in this example, would the effect of physiological renal compensatory changes become evident; these would include retention of HCO_3^-, resulting in a further rise in plasma $[HCO_3^-]$ related to the increased excretion of $[H^+]$ in the urine (p. 94).

ACID–BASE STATUS

The acid–base status of a patient is usually determined by measurements made on arterial or arterialized capillary samples of blood, and both the Henderson and Henderson–Hasselbalch equations provide a starting-point for discussing the measurements required. If two of the three variables in equations (B) or (D) are measured, the third variable can be calculated.

Plasma $[H^+]$, measured in practice with a pH meter, and $P\text{CO}_2$

determinations are regularly performed; together they provide partial characterization of the acid–base status of a blood sample. Whereas all laboratories report plasma $[H^+]$ or pH and P_{CO_2}, there is no uniformity of view as to which of the various ways of expressing and reporting data for the bicarbonate (HCO_3^-) term in equations (B) or (D) should be adopted. Some laboratories report two or more of the following variables, and leave it to clinicians to decide which of the data to use:

(1) *Plasma [HCO_3^-].* This is the concentration of bicarbonate in the blood specimen, as collected.
(2) *Standard bicarbonate.* This is the concentration of bicarbonate in plasma of a blood specimen that has, following collection, been equilibrated with O_2–CO_2 mixtures at 37°C. These *in vitro* manipulations allow the bicarbonate concentrations to be determined under precisely specified conditions (fully oxygenated blood, 37°C and P_{CO_2} of 5.3 kPa or 40 mmHg).
(3) *Base excess.* This is another measurement that involves manipulations *in vitro*. It is the amount of strong acid that would be required to titrate one litre of fully oxygenated blood to a $[H^+]$ of 40 nmol/L (pH 7.40) at 37°C and under conditions where the P_{CO_2} is 5.3 kPa or 40 mmHg. If alkali were to be required to perform this titration, the results would be expressed as negative quantities or else described as a *base deficit.*

The main advantage of plasma $[HCO_3^-]$ is that it is a measure of the actual composition of the blood specimen, as collected, without it having been subjected to *in vitro* manipulations. The principal disadvantage of plasma $[HCO_3^-]$ is that the results reflect primary changes in acid–base status that may be either respiratory or metabolic in nature, and sometimes both; it is not independent of changes in P_{CO_2}. Data for plasma $[HCO_3^-]$ cannot provide any 'rule of thumb' guide to the amount of acid or base that might be given to a patient to correct an acid–base disturbance.

In a physicochemical sense, base excess and standard bicarbonate are independent of P_{CO_2}. To a certain extent, therefore, these variables (especially base excess) provide an indication of the amount of acid or base required to correct an acid–base disturbance; some doctors find them useful for this reason. However, *in vitro* and *in vivo* acid–base conditions are not the same, and it has been shown that the laboratory equilibration procedure which yields the results for base excess and standard bicarbonate does not adequately reflect *in vivo* conditions. It is not possible, by *in vitro* manipulations on blood, to obtain an accurate measure of the deficit or excess of acid in the body.

In the past, acid–base equipment enabled measurements of standard bicarbonate and base excess to be made more easily than plasma [HCO_3^-]. However, plasma [HCO_3^-] can now be determined as readily and reliably as base excess or standard bicarbonate, and opinion seems to be swinging in favour of plasma [HCO_3^-], as being the more obviously physiological and less artificial measurement. Whatever measurement is requested, the collection of suitable specimens and care in their subsequent handling are considerations of prime importance.

Collection and transport of acid–base specimens

Arterial blood specimens are the most appropriate for full assessment of acid–base status. However, unless an arterial cannula is *in situ,* these specimens may be difficult to obtain for repeated assessments of a patient's rapidly changing condition.

Arterialized capillary blood specimens are widely used. They can most readily be obtained from the ear-lobe, the pulp of the fingers or, from infants, by heel-stab. It is essential for the blood to flow freely. Collection of satisfactory capillary samples may be impossible if there is peripheral vasoconstriction or the blood flow is sluggish.

There are still some who use venous blood for full acid–base studies. Venous samples can give results that approximate closely to values obtained with arterial samples, as long as steps are taken to minimize the effects of local metabolism, e.g. by warming the forearm to produce arterialization of the venous blood. However, arterial and arterialized capillary samples have been repeatedly shown to reflect the acid–base status of patients more closely than arterialized venous samples, and are much to be preferred when full acid–base assessment is required.

Patients must be relaxed and their breathing pattern should have settled after any temporary disturbance associated with the preparations for obtaining specimens before the specimens themselves are collected. Some patients, for instance, may hyperventilate temporarily because they feel apprehensive.

Blood is collected into syringes or capillary tubes that contain heparin as anticoagulant; glass or certain kinds of plastic syringes can both be used satisfactorily. Specimens must be free from air bubbles, and any air must be ejected from the syringe before the specimen is mixed with the heparin.

Arrangements should be made for the *immediate* performance of the acid–base measurements. Unless this can be done, the specimens *must* be chilled in iced water as otherwise glycolysis occurs and the acid–base composition of the specimen alters rapidly. Under no circumstances should specimens of blood for acid–base measurement be frozen as this causes haemolysis. Specimens chilled in ice water can have their analysis delayed as long as 4 h. However, the clinical reasons that give rise to requests for full acid–base studies usually demand much more rapid answers. This is reflected by the increasing numbers of automatic blood gas analysers that are being installed in intensive care units and other ward environments.

Temperature effects

Acid–base measurements are made at 37°C. However, some patients on whom these investigations are requested may have body temperatures that are much removed from 37°C, for instance severely hypothermic patients. Equations have been defined that relate $[H^+]$, $P\text{CO}_2$ and $P\text{O}_2$, determined at 37°C, to the equivalent values that correspond to the patient's body temperature. Questions are often raised, therefore, as to whether acid–base data obtained at 37°C should be corrected for such temperature effects before being reported.

Reference values for acid–base data have only been properly established in respect of observations made at 37°C. We cannot, therefore, recommend 'correction' of analytical results (obtained at 37°C) to values that would have been obtained, according to the equations used, at the temperature of the patient (e.g. 28°C). Since we do not know what 'normal' acid–base values are appropriate for a patient with a temperature of 28°C, for instance, we recommend that analytical results for blood acid–base measurements made at 37°C should be related to the reference values that have been established for 37°C, *whatever the temperature of the patient.*

This practice, if accepted, greatly simplifies the interpretation of results. Furthermore, on present evidence, it seems to be physiologically more valid than complex temperature-correction procedures. If treatment aimed at reducing an acid–base disturbance (e.g. $NaHCO_3$ infusion) is given to a severely hypothermic patient, the effect of the treatment should be monitored frequently by repeating the acid–base measurements.

Presentation of results

Some prefer to use the Henderson equation, equations (A) and (B).

Thereby, they avoid the confusion which can arise when using pH notation, where an *increase* in $[H^+]$ is denoted by a *decrease* in pH. On the other hand the Henderson–Hasselbalch equation, equations (C) and (D), will be more familiar to some readers. It is a matter of personal preference which equation to use. We shall base our discussion mainly on the Henderson equation.

Many different diagrams and graphic representations of acid–base balance have been proposed, e.g. Fig. 5.1 in which $[H^+]$ and pH are plotted on the appropriate scales against P_{CO_2} with lines of equal concentration of HCO_3^- radiating from the origin. The entry of results on diagrams such as Fig. 5.1 gives a visual indication of the acid–base status of a patient at the time a specimen was collected. Values can be

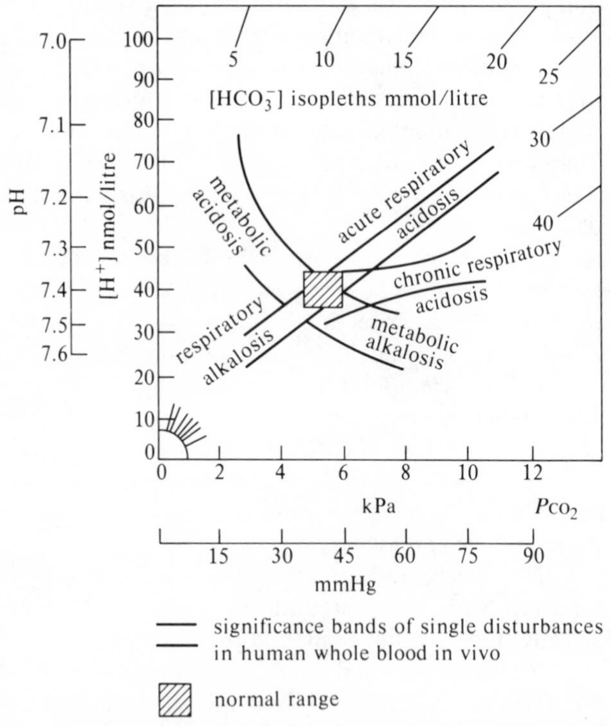

FIG. 5.1. *Acid-base diagram for arterial blood*, which can be used for plotting results for $[H^+]$ or pH, P_{CO_2} and plasma $[HCO_3^-]$; the points shown on the scale for $[HCO_3^-]$ indicate the places where a straight line from the origin to the limits of the figure cut the axes. The shaded rectangle indicates the reference values. The pairs of lines radiating from the shaded rectangle indicate the 95% confidence limits for acid-base data for the corresponding single disturbances. Conversion factor: 4 kPa is equivalent to 30 mmHg. After Flenley (1978); reproduced with permission.

related directly to the pattern of findings in various categories of acid–base disturbances. Plotting successive observations on a diagram provides one way of seeing how a patient's acid–base balance is altering.

Some people find all acid–base diagrams confusing. For them, we would emphasize that there is no need to use diagrams at all. The information to be derived from laboratory results of acid–base studies is all contained in the numerical data. The data can be satisfactorily interpreted without recourse to diagrams by first considering the patient's clinical condition and then examining the acid–base results in the light of the clinical findings, following a logical sequence that begins with the $P\text{CO}_2$ results.

To use the scheme that follows, first decide whether the $P\text{CO}_2$ is normal, increased or decreased. Next examine the data for plasma $[H^+]$ and thereafter the plasma $[HCO_3^-]$, or alternative measurement (e.g. standard bicarbonate) if such is preferred.

$P\text{CO}_2$ is normal

If plasma $[H^+]$ is also normal, there is no acid–base disturbance. Any departure of plasma $[H^+]$ from normality can be attributed to a disturbance of acid–base equilibrium due to a metabolic (as distinct from a respiratory) cause:

(1) *$[H^+]$ is increased.* The patient has a metabolic acidosis. This would be confirmed by finding a decreased plasma $[HCO_3^-]$.

(2) *$[H^+]$ is decreased.* The patient has a metabolic alkalosis. This would be confirmed by finding an increased plasma $[HCO_3^-]$.

$P\text{CO}_2$ is increased

This finding, on its own, allows one to conclude that the patient has a respiratory acidosis, since plasma $P\text{CO}_2$ is directly related to alveolar $P\text{CO}_2$. Clinical considerations and examination of the data for plasma $[H^+]$ and $[HCO_3^-]$ together amplify this first conclusion:

(1) *$[H^+]$ is increased.* The patient has a respiratory acidosis. If this is a simple disturbance, plasma $[HCO_3^-]$ will be increased. If the acid–base disturbance is mixed, the results for plasma $[HCO_3^-]$ will help to define its nature. For example, if there is a respiratory acidosis and, *in addition*, there is a metabolic acidosis, the latter will tend to lower plasma $[HCO_3^-]$ which may then be found to be normal or even decreased.

(2) *[H^+] is normal.* This indicates *either* that the respiratory acidosis has been fully compensated *or* that a metabolic alkalosis is present in addition to the respiratory acidosis. Both possibilities give rise to increased plasma [HCO_3^-], but they can usually be readily distinguished on clinical grounds, including consideration of treatment received (e.g. $NaHCO_3$).

(3) *[H^+] is decreased.* These patients have a mixed acid–base disturbance, consisting of a metabolic alkalosis and a respiratory acidosis. Plasma [HCO_3^-] is increased. Clinical considerations will determine which of the disturbances was the primary one.

P_{CO_2} is decreased

This finding, on its own, means that the patient has a respiratory alkalosis. Clinical considerations and examination of the other acid–base data will amplify this first conclusion:

(1) *[H^+] is decreased.* The patient has a respiratory alkalosis. If this is a simple disturbance, plasma [HCO_3^-] will be decreased. If the acid–base disturbance is mixed, the results for plasma [HCO_3^-] may help to define its nature. For example, if there is a respiratory alkalosis and, *in addition,* there is a metabolic alkalosis, the latter will tend to increase plasma [HCO_3^-], which may then be found to be normal or even increased.

(2) *[H^+] is normal.* The respiratory alkalosis is either fully compensated, or a metabolic acidosis is also present. Both possibilities give rise to decreased plasma [HCO_3^-], but the distinction can usually be made on clinical grounds and by considering the results of renal function tests.

(3) *[H^+] is increased.* There is a metabolic acidosis as well as a respiratory alkalosis. Often the metabolic acidosis is the *primary* disturbance, the reduced P_{CO_2} representing a physiological compensatory change occurring in response to the metabolic acidosis (e.g. the overbreathing which occurs in patients with diabetic ketoacidosis). Plasma [HCO_3^-] is decreased in these patients.

Simple and mixed disturbances

Tables 5.2a and 5.2b give a summary account of *simple* acid–base disturbances and Table 5.3 presents sets of illustrative data.

Mixed disturbances of acid–base balance show the combined effects of two primary disturbances, for instance effects due to a combination

TABLE 5.2a. Mechanisms, and causes of simple acid-base disturbances.

Descriptive terms and mechanism	Examples of causes
Respiratory acidosis	
Alveolar Pco_2 increased	Lung disease Weakness of respiratory muscles CNS disease Drug overdosage (e.g. hypnotics, anaesthetics)
Metabolic acidosis	
(1) Addition of H^+ to body fluids in excess of excretory capacity	(1) Disorders of metabolism (e.g. starvation ketosis, diabetic ketoacidosis, lactic acidosis) (2) Ingestion of substances that give rise to H^+ (e.g. NH_4Cl, methanol, paraldehyde, salicylate poisoning sometimes)
(2) Failure to excrete H^+ at normal rate	(1) Inadequate production of NH_3 by kidney (e.g. chronic renal failure) (2) Inability to maintain a gradient of H^+ between blood and urine (e.g. distal renal tubular acidosis) (3) Anuria (e.g. acute renal failure)
(3) Loss of HCO_3^- from the body	(1) From gastrointestinal tract (e.g. severe diarrhoea, fistulous drainage, ureterosigmoidostomy) (2) In urine (e.g. proximal renal tubular acidosis, carbonic anhydrase inhibitors)
Metabolic alkalosis	
(1) Loss of H^+ from the body	Loss of H^+ in vomit, diuretics (e.g. thiazides), mineralocorticoid excess, glucocorticoid excess, K^+ depletion (if severe)
(2) Addition of base to body fluids in excess of excretory capacity	Ingestion of alkali (e.g. $NaHCO_3$ in milk-alkali syndrome)
Respiratory alkalosis	
Alveolar Pco_2 lowered	Voluntary overbreathing Artificial ventilation Drug overdosage (e.g. salicylate poisoning, sometimes)

of metabolic acidosis and respiratory alkalosis. These patients need to be distinguished (in the example given) from patients with a primary metabolic acidosis accompanied by respiratory compensation, a distinction that can usually be made on the basis of clinical assessment, including the history of the illness, and confirmed by the results of

TABLE 5.2b. Effects of simple acid–base disturbances.

Type of disturbance	Effects on acid–base measurements			
	$P\text{CO}_2$	$[H^+]$	$[HCO_3^-]$	
Respiratory acidosis	↑	↑	↑	(1)
	↑	↑ or HN	↑	(2)
	↑	HN	↑ ↑	(3)
Metabolic acidosis	N	↑	↓	(1)
	↓	↑ or HN	↓ or ↓ ↓	(2)
Metabolic alkalosis	N	↓	↑	(1)
	↑	↓ or LN	↑	(2)
Respiratory alkalosis	↓	↓	N or ↓	(1)
	↓	↓ or LN	↓	(2)
	↓	LN	↓	(3)

Key to symbols

(1) = before compensation
(2) = partially compensated
(3) = fully compensated

N = within the range of reference values
LN = in the lower part of the reference range
HN = in the upper part of the reference range

↓ = low ↓ ↓ = very low ↑ = high ↑ ↑ = very high

acid–base measurements. For a disturbance to qualify for the description 'mixed', there must be evidence of more than one disturbance of a primary nature. This type of disorder may be difficult to treat properly, without full acid–base assessments that usually need to be repeated on several occasions.

Alternatives to full acid–base assessment

The full characterization of acid–base status requires a knowledge of all three variables in the Henderson or the Henderson–Hasselbalch equation. Ideally, this requires arterial or arterialized capillary blood samples since venous blood $P\text{CO}_2$ bears no constant relationship to the alveolar $P\text{CO}_2$. However, full acid–base assessment of all patients with acid–base disturbances would present difficulties in the collection of specimens and the numbers of such full assessments might present logistic problems to the laboratory.

Total CO_2

The most commonly performed acid–base measurement is one carried out on venous blood, plasma [total CO_2]. This is the CO_2 released from plasma (or serum) by strong acid, and includes contributions from

TABLE 5.3. Data for patients with simple disturbances of acid–base balance.

Nature of disturbance	$[H^+]$ (nmol/L)	P_{CO_2} (kPa)	Plasma $[HCO_3^-]$ (mmol/L)	pH	Standard bicarbonate (mmol/L)	Base excess (mmol/L)	Total CO_2 (mmol/L)
Reference values	36–44	4.5–6.1	21.0–27.5	7.36–7.45	22.0–26.0	−2.5 to +2.5	24–30
Metabolic acidosis							
Uncompensated	90	5.3	10	7.05	10	−20	14
Compensated (partial)	72	3.2	8	7.14	10	−20	11
Respiratory acidosis							
Uncompensated	58	9.3	29	7.24	24	0	32
Compensated (partial)	49	9.3	34	7.31	29	+ 6	37
Metabolic alkalosis							
Uncompensated	26	5.3	37	7.59	37	+16	40
Compensated (partial)	32	7.3	40	7.49	37	+16	44
Respiratory alkalosis							
Uncompensated	29	3.2	20	7.53	24	0	22
Compensated (partial)	32	3.2	18	7.49	22	− 2	20

Definitions of measurements: Appendix 2 (p. 522)

The data in the first 3 columns are all that need to be considered. Alternatively, for those who prefer pH to $[H^+]$, the data in columns 2–4 can be used. The remaining columns contain data for alternative ways of presenting the information contained in the plasma $[HCO_3^-]$ column. Readers may find this table less confusing if they cover columns that they are not accustomed to using.

HCO_3^-, H_2CO_3, dissolved CO_2, CO_3^{2-} and carbamino compounds; HCO_3^- is by far the largest component in this mixture since, at normal blood $[H^+]$, HCO_3^- contributes about 95% of the 'total CO_2'. In this book, we use the expression plasma [total CO_2] consistently to refer to the measurement of this mixture of compounds.

The measurement, plasma [total CO_2], has its limitations (see below) and these have to be accepted. It is, nevertheless, a determination that can be satisfactorily carried out on venous blood samples collected and prepared for analysis without all the special precautions required in order to obtain reliable data for full acid–base assessments, using arterial blood or arterialized capillary blood.

Some laboratories incorrectly describe plasma [total CO_2] as plasma $[HCO_3^-]$. Readers must beware of this potential source of confusion. In this book, we reserve plasma $[HCO_3^-]$ for measurements of bicarbonate (HCO_3^-) concentration, usually performed as part of full assessments of acid–base status.

Plasma [total CO_2], by itself, cannot provide a precise definition of a patient's acid–base status, since the values for $[H^+]$ and P_{CO_2} are unknown. However, combined with a knowledge of the patient's history and clinical condition, plasma [total CO_2] often enables an adequate initial assessment to be made of whether an acid–base disturbance is present and, if present, an indication of its severity. This is especially the case when a metabolic (non-respiratory) disturbance is present.

Full acid–base assessment, on arterial or arterialized capillary blood samples, is required on relatively few patients. These patients, who are presenting problems in diagnosis or management, usually have problems of *respiratory* origin.

Anion gap (AG)

This term describes a calculation that is frequently performed in clinical practice as a pointer towards the presence of an acid–base disturbance, especially a metabolic acidosis. This calculation, for which another name is *ion difference,* makes use of the results of plasma electrolyte determinations in equations such as the following:

$$AG_1 = ([Na^+] + [K^+]) - ([Cl^-] + [\text{total}\,CO_2])$$
$$AG_2 = [Na^+] - ([Cl^-] + [\text{total}\,CO_2])$$

The difference between the cation(s) and the anions in these equations represents the so-called unmeasured anions or anion gap. These anions include proteins, phosphate, sulphate and lactate. The average numerical value for AG_1 is 15 mmol/L and for AG_2 it is 11 mmol/L.

TABLE 5.4 Causes of an increased anion gap

Metabolic acidosis	
Uraemic acidosis	Salicylate toxicity
Ketoacidosis	Methanol toxicity
Lactic acidosis	Ethylene glycol toxicity
Increase in plasma [Na^+]	
Treatment with sodium salts, including some forms of high dose antibiotic therapy	
Loss of CO_2	
Improper handling of specimen after collection	

The main causes of an increased anion gap are listed in Table 5.4. Reduction in the anion gap occurs much less frequently; when it occurs, this can usually be traced to an error in the laboratory.

There are two opposing views as to the value of calculating the anion gap. Some are strong advocates, whereas others consider that the calculation provides no information that cannot be equally well obtained from properly assessing the results for plasma [total CO_2] or, in certain conditions (e.g. lactic acidosis), better obtained by making the appropriate additional measurements (e.g. plasma [lactate]).

Restricting the application of our comments to the clinical value of calculating the anion gap, we would take a middle view and point to its particular usefulness in patients with lactic acidosis. Plasma lactate measurements are not yet available from some laboratories, especially in an emergency, and the anion gap then provides a valuable pointer to this metabolic disturbance. The anion gap cannot be calculated by those laboratories that have stopped measuring plasma [Cl^-].

Intracellular [H^+]

Considerable interest exists in the relationship between extracellular and intracellular [H^+]. The acid–base determinations discussed so far only provide information about conditions in the ECF. However, in some disorders there is a marked dissociation between the [H^+] of the ECF and ICF, as frequently occurs in potassium deficiency (p. 53). Method of measuring intracellular [H^+] are still research procedures.

THE TRANSPORT OF OXYGEN

The properties of the dissociation curve for oxyhaemoglobin (HbO_2) enable O_2 to be carried from the lungs to the tissues (Fig. 5.2). It is

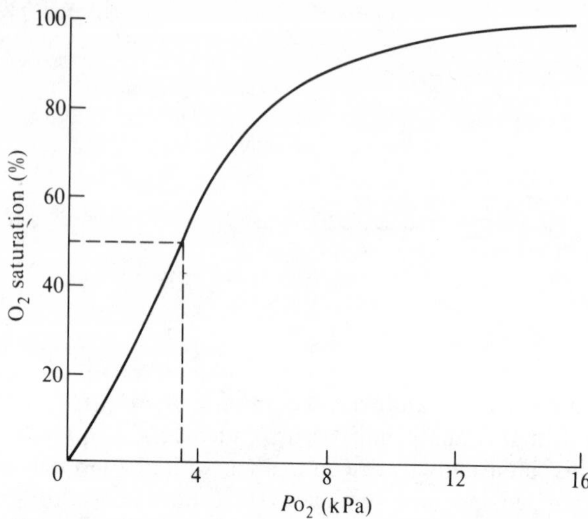

FIG. 5.2. *The oxygen dissociation curve of haemoglobin*. The Po_2 corresponding to an oxygen saturation of 50% (P_{50}) is indicated by the dotted lines. Conversion factor: 4kPa is equivalent to 30 mmHg.

convenient to define the position of the HbO_2 dissociation curve for a given sample of blood as the Po_2 required to cause 50% oxygen saturation; this value is called the P_{50}. The following factors are known to cause a shift to the right (increased P_{50}):

(1) An increase in $[H^+]$ in the ECF (the Bohr effect).
(2) A rise in Pco_2, partly due to a $[H^+]$ effect and partly due to formation of carbamino compounds with haemoglobin.
(3) A rise in temperature.
(4) Increased red cell 2,3-diphosphoglycerate concentration.

A shift of the HbO_2 dissociation curve to the right means that the per cent oxygen saturation at a given Po_2 will be lower. Proportionately less O_2 will then be taken up in the lungs, whereas more will be delivered to the tissues. The effect on the unloading of O_2 in the tissues, where the Po_2 corresponds to a very steep portion of the curve, will be much greater than any effect exerted on the O_2 uptake in the lungs where, at a Po_2 of 13.2 kPa (100 mmHg), the curve (Fig. 5.2) is very much flatter.

2,3-Diphosphoglycerate (2,3-DPG)

During glycolysis most tissues metabolize 1,3-diphosphoglycerate (1,3-DPG) to 3-phosphoglycerate (3-PG). In erythrocytes, however, there is an active shunt mechanism whereby 2,3-DPG is formed from 1,3-DPG; the 2,3-DPG can then be hydrolysed to 3-PG.

There is a high [2,3-DPG] in red cells, and this is able to shift the position of the HbO_2 dissociation curve by reacting with reduced Hb to increase the stability of the deoxygenated form. A higher Po_2 is then required to obtain a given per cent oxygen saturation. Fetal Hb does not bind 2,3-DPG as well as maternal Hb, and this helps to explain the greater affinity of fetal blood for oxygen.

Variations in red cell [2,3-DPG] seem to enable the body to effect alterations in the HbO_2 dissociation curve with changing physiological circumstances. For example, in hypoxia due to high altitude, there is an increase in the activity of diphosphoglyceromutase in the red cell. This increase directs 1,3-DPG metabolism along the shunt pathway, and there is a rise in red cell [2,3-DPG] with a consequent shift of the dissociation curve to the right. Similar findings occur in hypoxia due to chronic bronchitis or right to left cardiac shunts.

Differences in red cell [2,3-DPG] are thought to be partly responsible for the rapid component of an individual's response to changes in altitude, and for the compensatory effects for reduced oxygen-carrying capacity that occur in anaemia and in acidosis. In acidosis, for instance, red cell [2,3-DPG] falls and the P_{50} falls with a consequent shift of the oxygen dissociation curve to the left, thereby serving to counteract the Bohr effect; the net effect of these two factors is that the affinity of Hb for O_2 remains about normal.

Measurement of red cell [2,3-DPG] is only available as a research procedure, except in a very few centres.

Po_2 and oxygen saturation

The full characterization of the oxygen composition of a blood sample requires measurement of Po_2, [Hb] and per cent oxygen saturation. Above a Po_2 of 10.5 kPa (80 mmHg), however, Hb is almost fully saturated with O_2 so the per cent oxygen saturation is not often required at high levels of Po_2. Measurement of Po_2 is important, especially at levels above 10.5 kPa, and the results are often valuable in assessing the efficiency of oxygen therapy, when high Po_2 values may be found. On the other hand, the results of Po_2 measurements may be misleading in conditions where the oxygen-carrying capacity of blood is grossly impaired, as in severe anaemia, in carbon monoxide poisoning and in methaemoglobinaemia; measurements of the [Hb] and of the per cent oxygen saturation are required in addition to Po_2 under these circumstances.

Measurements of the oxygen-carrying capacity of blood, the arteriovenous difference in per cent oxygen saturation, and cardiac output are all needed if the amount of oxygen delivered to the tissues is to be calculated.

AVAILABILITY OF MEASUREMENTS

Of the principal investigations mentioned in this section, [Hb] is widely available and Po_2 is now usually one of the measurements automatically performed by blood gas analysers as part of full acid–base assessment of patients. Per cent oxygen saturation is much less frequently performed. In the foreseeable future, transcutaneous oximetry could progressively replace arterial Po_2 measurements, but cost considerations at present limit its use.

CLINICAL APPLICATIONS OF BLOOD ACID–BASE AND OXYGEN MEASUREMENTS

The main indications for full acid–base assessment, coupled with Po_2 or per cent oxygen saturation measurements, are in the investigation of patients with pulmonary disorders and in the control of their treatment. Other important applications include the investigation and management of patients with vascular abnormalities involving the shunting of blood. These groups of patients either have a simple respiratory acidosis or a mixed disturbance that includes a respiratory acidosis as one of the primary features. Some features of respiratory insufficiency are considered below.

Full acid–base assessment is much less often needed in patients with metabolic acidosis or alkalosis, for whom measurement of plasma [total CO_2] on venous blood usually gives sufficient information. Also, full assessments are not often needed in patients with respiratory alkalosis, except in patients being treated by assisted ventilation; by far the commonest cause of respiratory alkalosis is voluntary over-breathing.

Respiratory insufficiency

This term is applied to disorders in which lung function is impaired sufficiently to cause the Po_2 to become abnormally low, i.e. less than 8.0 kPa (60 mmHg). It is convenient to distinguish two types of respiratory insufficiency, Type I in which the Pco_2 is normal or low, and Type II in which the Pco_2 is raised, i.e. to greater than 6.6 kPa (49 mmHg), in addition to the low Po_2. Both types are caused primarily by abnormalities in ventilation/perfusion ratios.

TYPE I: LOW $P\text{O}_2$ WITH NORMAL OR LOW $P\text{CO}_2$

This combination of hypoxaemia without hypercapnia occurs in patients in whom there is a preponderance of alveoli that are adequately perfused with blood, but inadequately ventilated. It occurs, for example, in chronic bronchitis, pulmonary oedema, asthma or obesity.

In Type I respiratory insufficiency there is, in effect, a partial right to left 'shunt' bringing unoxygenated blood to the left side of the heart. Increased ventilation of those alveoli that are adequately perfused and ventilated is able to compensate for the tendency for the $P\text{CO}_2$ to rise. It cannot, however, restore the $P\text{O}_2$ to normal, since the blood perfusing these alveoli conveys Hb which is already nearly saturated with O_2; the Hb is on the upper, 'flat' portion of the O_2 saturation curve.

TYPE II: LOW $P\text{O}_2$ WITH HIGH $P\text{CO}_2$

This combination of hypoxaemia with hypercapnia means that there is hypoventilation. Altered ventilation/perfusion relationships result in an excessive number of alveoli being inadequately perfused. This causes 'wasted' ventilation and an increase in 'dead space'.

Chronic bronchitis is an important cause of Type II abnormality. It also occurs in patients with mechanical defects in ventilation caused, for example, by chest injuries, myasthenia gravis or polyneuritis. In patients with status asthmaticus, if serial blood gas observations show a rising $P\text{CO}_2$ and a falling $P\text{O}_2$, more intensive treatment is urgently needed.

Treatment of acid–base disturbances

It cannot be stressed too often that the reliability of acid–base measurements, and their value in the management of patients, are critically dependent on the collection of satisfactory specimens and their proper handling after collection. Also, the data *must* be interpreted in relation to the clinical information, if they are to help provide a rational basis for the treatment that is to be given. Having defined the nature of an acid–base disturbance, therapeutic measures should be aimed at correcting the primary disorder and at assisting the physiological compensatory mechanisms. In some cases, more active intervention may be considered necessary. This may involve measures aimed at reducing the acid–base disturbance, for instance by treatment with HCO_3^- (as $NaHCO_3$), if the patient has an acidosis.

Empirical formulae have been published for calculating the dose of $NaHCO_3$ on the basis of the acid–base measurements discussed in this chapter. The best-known formulae make use of the term base excess. One of these indicates that 0.3 mmol base (as HCO_3^-) should be given per kg body weight for each 1.0 mmol/L change in base excess that it is desired to produce. In another expression, the factor 0.3 is replaced by 0.2. Similar formulae can be written incorporating other indices of the HCO_3^- term in equations expressing acid–base balance (e.g. standard bicarbonate).

These formulae make it appear simple for patients to be 'titrated' back to normal on the basis of measurements of $[H^+]$ in the ECF, and on the assumption that only the ECF is involved. They represent over-simplifications, however, and we would warn against their uncritical application. If active therapeutic intervention is considered necessary, to reduce the degree of acid–base disturbance by giving $NaHCO_3$, formulae should at most be used to give an approximate indication of the dosage required. Thereafter, while administering the treatment, patients should be monitored by frequent measurement of acid–base variables, probably every 2 h, until the desired therapeutic effect has been achieved. Plasma $[K^+]$ should also be monitored because shifts of K^+ can occur acutely between the ECF and ICF, especially when acid–base status alters rapidly.

FURTHER READING

ASHWOOD, E.R., KOST, G. and KENNY, M. (1983). Temperature correction of blood-gas and pH measurements. *Clinical Chemistry,* **29,** 1877–1885.

BEETHAM, R. (1982). A review of blood pH and blood-gas analysis. *Annals of Clinical Biochemistry,* **19,** 198–213.

BISWAS, C.K., RAMOS, J.M., AGROYANNIS, B. and KERR, D.N.S. (1982). Blood gas analysis: effect of air bubbles in syringe and delay in estimation. *British Medical Journal,* **284,** 923–927.

BREMNER, B.M. and STEIN, J.H., editors (1978). *Acid–base and Potassium Homeostasis.* Edinburgh: Churchill Livingstone.

COHEN, R.D. (1982). Some acid problems. *Journal of the Royal College of Physicians of London,* **16,** 69–77.

FLENLEY, D.C. (1978). Interpretation of blood-gas and acid–base data. *British Journal of Hospital Medicine,* **20,** 384–394.

HUTCHISON, A.S., RALSTON, S.H., DRYBURGH, F.J., SMALL, M. and FOGELMAN, I. (1983). Too much heparin: possible source of error in blood-gas analysis. *British Medical Journal,* **287,** 1131–1132.

Leading Article (1977). The anion gap. *Lancet,* **1,** 785–786.

LOLEKHA, P.H. and LOLEKHA, S. (1983). Value of the anion gap in clinical diagnosis and laboratory evaluation. *Clinical Chemistry,* **29,** 279–282, and the addendum from S. NATELSON on pages 282–283.

SHAPIRO, B.A. (1973). *Clinical Applications of Blood Gases.* London: Lloyd-Luke.

Chapter 6

Renal Disease

Many diseases affect renal function. In some, several different functions are affected, whereas in others there is selective impairment of glomerular function or of one or more tubular functions.

Urine is produced as a result of approximately two million nephrons performing the following processes:

(1) Formation of an ultrafiltrate of plasma at the glomerulus.
(2) Tubular reabsorption.
(3) Tubular secretion.

Excretion of urine is the main route whereby non-volatile products of metabolism are eliminated from the body.

Most types of renal disease cause destruction of complete nephrons; this is particularly true for chronic renal disease. It has been suggested, therefore, that the resulting alterations in renal function can be most readily explained on the basis of the remaining nephrons being functionally normal but overloaded. This 'intact nephron hypothesis' can explain some of the disturbances of function that occur in renal disease, but it almost certainly represents an oversimplification.

Chemical investigations are mainly of value in detecting the presence of renal disease, by its effects on renal function, and in assessing its progress. They are of less value in determining the cause of the disease.

Side-room tests and glomerular function tests are essential steps in the initial assessment of renal function. Tubular function tests are less convenient, since their performance usually requires conditions in which the tubules are stressed by artificial loading. Divided renal function studies based on chemical measurements, and requiring ureteric catheterization, are obsolete.

GLOMERULAR FUNCTION

The glomerular filtration rate (GFR) depends on the net pressure exerted across the glomerular membrane, the physical nature of the membrane and its surface area, which reflects the number of functioning glomeruli. All these factors may be modified by disease. However, in the absence of large changes in blood pressure or gross changes in the structure of the glomerular membrane, the GFR provides an index of the number of functioning glomeruli.

Measurement of the glomerular filtration rate

The GFR is determined by measuring the concentrations in plasma and urine of a substance which ideally fulfils the following criteria:

(1) It is readily filtered from the plasma at the glomerulus.
(2) It is neither reabsorbed nor secreted by the tubules.
(3) Its concentration in plasma remains constant throughout the period of urine collection.
(4) The measurement of its concentration in plasma and urine is both analytically convenient and reliable.

There are considerable practical advantages in choosing a substance that is normally present in plasma when measuring the GFR.

If a substance (s) meets the four criteria listed, the amount excreted in the urine (U_s) in unit time equals the amount cleared from the plasma (P_s), and the clearance of s by the kidney is given by the formula:

$$\text{GFR} = \frac{U_S}{P_S} \cdot V$$

The units for V, the rate of formation of urine, and for GFR are mL/min, or other similar units (e.g. mL/s).

INULIN CLEARANCE

This is the reference procedure by which values for GFR have been established and against which other methods of measuring GFR are compared. Inulin meets the criteria listed above, but it is not suitable for routine diagnostic use since it is an exogenous material. It has to be given by intravenous infusion and the infusion monitored to check that plasma [inulin] is kept constant during the urine collection period.

CREATININE CLEARANCE

Creatinine is an end-product of nitrogen metabolism; it is an endogenous substance, derived from creatine. Creatine is formed mainly in the liver and kidney, and is an important constituent of muscle, which contains most of the body's creatine. The rate of creatinine excretion is related to the amount of cooked meat in the diet (cooking converts creatine to creatinine), and to an individual's muscle mass.

Determinations of GFR, based on measurements of plasma and urine [creatinine], are widely used. Urine collections are usually made over 24-h periods in the U.K., when determining creatinine clearance.

Creatinine clearance measurements correlate fairly closely with inulin clearance measurements, except in patients in whom the GFR is severely impaired. This good correlation is fortuitous, since there are two opposing potential sources of inaccuracy in the creatinine measurements that cancel one another out. There is normally some tubular secretion of creatinine; this would tend to yield creatinine clearance values which exceed the true GFR. On the other hand, the presence of certain chromogens in plasma (but not urine) that react in a manner similar to creatinine means that most laboratories tend to overestimate plasma [creatinine].

All clearance measurements are subject to inaccuracies due to failure to collect specimens of urine properly over precisely timed intervals. It is surprising how difficult it can be to obtain an accurate collection of urine. Collections made over short periods are liable to inaccuracies from difficulties that patients may experience in emptying the bladder completely, especially on the second occasion. The satisfactory performance of clearance tests is so dependent on adherence to simple rules that it is worth stating these:

(1) At the start of the collection, the patient is told to empty the bladder as completely as possible; this specimen is discarded. Thereafter, the patient (and nursing staff) must know that *all* specimens of urine passed during the collection period are to be collected up to and including the final specimen. The urine specimens are pooled in a suitable container and the volume measured.

(2) A specimen of blood is collected at a convenient time, at some stage during the period of the urine collection. It is advisable to send the blood and urine specimens to the laboratory together, to help with the association of analytical results for purposes of calculating the GFR.

Although 24-h urine collections are widely used in the U.K. because of the ease of arranging them in relation to the timing of nurses' shifts (e.g. start at 0800 h on Day 1, and finish at 0800 h on Day 2), creatinine clearance measurements made over much shorter periods (e.g. 2h) have been shown to correlate very highly with results for inulin clearance. The advantages of short collection periods are that the tests can be performed much more readily under the direct control of one group of staff. The results are less likely to be influenced by within-day fluctuations in plasma [creatinine]. The following are commonly encountered causes of unsatisfactory 24-h urine collections:

(1) Collections sometimes mistakenly include urine passed at 0800 h at the start of the collection period as well as urine passed at 0800 h at the end of the collection period.
(2) Often, one or more of the intermediate specimens is lost down the toilet, because the patient or the nurse forgot that a 24-h collection was in progress.
(3) Some ward staff are under the mistaken impression that it is only necessary to retain a small sample each time a urine specimen is passed, having first measured the volume of that specimen. These samples are pooled as the collection period goes on and the pooled sample sent to the laboratory, together with a note of the total volume of urine passed by the end of the collection period. Unless precisely the same measured fraction of each urine specimen is used every time for pooling purposes, the final 'pooled sample' will clearly be unrepresentative.
(4) Some urine collections are simply stopped if the container is filled before the 24-h period is over, without necessarily ensuring that the last specimen to be collected involved complete emptying of the bladder, and usually without a note being made of the time when the collection period ended.

It should be apparent, from the above discussion, that even procedures which are apparently simple, such as a creatinine clearance test, can all too readily give very variable and sometimes grossly misleading results because of factors that are outside the laboratory's control. From now on, however, we shall assume that the urine collections have been made satisfactorily.

Creatinine clearance in healthy adults is normally greater than 100 mL/min. When urine collections are made over short periods, within-day fluctuations of up to 10% in the GFR have been reported. With 24-h urine collections, between-day fluctuations in creatinine clearance may be as high as 30%. In children, values for the GFR should be related to surface area. The creatinine clearance falls in old age.

The GFR may be impaired in a wide range of disorders affecting the renal tract. These include:

(1) Any disease in which there is impaired renal perfusion, e.g. fall in blood pressure, fluid depletion and renal artery stenosis.

(2) Most types of disease where there is loss of functioning nephrons.
(3) Diseases where pressure is increased on the tubular side of the glomerulus (e.g. urinary obstruction due to prostatic enlargement).

When renal function becomes impaired and plasma [creatinine] is increased, measurements of creatinine clearance overestimate the true GFR, because tubular secretion of creatinine is greater in these patients. Creatinine clearance measurements also overestimate the GFR in patients with marked proteinuria. In patients with moderate degrees of renal failure or with marked proteinuria, the GFR as determined by creatinine clearance measurements may exceed inulin clearance values by as much as 40%. However, in severe renal failure (GFR less than 20 mL/min) creatinine clearance values again approach results for inulin clearance since the excretory capacity of the remaining nephrons becomes saturated.

The creatinine clearance may be impaired in patients who are taking drugs that increase plasma [creatinine], such as cimetidine, co-trimoxazole and salicylate. These have all been reported to increase plasma [creatinine] by as much as 40%, and to be associated with a reduction of about 30% in the creatinine clearance. These drugs, in therapeutic dosage, appear to inhibit the secretion of creatinine by the renal tubules.

UREA CLEARANCE

These measurements are not recommended for estimating GFR. Urea diffuses back from the glomerular filtrate into the tubular cells and is reabsorbed, the amount reabsorbed varying inversely with the rate of urine flow.

Plasma measurements of glomerular function

The body must eliminate the same amount of nitrogenous and other waste substances in renal disease as in health, if nitrogen balance is to be maintained. When the GFR falls, plasma concentrations of these waste substances begin to rise, and this helps to maintain the filtered load. If a patient's GFR has fallen to half its normal value, in order to maintain the normal rate of excretion of substances such as urea and creatinine, plasma [urea] and [creatinine] need to double; this statement assumes that there has been no change in their rate of formation, nor in their handling by the renal tubules.

These considerations explain why measurements of plasma [urea] and [creatinine] are so widely used to provide an assessment of glomerular function. It is easier and quicker just to have plasma concentrations measured rather than to have the trouble of making a timed collection of urine in order to determine clearance. However, various difficulties arise as a result of using only plasma measurements to assess GFR.

PLASMA UREA

The main advantage of determining plasma [urea] is that it can be easily, quickly and precisely measured. However, the reference range for plasma [urea] is wide, e.g. 2.5–6.6 mmol/L in males below 50. Significant changes can occur in some patients for pathological reasons but without the plasma [urea] necessarily rising above the upper reference value. There are also many causes of an increased plasma [urea], summarized in Table 6.1. Several of these are not primarily renal in nature, although their effects may be much greater if renal function is impaired.

Any condition in which there is increased delivery of amino acids to the liver will tend to cause an increase in plasma [urea]. Thus marked alterations in plasma [urea] may result from changes in dietary protein intake, especially in patients with impaired renal function. Large rises in plasma [urea] also occur when protein breakdown is increased, e.g. in the catabolic states that commonly follow recent trauma, whether this be accidental or due to surgery.

Plasma [urea] increases in any patient with fluid depletion, whether the deficit is of water, Na^+ or the much commoner mixed state of water and Na^+ depletion. Increases in plasma [urea] often provide an early indication that patients are becoming dehydrated.

Other pre-renal causes of increased plasma [urea] include fever, sepsis, internal haemorrhage and right-side cardiac failure. There are, therefore, many conditions that can give rise to increased plasma [urea], in addition to the various causes of acute and chronic renal failure or of urinary tract obstruction.

TABLE 6.1. Causes of increased plasma urea concentration.

Physiological	High protein diet
Pathological	
(1) Pre-renal	Catabolic states, dehydration, circulatory failure
(2) Renal	Renal failure, acute or chronic
(3) Post-renal	Obstructive nephropathy
Therapeutic (iatrogenic)	Urea infusion

Despite these reservations about using plasma [urea] as an index of glomerular function, its measurement is valuable in patients with suspected renal disease, for their initial assessment (as a means of excluding the presence of severe renal disease), and in patients with known disease for the subsequent monitoring of progress. Because of the precision with which plasma [urea] can be measured, it can provide a good guide to the effectiveness of therapeutic regimes.

PLASMA CREATININE

Much of what has been said about plasma urea would apply almost equally well to plasma creatinine, except that plasma [creatinine] is less affected by changes in diet or in the rate of protein catabolism. Plasma [creatinine] might, therefore, be expected to provide a somewhat better indication of changes in GFR than plasma [urea].

Plasma [creatinine] has not replaced plasma [urea] as an approximate measure of glomerular function, in practice, mainly because measurements of plasma [creatinine], when this is normal or only slightly increased, tend to be imprecise.

Nowadays, with the increasing use of multi-channel analysers, both plasma [urea] and [creatinine] are often measured. Both tests provide an indication of the GFR, but not as precise a measure as does the creatinine clearance. It remains true, however, that clearance tests are a nuisance; they can be particularly difficult to perform satisfactorily in children.

Plasma creatinine in children

Reference values have been established for both sexes, and for different ages. The relationship between weight and surface area has been used to establish the prediction that GFR, corrected for surface area, is proportional to the child's height divided by plasma [creatinine]. For values of GFR below 90 mL/min/1.73 m^2, one formula that has been developed is:

$$\text{GFR} = 40. \frac{\text{height (cm)}}{\text{plasma [creatinine]}}$$

The units for plasma [creatinine] are μmol/L. This formula is *not* appropriate for GFR studies in elderly patients.

TUBULAR FUNCTION

The healthy kidney has considerable reserve capacity for reabsorbing water and for the excretion of H^+; this reserve is only exceeded under exceptional physiological loads. Moderate impairment of renal function may reduce this capacity, but this is only apparent if loading tests are employed to stress the kidney. Tubular function tests are used much less often than glomerular function tests, which do not stress the patient.

The distal convoluted tubules and the collecting ducts fulfil a very important role in determining the concentration of urine that is finally excreted. This function can be investigated by renal concentration tests, which assess the capacity of the kidney to perform osmotic work, this capacity being reflected in the urine osmolality.

Renal concentration tests

The tests investigate the ability of the kidney to produce a concentrated urine in response to fluid deprivation or other appropriate stimuli. They may be indicated in patients with polyuria and polydipsia in whom hypothalamic or posterior pituitary disease or renal tubular dysfunction is thought to be the cause, and in whom other causes of these symptoms (e.g. diabetes mellitus, chronic renal failure, hyperparathyroidism) have been excluded before these tests are performed (Table 6.2). Indeed renal concentration tests can be positively dangerous in some of these other conditions.

Under most circumstances, urine sp. gr. (p. 29) is linearly related to osmolality. For screening purposes, urine sp. gr. is adequate provided that side-room tests show that there is neither marked glycosuria nor

TABLE 6.2. Causes of failure to concentrate urine.

Mechanism	Examples
Insufficient secretion of ADH	Lesions of supra-optic-hypothalmic-hypophyseal tract (e.g. trauma, neoplasms, infection)
Inhibition of ADH release	Increased thirst giving rise to polydipsia (e.g. psychogenic diabetes insipidus, lesions affecting thirst centre)
Inability to respond to ADH	Renal tubular defect, as in nephrogenic diabetes insipidus, Fanconi syndrome
Inability to maintain renal medullary hyperosmolality	Renal papillary necrosis (e.g. analgesic nephropathy), chronic renal failure, hydronephrosis, hypokalaemia, hypercalcaemia, lithium toxicity
Increased solute load per nephron	Chronic renal failure, diabetes mellitus

significant proteinuria. If the sp. gr. of an early morning specimen is greater than 1.020 or the osmolality greater than 800 mmol/kg, there is little point in performing further tests of renal concentrating ability.

Maximal concentration of urine may be produced in response either to fluid deprivation or to intramuscular injection of desaminocys-1-8-D-arginine vasopressin (DDAVP), a synthetic analogue of arginine vasopressin. Fluid deprivation stimulates the release of ADH, the natural hormone; DDAVP acts in exactly the same way as ADH. If a renal concentration test is to be performed, this is usually undertaken as an outpatient procedure in the first place. Only if the response to the outpatient test is abnormal does the inpatient test procedure need to be considered.

If a patient is receiving drugs such as carbamazepine, chlorpropamide,clofibrate or DDAVP, these should be stopped at least 48 h before any renal concentration test is performed.

RENAL CONCENTRATION TESTS (OUTPATIENT PROCEDURE)

In a co-operative patient, this tests the ability of the kidneys to produce a concentrated urine in response to fluid deprivation.

The patient is told to drink nothing from 1600 h on the day before the clinic appointment. At the clinic the following morning, a urine specimen is collected and its osmolality measured. In most healthy individuals, the osmolality will be greater than 800 mmol/kg; if this result is obtained, no further test of renal concentrating ability is required. Failure to concentrate the urine to 800 mmol/kg indicates the need for further testing, but no other conclusion can be drawn. In particular, it is wrong to interpret failure to concentrate urine under outpatient conditions as a sign of renal tubular dysfunction since patients sometimes do not co-operate fully even when given explicit instructions about the need for fluid deprivation.

RENAL CONCENTRATION TESTS (INPATIENT PROCEDURE)

These tests are performed in two stages. Initially, a fluid deprivation test is performed under more strictly controlled conditions than are possible with the outpatient procedure. If the renal concentrating ability in response to fluid deprivation is found to be inadequate, the second stage of the test follows on immediately, injection of DDAVP being used as the test stimulus. There are several ways in which these tests can be performed, so the description given here is only illustrative.

Fluid deprivation test

The patient is weighed before the test starts, and is instructed not to smoke until the test is over, sometime the following day. Beginning at 2200 h, the patient is told not to drink again until permitted to do so. However, formal supervision of the patient overnight is generally impracticable, so it can be difficult to be certain that the instruction not to drink is obeyed.

On the morning of the test, the patient is weighed again, and thereafter every 2 h until the test is ended. Weight loss of 3–5% indicates that severe dehydration has occurred, in which case the procedure should be stopped as soon as the next specimen of urine has been collected. The prohibition on smoking is reiterated.

The patient may have a light breakfast with a moderate amount of fluid, but no tea or coffee. Thereafter constant supervision is required, to ensure there is no surreptitious water drinking. If the timing of urine collections is started at 0800 h, the following specimens are obtained:

(1) Urine collected over the periods 0800–0900 h, 1100–1200 h and 1400–1500 h.

(2) Blood (lithium heparin anticoagulant) specimens obtained at 0830 h, 1130 h and 1430 h.

The osmolalities of the blood and urine specimens are measured as soon as possible. Normally, there is no increase in plasma osmolality (reference values 285–295 mmol/kg) but the urine osmolality rises to 800 mmol/kg or more. If the urine osmolality fails to reach 800 mmol/kg, it is convenient to proceed to the DDAVP test immediately.

DDAVP test

The patient is allowed a moderate amount of water to drink at 1500 h, to alleviate thirst. Then, preferably in the light of reports that the urine osmolality failed to reach 800 mmol/kg during the fluid deprivation test, the patient is given an intramuscular injection (4 μg) of DDAVP at 1600 h. Further specimens are collected at 1700 h, 1800 h, and at 1900 h and urine osmolality measured.

Interpretation of inpatient tests of renal concentrating ability

These tests are of most value in distinguishing between hypothalamic-pituitary, psychogenic and renal causes of polyuria (Table 6.2). These three categories will be discussed seriatim.

Patients with *diabetes insipidus of hypothalamic-pituitary origin* produce insufficient ADH; they should, therefore, respond to the DDAVP test but not to fluid deprivation. As a rule, these patients show an increase in plasma osmolality during the fluid deprivation test, to more than 300 mmol/kg, and a low urine osmolality (200–400 mmol/kg). There is a marked increase in urine osmolality, to 600 mmol/kg or more, in the DDAVP test.

Patients with *psychogenic diabetes insipidus,* assuming they co-operate in the performance of the tests, should respond to both fluid deprivation and DDAVP. In practice, renal medullary hypo-osmolality may prevent the urine osmolality reaching 800 mmol/kg in these patients after either fluid deprivation or DDAVP in tests carried out over a 24-h period, as described above. Also, the chronic suppression of the physiological mechanism that controls ADH release may impair the normal hypothalamic response to dehydration. These patients have a plasma osmolality that is initially low but which rises to normal during the tests. However, fluid deprivation may have to be continued for considerably longer than 24 h in patients with psychogenic diabetes insipidus before renal medullary hyperosmolality is restored; they then show normal responses to tests of renal concentrating ability.

Polyuria of renal origin may be due to inability of the renal tubule to respond to ADH, as in nephrogenic diabetes insipidus. In a number of other conditions, the kidney loses its ability to maintain medullary hyperosmolality (Table 6.2), and these need to be excluded before renal concentration tests are performed. In nephrogenic diabetes insipidus, there is failure to concentrate the urine in response both to fluid deprivation and to DDAVP injection, the urine osmolality usually remaining below 400 mmol/kg; in these patients, plasma osmolality increases as a result of fluid deprivation.

Urinary acidification tests

The hydrogen ion concentration in urine is normally much higher than blood $[H^+]$ or the $[H^+]$ in glomerular filtrate. This degree of acidification occurs as a result of the kidney reabsorbing large amounts of HCO_3^- filtered at the glomerulus, and excreting H^+ produced as non-volatile acids during tissue metabolism. The H^+ in urine is only partly eliminated as H^+, the rest being either NH_4^+ or H^+ combined with buffer ions, principally inorganic phosphate (Fig. 6.1).

The net excretion of H^+ by the kidney is the sum of the titratable acidity (the amount of base needed to titrate the specimen of urine to pH 7.4) *plus* the excretion of NH_4^+ *less* the excretion of HCO_3^-.

Titratable acidity (TA), NH_4^+ and HCO_3^- excretion can all be measured (results expressed in μmol/L urine); it is rarely necessary to measure all 3 quantities.

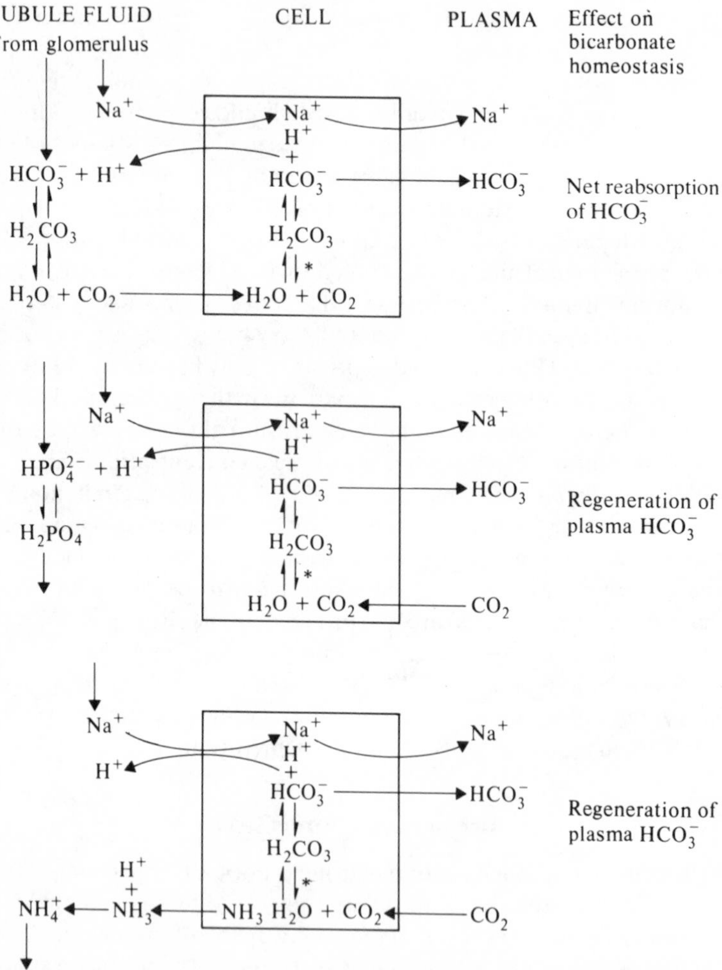

FIG. 6.1. *The reabsorption of filtered bicarbonate* (top diagram) *and the regeneration of bicarbonate*. Points of action of carbonic anhydrase are shown (*).

Reclamation of bicarbonate can only take place in the proximal tubule, where carbonic anhydrase is located on the brush border. Quantitatively, bicarbonate reclamation is normally far more important than bicarbonate regeneration. Regeneration of bicarbonate (middle and bottom diagrams) is effected by two mechanisms, both of which result in excretion of H^+ in exchange for Na^+; it is these latter mechanisms that are impaired in chronic renal failure.

SHORT TEST OF URINARY ACIDIFICATION (Wrong and Davies)

This assesses the capacity to produce an acid urine after a metabolic acidosis has been caused by the administration of ammonium chloride (NH_4Cl). It is a simple and widely performed test.

The patient fasts overnight before the test but is encouraged to drink water. The patient empties the bladder at the start of the test, i.e. at 0800 h. Urine specimens are then collected at hourly intervals until 1800 h. At 1000 h the patient is given NH_4Cl by mouth (0.1 g/kg body weight in gelatin capsules, *not* enteric-coated tablets). The test should be abandoned if the patient vomits. A portion of each urine specimen is transferred immediately to a glass syringe or container. This is capped, labelled and taken to the laboratory for measurement of pH; side-room test materials are not suitable for this purpose. Plasma [total CO_2] is measured on blood specimens collected before the test starts and 2 h after taking NH_4Cl.

In response to the NH_4Cl load, urine pH normally falls to below pH 5.3 in at least one specimen. However, it is always necessary to check that a satisfactory acidosis was induced by the dose of NH_4Cl; this is assumed to have occurred if plasma [total CO_2] falls by about 4 mmol/L in the 2-h period after NH_4Cl ingestion.

Renal tubular acidosis

Two distinct tubular abnormalities may be responsible for conditions in which there is acidosis of renal origin but little or no change in GFR.

Distal renal tubular acidosis is the commoner type and is due to an inability to maintain a gradient of $[H^+]$ across the distal tubule and collecting ducts. It is usually due to an inherited abnormality, but may occur in certain forms of acquired renal disease. Bone disease, commonly osteomalacia, results from buffering of H^+ by bone, and there is often hypercalciuria and nephrocalcinosis. There is loss of Na^+ and K^+ in the urine and hypokalaemia is common. Urinary pH rarely falls below pH 6.0 and never below pH 5.3 in the short test of urinary acidification.

Proximal renal tubular acidosis is much less common. It is due to proximal tubular loss of HCO_3^- caused by a low renal threshold for HCO_3^-. Occasionally this is an isolated abnormality but more often it occurs as one of the features in some patients with Fanconi syndrome; if these patients are given enough NH_4Cl to reduce plasma [total CO_2] below the renal threshold for HCO_3^-, the urinary pH can be made to fall below pH 5.3.

More elaborate tests of urinary acidification, e.g. determination of the renal threshold for HCO_3^-, are needed in order to differentiate between proximal and distal renal tubular acidosis.

Sodium and potassium excretion

SODIUM EXCRETION

The kidney normally conserves Na^+ very efficiently if dietary Na^+ is reduced (p. 44). In chronic renal failure, however, the kidney's capacity to adapt to changes in Na^+ intake is limited.

In most patients with chronic renal failure, Na^+ balance is maintained provided large changes in Na^+ intake are avoided but, when the limited ability to adapt to changes in Na^+ intake is exceeded, Na^+ and water tend to be retained and oedema tends to develop whenever dietary Na^+ is increased. By contrast, if dietary Na^+ is restricted, the diseased kidneys fail to conserve Na^+ and water, and depletion occurs; this in turn, may reduce the GFR even further. Occasionally, in chronic pyelonephritis or other disorders primarily affecting the renal tubules, large amounts of Na^+ are lost in the urine and severe Na^+ and water depletion can occur.

The ability to conserve Na^+ can be tested by giving a diet containing 20 mmol Na^+/day. Normally, urinary Na^+ excretion falls within a week to the amount present in the diet. This test should always be carefully monitored, by daily measurement of plasma [Na^+] and [urea], since severe Na^+ depletion can be induced.

POTASSIUM EXCRETION

About 90% of K^+ in the glomerular filtrate is normally reabsorbed in the proximal tubules. The distal tubules regulate the amount of K^+ excreted in the urine. The rate of secretion of K^+ by the distal tubules is influenced by the trans-tubular potential and by the tubular cell [K^+]. It is usually maintained adequately provided the urine flow rate is greater than one litre per day.

Retention of K^+ tends to occur late in the progress of renal disease, when oliguria and anuria supervene. However, in patients with oliguria due to acute or chronic renal failure, hyperkalaemia and K^+ retention can develop rapidly. Dangerous hyperkalaemia can also quickly follow the ingestion of K^+-containing foods in these patients.

Excessive renal losses of K^+ occur only rarely in chronic renal disease. However, the Na^+ depletion which sometimes occurs in renal disease may be associated with secondary aldosteronism, which causes excessive loss of K^+. Acid–base disturbances markedly affect renal output of K^+, excessive losses occurring particularly when there is a metabolic alkalosis; the kidney is unable to conserve K^+ efficiently in the presence of an alkalosis. Some primary tubular disorders are also associated with excessive losses of K^+. Treatment with diuretics can often cause K^+ depletion.

If dietary K^+ is reduced to 20 mmol/day, urinary output normally falls to this level within one week but occasionally takes 2 weeks. The persistence of a relatively high urinary K^+ output in the presence of hypokalaemia strongly suggests that the kidney is not able to conserve K^+ adequately. Measurement of urinary K^+ can prove very helpful in patients suspected of losing abnormal amounts of K^+.

Disorders of proximal tubular function

Specific disorders affecting the proximal tubules may cause impaired reabsorption of glucose, or phosphate, or amino acids, etc. In some conditions these defects occur singly; in others, multiple defects are present. Chemical investigations are needed for specific identification of these abnormalities and may include amino acid chromatography, measurement of plasma [calcium], [phosphate] and alkaline phosphatase activity, and an oral glucose tolerance test.

AMINO ACIDURIA

There are four groups of amino acids: (1) the neutral, (2) acidic and (3) basic amino acids, and (4) the imino acids proline and hydroxyproline. Each group has its own specific mechanism for transport across the proximal tubular cell. Normally, the renal tubules reabsorb all the filtered amino acids except for small amounts of glycine, serine, alanine and glutamine. Amino aciduria may be due to:

(1) Disease of the renal tubule *(renal or low threshold type).*
(2) Raised [amino acids] in blood *(overflow type).*

Renal amino aciduria may be due to impairment of one of the specific transport mechanisms (as in cystinuria, Hartnup disease). It may also occur as a non-specific renal amino aciduria due to generalized tubular damage, together with reabsorption defects affecting glucose or phosphate or both.

The overflow types of amino aciduria result when the renal threshold for amino acids is exceeded, due to overproduction or to accumulation of amino acids in the body (e.g. phenylketonuria, p. 419; acute hepatic necrosis, p. 182).

Hartnup disease

This is a rare hereditary disorder in which epithelial transport of the monoamino-monocarboxylic (neutral) amino acids is affected. There is an amino aciduria and intestinal malabsorption of the neutral amino acids. Symptoms, if they arise, are mostly attributable to tryptophan deficiency. They include a pellagra-like rash that responds to treatment with nicotinamide, which is normally derived partly from tryptophan (Fig. 13.2). Cerebellar ataxia may also be present, possibly caused by the toxic effects of substances such as indoles formed by bacterial metabolism of unabsorbed tryptophan.

The diagnosis is confirmed by demonstrating the presence of an amino aciduria affecting the neutral group of amino acids.

Cystinuria

This is a hereditary condition in which cystine and the basic amino acids (lysine, ornithine and arginine) are excreted in excessive amounts. It is due to a specific defect in epithelial transport of these amino acids, the defect being present in the intestinal mucosa as well as the renal tubular cell. Symptoms may arise due to the formation of renal stones composed of cystine; cystine is relatively insoluble. The diagnosis of cystinuria is established by demonstrating excessive amounts of these amino acids in the urine, usually by chromatography. Cystinuria should not be confused with cystinosis.

Cystinosis (Lignac–Fanconi disease)

This is one of the causes of the Fanconi syndrome. It is a rare inherited disease in which cystine is deposited in the cells of the lymphoreticular system. Renal tubular damage also occurs and there is a generalized amino aciduria, glycosuria and phosphaturia. Affected infants fail to thrive and develop vitamin D-resistant rickets; death usually occurs in early childhood.

Fanconi syndrome

This syndrome may be inherited or secondary to a number of uncommon disorders such as heavy metal poisoning or multiple myeloma. One of the inherited causes of the Fanconi syndrome is cystinosis.

The syndrome comprises multiple defects of proximal tubular function, there being excessive losses of amino acids (generalized amino aciduria), phosphate, glucose and sometimes HCO_3^- (proximal renal tubular acidosis). Distal tubular functions may also be affected. Sometimes globulins of low molecular mass may be detectable in urine, in addition to amino aciduria.

Hypophosphataemia

This can result from an isolated tubular defect of phosphate reabsorption. It occurs in association with vitamin D-resistant rickets (p. 300).

Renal glycosuria

This is a benign condition. The renal threshold for glucose is lowered giving rise to glycosuria in the presence of normal plasma [glucose]. Diagnosis requires the performance of an oral glucose tolerance test (p. 222). The condition may be associated with other tubular defects.

PROTEINURIA

The normal glomerular filtrate contains small amounts (about 30 mg/L) of protein; this represents a total filtered load of about 5 g/24 h. Since less than 100 mg protein is normally excreted in the urine each day, tubular reabsorption of protein must be very efficient in health.

Routine 'side-room' testing of urine for the presence of protein (p. 31) should be part of the full clinical assessment of patients when first examined. It is also an important test in the medical assessment of applicants for life insurance, pre-employment health checks and any other examination where it is particularly necessary to try to exclude the presence of asymptomatic and unsuspected disease in apparently healthy individuals. Routine testing for protein is usually best carried out on an early morning specimen of urine, as this is normally the most concentrated specimen that can be obtained.

Nearly all renal diseases give rise to proteinuria, but other causes need to be considered (Tables 6.3 and 6.4). Although Table 6.4 classifies glomerular and tubular proteinuria separately, many patients show features of both glomerular and tubular protein loss. Where quantitative measurements of urine protein loss are required (e.g. when monitoring treatment of patients with the nephrotic syndrome), side-room tests are usually insufficiently precise; 24-h collections of urine should be examined in the laboratory. Esbach's test is obsolete.

TABLE 6.3. Overflow proteinuria.

Protein	Molecular mass (daltons)	Cause
Amylase	45 000	Acute pancreatitis
Bence-Jones protein	44 000	Multiple myeloma
Haemoglobin	68 000	Haemolytic anaemia, March haemoglobinuria
Lysozyme	15 000	Myelomonocytic leukaemia
Myoglobin	17 000	Myoglobinuria, Crush injuries

OVERFLOW PROTEINURIA

Several conditions may give rise to abnormal amounts of low mol. mass proteins (i.e. less than 70 000 daltons or 70 kDa) in urine. These proteins are filtered at the glomerulus and may then be neither reabsorbed nor catabolized completely by the renal tubular cells. The principal examples are listed in Table 6.3.

Electrophoresis of urine, following preliminary concentration, may be valuable in the investigation of multiple myeloma (p. 134). By this technique, Bence–Jones protein can be demonstrated in over 70% of patients with this disease. Measurements of enzymic activity, or spectroscopic examination, can be used to detect the other proteins in Table 6.3.

GLOMERULAR PROTEINURIA

Orthostatic proteinuria is a benign condition that affects children and young adults, the patients only exhibiting proteinuria after they have been standing up. If the patient has orthostatic proteinuria, protein will not be detectable in an early morning urine specimen, when tested by normal side-room methods (i.e. urine then contains less than 100 mg/L). The patient is instructed to empty the bladder just before going to bed, and the test for protein is performed on a specimen of urine passed the following morning, before getting up.

Orthostatic proteinuria is usually observed in only some of the urine specimens passed when up and about, and for these individuals the prognosis is good. However, the prognosis is less good in patients in whom proteinuria is always detected when they are up and about.

Effort proteinuria and febrile proteinuria usually have no long-term pathological significance.

TABLE 6.4 Glomerular and tubular proteinuria (after Parfrey, 1982)

Glomerular proteinuria	*Tubular proteinuria*
(1) *Intermittent* Orthostatic (occasionally persistent)	Acute tubular necrosis Chronic nephritis, pyelonephritis
Effort	Renal transplantation
Febrile	Congentital disorders affecting the renal tubule (e.g. cystinosis, Fanconi syndrome, renal tubular acidosis)
(2) *Persistent* Glomerulonephritis (all forms)	
Pathological causes of altered haemodynamics (e.g. congestive cardiac failure, renal artery stenosis)	Heavy metal poisoning (e.g. lead, cadmium) and Wilson's disease

The persistent forms of glomerular proteinuria are commonly occurring pathological causes of persistent proteinuria (Table 6.4). Plasma proteins escape in varying amounts, depending on their mol. mass, on the amount of glomerular damage, and on the capacity of the renal tubule cells to reabsorb or metabolize the proteins that have passed the glomerulus.

TABLE 6.5. Causes of the nephrotic syndrome (after Parfrey, 1982).

Category	Examples
Renal disease	**Glomerulonephritis,** renal artery stenosis, renal vein thrombosis, transplant rejection
Systemic disease	**Amyloidosis,** systemic lupus erythematosus, polyarteritis nodosa, Henoch–Schoenlein purpura
Metabolic disease	**Diabetes mellitus,** hypothyroidism
Neoplastic disease	**Lymphoma,** leukaemia, plasma cell myeloma, carcinoma (e.g. lung, gastrointestinal tract)
Infections	Malaria *(Plasmodium malariae),* schistosomiasis, hepatitis B, bacterial endocarditis, cytomegalovirus, infectious mononucleosis
Drugs	Heavy metals, organic gold and mercury compounds, captopril, penicillinase, phenindione, probenecid, troxidone
Familial disorders	Congenital nephrotic syndrome, Alport's syndrome, sickle-cell anaemia
Allergy	Bee stings, poison oak, poison ivy, pollen, vaccines

Note: The principal causes of the nephrotic syndrome, in the U.K., are shown in bold type. Malaria is the most common cause, worldwide.

The degree of proteinuria is not an index of the severity of renal disease. However, it is convenient to distinguish *mild or moderate proteinuria,* where the loss is not sufficient to cause protein depletion, from *severe proteinuria,* in which the urinary protein loss exceeds the body's capacity to replace losses by synthesis (usually 5–10 g/24 h). Severe persistent proteinuria is one feature of the nephrotic syndrome (Table 6.5) in which protein loss is sometimes more than 30 g/24 h.

Differential protein clearance (selectivity)

These measurements are sometimes performed in patients with nephrotic syndrome, since patients who have a selective proteinuria (e.g. due to minimal change disease) are more likely to respond to steroid therapy than patients with unselective proteinuria.

Selectivity measurements are based on the fact that glomerular permeability to a plasma protein depends largely on its mol. mass, small molecules being cleared more rapidly than large. Differences in clearance are investigated by measuring simultaneously the clearance

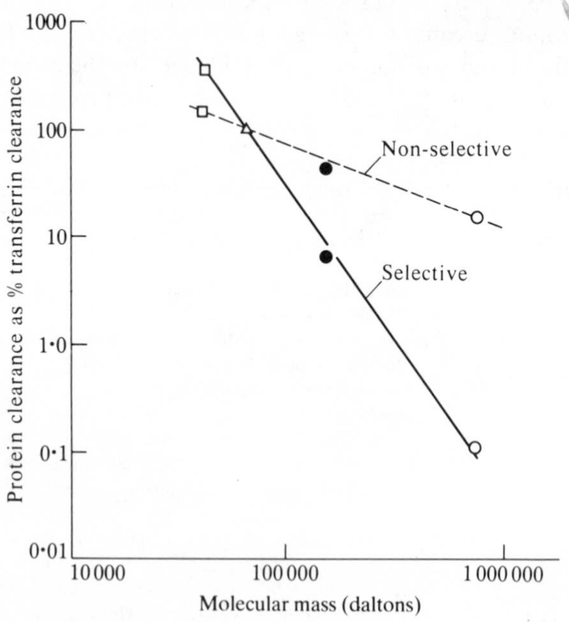

FIG. 6.2. *Selectivity of proteinuria.* Clearance of the following proteins is expressed as a percentage of the transferrin clearance: α_1-acid glycoprotein (mol. mass 44 100 Da) □; transferrin (mol. mass 80 000 Da) △; IgG (mol. mass 146 000 Da) ●; and α_2-macroglobulin (mol. mass 820 000 Da) ○. Note that both axes have logarithmic scales.

of two or more proteins of different mol. mass, e.g. transferrin and α_2-macroglobulin (Fig. 6.2). If the ratio is high, this means that large molecules are failing to pass into the glomerular filtrate and the proteinuria is called selective; if the ratio is low, the proteinuria is called unselective.

TUBULAR PROTEINURIA

This may be due to tubular or interstitial damage resulting from a variety of causes (Table 6.4), and the proteinuria results from failure of the tubules to reabsorb some of the plasma proteins that have been filtered by the normal glomerulus. Characteristically, the proteins that are mostly excreted in tubular proteinuria have low mol. mass, e.g. β_2-microglobulin (11.8 kDa) and lysozyme (15 kDa). The loss of protein is usually mild, rarely more than 2 g/24 h.

Urinary β_2-microglobulin excretion is normally very small in amount (under 0.4 mg/24 h). Its measurement has been used as a sensitive test of renal tubular damage. The test is of limited value for this purpose, however, if there is evidence of renal insufficiency, e.g. increased plasma [creatinine] or decreased creatinine clearance.

RENAL FAILURE

Acute renal failure

By definition, there is renal disease of acute onset and severe enough to cause failure of renal homeostasis. A few patients maintain a normal output of urine (in terms of volume) throughout the course of the illness, but usually oliguric, diuretic and recovery phases can be recognized. Chemical investigations help to determine the severity of the disease and to follow its course, but they do not help much in determining the cause. Renal ischaemia is the commonest cause of acute renal failure; primary renal disease, urinary tract obstruction and toxic renal damage account for most of the other cases.

Oliguric phase

Less than 400 mL urine is produced each day, and there may be anuria in renal failure due to outflow obstruction. The oliguria is mainly due to a fall in GFR. The urine that is formed usually has an osmolality similar to plasma and a relatively high [Na^+].

Plasma [Na^+] is usually low due to a combination of factors, including intake of water in excess of the amount able to be excreted, increase in metabolic water from increased tissue catabolism, and a shift of Na^+ from ECF to ICF. Plasma [K^+], on the other hand, is usually increased due to the impaired renal output and excess tissue catabolism, aggravated by the shift of K^+ out of cells that accompanies the metabolic acidosis; the acidosis develops due to failure to excrete H^+ and the increased formation of H^+ due to tissue catabolism.

Retention of urea, creatinine, sulphate and other waste products occurs. The rate at which plasma [urea] rises is affected by the rate of tissue catabolism; this, in turn, depends on the cause of the acute renal failure. In renal failure due to trauma (including failure developing after surgical operations), the plasma [urea] tends to rise more rapidly than in patients with renal failure due to medical causes such as acute glomerulonephritis.

To differentiate the low urinary output of suspected acute renal failure from that due to severe circulatory impairment with reduced blood volume, the tests summarized in Table 6.6 may be helpful. However, none of the chemical tests can be completely relied upon to make the important distinction between renal failure and hypovolaemia. Priority, therefore, must always be given to fluid administration, to prevent the progression of pre-renal uraemia to established renal failure.

For monitoring patients with acute renal failure, plasma [urea] and [K^+] are particularly important and are required at least once daily; plasma [creatinine] may also be of value. Decisions to use haemodialysis are reached at least partly on the results of these tests.

Diuretic phase

With the onset of this phase, urine volume increases but the clearance of urea, creatinine and other waste products may not improve to the same extent. Plasma [urea], [creatinine], etc. may therefore continue to rise, at least at the start of the diuretic phase. Large losses of electrolytes may occur in the urine and require to be replaced orally or

TABLE 6.6. Low urinary output: Recognition of cause.

	Simple hypovolaemia	Acute renal failure
Urine osmolality	Usually >600 mmol/kg	Usually <350 mmol/kg
Urine [urea]: plasma [urea]	Usually >10:1	Usually <5:1
Urinary [Na^+]	Usually <10 mmol/L	Usually >20 mmol/L
Effect of fluid plus diuretics	Flow increases	Little or no effect

parenterally. Measurement of these losses is needed so that correct replacement therapy can be given.

Plasma [K^+] tends to fall as the diuretic phase continues due to the shift of K^+ back into the cells and to marked losses in urine resulting from impaired conservation of K^+ by the still damaged tubules. Usually, Na^+ deficiency occurs also, due to failure of renal conservation. Throughout the diuretic phase, therefore, it is important to measure plasma [urea], [Na^+] and [K^+] at least once daily, and the output of Na^+ and K^+ in the urine.

Chronic renal failure

Most of the functional changes seen in chronic renal failure may be explained in terms of a full solute load falling on a reduced number of functionally normal nephrons. The GFR is invariably reduced, associated with retention of urea, creatinine, urate, various phenolic and indolic acids and other organic substances. The progress and severity of renal disease are usually monitored by measuring plasma [urea] periodically, but this is only a crude index of renal function (p.88).

Sodium, potassium and water

The increased solute load per nephron impairs the kidney's ability to produce concentrated urine, although the capacity to secrete dilute urine is often relatively unaffected. In less severe cases of failure, especially where the renal medulla is mainly affected, Na^+ tends to be lost due to impaired ability to lower [Na^+] in the tubules. Therefore, Na^+ and water depletion may occur if there is any reduction in Na^+ intake (e.g. due to vomiting), or increase in output as may occur with diuretic therapy. However, as the GFR falls to lower levels in more severe cases, retention of Na^+ occurs but is rarely accompanied by oedema. There is no consistent pattern of alteration in plasma [Na^+], but plasma [Cl^-] is often increased.

Potassium clearance may be increased and raised plasma [K^+] is uncommon in spite of the tendency for K^+ to come out of cells due to the metabolic acidosis that is usually present. However, patients with renal failure are unable to excrete large loads of K^+. Potentially fatal hyperkalaemia may follow ingestion of substances with a high K^+ content, e.g. certain fruit juices.

Patients with chronic renal failure can usually maintain Na^+, K^+ and water balance well, providing the dietary load remains close to normal daily requirements, and does not vary much.

Acid–base disturbances

The total excretion of H^+ is impaired mainly due to a fall in the renal capacity to form NH_4^+. There may also be a fall in titratable acidity due to a reduction in urinary buffering capacity. Metabolic acidosis is present in most patients but the severity of the acidosis remains fairly stable in spite of the reduced urinary H^+ excretion. There may be an extra-renal mechanism for H^+ elimination, possibly involving buffering of H^+ by calcium salts in bone; this would help to account for the demineralization that often occurs in chronic renal failure.

Calcium and phosphate

Plasma [calcium] is variable, but tends to be low, often due at least partly to reduced plasma [albumin]. Plasma [phosphate] is high, mainly due to a reduction of GFR.

Virtually all patients with chronic renal failure have secondary or tertiary hyperparathyroidism (p. 299), and may develop osteitis fibrosa. Plasma [calcium], which is decreased or close to the lower reference value in patients with secondary hyperparathyroidism, increases later if tertiary hyperparathyroidism develops. Many patients with a low plasma [calcium] have reduced activity of the renal enzyme responsible for the 1α-hydroxylation of 25-hydroxycholecalciferol (p. 265) and develop osteomalacia. They require treatment with 1α-hydroxycholecalciferol; neither vitamin D nor its 25-hydroxy derivative can be effective, at least not in normal doses.

A few patients show a third type of bone abnormality, with increased bone density (osteosclerosis). It is not clear why any particular one of these various types of uraemic osteodystrophy should develop in an individual patient.

Other metabolic abnormalities

Other findings in chronic renal failure may include impaired glucose tolerance and raised plasma [magnesium]. These are of no particular diagnostic significance, although the increased plasma [magnesium] may be responsible, at least in part, for the feeling of lethargy.

Uraemic toxins

Many substances accumulate in patients with uraemia, and many explanations have been offered for the symptoms and complications that develop. The subject remains confused, despite much study of the effects of maintenance dialysis (comparing the effects of haemodialysis and peritoneal dialysis), modifications to the diet, and renal transplantation.

At one time or another, substances suspected of being uraemic toxins have included urea, creatinine, amino acids, amines, guanidine derivatives, urate, phenols, indoles, magnesium, parathyroid hormone, glucagon, growth hormone, and 'middle molecules'. None of the substances proposed as a uraemic toxin has been generally accepted as having this property, but much interest has focussed recently on 'middle molecules'. The main point of agreement is that urea is one of the least toxic of the substances listed.

'Middle molecules' is a term first applied to substances with mol. mass 500–5000 daltons. More recently, as far as uraemic toxins are concerned, this range has been narrowed to substances with mol. mass 1000–2000 daltons. These have been detected in the plasma of severely uraemic patients but not in normal plasma, and have been found in particularly large amounts in patients with uraemic pericarditis or peripheral neuropathy. However, these findings require confirmation, and the chemical structures of the 'middle molecules' need to be identified; as well as their relationship (if any) to the symptoms of uraemia.

RENAL STONES

Physicochemical principles govern the formation of renal stones, and these are relevant to the choice of treatment aimed at preventing progression or recurrence. Stones may cause renal damage, and this is often progressive; renal function tests then show deterioration of function.

The solubility of a salt depends on the product of the activities of the ions which make up the salt. Frequently, the solubility product is exceeded without the formation of a stone, provided there is no 'seeding'. Seeding promotes crystal formation in relation to particles that may be present in the urine, such as debris or bacteria. There is also evidence that formation of stones can be prevented by inhibiting substances, normally present in the urine.

People living or working in hot conditions are liable to become dehydrated, and show a greater tendency to form renal stones. Urinary tract infection may be a precipitating factor, and there are

TABLE 6.7. Renal stones.

Type of stone	Frequency	Cause or relevant factors
Calcium oxalate, and mixed calcium oxalate and calcium phosphate	80–85%	Hypercalciuria (various causes, see text), hyperabsorption of oxalate from diet, primary hyperoxaluria
Magnesium ammonium phosphate and calcium phosphate ("triple phosphate")	5–10%	Urinary tract infection, leading to fall in $[H^+]$.
Uric acid	5–10%	Gout, myeloproliferative disorders, high protein diet, uricosuric drugs.
Cystine	1%	Cystinuria
Xanthine	Rare	Xanthinuria

several metabolic factors that can cause stones to form in the renal tract. However, in many patients, no cause can be found to explain why stones have formed. The main types of renal stone are described in Table 6.7. Uric acid and xanthine stones are radiolucent, unless calcium salts are deposited on them; other stones are mostly radio-opaque.

Hypercalciuria

Stones in the upper renal tract have been shown to occur in 5–10% of adults in Western Europe and the U.S.A. These stones are mostly either pure calcium oxalate or a mixture of calcium oxalate and phosphate. Not every patient with renal stones, however, has hypercalciuria since there is a considerable overlap between 24-h urinary calcium excretion of normal individuals on their usual diet (up to 12 mmol/24 h) and the urinary calcium excretion of stone-formers. Time-consuming and much more demanding studies of urinary calcium excretion, carried out on patients taking low-calcium diets (3.5 mmol/24 h), have been largely abandoned.

The causes of urinary calcium excretion in excess of 12 mmol/24 h, in patients taking their normal diet, include:

(1) Idiopathic hypercalciuria.
(2) Primary hyperparathyroidism (p. 284).
(3) Vitamin D overdosage and hypersensitivity to vitamin D (e.g. sarcoidosis).
(4) Prolonged immobilization.
(5) Renal tubular acidosis.

Idiopathic hypercalciuria is the commonest single cause of renal stones, but little is known about the basic abnormality in this disorder. Up to 10% of renal calculi, depending on the series, have been attributed to primary hyperparathyroidism. It is important to investigate patients with renal calculi for primary hyperparathyroidism since it is amenable to curative treatment.

Oxalate excretion

The majority of urinary calculi contain oxalate, but excessive excretion of oxalate is primarily responsible for the formation of stones in only a small percentage of cases. Primary hyperoxaluria is a rare condition in which there is increased excretion of oxalate and of glyoxylate, the latter due to deficiency of the enzyme responsibility for converting glyoxylate to glycine.

Patients with disease of the terminal ileum may have an increased tendency to form oxalate stones. This is due to hyperoxaluria caused by increased absorption of dietary oxalate.

Chemical investigation of patients with renal stones

The composition of stones should be determined, as this can be helpful. Care must be taken, however, to ensure that the specimen sent to the laboratory for analysis really is a stone that came from the patient — patients have been known to fool doctors, and to waste the laboratory's time, by producing small bits of gravel and saying they have passed them as stones. The following tests may be helpful in reaching a diagnosis:

(1) *Plasma* [calcium], [phosphate], [total CO_2], [urate], alkaline phosphatase activity. Full acid–base assessment is rarely needed.
(2) *Urine* side-room tests (pH and protein), and 24-h excretion of calcium, phosphate and urate. Occasionally, urinary excretion of oxalate, cystine or xanthine may be required, or urinary acidification tests.
(3) *Stone analysis* can be performed for calcium, phosphate, magnesium, oxalate, uric acid, cystine and xanthine, but these are not all required in every case.
(4) *Renal function tests,* usually creatinine clearance and plasma [urea].

In addition to chemical tests, microbiological examination of urine is usually performed in these patients, and radiological investigations including special investigations of the urinary tract.

FURTHER READING

BARRATT, M. (1983). Proteinuria. *British Medical Journal,* **287,** 1489–1490.

BLACK, D. and JONES, N.F., editors (1980). *Renal Disease,* 4th Edition. Oxford: Blackwell Scientific Publications.

BROD, J. (1971). Study of renal function in the differential diagnosis of kidney disease. *British Medical Journal,* **2,** 135–143.

HAYCOCK, G.B. (1981). Old and new tests of renal function. *Journal of Clinical Pathology,* **34,** 1276–1281.

JONES, N.F., editor (1975). *Recent Advances in Renal Disease,* **1.** Edinburgh: Churchill Livingstone.

Leading article (1977). What causes toxicity in uraemia? *British Medical Journal,* **2,** 143–144.

MORGAN, D.B. (1982). Assessment of renal tubular function and damage and their clinical significance. *Annals of Clinical Biochemistry,* **19,** 307–313.

PARFREY, P.S. (1982). The nephrotic syndrome. *British Journal of Hospital Medicine,* **27,** 155–162.

PARFREY, P.S. (1982). Proteinuria. *ibid*, **27,** 254–258.

SCHOOTS, A.C., MIKKERS, F.E.P., LAESSENS, H.A., SMET, R. de, LANDSCHOOT, N. van, and RINGOIR, S.M.G. (1982). Characterization of uremic 'middle molecular' fractions by gas chromatography, mass spectrometry, isotachophoresis, and liquid chromatography. *Clinical Chemistry,* **28,** 45–49.

WILLS, M.R. (1977). Biochemical aspects of urinary stones. *Proceedings of the Royal Society of Medicine,* **70,** 517–520.

WILLS, M.R., (1978). *Metabolic Consequences of Chronic Renal Failure,* 2nd Edition. Aylesbury: HM & M.

WRONG, O. and DAVIES, H.E.F. (1959). The excretion of acid in renal disease. *Quarterly Journal of Medicine,* **28,** 259–313.

Chapter 7

Plasma Protein Abnormalities

The diets of most people living in the 'developed' countries contain about 100 g protein/day, i.e. considerably more than the minimal requirements which are probably for about 20 g 'first class' protein/day. First class protein provides essential amino acids in approximately the same proportions as the body's composition.

Protein digestion and absorption of amino acids are discussed in Chapter 10. Amino acids derived from the diet join those formed by breakdown of endogenous protein to provide the constituents for synthesis of new protein. Amino acids also give rise, after deamination and other metabolic steps, to compounds which may be used to help meet energy requirements, or for gluconeogenesis. The amino nitrogen of amino acids is converted to urea. The liver is the key organ in these metabolic processes (Chapter 11).

Urea is the main nitrogen-containing compound excreted by man. Proteins, as such, are normally only excreted in small amounts. On an average diet, about 10–20 g N/day is excreted in the urine and about 1–2 g N/day in the faeces. On a low-protein diet, the urinary N falls to about 4 g/day provided that energy requirements are satisfactorily met by other dietary constituents. After major surgery or accidental trauma, urinary N excretion rises due to increased protein catabolism. If the output exceeds the dietary intake, there is a state of negative nitrogen balance. When body protein is being laid down, during growth or repair, there is a state of positive balance.

The rate of protein synthesis is increased by several hormones including thyroxine, cortisol and anabolic steroids. The rate is depressed in protein-energy malnutrition (p. 437), in some patients with malignant disease, and it tends to fall in a hot environment. The rate of protein catabolism is increased by thyroid hormones and the glucocorticoids (e.g. cortisol).

FUNCTIONS OF PLASMA PROTEINS

The plasma proteins are mostly synthesized in the hepatocytes, except for the immunoglobulins which are synthesized in the lymphoreticular system. The principal functions of the plasma proteins are:

(1) *Maintenance of colloid oncotic pressure,* a function mainly of albumin, quantitatively the most important plasma protein.
(2) *Transport by carrier proteins of:*
 (a) Essential metabolites: e.g. lipids.
 (b) Hormones: e.g. thyroxine, cortisol.
 (c) Metals: e.g. calcium, copper, iron.
 (d) Excretory products: e.g. bilirubin.
 (e) Drugs and various toxic ingested substances.
(3) *Defence reactions,* functions which depend on (a) immunoglobulins and (b) the complement system.
(4) *Coagulation and fibrinolysis,* which involve some proteins circulating in plasma and others liberated, for instance, from damaged platelets.
(5) *Buffering of* H^+, a minor function of plasma proteins (but a very important function of haemoglobin in the erythrocytes).
(6) *Specialized functions, including:*
 (a) Lecithin cholesterol acyl transferase (p. 245).
 (b) Proteinase inhibitors, e.g. α_1-antitrypsin, α_2-macroglobulin.
 (c) Renin-angiotensin system (p. 373).

Most plasma proteins are glycoproteins, i.e. they contain covalently linked carbohydrate. The amount varies considerably, e.g. IgG contains about 3% whereas α_1-acid glycoprotein contains about 40% carbohydrate.

The proteins in plasma are also found in smaller concentrations in extravascular fluid, the amount in the intravascular space depending partly on the molecular mass. Thus, about 40% of the total body albumin (mol. mass 66 kDa) is intravascular, whereas the corresponding figures for IgG (160 kDa) and IgM (1 MDa) are 50% and 80% respectively.

Methods of investigation

Several different techniques for measuring protein concentrations have been developed, including:

(1) *Direct chemical measurement,* e.g. biuret method for detecting the presence of peptide bonds. This measures the total concentration of proteins, singly or in mixtures.
(2) *Direct physical measurement,* e.g. albumin by dye-binding (the protein error of indicators).
(3) *Measurement after separation* by techniques such as electrophoresis or isoelectric focusing.

(4) *Measurement of biological activity,* e.g. enzymic activity (Chapter 8), coagulation properties.
(5) *Immunological methods.*

We shall consider briefly the electrophoretic and immunological methods.

SERUM PROTEIN ELECTROPHORESIS

This technique separates proteins mainly on the basis of their electric charge. It can be performed on a variety of materials (e.g. paper, cellulose acetate) or in various media (e.g. starch gel, polyacrylamide gel). *Serum* is normally required for this investigation, as fibrinogen and fibrin interfere with the interpretation of the separations. Electrophoresis on cellulose acetate (Fig. 7.1) is the most widely used technique. It separates the serum proteins into five distinct fractions: albumin and α_1-, α_2, β- and γ-globulins. Apart from albumin, these fractions all contain mixtures of proteins (Table 7.1). If plasma is used instead of serum for the electrophoretic separation, fibrinogen forms a discrete band between the β- and γ-globulin regions. This band may

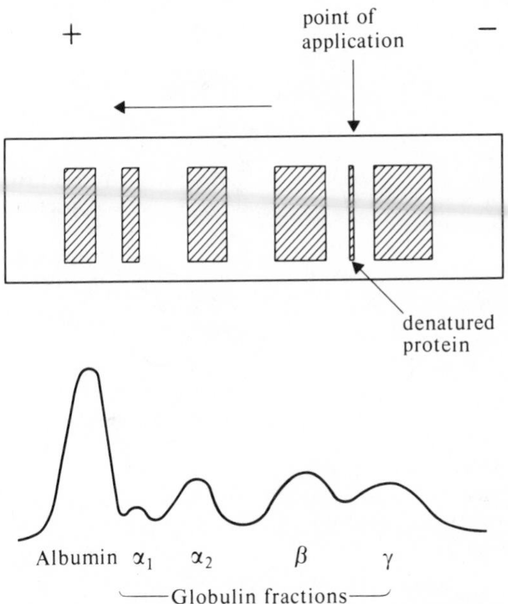

FIG. 7.1. *Serum protein electrophoresis on cellulose acetate*, showing the appearance of the strip and its corresponding densitometric scan after separation of the proteins and staining. The horizontal arrow shows the direction of flow of the electric current, towards the anode.

make it difficult to detect a band due to a paraprotein, e.g. a myeloma protein.

IMMUNOLOGICAL METHODS

The great advantage of immunological methods is their specificity. A

TABLE 7.1. Characteristics of plasma proteins. (Data from a table of about 100 'well-characterised' plasma proteins; Putnam, 1977)

Protein and electrophoretic mobility	Molecular mass (daltons)	Mean concentration (g/L)	Examples of functions
Pre-albumin	54 980	0.25	T4, T3 and retinol transport
Albumin	66 000	45	Colloid oncotic pressure, transport functions
α_1-*Globulins*			
α_1-Fetoprotein	74 000	Trace	
α_1-Acid glycoprotein	44 100	0.9	Binds cationic drugs
α_1-Antitrypsin	54 000	2.9	Antiproteinase
Prothrombin	72 000	0.06	Blood clotting
Transcortin	55 700	0.04	Cortisol transport
Vitamin D-binding globulin	52 800	0.005	Vitamin D transport
α_2-*Globulins*			
Thyroxine-binding globulin	60 700	0.015	T4 and T3 transport
Ceruloplasmin	132 000	0.35	Copper transport
Retinol-binding globulin	21 000	0.045	Retinol transport
Haptoglobin 1–1	100 000	2.0	Haemoglobin binding
α_2-Macroglobulin	820 000	2.4 (males) 2.9 (females)	Antiproteinase; also, transport functions
β-*Globulins*			
Transferrin	80 000	2.95	Iron transport
Plasminogen	81 000	0.12	Fibrinolysis
Pregnancy-specific β_1-glycoprotein	90 000	0.05–0.2	
β_2-Microglobulin	11 800	0.0015	
C-Reactive protein	140 000	Trace	
γ-*Globulins*			
IgG	143–149 000	13	Antibodies
IgA	158–162 000	2.7	Antibodies
IgM	800–950 000	1.5	Antibodies
IgD	175–180 000	0.1	Antibodies
IgE	185–190 000	Trace	Antibodies

wide variety of antibodies, each directed against a specific serum protein, can be raised moderately easily. The antigen–antibody reaction is relatively unaffected by the presence of other proteins in serum. Preliminary purification or other treatment of the serum is therefore unnecessary and it is possible, using immunological techniques, to measure the concentration in serum of any protein against which a suitable antiserum can be raised.

The most commonly used immunological methods are immunoelectrophoresis, immunodiffusion, immunoprecipitation and immunoassay. The first two are qualitative or can at best be made semi-quantitative. The latter two are quantitative procedures and have both proved capable of at least partial automation.

The specificity of immunological methods has allowed considerable advances to be made in the knowledge of variations in serum proteins that may occur in disease. It is possible to pinpoint abnormalities such as an isolated deficiency of IgA or IgM that cannot be detected by less sensitive and less specific techniques such as electrophoresis on cellulose acetate. It must be pointed out, however, that one of the reagents used in immunological techniques (the antiserum) is of biological origin. Since no two antisera are identical, variations in either their potency or specificity can cause anomalous results. Also, it is apparent that immunological reactivity does not always equate with biological activity, and this can considerably affect the interpretation of results (e.g. serum PTH, gastrin measurements).

Plasma protein concentrations in disease

The concentrations of proteins may be altered both in diseases that primarily affect protein metabolism and in diseases where there is dehydration or overhydration. Most diseases that alter plasma [proteins] affect the synthesis of proteins in the liver, or the distribution of proteins, or their rate of catabolism or rate of excretion. Albumin will be considered in detail, to illustrate these points. Similar principles apply to alterations in the concentration of many other proteins in plasma.

ALBUMIN

Albumin is quantitatively the most important of the plasma proteins, and has a low molecular mass compared with most of the other proteins listed in Table 7.1. These facts account for the importance of albumin's contribution towards maintaining the colloid oncotic pressure of plasma. Albumin is synthesized by the hepatocytes. Its half-life in plasma is 20 days.

Synthesis of albumin may be reduced in the following:

(1) *Hereditary defects.* Analbuminaemia is a rare disorder in which plasma [albumin] is usually less than 1.0 g/L. However symptoms, even oedema, may be absent due to compensatory increases in plasma [globulins].
(2) *Liver disease,* both acute and chronic, can cause defective albumin synthesis, but reductions in plasma [albumin] may not be detected in acute liver disease because of the long half-life of albumin.
(3) *Malnutrition* may cause a significant reduction in plasma [albumin], and this serves as a useful index of nutritional status in children.
(4) *Malabsorptive disease* has to be fairly severe before plasma [albumin] is reduced.

Distribution of albumin may be altered whenever there is increased capillary permeability. This allows proteins to leak into the extravascular fluid, and plasma [albumin] may be considerably reduced. This is a prominent feature in severely burned patients, and may be present with shock or chronic liver disease. Sequestration of protein in a serous effusion can have a similar effect.

Catabolism of albumin may be increased as a result of injury, e.g. major surgery or accidental trauma, or infection. The fall in plasma [albumin] is usually small.

Abnormalities of albumin excretion may give rise to increased loss in the urine, or may occur as a result of losses into the gastrointestinal tract or through the skin. Normally, little or no albumin is excreted by any of these routes.

Proteinuria is a feature of most forms of renal disease. However, it is only in the nephrotic syndrome that plasma [albumin] falls markedly. In these patients, losses usually exceed 5 g/day; the liver is unable to compensate fully for continuing losses of this size.

Gastrointestinal tract diseases associated with large losses of protein from the gut (protein-losing gastroenteropathy) are uncommon, e.g. intestinal lymphangiectasia, gastric mucosal hypertrophy. Smaller losses sometimes occur in ulcerative colitis, Crohn's disease and other gastrointestinal disorders. Demonstration of the cause of a low plasma [albumin] in these patients may be difficult; the commonest method is to inject ^{51}Cr-labelled albumin and measure faecal radioactivity over the next several days.

Burns and exfoliative dermatitis may cause large losses of albumin and other plasma proteins through the skin.

Transport functions of albumin

Albumin acts as a non-specific transport vehicle for a large number of endogenous substances. These include fatty acids, calcium, unconjugated bilirubin, thyroxine and urate. In addition, many drugs are transported bound to albumin, and in some instances drugs may displace endogenous substances, e.g. bilirubin can be displaced from albumin by salicylates and sulphonamides.

Bisalbuminaemia

Albumin is not a single protein, and over twenty structural variants have been described. None of these variants is associated with disease, but they do give rise to a number of different electrophoretic patterns. The most frequent appearance is a single, fairly discrete band (Fig. 7.5). Heterozygotes for some of the variant albumins may, however, show a double band, usually staining with equal intensity on the electrophoretic strip. This condition, called bisalbuminaemia, represents the products of two genes.

PRE-ALBUMIN

This protein, normally present in plasma in small amounts, is synthesized by the hepatocytes. Its functions are to act as one of the transport proteins for vitamin A and thyroxine. The plasma [pre-albumin] falls rapidly in response to injury (p. 124), and is decreased in both acute and chronic liver disease because its half-life is much shorter than albumin.

TRANSFERRIN

This, the major iron-binding protein in plasma, transports iron from the sites of absorption and red cell breakdown to the developing red cells in the bone marrow. Normally its iron-binding sites are about 30% saturated, although the wide fluctuations of plasma [iron], diurnal or with the phase of the menstrual cycle, render this percentage very variable. Determination of plasma total iron-binding capacity (TIBC) is widely used as a method for measuring plasma [transferrin].

Atransferrinaemia is a very rare congenital deficiency of transferrin.

Acquired deficiency occurs in protein-losing conditions, infections and neoplastic disease.

Increased plasma [transferrin] occurs in iron-deficiency states and in women taking oestrogen-containing oral contraceptives.

THE HAPTOGLOBINS

This is a group of proteins, all α_2-globulins, that bind haemoglobin to form haptoglobin/haemoglobin complexes; the complexes are then rapidly broken down in the lymphoreticular system.

Only a small proportion of red cell destruction occurs inside the vascular system, and it seems likely that only this component leads to the formation of haptoglobin/haemoglobin complexes. Nevertheless, even this small amount may, cumulatively, fulfil a useful function by helping to conserve the body's iron stores. Uncomplexed haemoglobin (mol. mass 68 kDa) can pass the glomerular filter, to a limited extent, with consequent loss of its iron from the body. By combining with haptoglobin, the small amount of haemoglobin normally released into the circulation is conserved and its iron is not lost in the urine.

Whenever intravascular haemolysis is increased, e.g. in haemolytic anaemia, plasma [haptoglobin] falls and free haptoglobin may be undetectable. Decreased plasma [haptoglobin] is found also in liver disease, and rarely as a congenital abnormality.

Plasma [haptoglobin] is increased in acute infections, in the nephrotic syndrome and following trauma. Haptoglobin is one of the acute phase reactants (p. 123).

Genetic polymorphism

The structure of haptoglobin varies between individuals. The basic molecule comprises two types of chain, α and β, giving rise to the structure $\alpha_2\beta_2$. The β chains (mol. mass 40 kDa) are the same in all haptoglobins, but the α chain may be of three types:

(1) α^{IS}, 9 kDa.
(2) α^{IF}, 9 kDa.
(3) α^{2}, 17 kDa.

The α^{IS} and α^{IF} polypeptide chains differ by only a single amino acid. The α^2 chain probably arose by partial duplication of the genes coding for the α^{IF} or α^{IS} chains. Three allelic genes Hp^{IF}, Hp^{IS} and Hp^2 code for the α chains in each $\alpha\beta$ pair, so several phenotypes are possible (Table 7.2). On electrophoresis of haptoglobin, multiple bands are seen, due to polymerization of those haptoglobins that contain the α^2 chain.

TABLE 7.2. Characteristics of haptoglobin phenotypes

Phenotype	Subunit structure	Molecular mass (daltons)	Electrophoretic appearance
Hp 1–1	$\alpha^{1F}\alpha^{1F}\beta_2$ $\alpha^{1F}\alpha^{1S}\beta_2$ $\alpha^{1S}\alpha^{1S}\beta_2$	100 000	Single band
Hp 2–1	$\alpha^{1S}\alpha^{2}\beta_2$ $\alpha^{1F}\alpha^{2}\beta_2$	105 000 210 000 315 000 etc.	Multiple bands
Hp 2–2	$\alpha_n^2\beta_n$, where n = 3–8	220 000 300 000 370 000 etc.	Multiple bands

Individuals inherit a characteristic genotype. This determines the nature of their plasma haptoglobin, and is an example of the phenomenon known as genetic polymorphism. A number of other plasma proteins show similar inherited variations, e.g. IgG, β-lipoprotein, α_1-antitrypsin. With the haptoglobins these are of considerable genetic interest but are not of pathological significance.

CERULOPLASMIN

This is a copper-containing protein that has oxidase properties. It normally binds about 90% of the copper present in plasma. However, it is not known whether ceruloplasmin is primarily concerned with plasma copper transport, nor whether its enzymic properties as an oxidase are of physiological importance.

Plasma [ceruloplasmin] is reduced in Wilson's disease (p. 187), in patients with malnutrition, and in the nephrotic syndrome.

Plasma [ceruloplasmin] is increased in pregnancy and in women taking oestrogen-containing oral contraceptives. It is also raised in acute infections, in some types of chronic liver disease and in neoplastic disease.

α_1-ANTITRYPSIN

This is another example of a plasma protein that exhibits genetic polymorphism. The physiological importance of α_1-antitrypsin, as an antiproteinase, is unknown. Interest principally relates to the association between certain diseases of the lung and liver and deficiency of α_1-antitrypsin.

α_1-Antitrypsin deficiency can usually be recognized by finding a greatly reduced α_1-globulin band on an electrophoretic strip stained for serum proteins. Specific immunological or enzymic methods can be used to measure [α_1-antitrypsin]. Phenotyping of α_1-antitrypism is best performed by isoelectric focusing, a very sensitive technique for separating proteins by means of differences in their isoelectric points. Phenotyping is desirable in all cases where plasma [α_1-antitrypsin] is low or borderline since the results are required before appropriate genetic counselling can be given to affected individuals or parents.

Several allelic genes code for α_1-antitrypsin, the alleles being given the general description Pi (Protease inhibitor). Besides having structurally different α_1-antitrypsins, individuals with different phenotypes also produce varying amounts of α_1-antitrypsin. The most striking example is provided by men and women who are homozygous for the Pi^Z allele (the ZZ type). They produce only about 15% of the normal amount of α_1-antitrypsin, and have a high incidence of the following disorders:

(1) *Pulmonary emphysema.* About 1% of patients with emphysema are found to have α_1-antitrypsin deficiency. However, this percentage is much higher in young patients with emphysema; when associated with α_1-antitrypsin deficiency, the disease tends to manifest itself in the third of fourth decades. Smoking seems to be a strong predisposing factor in the development of emphysema.

(2) *Hepatic disorders.* Neonatal jaundice, usually presenting a predominantly cholestatic picture, is common in type ZZ individuals. Although the jaundice may resolve, there is usually progression to hepatic cirrhosis. In about 20% of children with cirrhosis, the hepatic disorder can probably be attributed to α_1-antitrypsin deficiency.

Not all type ZZ patients develop lung or liver disease. This suggests that additional factors may be required as triggers. For example, mild hepatic injury may initiate release of leucocyte proteases which, in the absence of α_1-antitrypsin, cause liver damage.

Increased plasma [α_1-antitrypsin] occurs in pregnancy, in acute infections and following trauma. α_1-Antitrypsin is one of the acute phase reactants (p. 123). These increases in plasma [α_1-antitrypsin] are not important in themselves. However, α_1-antitrypsin-deficient individuals also show increases in plasma [α_1-antitrypsin] in these conditions, and these increases may be sufficient to bring their levels into the reference range. Delay in diagnosis of α_1-antitrypsin deficiency may occur as a result.

α_2-MACROGLOBULIN

This is the major α_2-globulin. It binds endopeptidases such as trypsin and chymotrypsin; the resulting complex has no endopeptidase activity. It is not known whether this antiproteinase activity represents the major biological function of α_2-macroglobulin.

Marked increases in plasma [α_2-macroglobulin] occur in the nephrotic syndrome, in some patients with cirrhosis and in some of the collagen disorders. There have been no descriptions of patients with significant decreases in plasma [α_2-macroglobulin].

α_1-FETOPROTEIN (AFP)

This protein is present in the tissues and plasma of the fetus. Its concentration falls very rapidly after birth, but minute amounts (up to 15 μg/L) can still be detected in plasma from adults, if sensitive radioimmunoassay methods are used. The functions of AFP are unknown but its measurement has important applications in the investigation of disease in adults:

(1) Pregnant women carrying fetuses with open neural tube defects have raised levels of plasma [AFP]. Screening programmes involving measurement of maternal serum [AFP] at 16–18 weeks' gestation are now widely practised (p. 410).

(2) Gross increases in serum [AFP] occur in about 50% of patients with hepatocellular carcinoma, and AFP is used as a tumour marker (p. 461). Smaller increases occur in many of the remaining patients with hepatocellular carcinoma, but are of less significance since a number of other liver diseases show similar changes. Other conditions showing large increases in plasma [AFP] include gonadal neoplasms and, rarely, gastric carcinoma.

CARCINOEMBRYONIC ANTIGEN (CEA)

This protein is normally present in fetal intestinal cells but is not detectable in adult cells, in health. It reappears in the cells of intestinal neoplasms, a process which may be associated with the loss of a normal membrane protein. Small amounts of CEA can normally be detected in plasma (up to 2–5 μg/L) by sensitive immunoassay techniques. Measurement of plasma [CEA] is useful as a tumour marker, in definable circumstances (p. 460).

At one time it was thought that the appearance of CEA in plasma in adults was diagnostic of intestinal carcinoma, particularly carcinoma of the colon. However, the antigen is also released into plasma by carcinoma occurring in other sites, and from non-carcinomatous lesions of the intestine.

Screening patients for the possible presence of neoplasms by measurement of plasma [CEA] is not warranted. However, the present international consensus recommendation is that plasma [CEA] should be measured preoperatively in patients with either colorectal or bronchial carcinoma, and used as an adjunct to clinical and pathological staging methods. Postoperatively, measurement of plasma [CEA] provides a way of assessing whether a tumour has been removed; the value should return to normal within 6 weeks. Thereafter periodic measurement of plasma [CEA] provides a non-invasive method of surveillance and a means of detecting recurrence, as well as a way of assessing response to anti-tumour therapy.

FIBRINOGEN

This is detected as a discrete band in the β-γ-globulin region if *plasma* is examined by electrophoresis. Conversion of fibrinogen to fibrin is effected by thrombin, which splits two polypeptides from fibrinogen with the formation of fibrin monomer. Fibrin monomer polymerizes spontaneously to form the fibrin clot.

Congenital fibrinogen deficiency is a rare hereditary disorder of clotting; its clinical manifestations are relatively mild. Acquired fibrinogen deficiency, however, is an acute bleeding disorder. It is usually caused by excessive consumption of fibrinogen which gives rise to the defibrination syndrome. The plasma [fibrinogen] is very low, usually due to excessive intravascular coagulation probably caused by release of thromboplastins into the circulation. Some of the commoner causes of acquired fibrinogen deficiency are:

(1) Complications of pregnancy, most often abruptio placentae or amniotic fluid embolism.
(2) Shock, which may be haemorrhagic, septic or anaphylactic.
(3) Following major surgery, especially thoracic operations.
(4) Snake bites.
(5) Disseminated carcinoma.

The coagulation disorder is usually compounded by the presence in the blood of fibrin degradation products (FDP). These are formed by the proteolytic breakdown of fibrin by plasmin (see below). The FDP inhibit the formation of a fibrin clot, probably by affecting both the thrombin/fibrinogen reaction and the subsequent polymerization of fibrin monomer.

Acquired fibrinogen deficiency is occasionally due to decreased synthesis; this may occur in severe hepatic disease. Deficiencies of prothrombin and other coagulation factors may also develop under these circumstances. Prothrombin synthesis and the effects of vitamin K deficiency are discussed elsewhere (p. 179).

Cryofibrinogen

Fibrinogen sometimes forms fibrin-like strands if plasma is cooled to 4°C; usually these strands redissolve on warming. This phenomenon tends to occur in patients with raised plasma [fibrinogen], particularly in patients with malignant neoplasms. It may also occur in patients with the defibrination syndrome.

If the presence of cryofibrinogen is suspected, blood should be collected into a warmed syringe containing oxalate as anticoagulant. The specimen should be kept at 37°C until the plasma has been separated from the cells. Unless this procedure is followed, cryofibrinogen may be centrifuged down with the cells and escape detection.

PLASMINOGEN AND PLASMIN

Plasminogen circulates in plasma as the inactive precursor of plasmin to which it is converted by splitting off a small polypeptide. Plasmin is usually not detectable in blood. It is a proteolytic enzyme with a special affinity for fibrinogen and fibrin. It splits these into smaller fragments (fibrin degradation products or FDP).

OTHER PLASMA PROTEINS

Lipoproteins and apolipoproteins, hormones, plasma enzymes, and some transport proteins such as thyroxine-binding globulin are considered elsewhere.

ACUTE PHASE REACTANTS

This is a group of proteins that are synthesized in greater amounts as a reaction to 'injury'. They include α_1-antitrypsin, α_1-acid glycoprotein, ceruloplasmin, C-reactive protein, fibrinogen and the haptoglobins. 'Injury' is here used as a general term to include:

(1) Acute tissue damage, due to trauma (accidents, major surgery),

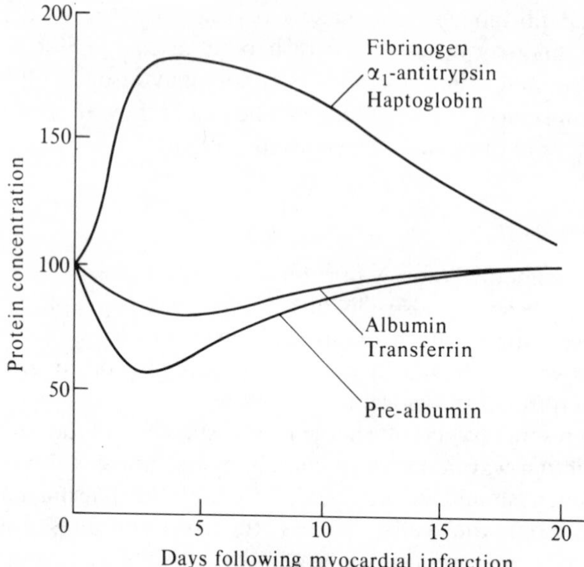

FIG. 7.2. This figure shows changes in plasma protein concentrations which occur following myocardial infarction, an example of an acute injury, expressed as a percentage of the values observed on the day of infarction, i.e. before the onset of changes. The acute phase reactants increase in concentration in response to injury.

infarction of tissue (e.g. myocardial infarction), burns, acute infections, etc.

(2) Chronic inflammation, e.g. chronic infections, collagen diseases.

(3) Malignant disease.

Acute phase reactants are also synthesized in increased amounts in pregnancy and in the puerperium.

Following an acute injury, the plasma concentrations of the acute phase reactants start to increase within 24 h, and usually reach their peak concentration after 3–5 days. The increase amounts, in most cases, to not more than an approximate doubling of concentration (Fig. 7.2), but C-reactive protein sometimes shows a 20–30 fold increase. Some of the other plasma proteins show a fall in concentration, e.g. pre-albumin and cholinesterase, after the injury.

THE IMMUNOGLOBULINS

The immunoglobulins are a group of structurally related proteins that function as antibodies; they are synthesized by the cells of the lymphoreticular system. Although predominantly moving with the γ-globulins on serum protein electrophoresis, some immunoglobulins (especially IgA and IgM) may migrate with the β-, or with the α_2-globulins.

The basic immunoglobulin molecule is made up of four polypeptide chains consisting of a pair of heavy chains, each having a mol. mass of 50–75 kDa, and a pair of light chains that each have a mol. mass of 22 kDa. There are five principal types of heavy chain (γ, α, μ, δ and ε) and two types of light chains (κ and λ). Every immunoglobulin can be assigned a formula that indicates its composition, according to its type of chain (e.g. $\alpha_2\lambda_2$, $\gamma_2\kappa_2$, etc.). The antigen-combining site is formed between the adjacent light and heavy chains (Fig. 7.3).

Three major (IgG, IgA and IgM) and two minor (IgD and IgE) classes of immunoglobulin have been recognized, the type of heavy chain determining the class to which the immunoglobulin belongs. Table 7.3 lists several properties of the major classes. Both light and heavy chains have 'constant' and 'variable' sections. The constant portion varies little within each particular chain type, whereas the variable portion, which is associated with the antigen-combining site, is different for each immunoglobulin even within a single chain type. The variable portion is responsible for the specificity of the antibody.

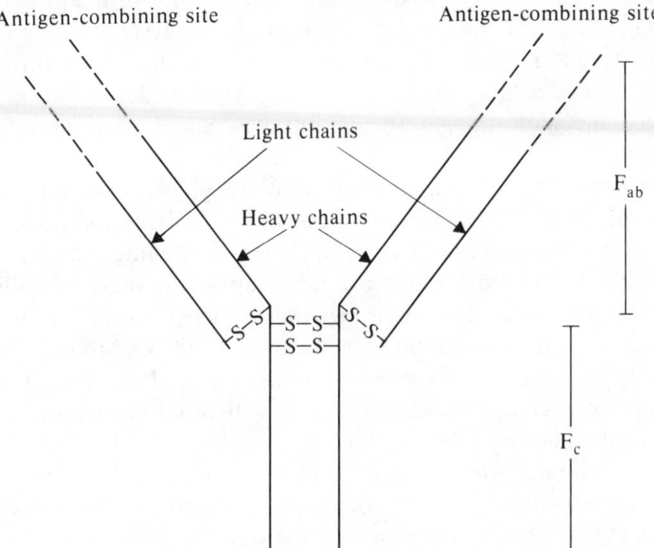

FIG. 7.3. *Diagrammatic representation of the immunoglobulin molecule.* The variable regions of the polypeptide chains are shown as interrupted lines. Heavy chains may be one of 5 types, and light chains one of 2 types (see text).

TABLE 7.3. Features of the major classes of immunoglobulins

	IgG	IgA	IgM
Average molecular mass (Da)	146 000	160 000	875 000
Plasma concentration (g/L)	7–15	1.5–2.5	0.6–1.7
Plasma half-life (days)	21	6	5
Percentage of protein in the vascular compartment	50	70	80
Present in secretions	Trace	Yes	Trace
Transplacental passage	Yes	No	No
Composition			
Light chain type	κ or λ	κ or λ	κ or λ
Heavy chain type	γ	α	μ
Structure	$\gamma_2\kappa_2$ or $\gamma_2\lambda_2$	$\alpha_2\kappa_2$ or $\alpha_2\lambda_2$	$(\mu_2\kappa_2)_5$ or $(\mu_2\lambda_2)_5$

IgG immunoglobulins, quantitatively the most important group, are formed particularly in response to soluble antigens such as toxins and the products of bacterial lysis. They are widely distributed in the ECF and cross the fetoplacental barrier.

IgM immunoglobulins are pentamers of the basic immunoglobulin structure. They tend to be formed especially in response to particulate antigens, such as those on the surface of bacteria. In the presence of complement, IgM is very effective in producing lysis of the cells with which it reacts. Following an antigenic stimulus, IgM formation usually precedes IgG formation. Thus, IgM is considered to be an early defence mechanism against the intravascular spread of infecting organisms.

IgA immunoglobulins, as they occur in plasma, are monomers. However, over 50% of IgA synthesis occurs in lymphoreticular cells lying under the mucosa of the respiratory and alimentary tracts. In these sites the IgA molecules are taken up by the mucosal epithelial cells with the formation of an IgA dimer such as $(\alpha_2\lambda_2)_2$; a secretory piece is also added which protects the molecule from proteolysis. The resulting protein, called secretory IgA, is secreted into the gut or the respiratory tract, and may form part of the defence mechanism against local viral and bacterial infections.

IgD immunoglobulins are present in minute amounts in plasma. They are also often present, with monomer IgM, on the surface of B lymphocytes. Their function in plasma is unknown, but on lymphocytes they are probably concerned with antigen recognition and with the development of tolerance.

IgE immunoglobulins include the reagins, which bind to cells such as the mast cells of the nasopharynx. In the presence of antigen (allergen), one result of the antigen–antibody reaction is the release of histamine and other amines and polypeptides from the cell, giving rise to local hypersensitivity reactions.

IMMUNOGLOBULINS IN CHILDREN

In neonates, IgG is present in plasma in relatively high concentration, having been transferred across the placenta, whereas IgM is present at birth in very small quantities; these small amounts have been synthesized by the fetus *in utero*. In the healthy neonate, IgA is undetectable in plasma.

During infancy and childhood, immunoglobulin concentrations rise towards the values observed in healthy adults, [IgM] increasing fairly rapidly; [IgG] and [IgA] increase more slowly (Fig. 7.4). In some children, immunoglobulin development may be delayed. A wide range of plasma [immunoglobulins] is found in apparently healthy adults (Table 7.3).

Disorders of immunoglobulin synthesis

There are two broad categories of disorder. In one, plasma [immunoglobulins] are decreased; in the other, plasma [immunoglobulins] are increased. The two categories are not mutually

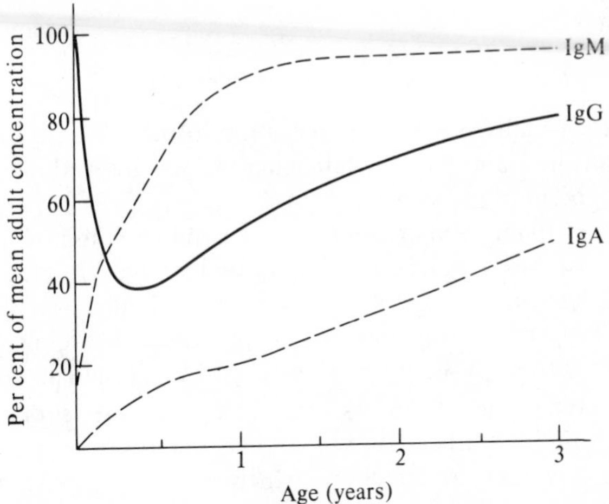

FIG. 7.4. The average concentration in plasma of the major immunoglobulin classes, in early childhood, expressed as a percentage of the mean concentration in healthy adults.

exclusive, since reductions in one immunoglobulin class may be accompanied by increases in another. Each of the categories subdivides into two groupings, and these are discussed separately.

Inherited deficiencies of immunoglobulin synthesis

Hypogammaglobulinaema, and rarely agammaglobulinaema, are conditions in which there is defective production of IgG, IgA and IgM. Children begin to develop severe, recurrent bacterial infections when over the age of 1 year.

Dysgammaglobulinaema is a condition in which there is a deficiency of one or two of the immunoglobulin classes; nearly all the possible combinations of deficiencies of IgG, IgA and IgM have been described.

These disorders of immunoglobulin synthesis may be associated in some syndromes with defective cell-mediated immunity, in which there is a deficiency of lymphocytes. Affected children are prone to viral infections. As a result, normally minor viral illnesses, such as vaccinia due to smallpox vaccination, may prove fatal.

These disorders are all very rare. They have been given eponymous names and sometimes have apparently bizarre clinical features associated with them. Serum protein electrophoresis should not be used for their detection, since deficiency of one immunoglobulin class may be masked by overproduction of another.

Acquired deficiencies of immunoglobin synthesis

These are usually secondary to some other disease. However, occasionally immunoglobulin deficiency may occur in adult life with no apparent precipitating cause.

Secondary hypogammaglobulinaemia is much commoner than the group of inherited deficiencies described above. It may occur in lymphoid neoplasia (e.g. chronic lymphatic leukaemia, Hodgkin's disease and multiple myelomatosis), in 'toxic' disorders or certain types of drug therapy, in protein-losing syndromes, and in prematurity.

In the lymphoid disorders, the cause of the suppression of immunoglobulin synthesis is not known but there is a consequent increased tendency to infection, which may be aggravated by the treatment adopted. Severe infections or treatment with cytotoxic drugs may themselves adversely affect immunoglobulin synthesis.

The loss of protein which occurs in the nephrotic syndrome, or in protein-losing gastroenteropathy, may occasionally be so severe as to cause significant hypogammaglobulinaemia.

Diffuse hypergammaglobulinaemia

Serum protein electrophoresis reveals a broad band, due to increased amounts of γ-globulin. The increase may affect all the immunoglobulin classes or it may affect predominantly one class. The antibodies produced are heterogeneous. The causes include:

(1) *Infections* — both acute and chronic. The increases are a part of the body's physiological response to infection, larger increases in plasma [immunoglobulins] generally being seen in chronic infection. All three major immunoglobulin classes are commonly affected, but sometimes a particular class tends to be associated with certain diseases, e.g. IgA is often increased in diseases of the skin and lungs and IgM in tropical diseases.

(2) *Chronic liver disease.* Most patients with cirrhosis show a diffuse increase in γ-globulin on serum protein electrophoresis. All the major immunoglobulin classes are affected, but IgA characteristically shows the greatest increase except in primary biliary cirrhosis where there is a predominant increase in IgM in over 90% of patients, and in chronic active hepatitis where IgG tends to be most affected.

(3) *Autoimmune diseases* such as lupus erythematosus, Sjögren's syndrome and chronic thyroiditis tend to show an increase in IgG, although other immunoglobulin classes are also affected.

Quantitation of the separate immunoglobulin classes is occasionally helpful in diagnosis, as part of the investigation of diffuse hypergammaglobulinaemia. In most cases, however, the cause of the diffuse increase in plasma [immunoglobulins] is apparent.

Discrete hypergammaglobulinaemia (paraproteinaemia)

These conditions are all uncommon or rare disorders, each thought to be due to a single clone of antibody-producing cells synthesizing excessive amounts of an individual immunoglobulin. These are often referred to as paraproteins or as 'M' (monoclonal) components. Two categories need to be distinguished:

(1) *Malignant paraproteinaemia,* which includes multiple myeloma, macroglobulinaemia and a few cases of lymphoma or lymphatic leukaemia (Table 7.4).

TABLE 7.4. Features of malignant paraproteinaemias

Disease	Relative incidence (approximate percentage)	Serum 'M' component		Urine protein	
		Nature	Frequency of occurrence (percentage of cases)	Nature	Frequency of occurrence (percentage of cases)
IgG myeloma	50	IgG	Over 95	Bence–Jones	50–90
IgA myeloma	25	IgA	Over 95	Bence–Jones	50–90
IgD myeloma	1–3	IgD	Over 50	Bence–Jones	Over 90
Light chain disease	10–25	κ or λ chains	Variable	Bence–Jones	100
Macroglobulinaemia	5–10	IgM	Over 95	Bence–Jones	10–60
Heavy chain disease	Very rare	α, γ, μ chain fragments	Variable	Heavy chain fragments	Variable

Notes

(1) Urine protein figures refer to urine protein electrophoresis after preliminary concentration.
(2) In a variable proportion of cases, urine may also contain albumin and other proteins.
(3) The very rare conditions of IgE myeloma and IgM monomer disease are not included.

(2) *Benign paraproteinaemia,* in which there is an isolated or incidental abnormality that does not progress, on prolonged follow-up, to one of the conditions listed in the malignant group.

Multiple myeloma is a malignant disease in which there is proliferation of plasma cells and plasma cell precursors. Most myelomas produce complete immunoglobulin molecules, usually either IgG or IgA, together with excessive amounts of immunoglobulin fragments which may be light chains or parts of heavy chains or other fragments.

About 10–20% of myelomas, usually the less differentiated neoplasms, only secrete insignificant amounts of complete immunoglobulins. Instead, they produce large quantities of immunoglobulin fragments consisting mainly of dimers of light chains (mol. mass 44 kDa), when the disease is called Bence–Jones myeloma or light chain disease. Occasionally, the fragments consist of heavy chains, a condition called Franklin's or heavy chain disease.

Waldenstrom's macroglobulinaemia usually follows a more prolonged course than multiple myeloma. There is a proliferation of cells that resemble lymphocytes rather than plasma cells. They produce complete IgM molecules and often an excess of light chains. Increased plasma [IgM] causes increased plasma viscosity, which tends to make the circulation sluggish and thromboses are common.

Cryoglobulins

These are immunoglobulins that precipitate when cooled to 4°C and redissolve when warmed to 37°C; sometimes they precipitate at temperatures intermediate between 37°C and 4°C. They occur in a number of diseases associated with hypergammaglobulinaemia, both the diffuse and the discrete forms. Their detection is of little value in determining the aetiology of the conditions in which they are formed. However, their importance lies in the fact that they may be associated with Raynaud's phenomenon. If cryoglobulin determinations are to be performed, the blood specimen needs to be collected into a warmed syringe (no anticoagulant) and maintained at 37°C until the serum for cryoglobulin investigation has been separated from the red cells.

COMMONLY PERFORMED INVESTIGATIONS OF PLASMA PROTEIN ABNORMALITIES

The concentrations of proteins and of protein-bound constituents in the plasma of healthy adults, when standing, may be up to 13% more than when recumbent. Effects of alterations in posture may be greater in oedematous patients, with changes in concentration of as much as 20% being observed. These are due to alterations in the distribution of the ECF, and are accompanied by changes in the PCV. Prolonged venous stasis before blood sampling can have similar effects. It is important for doctors to standardize their technique in the collection of blood specimens. Otherwise, results for plasma [protein] determinations, and measurements affected by changes in plasma [protein] such as calcium, cholesterol or cortisol, may not be directly comparable in specimens obtained on different occasions.

TOTAL PROTEIN

This measurement is frequently carried out on plasma or serum. It is, however, at best a screening investigation since the concentrations of individual plasma proteins do not all rise or fall in parallel with one another. Significant alterations in the concentration of one protein, or of a group of proteins, may be masked by coincident or compensatory changes in the concentrations of other proteins. For example, a fall in plasma [albumin] may be accompanied by an increase in plasma [immunoglobulins].

Increased plasma [total protein] may be due to dehydration, to increased plasma [immunoglobulins], monoclonal or polyclonal, or to faulty techniques in blood sampling.

Reduced plasma [total protein] may occur due to overhydration, including increased plasma volume due to pregnancy, impaired protein synthesis (as in cases of malnutrition), malabsorption, liver disease, hypogammaglobulinaemia, or increased protein loss as a consequence of renal, gastrointestinal or skin disorders.

ALBUMIN

Alterations in plasma [albumin] can occur for many reasons (p. 116). In general, measurement of [albumin] in plasma (or serum) is of much greater value diagnostically than plasma [total protein]. In many patients, decreased plasma [albumin] is a non-specific index either of poor nutrition, leading to reduced synthesis, or of the increased breakdown of tissue that occurs in association with pyrexia or trauma.

SERUM PROTEIN ELECTROPHORESIS

When this technique was introduced into routine use about 30 years ago, it was thought that different patterns of alterations in the various protein fractions would be of great value in differential diagnosis. These hopes have not been realized. 'Characteristic' electrophoretic patterns do appear, in the nephrotic syndrome for example (Fig. 7.5), but these changes often do not occur until so late in the disease that the diagnosis has long been apparent. Even then, the electrophoretic pattern is rarely pathognomonic. The pattern associated with the nephrotic syndrome, for instance, may be similar in many respects to the pattern seen in some of the collagen disorders. Earlier changes, due to alterations in individual proteins which form part of one of the globulin peaks, may be masked by other proteins in that peak which either show no change in concentration or which show changes in concentration in the opposite direction.

Examples of electrophoretic patterns are shown in Fig. 7.5, and densitometric findings are summarized in Table 7.5. Serum protein electrophoresis is principally of value in the detection of the monoclonal immunoglobulins synthesized in the paraproteinaemias; it is also useful in screening for α_1-antitrypsin deficiency.

It would clearly be more informative to measure quantitatively, by immunological or other means, those proteins which show characteristic changes in the disease under investigation. The advantage of serum protein electrophoresis (on cellulose acetate,

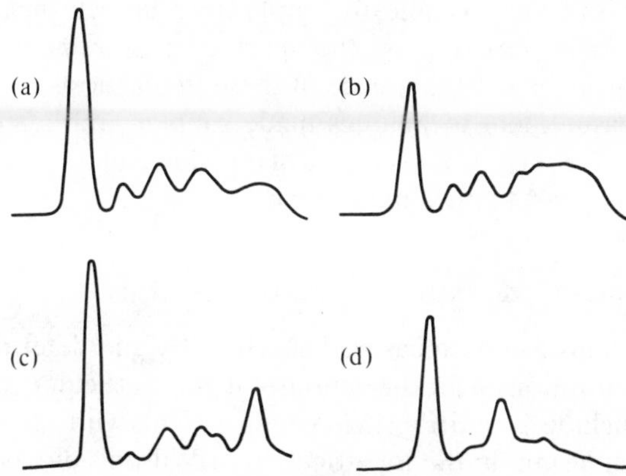

FIG. 7.5. *Densitometric scans of serum protein electrophoretic strips:*

(a) Normal scan.

(b) Reduced albumin, diffuse increase in γ-globulin, and the trough normally present between the β- and γ-globulins filled in (β-γ fusion) due to the predominant increase in IgA. This pattern is seen in cirrhosis of the liver.

(c) Monoclonal increase in γ-globulin, as seen in multiple myeloma.

(d) Reduced albumin, raised α_2-globulin and reduced γ-globulin, as seen, for instance, in the nephrotic syndrome.

TABLE 7.5. Alterations in serum protein electrophoretic patterns in disease.

Disease	Albumin	Globulin fractions			
		α_1	α_2	β	γ
Obstructive jaundice	N	N	↑	N	N
Chronic hepatocellular jaundice	↓	N	N	N	↑
Cirrhosis of the liver	↓↓	N	N	N	↑↑
Nephrotic syndrome	↓↓	N	↑↑	N	↑, N or ↓
Acute infections	↓	N	↑	N	(↑)
Chronic infections	↓	N	↑	N	↑
Collagen diseases	↓	N	↑	N	↑
Acute injury	↓	(↑)	↑	N	N
Severe diabetes	↓	N	↑	N	↑
Myeloma and macroglobulinaemia	N	N	Abnormal monoclonal band normal immunoglobulins ↓		

N = normal, ↓ = decreased, ↓↓ = considerably decreased, (↑) = occasionally or slightly increased, ↑ = increased, ↑↑ = considerably increased.

Notes (1) These features are not necessarily all present on every occasion.
(2) In most instances, the patterns listed do not appear until relatively late in the time-course of the disease.

paper, etc.) is that it is technically simple and relatively inexpensive. It is unlikely to be replaced by the specific measurement of several individual proteins until the place of these in diagnosis is established and the measurement techniques made simpler and less expensive. There are, however, a few examples of individual plasma proteins that have a specific part to play in diagnosis.

MEASUREMENT OF INDIVIDUAL GLOBULINS

Several proteins can be measured specifically, and determination of their concentrations can be important in particular conditions. Examples include α_1-antitrypsin in young patients with emphysema (p. 120), ceruloplasmin in the investigation of Wilson's disease (p. 187), α_1-fetoprotein (AFP) in the screening of pregnant women for open neural tube defects (p. 410), and tumour markers such as AFP and CEA (Chapter 24).

Investigation of paraproteinaemia

Chemical investigations are only one of several ways of investigating suspected cases of paraproteinaemia. Others include haematological and radiological investigations, and lymph node biopsy. The chemical tests can be used to detect the paraprotein, to determine its concentration and type, and to follow the progress of the disease.

INITIAL INVESTIGATIONS

Serum protein electrophoresis and urine protein electrophoresis should be carried out in all cases of suspected paraproteinaemia.

Serum protein electrophoresis will show a single discrete band, usually in the γ-globulin region but occasionally in the β, or α_2-globulin region, in over 90% of patients in whom there is overproduction of complete immunoglobulin molecules (Table 7.4), and it will occasionally show a band due to the presence of light chains. It is the most sensitive routinely available test for paraproteins.

Urine protein electrophoresis is usually needed to demonstrate Bence–Jones protein, as its small mol. mass (44 kDa) means that it is cleared rapidly by the kidney. It is frequently necessary to concentrate the specimen before examining it for paraproteins. If Bence–Jones protein is detected in urine, the monoclonal nature of the light chains can be confirmed by immunoelectrophoresis. In multiple myeloma the light chains are nearly always dimers of type λ or type κ, but not a mixture of the two. Protein electrophoresis is much more sensitive than side-room tests as a means of detecting Bence–Jones protein in urine.

Most cases of myeloma and many cases of macroglobulinaemia have Bence–Jones proteinuria. In light chain disease there is Bence–Jones proteinuria but usually no serum 'M' component. It is for this reason that all patients suspected of having a diagnosis of myeloma should have both serum and urine protein electrophoresis performed. The principal protein findings in myeloma and macroglobulinaemia are summarized in Table 7.4.

FURTHER INVESTIGATIONS

The type and amount of paraprotein and the concentration of normal immunoglobulins can be determined. The type (IgG, IgA, etc.) is best identified by immunoelectrophoresis, while the amount of paraprotein is most reliably determined by integration of the area under the 'M' peak on the serum electrophoretic scan. The concentration of normal immunoglobulins is measured using an immunological method. In patients with a malignant paraproteinaemia, determination of the amount of the normal immunoglobulins may help to assess the likelihood that infection will develop and this may influence the choice of treatment.

The most important diagnostic decision to be made, if a paraprotein is detected, is whether the condition is benign or malignant. The following chemical features point to the condition being benign:

(1) The other immunoglobulin classes are normal.
(2) Bence–Jones proteinuria is absent.
(3) The serum ['M' component] is less than 10 g/L (1 g/100 mL).
(4) The serum [paraprotein] does not increase with time.

OTHER CHEMICAL ABNORMALITIES

Renal disease in myeloma may be due to many factors including tubular damage caused by Bence–Jones protein, renal amyloid, hypercalcaemia or hyperuricaemia. There may be impaired renal function with a fall in GFR and a rise in plasma [urea].

Increased plasma [calcium] occurs frequently in multiple myeloma, probably due to release of calcium from bone rather than to increased binding of calcium by myeloma proteins. Plasma alkaline phosphatase is usually normal or only slightly raised; increased osteoblastic activity is not a feature of bone disease in multiple myeloma.

Increased plasma [urate] is liable to occur whenever cellular breakdown is rapid, e.g. following cytotoxic therapy. Renal damage may follow the formation of urate stones in these circumstances.

Similar abnormalities may be detected in the other paraproteinaemias, particularly in Waldenstrom's macroglobulinaemia.

PROGRESS OF PARAPROTEINAEMIA

The progress of both the malignant paraproteinaemias and the benign paraproteinaemias can be determined by measuring serum ['M' component] and serum [normal immunoglobulins]. These measurements may need to be repeated several times before the diagnosis is established. Monitoring the progress of benign paraproteinaemias indicates that these conditions only rarely become malignant.

The efficacy of treatment for the malignant paraproteinaemias can be assessed by measuring serum ['M' component], plasma [calcium], creatinine clearance, etc., depending on how many of these were abnormal before treatment started. Potentially adverse effects of the treatment may require periodical assessment of plasma [urate], hepatic function and other measurements, depending on the nature of the treatment.

FURTHER READING

CUTHBERTSON, D.P. (1982). The metabolic response to injury and other related explorations in the field of protein metabolism. *Scottish Medical Journal,* **27,** 158–171.

DURIE, B.G.M. and SALMON, S.E. (1977). Multiple myeloma, macroglobulinaemia and monoclonal gammopathies. In *Recent Advances in Haematology,* **2,** 243–261. Ed. A.V. Hoffbrand. Edinburgh: Churchill Livingstone.

HOBBS, J.R. (1971). Immunoglobulins in clinical chemistry. *Advances in Clinical Chemistry,* **14,** 219–317. New York: Academic Press.

PUTNAM, F.W., editor (1975 and 1977). *The Plasma Proteins. Structure, Function and Genetic Control,* 2nd Edition. New York: Academic Press.

WHICHER, J.T. (1980). The interpretation of electrophoresis. *British Journal of Hospital Medicine,* **24,** 348–360.

WHICHER, J.T. (1983). Abnormalities of plasma proteins. In *Biochemistry in Clinical Practice*, pp. 221–251. Ed. D.L. Williams and V. Marks. London: Heinemann.

Chapter 8

Enzyme Tests in Diagnosis

There are a number of circumstances in which measurement of enzyme activities in body fluids or tissue extracts may be of diagnostic value. In practice, the vast majority of enzyme tests required for diagnostic purposes measure the enzyme activities in plasma or serum, and this chapter will mostly be concerned with these. There are two main groups of enzymes in plasma:

(1) Enzymes which have been released from cells as a result of leakage or cell death in the normal process of 'wear and tear', or because of cell damage or death caused by disease. In general, this group of enzymes has no known function in the blood.
(2) Enzymes with clearly defined actions in the blood, e.g. the coagulation factors, renin-angiotensin system.

This chapter is concerned with the *first* category of plasma or serum enzyme activity.

ENZYME ACTIVITIES IN PLASMA

The activity of most enzymes normally detectable in plasma remains fairly constant in health, although some may show temporary increases after severe muscular exercise (e.g. creatine kinase), or after a meal (e.g. intestinal isoenzyme of alkaline phosphatase). In general, plasma enzyme activity represents a steady state in which the rate of release from cells into plasma and the rate of removal from plasma are equal (Fig. 4.3). The activity observed is governed by the following factors:

(1) The rate of release of the enzyme from cells.
(2) The volume of distribution of the enzyme in the ECF.
(3) The rate of removal of the enzyme from plasma, by catabolism or excretion.
(4) The presence in the plasma of factors which may affect the method of assay, i.e. of inhibitors or activators of enzyme activity.

Changes in plasma enzyme activity can nearly always be attributed to increases in the rate of release of enzyme into plasma. Such increases are most commonly caused by:

(1) *Necrosis or severe damage to cells.* This type of damage is usually caused by ischaemia or by toxic substances. The enzymes released are principally those present in the cytoplasm of the cell.
(2) *Increased rate of cell turnover.* This occurs normally during periods of active growth (e.g. alkaline phosphatase in the first year of life and at puberty), or tissue repair (e.g. alkaline phosphatase in patients who are recovering from multiple fractures), and in association with several forms of malignant disease.
(3) *Increased concentration of enzyme within the cell.* The synthesis of some intracellular enzymes is induced by certain diseases or drugs (e.g. gamma-glutamyl transferase by ethanol).
(4) *Duct obstruction.* Enzymes normally present in exocrine secretions (e.g. amylase) may be regurgitated into blood.

There is considerable variation in the rate of removal of enzymes from the circulation, by uptake into the cells of the lymphoreticular system and, in a few instances, by excretion in the urine. As a result, there is a wide range of half-lives of tissue enzymes following their release into plasma. The plasma half-lives of enzymes most often measured for diagnostic purposes vary from about 10 hours to more than 5 days.

The selection of plasma enzyme tests, and methods of reporting

Organ damage causes the release of increased amounts of many enzymes into the bloodstream. The choice of enzyme activity measurement depends on the nature and site of the disease, and the following factors are important in making this choice:

(1) *Sensitivity,* i.e. the ability to detect small amounts of tissue damage by measuring the activity of the enzyme. In many disorders it can be very difficult to make a clinical diagnosis when tissue damage is small.
(2) *Specificity,* i.e. the ability to identify which tissue has been damaged.

In general, enzymes provide very sensitive indicators of tissue damage. Since many enzymes are present in cells in much higher concentrations than in plasma, even small amounts of tissue damage can cause measurable increases in plasma enzyme activity (Table 8.1). The

TABLE 8.1. Comparison of the relative activities of 4 enzymes in tissue homogenates. The activities are all expressed in relation to each enzyme's activity in plasma, taken as unity.

	AST	ALT	LD	CK
Heart	8000	400	1000	10 000
Liver	7000	3000	1500	<10
Skeletal muscle	5000	300	700	50 000
Red blood cell	15	7	300	<1
Plasma	1	1	1	1

AST = Aspartate aminotransferase
ALT = Alanine aminotransferase
LD = Lactate dehydrogenase
CK = Creatine kinase

sensitivity of the assay techniques allows very small amounts of enzyme activity to be detected, following release from the tissues.

The biggest problem in clinical enzymology is one of specificity. Most enzymes are widely distributed throughout the body, although the relative amounts in different tissues vary considerably (Table 8.1). Elevation of the activity in plasma of each of these tissue enzymes may be seen in disorders affecting a number of different tissues. The problem can be partly overcome by measuring several different enzyme activities or by the study of isoenzymes, as described below.

Two other factors relevant to the choice of plasma enzyme activity measurements are:

(3) *Time-course of enzyme elevation.* The enzyme activity should be raised soon after the onset of the disease and should remain raised for an appreciable period. The time-course depends on the duration of enzyme release and on the half-life of the enzyme in plasma.

(4) *Technical factors.* Enzyme tests that are accurate and precise, easy to perform and inexpensive are the ideal.

ENZYME UNITS

The results of enzyme measurements are usually reported in units of activity since, for technical reasons, it is not often practicable to measure enzyme concentration directly. This use of activity measurements tacitly assumes that enzyme activity is directly proportional to enzyme concentration, an assumption which may only be valid if correct conditions are selected under which to perform the assay.

Some of the complexities relating to enzyme units are described in Appendix 1 (p. 517). In the main text, however, we have adopted a pragmatic approach, hoping thereby to avoid confusion when readers compare the results of enzyme activity measurements reported by different laboratories, each of which may have its own sets of reference values. In most cases, when describing quantitative data for enzyme activities, we shall express the activity as the ratio between the result observed on the specimen under examination and the upper reference value for the laboratory making the measurement. We would emphasize that we shall be doing this for the sake of clarity in a discussion intended to have general application. We do not necessarily advise laboratories to adopt a similar approach when reporting results for their enzyme activity measurements.

EXAMPLES OF CLINICALLY IMPORTANT PLASMA ENZYMES

The Nomenclature Commission of the International Union of Biochemistry has recommended that enzyme names be based on the substrate and the type of reaction catalysed. Using the following reaction to illustrate the recommended terminology:

$$\text{ATP} + \text{creatine} \rightleftharpoons \text{ADP} + \text{creatine phosphate},$$

the Nomenclature Commission would describe the enzyme that catalyses this particular reaction in the following ways:

(1) *Systematic name* (i.e. ATP: creatine phosphotransferase). Although systematic names fully characterize the reactions catalysed, they are often cumbersome and their use is often confined to when the enzyme is first mentioned.

(2) *Enzyme Commission code number* (i.e. EC 2.7.3.2.). This classifies the type of reaction catalysed. They are not needed for clinical practice.

(3) *Recommended name* (i.e. creatine kinase). These are the names in common use, and recommended names are the ones used in this book.

LACTATE DEHYDROGENASE (LD). EC 1.1.1.27

Human tissues contain five major proteins, all designated lactate dehydrogenase (LD). An additional minor LD is present in testes. Proteins that possess similar catalytic activity, but which differ in their structure and in certain other properties, are known as isoenzymes.

Lactate dehydrogenase (mol. mass 135 kDa) is a tetramer consisting of four polypeptide chains; each chain may be one of two types called H or M. The various combinations of these two types of subunit give rise to five distinct isoenzymes, having subunit structures H_4, H_3M, H_2M_2, HM_3 and M_4 They can also be characterized on the basis of electrophoretic mobility. The isoenzyme that migrates most rapidly towards the anode (H_4) is termed LD_1, and the slowest moving isoenzyme (M_4) is called LD_5. It is possible to divide most tissues into one of three categories:

(1) Tissues in which the fast-moving or anodal (LD_1, LD_2) isoenzymes predominate (e.g. heart, red blood cells).
(2) Tissues in which the slow-moving or cathodal (LD_5, LD_4) isoenzymes predominate (e.g. liver and most skeletal muscles).
(3) Tissues in which none of the five isoenzyme components predominates (e.g. lung and spleen).

Total LD activity, i.e. the combined activity of the isoenzymes of LD, is often measured in plasma. Alternatively, or additionally, LD isoenzyme studies are frequently performed. All five isoenzymes of LD can normally be detected in plasma although LD_2 and LD_1 are present in considerably larger amounts than LD_3, and normally there are only small amounts of LD_4 and LD_5 (Fig. 8.1).

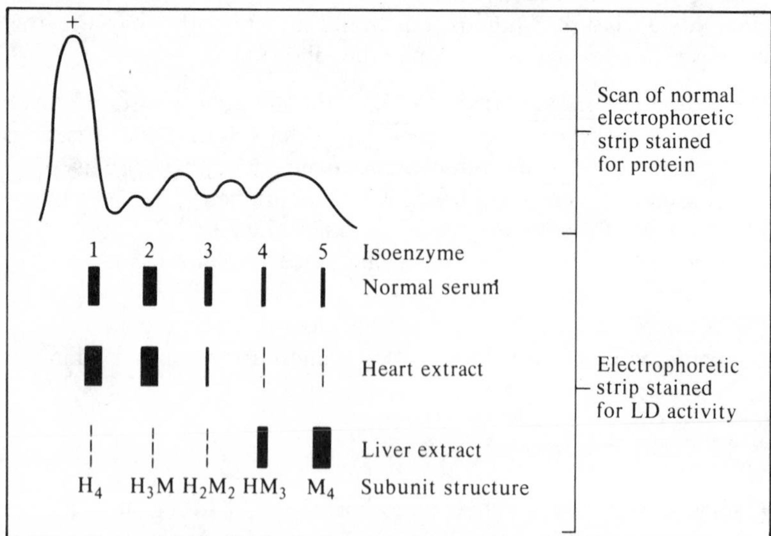

FIG. 8.1. Diagrammatic representation of the approximate relative proportions of the five principal isoenzymes of lactate dehydrogenase, after electrophoretic separation on cellulose acetate.

When LD is released from damaged tissue, the isoenzyme pattern of the damaged organ is reflected by a corresponding change in the isoenzyme pattern detectable in plasma. Isoenzyme studies may therefore help to localize the tissue of origin of an increased plasma enzyme activity and are used, with LD measurements, to distinguish between damage to the myocardium and damage to liver or skeletal muscle.

Isoenzymes of LD differ in their biological, chemical and physical properties and show a gradation of properties, with LD_1 and LD_5 representing the extremes, e.g. the half-life of LD_1 in plasma is about 100 h and the half-life of LD_5 is about 10 h. Chemical and physical methods used to distinguish LD isoenzymes are listed in Table 8.2.

Clinical applications

The activity of total LD in plasma may be increased in a wide variety of disorders. Examples include myocardial infarction, megaloblastic anaemia, acute hepatic disease, acute renal disease, pulmonary infarction, a variety of muscle diseases and many types of malignant disease.

In practice, LD activity is most often measured in the investigation of myocardial infarction. As LD_1 and LD_2 are the main isoenzymes released in this condition, one of the measurements for these 'heart-type' or so-called 'heart-specific' isoenzymes in plasma is usually employed, to make the test more specific for heart damage.

Very high activities of so-called 'heart-specific' LD are found in patients with megaloblastic anaemia, presumably due to the release of enzymes from abnormal red cell precursors that are being destroyed in

TABLE 8.2. Differentiation of lactate dehydrogenase isoenzymes.

	Isoenzyme activity	
Method of differentiation	$LD_1(H_4)$	$LD_5(M_4)$
Electrophoretic mobility	α_1-Globulin	γ-Globulin
Use of 2-oxobutyrate as substrate	Relatively high activity	Relatively low activity
Heating for 30 min at 60°C	Unaffected	Destroyed
Assay in presence of urea (2.0 mol/L)	Slight inhibition	Complete inhibition
Assay in presence of oxalate (2.0 mmol/L)	Marked inhibition	Slight inhibition
Assay in presence of anti-M subunit antibody	No inhibition	Complete inhibition

the bone marrow. Increased plasma 'heart-specific' LD activity may be found in patients with haemolytic anaemia, whatever the cause.

As red cells contain high concentrations of LD, care must be taken to avoid producing haemolysis when blood specimens are collected, and when they are being prepared for analysis. Otherwise, results for plasma LD measurements could be misleading.

It will be evident that the designation 'heart-specific' LD activity is a misnomer.

ASPARTATE AMINOTRANSFERASE (AST). EC 2.6.1.1

This enzyme, also known as glutamic oxaloacetic transaminase (GOT), is present in most tissues, but especially in skeletal and cardiac muscle, liver and kidney. There are two major isoenzymes, one cytoplasmic and the other mitochondrial. Both isoenzymes have been demonstrated in plasma following tissue damage, but their differentiation has not been shown to be of much diagnostic value.

Clinical applications

Increased activities of AST in plasma are of considerable diagnostic help in the recognition of myocardial infarction and other conditions associated with myocardial damage (e.g. rheumatic carditis). Peak levels of plasma AST activity observed after myocardial infarction are usually 2–10 times the upper reference value (Fig. 8.3, p. 155).

Measurement of plasma AST may be of value when investigating patients with acute liver damage. In these patients plasma AST activity is usually more than five times the upper reference value, if the damage is severe enough to cause jaundice; the peak may occur in the prodromal stage. In chronic liver disease much smaller elevations are often observed; these may indicate continuing hepatocellular damage.

Increased activity of plasma AST occurs in patients with muscular dystrophy, but the extent of the rise is relatively much smaller than the increase in creatine kinase activity. High plasma AST activities may also be found in patients with acute renal disease, acute pancreatitis, or when liver damage occurs secondary to heart failure or pulmonary infarction.

ALANINE AMINOTRANSFERASE (ALT). EC 2.6.1.2

This enzyme, often called glutamic pyruvic transaminase (GPT), is widely distributed. Its concentration in most tissues is considerably less than AST, but in liver the activities of the two enzymes are the same order of magnitude (Table 8.1).

Clinical applications

Plasma ALT activity is widely used as a test for hepatocellular damage. In early viral hepatitis and other types of acute liver cell injury, it is usually increased to a greater extent than AST, whereas plasma AST activity tends to be higher than ALT activity in patients with chronic hepatic disease.

Small increases in plasma ALT activity occur in uncomplicated myocardial infarction. Larger increases may occur when there is cardiac failure, presumably as a result of hepatic venous congestion.

Since ALT is widely distributed, increased plasma activity may be observed in many other conditions associated with tissue injury. For instance, plasma ALT activity sometimes rises in acute pancreatitis, acute renal disease, muscle disease and disseminated carcinoma, but these findings are not specific enough to be much help in diagnosis.

CREATINE KINASE (CK). EC 2.7.3.2

Creatine kinase has three principal isoenzymes, each composed of two polypeptide chains, denoted B or M; these give dimers with the composition MM, MB and BB. The MM isoenzyme (CK–MM) is present in high concentration in skeletal muscle and myocardium. The MB isoenzyme (CK–MB) is present as a minor component in myocardium (about 30% of total CK activity), but is virtually absent from most skeletal muscles. The BB isoenzyme is present in brain, thyroid and some other tissues.

In normal plasma over 95% of total CK activity is due to CK–MM; increases in activity are nearly always due mainly to CK–MM. There is also a variable contribution from CK–MB and, rarely, CK–BB depending on the organ from which CK has been released. It is possible to measure CK–MB separately. Since this isoenzyme has a high specificity for myocardium, CK–MB measurements are the most sensitive and specific chemical test for myocardial infarction presently available.

The BB isoenzyme can also be measured separately, but is rarely demonstrable in plasma. Raised plasma CK–BB activity has been reported in patients with metastatic carcinoma of the prostate, and in patients who have suffered brain damage due to accidental trauma, surgery or boxing injuries.

Prolonged muscular exercise can cause large increases in plasma total CK activity. Small increases may occur after relatively minor amounts of unaccustomed exercise. Increases in plasma total CK activity often follow intramuscular injections, especially of drugs that cause local inflammation at the injection site.

Clinical applications

Measurement of plasma total CK activity is useful in the diagnosis of myocardial infarction (Fig. 8.3) and plasma CK–MB activity may be particularly valuable in certain circumstances (p. 156). Plasma total CK activity measurements may also help with the recognition of certain types of overt muscle disease (e.g. muscular dystrophy), and in investigating possible carriers of the gene for Duchenne-type muscular dystrophy.

Several other diseases are associated with increased plasma total CK activity, probably due to secondary release of the enzyme from skeletal muscle. These include acute renal failure, coma due to poisoning or cerebral thrombosis, and diabetic ketoacidosis. Increases often occur after surgery. Raised plasma total CK activity may also be observed in myxoedema.

ALKALINE PHOSPHATASE. EC 3.1.3.1

This is the generic name for a group of enzymes that display maximum activity in the range of pH 9.0–10.5. They are widely distributed, different tissues possessing one, or occasionally more, characteristic and analytically distinguishable form. Liver, bone, placenta and intestine are clinically important sources of plasma alkaline phosphatase activity.

The precise biochemical role of alkaline phosphatase is not known. In many tissues it is attached to cell membranes, suggesting an association between alkaline phosphatase activity and membrane transport. In the liver, for example, activity is localized on (1) the cell membrane that adjoins the biliary canaliculus, and (2) close to the sinusoidal border of the parenchymal cell.

It is possible, by electrophoresis, to determine the tissue of origin of increased alkaline phosphatase activity present in a patient's serum. Alternative, and simpler, quantitative means of differentiation have also been described. For instance, sensitivity of intestinal alkaline phosphatase to inhibition by L-phenylalanine, and the heat-stability of placental alkaline phosphatase can both be made the basis of practicable routine procedures. Unfortunately, simple tests capable of making the most important clinical distinction, between liver and bone alkaline phosphatases, are less satisfactory, although differences in heat stability and sensitivity to urea (2.0 mol/L) have been used with some success.

Alkaline phosphatase is removed from the plasma by the cells of the lymphoreticular system, like other plasma enzymes. Although alkaline phosphatase is present in bile, this has not come from the plasma but has rather been produced in the liver, by the cells lining the biliary canaliculi.

Physiological changes in activity

Plasma alkaline phosphatase activity may vary considerably for entirely physiological reasons. It is important to have taken account of these before attributing a pathological reason to what at first sight might seem to be an abnormal result.

In *normal pregnancy,* release of alkaline phosphatase from placenta may cause plasma activity to rise in the second and third trimesters to about twice the normal adult levels.

In *infancy and childhood,* the upper reference value is as much as three times the adult value (Fig. 8.2). This increase is a direct result of the intensive osteoblastic activity associated with normal bone growth, e.g. at puberty, since osteoblasts secrete alkaline phosphatase.

After a *meal containing fat,* release of intestinal alkaline phosphatase may cause a transient small increase in plasma enzyme activity. Blood for measurement of plasma alkaline phosphatase activity should preferably be collected from fasting subjects.

Clinical applications

Increased activity of plasma alkaline phosphatase is found in many conditions, the two principal categories being liver and bone disease.

Bone disease. Plasma alkaline phosphatase activity is increased in patients with diseases in which there is increased osteoblastic activity such as Paget's disease, hyperparathyroidism, rickets and osteomalacia, and carcinoma with osteoblastic metastases.

Liver disease. Cholestasis has a dual effect since it causes increased synthesis of hepatic alkaline phosphatase and regurgitation of the hepatic enzyme into plasma. In the absence of cholestasis, hepatocellular damage causes relatively little release of alkaline phosphatase, presumably because the enzyme is firmly bound to the membrane of the liver cell and therefore does not 'leak' from damaged cells as readily as do soluble cytoplasmic enzymes.

In patients with *carcinoma,* increased plasma alkaline phosphatase activity is usually due to osteoblastic secondary deposits in bone or to hepatic metastases giving rise to cholestasis. Occasionally, atypical isoenzymes may be found, of which the commonest is the Regan isoenzyme, named after the patient in whom it was first detected. The Regan isoenzyme is very stable to heat, and very similar to the placental isoenzyme. There are several other atypical 'tumour marker' isoenzymes of alkaline phosphatase, sometimes present in amounts insufficient to raise total plasma alkaline phosphatase activity; special techniques are required to demonstrate them.

Lowered activity of plasma alkaline phosphatase is rare. It occurs in hypophosphatasia, a hereditary disease in which vitamin D-resistant

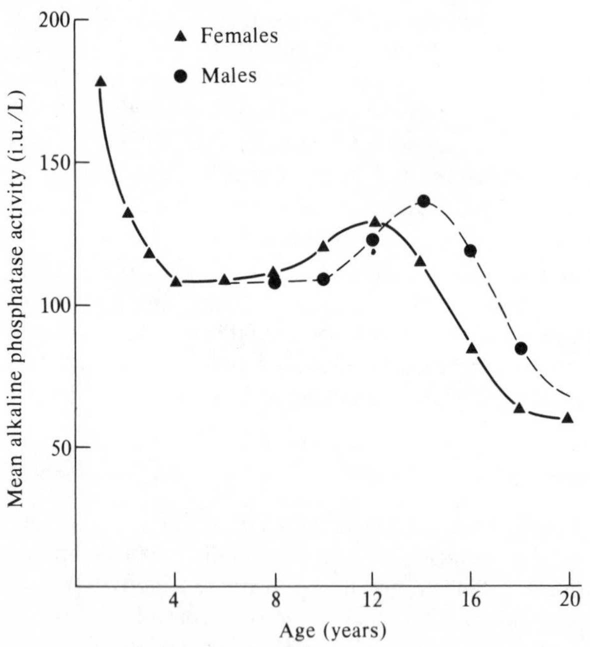

FIG. 8.2. *Variation in mean plasma alkaline phosphatase activity with age.* The reference values for each group are wide. With adults, for instance, the mean value of 70 i.u./L shown in the figure corresponds to a reference range of 40–100 i.u./L. In the first year of life, the mean activity rises within about one month from about 70 i.u./L to about 180 i.u./L. Below 6, there is no sex difference.

These figures are illustrative of the variations that occur. Values for individual laboratories are affected by the method used, but the overall pattern of changes in activity with age remains the same.

rickets is the most prominent finding. Both tissue and plasma alkaline phosphatase activities are usually low, and excessive amounts of phosphoryl-ethanolamine are present in the urine.

GAMMA-GLUTAMYL TRANSFERASE (GGT). EC 2.3.2.2

This enzyme, sometimes called gamma-glutamyl transpeptidase, is found in a number of tissues, but mainly in the kidney, liver, biliary tract and pancreas. The kidney contains by far the largest amount of GGT of any of the tissues, including the liver. However, renal GGT is not released into plasma and is of little diagnostic importance.

Much of the GGT activity in liver, as with alkaline phosphatase, is associated with the cell membrane adjoining the biliary canaliculus. There is also a soluble cytoplasmic form of GGT in liver.

Clinical applications

Raised plasma GGT activity occurs almost exclusively in diseases of the liver, biliary tract and pancreas. It is increased in all types of liver disorder, both acute and chronic, but the largest increases occur in cholestasis. The magnitude of the changes parallel those of alkaline phosphatase (p. 184) except that GGT is a more sensitive index of cholestasis and is more likely to be raised when there is hepatocellular damage accompanied by only minimal cholestasis. It is also increased in patients treated with phenytoin, phenobarbitone and other barbiturates and certain other drugs, and in 70–80% of chronic alcoholics. These increases are all due, at least partly, to liver enzyme induction (p. 181).

Plasma GGT activity is sometimes increased after myocardial infarction, and occasionally in other diseases that do not primarily involve the liver, biliary tract or pancreas. In all these cases, the increased GGT activity can usually be attributed to secondary hepatic involvement, e.g. due to venous congestion.

Measurement of plasma GGT is probably the most sensitive test for liver disease routinely available, particularly when there is cholestasis. It is, however, less effective than alkaline phosphatase in distinguishing hepatocellular from cholestatic disease. Plasma GGT activity is not raised in bone disease so GGT measurements can help to identify the origin of a raised plasma alkaline phosphatase activity; the activity of both enzymes is increased if there is hepatic disease.

The increasingly wide availability of plasma GGT measurements has rendered obsolete the measurement of aminopeptidase (arylamidase) and 5′-nucleotidase activities, two other enzymes closely associated with the cell membrane adjoining the biliary canaliculus.

ACID PHOSPHATASE. EC 3.1.3.2

The prostate contains high concentrations of acid phosphatase. Other acid phosphatase isoenzymes are present in small concentrations in liver, spleen, red cells and platelets. The prostatic and non-prostatic isoenzymes may be distinguished from one another by using a substrate that is preferentially hydrolysed by one of the isoenzymes, or by measuring enzymic activity in the presence of a selective inhibitor (Table 8.3).

Many laboratories report both plasma total acid phosphatase and prostatic or 'tartrate-labile' or 'formaldehyde-stable' acid phosphatase activities. The prostatic isoenzyme is the component of principal diagnostic interest. Radioimmunoassay (RIA) methods for determining the concentration of prostatic acid phosphatase have been described. These are more sensitive than enzyme activity measurements and are probably more specific for the prostatic isoenzyme, but the diagnostic gains from using RIA methods have not been as great as they appeared when the RIA tests were introduced.

Clinical applications

Acid phosphatase measurements are used mainly for diagnosis and monitoring of metastatic or invasive prostatic carcinoma. While the carcinoma is confined within the prostate, plasma prostatic acid phosphatase activity is raised in only about 20% of the patients. However, local or distant spread of the neoplasm causes increased activity, often to high levels, in up to 80% of patients. Enzyme activity falls rapidly after successful oestrogen therapy, but quickly rises again if relapse occurs.

TABLE 8.3.

Source of acid phosphatase	Degree of inhibition by	
	L(+) tartrate	Formaldehyde
Prostate	Almost complete	Minimal
Red cells	Minimal	Marked
Other tissues (e.g. liver, spleen)	Marked	Moderate

When investigating patients with prostatic disease by measuring plasma acid phosphatase activity, or monitoring treatment of prostatic carcinoma, the following points should be noted:

(1) The prostatic isoenzyme is unstable and specimens should therefore be sent to the laboratory, for assay or storage under correct conditions, as soon as possible after collection.
(2) It seems advisable, where possible, to collect blood samples before a rectal examination since temporary (and occasionally large) increases in plasma acid phosphatase activity have been reported after this procedure. However, this effect must usually be small, and some recent reports have stated that rectal examination has no effect on plasma acid phosphatase activity.

Increased plasma acid phosphatase activity occurs in several other disorders. The prostatic isoenzyme is released when there is prostatic inflammation, trauma or necrosis. A tartrate-stable (non-prostatic) isoenzyme is often increased in patients with bone disease, especially Paget's disease, but also in patients with hyperparathyroidism and metastatic breast carcinoma. Plasma total acid phosphatase may be increased in liver disease, in Gaucher's disease, and in thrombocytopenia when this is due to an increased rate of platelet lysis.

α-AMYLASE (Amylase). EC3.2.1.1

Large amounts of amylase are present in the pancreas and salivary glands, and smaller amounts in other tissues. Both the salivary and pancreatic isoenzymes have a mol. mass of 45 kDa, and are partially filtered from plasma at the glomerulus. The isoenzymes can be differentiated by electrophoretic methods, or by means of an inhibitor obtained from wheat. This inhibitor preferentially inhibits salivary and other non-pancreatic isoenzymes of amylase.

Clinical applications

Increased plasma amylase activity occurs in several conditions:

(1) *Acute pancreatitis.* Plasma amylase activity rises early, with a peak activity usually more than five times the upper reference value about 24 h after the onset of pain. Activity returns to normal between 2 and 7 days after the onset.
(2) *Other acute abdominal conditions.* Perforated peptic ulcer, acute

biliary obstruction and acute intestinal obstruction are often accompanied by increased plasma amylase activities but the peak values are not usually as high as in acute pancreatitis.

(3) *Salivary gland disease,* especially acute inflammation (e.g. mumps).

(4) *Renal disease,* due to reduced amylase clearance.

(5) *Drugs.* Morphine and other drugs that cause spasm of the sphincter of Oddi may give rise to an increase in plasma activity that is usually only transient.

(6) *Macro-amylasaemia.*

Plasma amylase measurements are mostly used to help distinguish acute pancreatitis, in which surgery is not indicated, from other acute abdominal disorders many of which require immediate operation. Although plasma amylase activity tends to rise to higher levels in acute pancreatitis than in these other conditions, this is not always so and the peak level of plasma amylase is not an absolute guide to diagnosis in these cases.

The specificity of plasma amylase measurements can be improved by measuring both total activity and the activity that persists in the presence of a wheat-protein inhibitor. This method is much quicker than electrophoresis.

Urinary amylase measurements offer no advantages over plasma amylase measurements for the diagnosis of acute pancreatitis. They are, however, of help in the recognition of *macro-amylasaemia*, an uncommon disorder in which part of the plasma amylase activity is attributable to an enzyme molecule with a much larger mol. mass than the usual plasma amylase. The increased plasma activity is due to reduced renal clearance of the larger molecule. In some cases, macro-amylase is probably a polymer but in others it is due to complex formation between amylase and an immunoglobulin. The disorder, although rare, may cause diagnostic difficulty due to the presence of an unexplained and persistently high plasma amylase activity. The explanation may be suspected when the increased plasma amylase activity is found to be persistent and accompanied by a normal urinary amylase activity. Confirmation of the diagnosis requires more complex tests.

Lipase and trypsin are two other enzymes of pancreatic origin that have increased plasma activities in patients with acute pancreatitis. However, these measurements are more difficult than amylase, and are much less frequently performed. The main value of determining trypsin activity in blood is for the detection of cystic fibrosis (p. 439).

CHOLINESTERASE (CHE). EC 3.1.1.8

There are two principal cholinesterases in human tissues. The one present in plasma and liver used to be known as pseudocholinesterase. The other, acetylcholinesterase or 'true' cholinesterase, is present at nerve endings and in erythrocytes, but not in plasma.

Plasma CHE is synthesized in the liver, and its activity tends to be reduced whenever plasma [albumin] is reduced. Table 8.4 gives examples of the many causes of decreased plasma CHE activity. These need to be borne in mind when interpreting results for CHE investigations in the two main groups of circumstances where these measurements are of particular value.

Scoline apnoea

Some patients develop prolonged apnoea, often lasting for several hours, after administration of succinyl dicholine (scoline) as a muscle relaxant. This drug is normally hydrolysed in plasma by CHE. Over 50% of patients who are sensitive to scoline have genetically determined abnormalities in the CHE enzyme protein, which is coded by at least four allelic genes at the first locus:

- E_1^u codes for the usual form of CHE, present in over 95% of the U.K. population.
- E_1^a codes for an atypical CHE, which is resistant to inhibition by dibucaine.
- E_1^f codes for an atypical CHE, which is resistant to inhibition by fluoride.
- E_1^s codes for a protein that has little or no CHE activity.

TABLE 8.4. Causes of low plasma cholinesterase activity. (Modified from Whittaker, 1977)

Category	Examples
Physiological	Pregnancy, third trimester; infancy
Inherited	Rare CHE variants (Table 8.5)
Acquired	
(1) Disease	Liver disease; malnutrition; acute injury states (p. 124); chronic renal failure; collagen diseases
(2) Poisoning, and effects of various drugs	Organophosphorus insecticides; oral contraceptives; monoamine oxidase inhibitors; cytotoxic drugs; X-ray therapy

TABLE 8.5. Results of inhibitor studies of plasma cholinesterase in ten genotypes, with their approximate frequency in the U.K. population. (Modified from Whittaker, 1977)

Genotype	Frequency (%)	Dibucaine number	Fluoride number	Scoline sensitivity
$E_1{}^uE_1{}^u$	95	77–83	50–68	Not sensitive
$E_1{}^uE_1{}^a$	4	48–69	44–54	Rarely sensitive
$E_1{}^uE_1{}^f$	0.5	70–83	48–54	Rarely sensitive
$E_1{}^uE_1{}^s$	0.5	77–83	50–68	Rarely sensitive
$E_1{}^aE_1{}^a$	0.05	8–28	10–28	Very sensitive
$E_1{}^aE_1{}^f$	0.005	45–59	28–39	Moderately sensitive
$E_1{}^aE_1{}^s$	0.003	8–28	10–28	Very sensitive
$E_1{}^fE_1{}^f$	0.0007	64–69	34–43	Moderately sensitive
$E_1{}^fE_1{}^s$	0.0007	64–69	34–43	Moderately sensitive
$E_1{}^sE_1{}^s$	0.001	No or very little activity		Very sensitive

Table 8.5 summarizes the results of inhibitor studies in the 10 genotypes that derive from these four allelic genes.

Most individuals with abnormal variants have low plasma CHE activity, but the only reliable way of demonstrating the variants is by means of inhibitor studies. The abnormal enzymes are less affected by some inhibitors than is the normal enzyme. Dibucaine and fluoride are the two most widely used inhibitors. The ranges of values in Table 8.5 are the percentage inhibition of the observed CHE activity when the enzyme measurement is repeated in the presence of dibucaine or fluoride. Thus, the $E_1{}^a$ $E_1{}^a$ genotypes, which show great sensitivity to scoline, usually show (1) low plasma CHE activity, but this low activity is (2) only inhibited to a small extent by dibucaine (8–29%) or fluoride (10–28%). On the other hand, the $E_1{}^uE_1{}^u$ genotypes (the main group of normal individuals) do not show sensitivity to scoline and have (1) normal plasma CHE activity which is, however, (2) strongly inhibited by dibucaine (77–83%) and by fluoride (50–68%). Results of inhibitor studies are reported simply as numbers.

Detection of individuals liable to scoline apnoea and identification of heterozygotes is important. Affected relatives can then be traced, and warnings given to anaesthetists not to use certain muscle relaxants, especially scoline, in future operations.

Organophosphorus poisoning

Several chemicals cause reduction both in plasma CHE and erythrocyte acetylcholinesterase activity; a particularly important group is the organophosphorus insecticides. Measurements of plasma

CHE activity should be made at regular intervals on workers at risk from exposure, so as to detect toxic effects early, preferably before symptoms develop. These measurements are also useful for following progress of patients recovering from organophosphorus poisoning.

PLASMA ENZYME TESTS IN VARIOUS CLINICAL CONDITIONS

Myocardial infarction

Characteristic changes in plasma enzyme activity occur after myocardial infarction. Activities of many enzymes may become raised, but only four enzyme tests are used with any frequency. These are AST, 'heart-specific' LD, total CK and CK–MB.

The time-course of plasma enzyme changes always follows the same general pattern (Fig. 8.3). After an initial 'lag' phase of at least 3 h, during which activities remain normal, they rise rapidly to a peak. Activities then return to normal at rates that depend on the half-life of each enzyme in plasma (Table 8.6). The rapid rise and fall of CK–MB activity should be noted, and the fact that LD activity remains raised for considerably longer than the other enzymes.

Enzyme activity measurements, especially plasma CK–MB, are very sensitive tests of myocardial damage. Probably over 95% of patients with myocardial infarction show detectable increases in

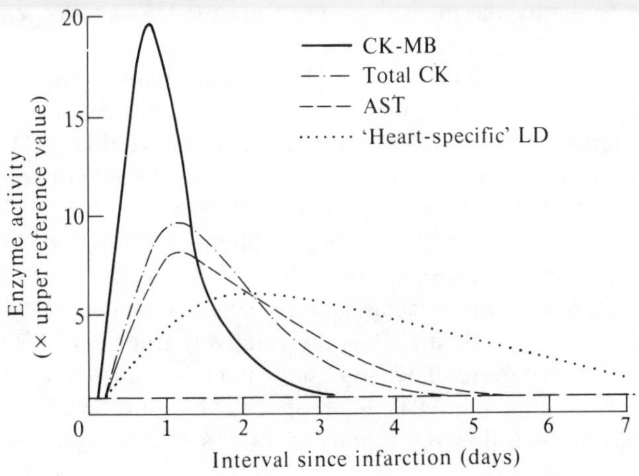

FIG. 8.3. Typical patterns of enzyme elevation following uncomplicated myocardial infarction.

TABLE 8.6. Time-course of plasma enzyme activities after myocardial infarction.

Enzyme	Onset (h)	Peak (h)	Duration (days)
Creatine kinase (MB isoenzyme)	3–10	12–24	1½–3
Creatine kinase (total)	5–12	18–30	2–5
Aspartate aminotransferase	6–12	20–30	2–6
'Heart-specific' lactate dehydrogenase	8–16	30–48	5–14

plasma enzyme activity, *provided that* specimens are taken at the optimal times (see below). The magnitude of the rise in activity is strongly correlated with infarct size. Patients with large increases, for example plasma AST activity more than ten times the upper reference value, have a much poorer prognosis than those with small increases in activity.

Selection of tests

Plasma AST and 'heart-specific' LD activities are the enzyme measurements that are requested most often for diagnosis of myocardial infarction or confirmation of the diagnosis. Because these enzymes have different tissue specificities, we consider it advisable to request both tests as a matter of routine. Provided that specimens are obtained at the optimal time, this combination of tests usually indicates whether or not infarction has occurred. However, it is possible to justify the measurement of plasma CK activity, either in addition to AST and 'heart-specific' LD or instead of one of these tests (usually instead of AST), because (1) CK is probably a more sensitive indicator of myocardial damage, and (2) CK is not released in liver disease (unlike AST) nor as a result of haemolysis (unlike LD). These advantages of CK measurements are, however, largely offset by the ease with which it is released from skeletal muscle (p. 146).

The place of plasma CK–MB measurements in routine diagnosis is established, and there is little doubt that CK–MB is the most sensitive and specific test for myocardial damage currently available. However, raised plasma CK–MB activities are relatively transient. Also, CK–MB methods are more expensive, slow and time-consuming than the methods for measuring AST, 'heart-specific' LD or total CK activities in plasma. In the following circumstances, CK–MB measurements are indicated:

(1) When very early confirmation of the diagnosis is required, i.e.

within 6 h of the incident. This is before either plasma total CK or AST activity might be expected to be significantly elevated.

(2) When it is suspected that an increased plasma total CK activity may be due to release of enzyme from skeletal muscle (e.g. after an intramuscular injection).

(3) When investigating post-operative patients for suspected myocardial infarction. In these patients, plasma CK–MB remains normal in the absence of myocardial damage, whereas AST, total CK and LD activities are often increased for non-cardiac causes.

(4) In patients suspected of having had a second infarct within a few days of the first. It is easier to show that a second rise in plasma enzyme activity has occurred if the activity of the enzyme that is being measured rises and falls rapidly after each incident involving myocardial damage (Fig. 8.3).

In many patients with suspected myocardial infarction it might seem unnecessary to measure plasma enzymes, since the ECG often provides unequivocal evidence of infarction. However, it is possible to misinterpret ECG traces, especially in the presence of arrhythmias, and the ECG is by no means always abnormal in patients who have recently had a myocardial infarction.

Plasma enzyme activity measurements can provide confirmation of the diagnosis of myocardial infarction, independent of the ECG findings. They can also provide information about the size of an infarct and give an indication of prognosis. We would therefore recommend that plasma enzyme tests, normally AST and 'heart-specific' LD, should be requested on all patients suspected of having had a myocardial infarction within the previous 48 h. Requests should be made at the optimal times for blood sampling.

Optimal times for blood sampling

Small increases in plasma enzyme activity following an infarct are liable to be missed unless care is taken over timing the collection of blood samples. If a single sample is to be taken, its collection should be timed to coincide with peak enzyme activity, i.e. between 18–30 h after the onset of symptoms if AST or CK activity is to be measured (Table 8.6). Samples taken at other times, e.g. during the first 12 h or between 30 and 48 h, may provide useful additional information, by helping to show whether the progress of enzyme activity changes is consistent with the provisional diagnosis. In the absence of further symptoms, it is neither necessary nor is it useful to take further specimens after 48 h.

This does not apply, of course, if the patient is not seen until 2 or more days after the onset of symptoms. In these circumstances, 'heart-specific' LD measurements may still be of value, and are certainly potentially of more value than other plasma enzyme tests since LD activity remains raised for longer (Fig. 8.3).

Sensitivity and specificity

Plasma enzyme activity measurements should be treated as complementary to ECG recordings, in the investigation of patients with suspected myocardial infarction. Assuming that the correct combination of enzyme tests has been performed, at the appropriate time, and the ECG recorded under optimal conditions, the sensitivity and specificity of the two methods of investigation are approximately:

	Sensitivity	*Specificity*
ECG	70%	100%
Plasma enzymes	95%	90%

By performing both types of investigation correctly, few wrong diagnoses of myocardial infarction will be made.

Trauma and operations

The activity of several enzymes is increased in plasma shortly after major operations, or as a result of severe trauma (e.g. road traffic accident). The changes are similar to those observed after myocardial infarction (Fig. 8.3), except that plasma CK–MB activity is not increased.

Muscle disease

Several enzyme activities can be measured in plasma in order to detect muscle damage. Plasma *total* CK activity is usually the measurement of choice, irrespective of the aetiology of the disorder; it is increased in the greatest number of cases and shows the largest change. Plasma AST, LD and ALT activities may be increased also. The enzyme tests may be considered under several disease categories.

Muscular dystrophy

This group of disorders includes the Duchenne, limb girdle and facioscapulohumeral types. Duchenne dystrophy is usually transmitted as a sex-linked recessive disorder and predominantly affects males. High levels of plasma CK activity may be detected while the child is still only a few months old, and before the onset of clinical signs. During the early clinical stages of the disease very high plasma CK activities are usually present, but these tend to fall as the terminal stages of the disease are reached.

About 75% of female carriers of the Duchenne dystrophy gene have raised plasma CK activities, but the increases are usually relatively small. The incidence of raised CK activities in more benign forms of muscular dystrophy is about 70%; the increases tend to be much smaller than in the Duchenne type.

Malignant hyperpyrexia

This is a rare but serious disorder characterized by raised body temperature, convulsions and shock following general anaesthesia. Many of the patients show evidence of myopathy. Extremely high plasma CK activities are seen in the acute, post-anaesthetic stage of the disorder, but smaller increases often persist and are also present in relatives of affected patients. It has been suggested that all patients might be screened preoperatively in order to detect people at risk, but preoperative plasma CK activities are not sufficiently reliable to detect all patients likely to develop malignant hyperpyrexia.

Preoperative screening for susceptibility to malignant hyperpyrexia should be limited to patients with a family history of anaesthetic deaths or of malignant hyperpyrexia. This involves pharmacological tests carried out on muscle biopsy specimens, and histological examination of muscle.

Other myopathies

Plasma CK activity is raised in most patients with alcoholic myopathy. There is also a large group of rare, genetically determined myopathies and myotonias. Plasma CK activity may be raised in the myopathies, but rarely in the myotonias, the increases observed usually being small.

Polymyositis

This may be due either to infective agents or to less well-defined causes (e.g. collagen disease). Plasma CK activity is often raised, the amount of the increase reflecting the activity of the disease.

Neurogenic muscle disease

In peripheral neuritis, poliomyelitis and motor neurone disease, plasma CK activity is usually normal.

Lung disease

Increases in plasma enzyme activity sometimes occur after pulmonary embolism, e.g. plasma LD activity is quite often increased. The isoenzyme pattern in serum is different in these patients from the pattern produced after myocardial infarction. However, the plasma enzyme changes after pulmonary embolism have not been found to be of diagnostic value, and they do not offer a reliable means of differentiating between pulmonary embolism and pneumonia.

Liver disease

Three patterns of altered plasma enzyme activity may be seen in patients with liver disease, due to the following derangements:

(1) Release of soluble cytoplasmic enzymes and, to a lesser extent, mitochondrial enzymes. This occurs in hepatocellular damage.
(2) Release of membrane-associated enzymes. This occurs in cholestasis.
(3) Impaired synthesis of certain enzymes (e.g. cholinesterase). This occurs in severe impairment of hepatocellular function.

Only the first two patterns will be further considered here. The third group is rarely used for assessing the severity of liver disease.

Soluble cytoplasmic enzymes

Aminotransferase (ALT and AST) activity measurements are the standard tests of hepatocellular damage in most hospital laboratories, even though some other plasma enzymes have been shown to be more specific and, in some reports, more sensitive indices of liver damage. Aminotransferases continue to be measured mainly because of their analytical convenience and simplicity.

Plasma ALT measurements are more liver-specific than AST (Table 8.1), and increases in plasma ALT are usually greater than AST in early hepatocellular disease. Aspartate aminotransferase has both cytoplasmic and mitochondrial isoenzymes and tends to be released more than ALT in chronic hepatocellular disease (e.g. cirrhosis).

Membrane-associated enzymes

Many enzymes are located close to the biliary canaliculus, including alkaline phosphatase and gamma-glutamyl transferase (GGT). These tend to be released in only small amounts following hepatocellular damage, but are released in much greater amounts when there is cholestasis.

Alkaline phosphatase is used as one of the routine tests of liver function in almost all hospital laboratories. Changes in the activities of most of the other membrane-associated enzymes usually accompany the changes in plasma alkaline phosphatase in cholestatic liver disease. Plasma GGT activity, however, provides a more sensitive index of cholestasis than alkaline phosphatase, and has the advantage of being more specific (e.g. plasma alkaline phosphatase may also be increased if bone disease is present). Plasma GGT measurements have become much more generally available than a few years ago.

ROUTINE ENZYME MEASUREMENTS IN LIVER DISEASE

Most laboratories select one plasma enzyme test as an index of hepatocellular damage, usually ALT or AST, and another as an index of the presence of cholestasis, usually alkaline phosphatase. These measurements may help to distinguish between the various types of jaundice, and are particularly valuable in detecting liver disease in unjaundiced patients. Other enzyme tests such as plasma GGT and alkaline phosphatase isoenzyme studies are routinely performed in some laboratories.

The jaundiced patient

In haemolytic jaundice, plasma ALT and AST activities are usually normal but 'heart-specific' LD may be increased due to its release from erythrocytes. Plasma alkaline phosphatase is usually normal in relation to the patient's age (Fig. 8.2).

In hepatocellular jaundice, plasma ALT and AST activities are frequently considerably increased, to more than six times the upper reference value, whereas alkaline phosphatase activities are usually less than twice the upper reference value.

In cholestatic jaundice, plasma ALT and AST activities are usually only slightly raised, whereas alkaline phosphatase activities are usually more than twice the upper reference value.

Many patients with jaundice conform to the simple criteria outlined above, but the more difficult cases often have atypical plasma enzyme activity patterns. In long-standing cholestatic jaundice, for instance, alkaline phosphatase activities may fall due to impairment of hepatic protein synthesis and return to within the range of reference values; this emphasizes the importance of obtaining the first set of chemical investigations as early as possible in patients with liver disease. Difficulty in interpreting results for plasma enzyme activity measurements may also occur in patients in whom cholestasis develops when the jaundice is thought to be primarily due to hepatocellular damage, and *vice versa*.

The unjaundiced patient

Plasma enzyme activity measurements provide a very sensitive means of recognizing hepatocellular damage. In early cases of hepatitis, increased plasma ALT and AST activities may be the only abnormal chemical findings. These measurements can also be used to determine whether liver cell destruction is continuing, and to provide an index of the activity of cirrhosis.

Alkaline phosphatase and GGT activities may be increased in plasma in the presence of only minor degrees of cholestasis, for instance in patients with partial biliary obstruction due to obstruction to one of the smaller biliary ducts such as often occurs in both primary and secondary carcinoma of the liver. Partial biliary obstruction may have little or no effect on the capacity of the liver to excrete bilirubin, so there is frequently no evidence of jaundice. Plasma alkaline phosphatase and GGT activities, however, may become greatly increased in association with higher concentrations of alkaline phosphatase and GGT in the part of the liver normally drained by the obstructed bile duct, due to induction of enzyme synthesis.

Drug therapy with potentially hepatotoxic agents may give rise to increased plasma enzyme activities before any other abnormalities are apparent. This applies to hepatotoxic drugs such as iproniazid, which cause raised ALT and AST activities. It also applies to drugs which may cause cholestasis, often of long duration, such as chlorpromazine, but in this case it is the plasma activities of alkaline phosphatase and GGT that are particularly increased.

Enzyme tests in malignant disease

Cancer cells show altered patterns of metabolic activity, which are accompanied by changes in the relative activity of various enzymes within the cell. For example, enzymes concerned with carbohydrate catabolism tend to be more active in neoplastic than in normal tissues whereas the opposite applies to enzymes concerned with carbohydrate synthesis. There is also a change in the pattern of isoenzymes within the cell. For example, LD_5 tends to become more prominent in neoplastic cells. Several different measurements of plasma enzyme activity have a place in the investigation of particular forms of malignant disease.

Acid phosphatase

Acid phosphatase activity in plasma is raised in about 80% of patients with prostatic carcinoma when this has spread beyond the prostatic capsule, but is raised in only about 20% of patients with non-invasive carcinoma. The main value of the test is in confirming the likely origin and existence of spread of the tumour, and in assessing the response to therapy.

The relatively new RIA methods offer few advantages over the older methods based on enzyme activity measurements and the use of inhibitors. The RIA methods are not significantly more sensitive at detecting intracapsular prostatic carcinoma, nor are they much better at detecting and following the course of more extensive disease.

The BB isoenzyme of creatine kinase, CK–BB, has recently been detected in plasma from patients with prostatic carcinoma. It may prove to be an alternative to acid phosphatase, or an additional tumour marker, for detecting and monitoring carcinoma of the prostate.

Alkaline phosphatase

The causes of an increased plasma alkaline phosphatase activity in patients with malignant disease can be grouped as follows:

(1) *Bone alkaline phosphatase* activity is increased in osteoblastic bone tumours, either primary or secondary, for instance in osteogenic sarcoma. It is not raised as a rule in lesions that are primarily osteolytic, such as those found in multiple myeloma.

(2) *Liver alkaline phosphatase* activity is increased in a high proportion of patients with either primary or secondary neoplasms in the liver, usually associated with areas of obstruction to biliary flow caused by these tumours. Other membrane-associated hepatic enzymes (e.g. GGT) may be similarly affected.

(3) *Tumour-specific alkaline phosphatase isoenzymes,* of which several have been described. The commonest is the Regan isoenzyme. This has the heat-stability properties of placental alkaline phosphatase. It is found in the plasma of some patients with various types of carcinoma, e.g. bronchus. There may be no evidence of hepatic or bony involvement, but the Regan isoenzyme may be detectable even though total plasma alkaline phosphatase activity is still normal.

'Ubiquitous enzymes'

These include glucosephosphate isomerase (EC 5.3.1.9), LD and the aminotransferases. All these enzymes are found in most tissues, including tumour tissue. Their activities in plasma tend to increase in disseminated carcinoma irrespective of the origin of the tumour, presumably due to increased cell turnover. The most sensitive is probably glucosephosphate isomerase, the activity of which has been reported as being increased in about 80% of patients with disseminated carcinoma.

Alterations in the activity of 'ubiquitous enzymes' tend to occur late in the disease and their measurements have not lived up to initial forecasts that they would be of value in the early and specific diagnosis of cancer. However, these tests can be used to follow the progress of the disease and the effect of therapy.

RED CELL ENZYMES

Erythrocytes contain large numbers of enzymes, many of them involved in glycolysis or with maintaining cellular levels of glutathione. Even in health, red cells contain different enzyme activities depending on the age of the cells, the less stable enzymes becoming inactivated as the cells get older.

HAEMOLYTIC ANAEMIA

Several inherited abnormalities have been described, most of which are rare. The disorders are usually due to abnormal function of an enzyme molecule rather then to a reduction in protein synthesis.

Glucose-6-phosphate dehydrogenase (G6PD), EC 1.1.1.49. Deficiency of G6PD is by far the most common of the rare inherited red cell enzyme disorders. It is transmitted by a sex-linked gene and therefore principally affects males. Female heterozygotes produce a mixed population of red cells, half of which are normal and the other half enzyme-deficient. The enzyme-deficient cells are more prone to haemolysis.

The incidence of the disorder varies, but is relatively high among American negroes and Mediterranean populations. A large number of different abnormal enzyme variants have been described. There may be no clinical effects, but in other individuals the defect may cause one of the following clinical syndromes:

(1) Hereditary non-spherocytic haemolytic anaemia.
(2) Drug-induced haemolytic anaemia. Primaquine and the sulphonamides are examples of drugs that have been implicated.
(3) Favism, an acute, severe haemolytic disorder which follows ingestion of the fava bean. Only a minority of individuals with G6PD deficiency is susceptible; this suggests that another defect may have to be present as well.

Pyruvate kinase, EC 2.7.1.40. Deficiency of pyruvate kinase is one of the more common, known causes of non-spherocytic haemolytic anaemia, itself a rare condition. The enzyme deficiency causes accumulation of phosphoenolpyruvate and 2:3-diphosphoglycerate (2:3-DPG). The latter, by shifting the oxygen dissociation curve to the right (p. 78), helps to compensate for the haemolytic anaemia.

Other inherited abnormalities of erythrocyte enzymes, rare causes of hereditary non-spherocytic haemolytic anaemia, include deficiencies of glucosephosphate isomerase, hexokinase, phosphofructokinase and aldolase.

Methaemoglobinaemia may be caused by deficiency of erythrocyte NADH-methaemoglobin reductase (p. 324).

VITAMIN DEFICIENCIES

Tests for deficiency of some of the B vitamins have been described, based on the activity of a particular enzyme in haemolysed erythrocytes. The enzymic activity is determined before and after the addition of the appropriate prosthetic group (Table 13.1, p. 269). The effect of adding the prosthetic group (e.g. the thiamin pyrophosphate or TPP effect) gives a direct indication of the degree of vitamin deficiency.

MISCELLANEOUS ENZYME TESTS

URINE

Most of the enzyme activity detectable in normal urine is derived from cells lining the urinary tract. Only a few enzymes normally present in plasma have a low enough mol. mass (less than about 60 kDa) to allow significant filtration to occur at the glomerulus.

Amylase (mol. mass 45 kDa) is normally detectable in urine. Excretion is increased in patients with acute pancreatitis and other causes of increased plasma amylase activity. However, urinary amylase adds nothing to measurements of plasma amylase activity except when macro-amylasaemia is suspected (p. 152).

N-Acetyl-β-glucosaminidase (NAG) activity derives from renal tubule cells. Its activity in urine, if measured serially in patients after renal transplantation, may be used to predict the onset of rejection episodes. Total urinary enzyme activity rises, and the isoenzyme pattern changes, prior to rejection.

GASTRIC AND PANCREATIC JUICE

Measurement of enzyme activities in gastric juice is rarely useful. However, determination of amylase, lipase, trypsin or chymotrypsin activities in duodenal juice, as part of a pancreatic function test (p. 201) can be of value; usually only amylase or trypsin activity is measured.

FAECAL ENZYMES

Measurement of tryptic activity in faeces may help in the diagnosis of cystic fibrosis (p. 439). Low levels are found in this disorder.

VAGINAL FLUID

Fluid obtained from patients with carcinoma of the cervix often contains increased activities of G6PD, LD, glucosephosphate isomerase and several other enzymes. However, similar increases occur in many other gynaecological disorders, e.g. vaginitis. Also, some patients with carcinoma-in-situ do not show any increase in enzyme activity. These tests are not suitable for screening for cervical carcinoma.

AMNIOTIC FLUID

Amniocentesis can be carried out from about the 15th week of pregnancy, and enzyme tests on amniotic fluid specimens can be used for two entirely different purposes:

(1) Some rare untreatable metabolic disorders can be recognized by culturing fibroblasts from amniotic fluid specimens obtained as early as possible, i.e. about the 15th week, in the case of pregnancies where there is a known high risk. Many of these disorders can be identified by measuring in the cultured fibroblasts the activity of the enzyme specifically associated with the disease (p. 429).
(2) Women who have previously given birth to a child with a neural tube defect, and women found on routine screening at the 16th–18th week of pregnancy to have abnormally high serum [AFP] are advised to undergo amniocentesis. If this examination is carried out, amniotic fluid [AFP] should be measured and amniotic fluid acetylcholinesterase isoenzyme patterns studied if the AFP result is abnormal (p. 412).

OTHER FLUIDS

Enzyme and isoenzyme investigations carried out on pleural and ascitic fluids are rarely of diagnostic value. Occasionally, measurement of amylase activity in pleural or peritoneal fluid may be of use in patients suspected of having fistulous tracts leading from the pancreas. Pancreatic fistulae may communicate with either the pleural or the peritoneal cavity.

Enzyme measurements on CSF are very rarely of value (p. 470).

Tissue enzymes

At present the study of enzymes in biopsy specimens is rarely performed on a routine basis. The main place of these investigations is to enable a precise diagnosis to be made in certain inborn errors of metabolism. For example, measurement of the activity of glucose-6-phosphatase and other enzymes related to glycogen metabolism, in liver biopsy specimens from patients with glycogen storage disease, is needed to identify the type of disorder on a biochemical basis (p. 238).

Alterations in cellular enzyme patterns may occur in malignant and pre-malignant disease. For example, the tissue LD isoenzyme pattern alters in carcinoma of the bronchus and of the colon. Relatively greater amounts of LD_4 and LD_5 are present than in the normal tissues. These measurements have not yet been shown to have enough diagnostic value to merit introduction into routine use.

FURTHER READING

FRIDHANDLER, L. and BERK, J.E. (1978). Macroamylasaemia. *Advances in Clinical Chemistry,* **20,** 267–286. New York: Academic Press.

HEARSE, D.J. and LERIS, J.de, editors (1979). *Enzymes in Cardiology: Diagnosis and Research.* Chichester: John Wiley.

MARSHALL, W.J. (1978). Enzyme induction by drugs. *Annals of Clinical Biochemistry,* **15,** 55–64.

MOSS, D.W. (1982). *Isoenzymes.* London: Chapman and Hall.

ROSALKI, S.B. (1975). Gamma-glutamyl transpeptidase. *Advances in Clinical Chemistry,* **17,** 53–107. New York: Academic Press.

SCHWARTZ, M.K. (1973). Enzymes in cancer. *Clinical Chemistry,* **19,** 10–22.

WHITTAKER, M. (1977). Estimation of plasma cholinesterase activity and the use of inhibitors for the determination of phenotypes. *Association of Clinical Pathologists, Broadsheet,* **87.** London: BMA House.

WILKINSON, J.H. (1976). *The Principles and Practice of Diagnostic Enzymology.* London: Arnold.

Chapter 9
Liver Disease

The liver plays a key role in many of the processes of intermediary metabolism. Its functions include the synthesis of carbohydrates, lipids and a wide range of proteins. It exchanges substances with the plasma, adding some for distribution in the body and removing others, often with subsequent metabolism. Bile is formed by the liver and it is the organ mainly responsible for the detoxication of many drugs and carcinogens. The liver excretes a wide range of compounds into bile.

Structure of the liver

It is a common fallacy to think of the liver as being composed almost exclusively of a single type of cell, the hepatocyte or parenchymal cell. In fact, only about 60% of the cells are parenchymal cells, and 30% are endothelial cells (Kupffer cells) lining the hepatic sinusoids. The rest consists of vascular and supporting tissue and bile ducts.

The parenchymal cells are arranged in sheets which are effectively only one cell in thickness. The membrane of the parenchymal cells is raised into microvilli both at the surface related to the sinusoidal cells and within the biliary canaliculi (Fig. 9.1), and the parenchymal cells thus have a direct relationship with plasma across the large surface area of membrane produced by the microvilli. A rapid, two-way exchange between the hepatocytes and plasma probably occurs through pores in the lining of the sinusoids.

The parenchymal cells are responsible for the formation of canalicular (hepatic) bile. The smallest radicles of the biliary tree have no separate structure of their own but are formed by fusion of the membranes of adjacent liver cells, and the biliary canaliculi have a rich network of microtubules and microfilaments surrounding them. There is a rapid one-way exchange between parenchymal cells and newly secreted canalicular bile, promoted by the formation of microvilli.

Carbohydrate metabolism

The liver is an important site of glycogen storage, which provides a rapidly available source of glucose during fasting. Glucose can also be formed in the liver by gluconeogenesis from amino acids such as alanine and aspartic acid. The liver and, to a lesser extent, the kidney are the only organs capable of releasing glucose into plasma.

Lipid metabolism

Cholesterol is synthesized in the liver, and is taken up by the liver from the circulation in the form of chylomicron remnants and LDL (p. 248). In the liver, newly synthesized cholesterol is esterified with fatty acids; both free and esterified cholesterol are released into the circulation.

Fatty acids and phospholipids are taken up from the plasma and metabolized by the liver, and it incorporates lipids into lipoproteins before their transfer into the plasma. It is an important storage site for fat-soluble vitamins and is the organ where 25-hydroxylation of vitamin D occurs.

Protein metabolism

Most of the plasma proteins, including the coagulation factors, are synthesized in the liver, the principal exception being the

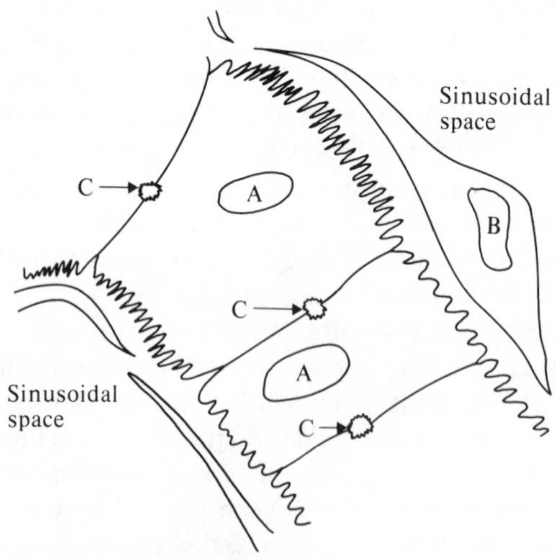

A - Nucleus of parenchymal cell
B - Nucleus of Kupffer cell
C - Biliary canaliculus

FIG. 9.1. *Diagrammatic representation of the microscopic structure of the liver*, to show the relationship between the hepatocytes and (i) the sinusoidal spaces, (ii) the biliary canaliculi.

immunoglobulins. It is also responsible for the synthesis of urea, via the urea cycle.

Urea is the compound in which most of the amino nitrogen of amino acids is eliminated from the body. Although creatinine, creatine, urate, NH_4^+, free amino acids and a number of other nitrogen-containing substances are also present in urine, these are quantitatively much less important than urea as means of nitrogen excretion in man.

Hepatocellular transport systems

The liver removes many substances from plasma and in some cases at least partly excretes them into bile. Examples include bilirubin, the primary and secondary bile acids (the latter returning to the liver as a result of the enterohepatic circulation; see below), and cholesterol (transported to the liver in chylomicron remnants). During their passage across the hepatocyte, many organic substances, endogenous and exogenous, are metabolized or conjugated (see below).

The transport systems for organic anions are very important. There are at least two of these systems, (1) for the excretion of bile acids and (2) for the excretion of other organic anions, of which bilirubin is the principal physiological example. Quantitatively, the system that transports the primary and secondary bile acids is the most important. The other system (or systems) removes a wide range of drugs and dyes (e.g. bromsulphthalein, indocyanine green) as well as bilirubin. The organic anions transported by these systems mostly have a mol. mass greater than 400 Da.

Enterohepatic circulation

Bile acids are formed in the liver from cholesterol. After conjugation with glycine or taurine, these primary bile acids (cholic acid and chenodeoxycholic acid conjugates mainly) are secreted into bile together with cholesterol and phospholipids. During the digestion and absorption of a lipid-rich meal, the absorption of primary bile acids from the terminal ileum produces a large increase in portal blood [bile acids]. There is a very efficient hepatic mechanism for removing bile acids from portal blood and, because of this mechanism, there is only a small increase in systemic blood [bile acids]; bile acids removed by the liver from the bloodstream are then re-excreted into bile. This cycle of excretion into bile, absorption from the intestine and re-excretion into bile is known as the enterohepatic circulation of bile acids.

Other substances that undergo an enterohepatic circulation include urobilinogen (stercobilinogen), cholesterol, secondary bile acids (formed in the intestine by bacterial metabolism of the primary hepatic bile acids) and thyroid hormones.

Detoxication and conjugation

Many different substances including bilirubin, many drugs, some hormones and a number of carcinogens are detoxicated, inactivated and in some cases excreted by the liver. For example, most steroid hormones are taken up by the liver and converted into inactive, water-soluble substances. The detoxication of some carcinogens proceeds through the formation of glutathione conjugates to a series of mercapturic acid derivatives. Some of the products are excreted in the urine, others in bile.

GROUPINGS OF CHEMICAL TESTS USED IN THE DIAGNOSIS OF LIVER DISEASE

The history of the patient's illness and the findings on clinical examination, including the results of side-room tests, provide an essential preliminary basis on which to select further investigations and to interpret their results. The relative value of different sorts of investigation changes from time to time, because of technical advances, and at present both Clinical Chemistry and Radiology have much to offer in making diagnoses of liver disease. Chemical investigations are particularly valuable in the following respects:

(1) Detecting the presence of liver disease.
(2) Placing the disease into the appropriate broad category.
(3) Following progress.

However, chemical investigations are only rarely instrumental in pin-pointing the diagnosis. Chemical tests for the diagnosis and follow-up of liver disease are discussed under the following headings:

(1) *Hepatic anion transport:*
- **(a)** Bilirubin excretion.
- **(b)** Other tests of hepatic anion transport function.

(2) *Plasma protein abnormalities:*
- **(a)** Proteins synthesized in hepatic parenchymal cells.
- **(b)** Immunoglobulins.

(3) *Plasma enzyme tests:*
  (a) Tests indicative of hepatocellular damage.
  (b) Tests indicative of cholestasis.
(4) *Other tests*

Hepatic anion transport: Bilirubin excretion

About 80% of the total amount of bilirubin formed each day (400–500 μmol, 240–300 mg) arises from red cells, broken down in the lymphoreticular system at the end of their life-span of approximately 120 days. The remaining 20% arises from red cell precursors destroyed in the bone marrow (so-called 'ineffective erythropoiesis'), and from other haem proteins such as myoglobin, the cytochromes and peroxidase which are all widely distributed in the body. Regardless of the origin of the haem, iron is removed from the molecule and the porphyrin ring opened to form bilirubin, a linear tetrapyrrole (p. 311). Bilirubin is soluble in lipid solvents but almost insoluble in water. These characteristics enable it to cross cell membranes readily, but special mechanisms are needed to make it water-soluble for carriage in plasma and excretion in bile.

Transport in plasma

Bilirubin is rendered soluble in plasma by protein-binding, mainly to albumin. In this form it does not readily enter most tissues, nor is it filtered at the glomerulus unless there is glomerular proteinuria.

Transport across the hepatocyte

The overall process of bilirubin transport by the liver is rapid. After intravenous injection, as a single bolus, bilirubin can be detected in bile within 1 min and the maximum concentration appears in bile between 2 and 3 min. The process of bilirubin transport across the liver is usually described as occurring in four distinct but related steps:

(1) Hepatic uptake of bilirubin.
(2) Binding of bilirubin to intracellular binding proteins.
(3) Conjugation of bilirubin with glucuronic acid.
(4) Secretion of bilirubin glucuronides into bile.

Hepatic uptake of bilirubin is poorly understood. The bilirubin-albumin complex in plasma appears to be dissociated at the liver cell membrane, leaving albumin in the plasma space. It is not known whether there are specific membrane-bound receptors for bilirubin uptake, but carrier-mediated uptake of bilirubin has been demonstrated.

Binding of bilirubin. Liver cells contain a series of relatively non-specific anion-binding proteins. One of these proteins, ligandin, may be involved in the uptake phase of bilirubin transport.

Conjugation of bilirubin within the hepatocyte makes it water-soluble by enzymic conjugation with glucuronic acid. Three bilirubin glucuronides (glucosiduronates) are formed, two monoconjugates and one diconjugate; the link is with the propionic acid side-chains (p. 311).

Secretion of bilirubin glucuronides into bile occurs by a process which is still poorly understood. Secretion is against a high concentration gradient, so a carrier-mediated energy-dependent process seems likely. In man, the bilirubin glucuronides in bile consist of about 75% as diglucuronide and 25% as a mixture of the two monoglucuronides.

Further metabolism of bilirubin in the gut

After secretion into the upper small intestine, bilirubin glucuronides are not reabsorbed to any great extent. Instead, they are degraded by bacterial action, mainly in the colon, being deconjugated and then converted into compounds collectively termed urobilinogen (sometimes called stercobilinogen). Most of the urobilinogen so formed is excreted in the faeces but a small proportion is reabsorbed and then mostly excreted by the liver, providing one example of an enterohepatic circulation.

Urobilinogen is water-soluble and has a low mol. mass (593 Da). Some of the reabsorbed material passes through the liver into the systemic circulation and is then excreted in the urine; it can be detected by side-room tests. After excretion, urobilinogen (which is colourless) oxidizes spontaneously to a dark brown pigment called urobilin.

MEASUREMENTS OF PLASMA BILIRUBIN

Plasma contains two different forms of bilirubin, (1) the unconjugated lipid-soluble form which is being transported, bound to albumin, from the lymphoreticular system to the liver, and (2) the conjugated water-soluble forms which have been regurgitated from the liver to the plasma. Normally, most of the bilirubin in plasma is unconjugated.

The chemical measurement that is normally performed is plasma or serum [total bilirubin], which measures the sum of the unconjugated and conjugated forms (reference values 2–20 μmol/L, 0.1–1.2 mg/100 mL). It is possible to measure plasma [conjugated bilirubin] and plasma [unconjugated bilirubin] separately, but for most purposes plasma [total bilirubin] is sufficient, when interpreted in relation to the results of side-room tests and other chemical investigations.

Abnormalities of bilirubin metabolism

Clinical jaundice is usually attributable to hyperbilirubinaemia. It becomes apparent when the plasma [bilirubin] exceeds 50 μmol/L (3 mg/100 mL) but smaller degrees of hyperbilirubinaemia may be of diagnostic significance. It is convenient to consider disorders of bilirubin metabolism under 3 headings (Fig. 9.2):

Pre-hepatic hyperbilirubinaemia

Over-production of bilirubin occurs in all forms of haemolytic anaemia and, less commonly, in conditions where there is a large amount of ineffective erythropoiesis, e.g. pernicious anaemia. In haemolytic disease of the newborn due to Rhesus incompatibility, if untreated, the concentration of lipid-soluble unconjugated bilirubin in plasma may

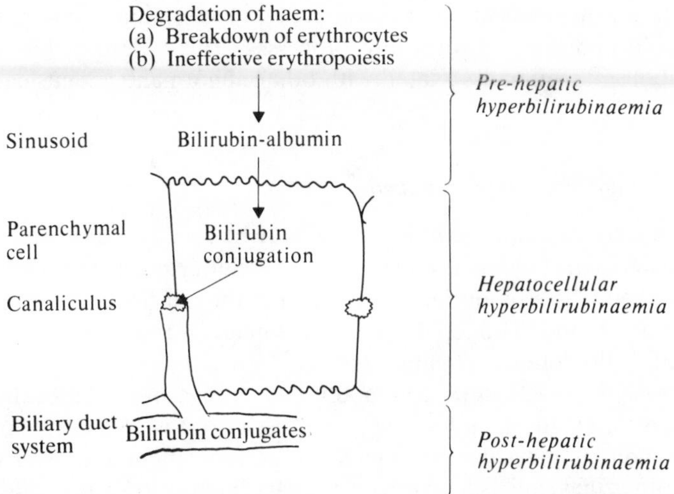

FIG. 9.2. *Formation, metabolism and excretion of bilirubin.* Diagrammatic representation of the sequence of stages, beginning with the degradation of haem and ending with the appearance of bilirubin conjugates in the bile.

exceed 340 μmol/L (20 mg/100 mL); appreciable amounts then cross the blood–brain barrier with resulting kernicterus. Pre-hepatic hyperbilirubinaemia is also sometimes due to bleeding into the tissues (e.g. sports injuries).

Hepatocellular (hepatic) hyperbilirubinaemia

Hepatic uptake of bilirubin may be abnormal in the common, mild unconjugated hyperbilirubinaemia known as Gilbert's syndrome (see below). Uptake may also be abnormal in patients with hepatitis or cirrhosis; plasma [unconjugated bilirubin] is then increased.

Conjugation of bilirubin by the liver, if impaired, leads to failure of excretion and hence to hyperbilirubinaemia. This may occur in the newborn, especially in premature infants; normally the microsomal conjugating enzyme system develops at about full term. Hepatic conjugation may also be abnormal in generalized hepatocellular jaundice and in congenital deficiency of the conjugating enzyme (Gilbert's and Crigler–Najjar syndromes), and occur as a result of competitive inhibition of the conjugating enzyme (e.g. due to novobiocin).

Secretion of conjugated bilirubin into bile is abnormal in the Rotor and the Dubin–Johnson syndromes (see below).

Hepatocellular damage, e.g. due to viral hepatitis or cirrhosis, may interfere with conjugation of bilirubin or with secretion of conjugated bilirubin into bile, or with both processes. This failure of bilirubin metabolism is often associated with other abnormalities that lead to cholestasis.

Post-hepatic hyperbilirubinaemia

Cholestasis, or stoppage of bile flow, may be intrahepatic ('medical') or extrahepatic ('surgical'). In both there is conjugated hyperbilirubinaemia and bilirubinuria. In general, the distinction between intrahepatic and extrahepatic jaundice is made by radiological investigations rather than by chemical tests, or by liver biopsy.

Intrahepatic cholestasis commonly occurs in acute hepatocellular damage, e.g. due to infectious hepatitis, cirrhosis and intrahepatic carcinoma. It may also be due to drugs (e.g. methyltestosterone, phenothiazines) and is a feature of primary biliary cirrhosis.

Post-hepatic cholestasis, caused by mechanical obstruction of the biliary tree, is most often due to gallstones or carcinoma of the head of the pancreas.

THE CONGENITAL HYPERBILIRUBINAEMIAS

These disorders are all due to inherited defects in the mechanism of bilirubin transport.

Gilbert's syndrome

This condition is characterized by an asymptomatic unconjugated hyperbilirubinaemia. It is probably present in 2–3% of the population. The plasma [bilirubin] fluctuates, with higher concentrations tending to occur during intercurrent illness. It is unusual for the concentration to exceed 50 μmol/L (3 mg/100mL).

The defect has not been defined precisely, and no morphological abnormalities of the liver have been shown. However, there is both an abnormality of hepatic uptake of bilirubin from the plasma, and a mild deficiency of hepatic bilirubin-UDP-glucuronyl-transferase in the endoplasmic reticulum. The uptake defect is probably secondary to the enzyme deficiency. Patients with Gilbert's syndrome are unable to metabolize some drugs normally. Plasma [bile acids], however, are normal in these patients.

Gilbert's syndrome can be differentiated from the mild degree of hyperbilirubinaemia of haemolytic anaemia, or of some patients with hepatic disease, by observing the effect on plasma [unconjugated bilirubin] of a reduced energy intake or of an injection of nicotinic acid. These tests are not often performed.

Reduced energy intake (1.67 MJ/day; 400 Cal/day) for 72 h results in at least a doubling of plasma [unconjugated bilirubin] in patients with Gilbert's syndrome. In normal individuals, it does not rise above 25 μmol/L (1.5 mg/100mL). Usually no increase is seen in patients with haemolytic anaemia, but in hepatitis there is an increase similar to the rise seen in Gilbert's syndrome. In this test, it appears that the increase in plasma [unconjugated bilirubin] is caused by the reduction in lipid intake; a reduced carbohydrate intake has no effect.

Nicotinic acid test. Following intravenous injection of nicotinic acid (50 mg), plasma [unconjugated bilirubin] rises to a maximum value after 90 min in normal individuals. In patients with Gilbert's syndrome, plasma [unconjugated bilirubin] rises to a peak that is higher than in normal individuals, and occurs later (after 2–3 h).

Crigler–Najjar syndrome

This is a rare condition that gives rise to severe hyperbilirubinaemia in neonates, leading to kernicterus and often to early death. The unconjugated hyperbilirubinaemia is due to marked reduction in the activity of bilirubin-UDP-glucuronyltransferase.

Dubin–Johnson and Rotor syndromes

These syndromes are both characterized by a benign conjugated hyperbilirubinaemia. There is a defect in the transfer of conjugated bilirubin into the biliary canaliculus. The mechanism of these defects is not known.

Other tests of hepatic anion transport function

Plasma bile acids

Conjugated bile acids are secreted in bile. They are absorbed from the small intestine mainly into the portal bloodstream from which they are rapidly cleared by the liver. Bile acids have a rapid enterohepatic circulation round which approximately 60 mmol (30 g) bile acids circulate each day.

Plasma [bile acids] rise in normal individuals from a fasting value of less than 1 μmol/L to 4–6 μmol/L between 90 and 120 min after a fat-containing meal. This increase is exaggerated in patients with abnormalities of bile acid clearance, and is often associated with an increased fasting plasma [bile acids]. The measurement of fasting plasma [bile acids] is probably the most sensitive test of hepatic anion transport function presently available.

Bromsulphthalein excretion and related tests

Bromsulphthalein (BSP) is an anionic dye still occasionally used to test hepatic anion transport function in unjaundiced patients. After intravenous injection BSP is bound to albumin, normally cleared rapidly from the circulation by the liver and excreted into bile. Passage across the liver involves conjugation of BSP, mainly with glutathione. The test is not entirely free from risk, as severe systemic reactions have followed BSP injection. Some hospitals no longer use it but, where the BSP test is still employed, local details of its performance should be followed.

Dibromsulphthalein would seem to be a safer dye for use in this type of clearance test, but unfortunately it is very difficult to obtain in the U.K. Another dye with apparently a low incidence of side-effects, as compared with BSP, is indocyanine green but again it has been little used in the U.K.

Plasma protein abnormalities

Plasma proteins are considered in Chapter 6, but certain features need to be recapitulated here. Measurements of plasma proteins synthesized by hepatocytes will be discussed first, followed by tests which reveal changes in immunoglobulin concentrations.

ALBUMIN

The capacity to synthesize albumin is impaired in many patients with liver disease. In chronic hepatocellular damage, plasma [albumin] falls and, in the later stages, provides a fairly good index of the progress of the disease. In acute liver disease, however, there may be little or no reduction in plasma [albumin] as the biological half-life of albumin is about 20 days and the fractional clearance rate is therefore low.

Increased portal venous pressure combined with a low plasma colloid oncotic pressure, together with Na^+ retention due to secondary hyperaldosteronism, give rise to the formation of ascites in cirrhotic patients. Ascites usually forms when plasma [albumin] falls below 30 g/L (3 g/100 mL).

COAGULATION FACTORS

Multiple defects of these factors occur in liver disease. In patients with hepatocellular damage, for instance, plasma [prothrombin] is often decreased. It is, however, unusual to measure coagulation factors individually in these patients. Instead, measurement of the *prothrombin time,* which depends on a number of coagulation factors, often suffices. The prothrombin time may be prolonged due to:

(1) Deficiency of vitamin K due to failure of absorption, caused by cholestasis.
(2) Failure to synthesize prothrombin and other factors measured by the prothrombin time, due to hepatocellular damage.

In vitamin K deficiency the coagulation defect can often be corrected by parenteral administration of vitamin K, but this is without effect in hepatocellular damage. The prothrombin time becomes prolonged relatively late in the hepatic disease, but then alters rapidly as the fractional turnover rates of the proteins concerned are short. It is used to assess progress and prognosis in patients with acute hepatocellular failure.

OTHER INDIVIDUAL PLASMA PROTEINS

Haptoglobin (p. 118), ceruloplasmin (p. 119) and cholinesterase (p. 153) are examples of plasma proteins synthesized in the liver and able to be measured specifically. Their concentrations alter when there is hepatocellular dysfunction. As tests of the liver's ability to synthesize proteins their measurement offers no advantage over the other tests already mentioned.

α_1-Fetoprotein (p. 121) is normally synthesized in minute amounts by adult hepatocytes. However, most patients with hepatocellular carcinoma and some with teratomas and certain other tumours have increased plasma [AFP], and about 50% of patients with hepatocellular carcinoma have grossly elevated plasma [AFP].

SERUM PROTEIN ELECTROPHORESIS

Abnormal patterns often occur in liver disease (p. 133); these reflect the composite effect of changes in individual serum proteins. In cirrhosis, for example, serum [albumin] is reduced and there is a polyclonal increase in [γ-globulin]; there is also filling of the trough between the β- and the γ-globulins, due to the predominant increase in the electrophoretically fast-moving IgA fraction (Fig. 7.5b). In cholestasis, increases in serum [γ-globulin] are smaller, but there are increases in the concentrations of α- and β-globulins due in part to increased serum [lipoproteins]. Although frequently requested in patients with liver disease, it is doubtful whether much specific diagnostic information derives from serum protein electrophoresis, except in the case of α_1-antitrypsin deficiency. This is one cause of chronic liver disease that is detectable in most cases by electrophoresis.

IMMUNOGLOBULINS

The diagnostic value of plasma immunoglobulin measurements in liver disease is uncertain. In acute liver disease abnormalities are inconstant, but in chronic liver disease there is usually an overall increase in plasma [immunoglobulins]. In most types of cirrhosis, plasma [IgA] is increased to about 2–3 times the mean value observed in health; there are usually smaller increases in plasma [IgG] and [IgM], but in primary biliary cirrhosis plasma [IgM] increases greatly and there are much smaller increases in plasma [IgG] and [IgA].

OBSOLETE TESTS OF LIVER FUNCTION

These include several empirical turbidity and flocculation tests such as the zinc sulphate turbidity, thymol turbidity and flocculation, and colloidal gold tests. These depend upon differences in the relative concentrations of serum proteins that can now be measured individually, specifically and quantitatively by other means.

The albumin : globulin ratio is also obsolete. There is little merit in expressing these two results of quantitative plasma measurements as an arbitrary ratio.

Plasma enzyme tests in liver disease

Individual plasma enzymes and alterations in their activity in liver disease are considered in detail elsewhere (p. 160). There are two principal groups of measurement used to help in the diagnosis of hepatic disorder.

Hepatocellular damage. Plasma alanine and aspartate aminotransferase (ALT and AST) activities are most often measured. These two enzymes are about equally sensitive as indices of hepatocellular damage, but plasma ALT is more liver-specific. Lactate dehydrogenase activity, especially LD_5, is also increased in plasma in many of these patients.

Cholestasis. Plasma alkaline phosphatase and GGT activities are the measurements most widely used. It should be noted that the hepatic synthesis of certain enzymes may be induced in response to biliary obstruction, to cholestasis caused by drugs, to ingestion of a variety of drugs shown to produce enzyme induction, or to the ingestion of large amounts of alcohol. Enzyme induction leads to an increase in the amount of enzyme protein in the hepatocytes, associated with increased enzyme activity in plasma. Both alkaline phosphatase activity and GGT activity in plasma are increased by cholestasis. Induction of GGT by drugs or alcohol can cause difficulty in interpreting the results of plasma GGT activity measurements.

Most cases of liver disease are neither entirely 'hepatocellular' nor entirely 'cholestatic'. The categorization of enzyme activity measurements in this way is thus a considerable oversimplification.

Other tests in the diagnosis of liver disease

Disordered amino acid metabolism

In advanced liver disease there is ineffective conversion of amino acids and ammonia to urea. The consequences of these abnormalities are as follows:

Amino acids. Plasma [amino acids] are increased and there is an amino aciduria of the generalized (overflow) type. These findings are rarely of sufficient diagnostic or prognostic value to justify measurement of plasma or urinary amino acids in these patients.

Ammonia. Plasma [NH_3] increases in patients with acute hepatocellular failure and with portal-systemic encephalopathy complicating hepatic cirrhosis. Technical difficulties in the reliable measurement of plasma [NH_3] have limited the use of this test, and results tend to correlate poorly with other indices of acute hepatic decompensation.

Urea. Significant decreases in plasma [urea] may occur late in hepatic disease, due to failure of the liver to convert amino acids and NH_3 to urea. There are, however, other causes of a reduction in plasma [urea], some of which may need to be excluded (Table 9.1) before a fall in plasma [urea] in a patient with liver disease should be interpreted as a sign of seriously deteriorating hepatic function. These patients, for instance, are sometimes maintained for long periods on diets low in protein and high in carbohydrate. Also, one of the commonest reasons for laboratories reporting low values for plasma [urea] is the collection of blood from a limb into which an intravenous infusion is being given.

TABLE 9.1. Causes of a decreased plasma urea concentration.

Cause	Examples
Non-pathological	Low protein, high carbohydrate diet
	Pregnancy
	Faulty collection of blood specimen
Pathological	Hepatic failure
	Inappropriate ADH secretion
Iatrogenic	Prolonged intravenous fluid treatment

Disordered carbohydrate metabolism

Hypoglycaemia may occur in a number of liver diseases, due to impaired gluconeogenesis or glycogen breakdown, or both. However, plasma [glucose] measurements are of little value in diagnosis of liver disease.

The *galactose tolerance test,* based on the fact that the liver is the main site of conversion of galactose to glucose, is little used in Britain. It is strongly contra-indicated in patients with galactosaemia (p. 239), in whom the conversion of galactose to glucose is markedly abnormal.

Disordered lipid metabolism

Plasma [free fatty acids] are increased in many types of liver disease. In cholestatic syndromes the concentrations of all the plasma lipid fractions are frequently raised. In contrast to most other causes of hyperlipidaemia, plasma [phospholipids] are increased at least as much as plasma [cholesterol] and [triglycerides].

An abnormal lipoprotein, lipoprotein X, is present in the plasma in nearly all cases of cholestasis. The mechanism of its production is poorly understood. It is not found in healthy people and is rarely present in patients with hepatocellular damage unless there is cholestasis also.

Serological tests

Several potentially diagnostic serological abnormalities may be present in liver disease, including:

Anti-mitochondrial antibodies. Antibodies directed against mitochondrial membranes of cells from many tissues are present in the serum of over 95% of patients with primary biliary cirrhosis. They are occasionally detected in the serum of patients with obstruction of the main bile ducts, but more often in serum of patients with cirrhosis or active chronic hepatitis.

Anti-smooth muscle antibodies. These have been found in the serum of about 50% of patients with active chronic hepatitis. In most of these patients the LE cell test is also positive.

THE PLACE OF CHEMICAL TESTS IN THE DIAGNOSIS OF LIVER DISEASE

Side-room tests can be a great help in the initial investigation of patients suspected of having liver disease and should always be performed as early in the illness as possible. Urine specimens should be examined for the presence of bilirubin and urobilinogen. The results of these examinations should be interpreted in relation to plasma [bilirubin] and the results of other tests, particularly aminotransferase and alkaline phosphatase activities. Where hyperbilirubinaemia is present, these results help to place the cause of the hyperbilirubinaemia into one of the three broad categories already discussed – pre-hepatic, hepatocellular and post-hepatic or cholestatic.

Table 9.2 provides examples of findings obtained by examining urine specimens and measuring plasma [bilirubin] in some of the

TABLE 9.2. Examples of results for bilirubin and urobilinogen investigations.

	Urine		Plasma	
			Bilirubin	
	Bilirubin	**Urobilinogen**	(μmol/L)	(mg/100mL)
Healthy individuals	Absent	Trace amount	2–20	0.1–1.2
Patients with:			*Increases up to:*	
Haemolytic disease	Absent	Increased	60	3.5
Hepatitis	Present	Varies with stage of disease	250	15
Biliary obstruction	Present	Decreased	400	24
Gilbert's syndrome	Absent	Decreased or normal	50	3

Notes

(1) Values for plasma [bilirubin], with reference values, are included so as to give an indication of the order of severity of the hyperbilirubinaemia observed.

(2) Gilbert's syndrome is included as an example of a common congenital hyperbilirubinaemia.

commoner conditions that give rise to hyperbilirubinaemia. With some of the patients categorized as having hepatic jaundice (e.g. cases of Gilbert's syndrome), determination of both plasma [unconjugated bilirubin] and plasma [conjugated bilirubin] may be helpful.

ACUTE HEPATITIS

In the pre-icteric phase of serum and of infectious hepatitis, plasma aminotransferase (ALT and AST) activities are increased and urobilinogen and bilirubin are present in the urine. By the time clinical jaundice appears, aminotransferase activities are usually more than five times, and occasionally more than 100 times, the upper reference value. The stools may be very pale, due to impaired biliary excretion of bilirubin, and urobilinogen disappears more or less completely from the urine. Alkaline phosphatase activity is usually slightly increased, up to about twice the upper reference value, but it may be considerably raised in those relatively uncommon cases where there is a marked cholestatic element.

Chronic persistent hepatitis continues in a few patients. In these, plasma ALT and AST activities and plasma [bile acids] may remain high for many months.

Similar findings to those in acute hepatitis are observed in patients with hepatocellular toxicity due to drugs, e.g. in paracetamol overdosage, halothane jaundice or carbon tetrachloride poisoning.

CHOLESTASIS

This is most commonly due to extrahepatic obstruction of the bile ducts. However, intrahepatic cholestasis can be prolonged and severe, especially when due to primary biliary cirrhosis or drugs such as the phenothiazines. In other patients, intrahepatic cholestasis may be relatively benign and short-lived, as is usually the case in the cholestasis sometimes associated with pregnancy.

Plasma [bilirubin] and [bile acids] are usually increased, often to high levels, and both are present in the urine. Urobilinogen may be present in urine, depending on the degree of biliary obstruction, but frequently it is undetectable.

Plasma alkaline phosphatase activity is considerably increased, often to more than three times the upper reference value, at least initially. Plasma AST and ALT activities, however, are usually only *moderately* raised by comparison with acute hepatitis, to values less than five times the upper reference value, unless complications such as ascending infection of the biliary tract develop. Plasma protein concentrations do not show characteristic changes except for the increase in plasma [IgM] seen in primary biliary cirrhosis. Plasma lipids show a variable increase in [cholesterol], [triglycerides] and [phospholipids]; lipoprotein X is usually demonstrable.

Chemical features that may help to distinguish cholestasis from hepatocellular damage are summarized in Table 9.3. It should be emphasized that these are 'typical' findings. Many cases do not follow these patterns exactly, and the distinction between intrahepatic and extrahepatic cholestasis can rarely be made by chemical tests.

INFILTRATIONS OF THE LIVER

The liver parenchyma may be progressively disorganized and destroyed by a wide variety of pathological conditions that can loosely be described as 'infiltrations'. These include multiple deposits of secondary carcinoma (e.g. lung, stomach), amyloidosis, the reticuloses, tuberculosis, sarcoidosis and abscesses.

In terms of their effects on the liver, 'infiltrations' are often characterized by the finding of considerably increased plasma alkaline phosphatase (hepatic isoenzyme) and GGT activities in the presence of normal or only slightly increased plasma [bilirubin]. Plasma ALT

TABLE 9.3. Distinction between hepatocellular damage and cholestasis.

Plasma investigation	Hepatocellular damage: Acute	Hepatocellular damage: Chronic	Cholestasis
Bilirubin	↑ or ↑ ↑	N or ↑	↑ or ↑ ↑
Aminotransferases	↑ ↑	N or ↑	↑
Alkaline phosphatase	N or ↑	N or ↑	↑ ↑
Gamma-glutamyl transferase	N or ↑	N or ↑	↑ ↑
Albumin	N or ↓	N or ↓ or ↓ ↓	N or ↓
Immunoglobulins	N or ↑	↑ *	N†
Prothrombin time	N or ↑	N or ↑	N or ↑
Vitamin K effect on prothrombin time	None	None	May correct abnormality

N = normal, ↑ = increased, ↑ ↑ = much increased,
↓ = decreased, ↓ ↓ = much decreased.

* Plasma [IgA] is particularly increased in cirrhosis, and plasma [IgG] in chronic active hepatitis.

†Plasma [IgM] is increased in primary biliary cirrhosis

and AST activities may be normal or slightly increased. The increase in plasma alkaline phosphatase and GGT activities is due to enzyme induction, caused by cholestasis in some parts of the liver.

CIRRHOSIS OF THE LIVER

Alcoholism, viral hepatitis and prolonged cholestasis are thought to be the most frequent causes of cirrhosis in Britain, but in half the cases no obvious cause is found. Less often, cirrhosis is associated with congenital metabolic disorders such as Wilson's disease (see below), cystic fibrosis (p. 438), haemochromatosis (p. 310), or galactosaemia (p. 239).

It is convenient to divide cirrhosis into mild and severe cases. Mild cases show minimal clinical abnormalities. In severe cases, ascites, haematemesis and other features are present; acute hepatic decompensation may arise and often proves fatal.

In *mild or latent cirrhosis,* plasma [bile acids] are likely to be abnormal more often than other tests. Measurements of plasma GGT activity are also very sensitive in detecting mild cirrhosis. Plasma ALT, AST and alkaline phosphatase activities are less often increased at this stage, and plasma [bilirubin] is usually below 25 μmol/L (1.5 mg/100 mL). Alterations in plasma [albumin] and in plasma [immunoglobulins] are rarely present at this stage.

In *more severe cases,* jaundice may develop; plasma [albumin] falls and plasma [immunoglobulins] increases. Clinical deterioration accompanied by prolonged prothrombin time, increased plasma [NH_3] and reduced plasma [urea] may herald the development of acute hepatic failure.

COPPER IN LIVER DISEASE

The liver is the principal organ involved in copper metabolism. The amount contained in the liver is maintained at normal levels by excretion of copper in bile and by incorporation into ceruloplasmin (p. 119).

The quantity of copper in the liver is increased in Wilson's disease, primary biliary cirrhosis, prolonged extrahepatic cholestasis, and intrahepatic bile duct atresia in the neonate. Increased liver [copper] may be a factor in the continuing hepatic damage that occurs in primary biliary cirrhosis.

Wilson's disease (hepatolenticular degeneration)

This is a rare hereditary disorder in which copper is deposited in many tissues, including the liver, brain and kidney. Symptoms usually appear in childhood and are mainly due to liver disease and to degenerative changes in the basal ganglia. The biochemical defect has not been precisely defined but it seems likely that there is a primary genetically determined abnormality of ceruloplasmin synthesis.

A diagnosis of Wilson's disease may be made on the basis of clinical findings including the demonstration of Kayser–Fleischer rings, which are due to the deposition of copper in the cornea. Chemical tests usually show a reduction in plasma [total copper] and [ceruloplasmin], and an increased plasma [free copper], i.e. copper not bound to ceruloplasmin. There is always an increased amount of copper in the liver and increased excretion in the urine. It should be noted that patients with cirrhosis due to causes other than Wilson's disease often have an increased plasma [ceruloplasmin].

Abnormalities of other chemical tests are often present, e.g. increased plasma [bilirubin] and ALT and AST activities, and there is usually evidence of renal tubular damage. There is a generalised renal amino aciduria and there may be glycosuria and phosphaturia and, in advanced cases, renal tubular acidosis.

ASCITES

Several chemical tests on blood (plasma [albumin], [bilirubin] and enzyme activity measurements) are performed routinely in any patient who develops ascites, as liver disease is the commonest cause. Other causes include congestive cardiac failure, malignant disease, chronic pancreatitis, tuberculosis and the nephrotic syndrome. If a diagnostic paracentesis is performed, the appearance of the fluid (blood-stained, bile-stained, milky, etc.) should be noted and ascitic fluid [protein] should be determined.

Ascitic fluid with a [protein] less than 30 g/L (3 g/100 mL) is called a *transudate*, and is usually associated with non-infective causes such as uncomplicated cirrhosis, in which there is a combination of back-pressure effects and low plasma [albumin]. However, ascitic fluid [protein] may be greater than 30 g/L in some patients, and 30 g/L cannot be designated as a diagnostic cut-off point.

Ascitic fluid [protein] much in excess of 30 g/L is called an *exudate*. It usually indicates the presence of infective conditions such as tuberculous peritonitis, malignant disease or pancreatic disease. If pancreatic disease is thought to be the cause, fluid amylase activity should be measured and the finding of a high activity will help to confirm the diagnosis. If hepatoma is suspected, plasma and ascitic fluid [α_1-fetoprotein] may both be considerably increased.

Chemical investigations are relatively non-specific and of minor importance in the examination of ascitic fluid. Bacteriological and cytological examinations may be much more informative. The same is generally true for the examination of pleural fluid, where again the most frequently performed chemical measurement is fluid [protein], to help differentiate between a transudate and an exudate.

GALLSTONES

Most gallstones are composed of cholesterol and insoluble salts of calcium along with trace amounts of other substances such as bilirubin. Cholesterol is virtually insoluble in water and is normally maintained in solution in bile by an interaction with bile acids and lecithin. Experimental conditions that lead to the adequate solubilization of cholesterol have been described. By determining the concentrations of cholesterol, bile acids and lecithin in bile, it is possible to say whether or not individual bile samples are 'lithogenic'.

Formation of gallstones may be partly due to disturbances in the relative concentrations of the biliary lipids leading to a reduction in the solubility of cholesterol. However, factors which promote crystallization of cholesterol from bile are not well understood and chemical examination of specimens of bile has no part to play in the routine diagnosis and surgical treatment of gallstones.

Chemical tests, such as plasma ALT or AST and GGT activities, are important in assessing whether gallstones have caused hepatocellular damage or cholestasis, due to complications such as ascending cholangitis or biliary tract obstruction.

Treatment of gallstones with oral chenodeoxycholic acid is sometimes prescribed. In these patients, measurement of biliary lipids as a check on the effectiveness of treatment, in changing the composition of bile, may be appropriate but are not widely available.

Patients with haemolytic anaemia may develop gallstones that consist wholly or almost entirely of bilirubin glucuronides. Stones form because of overproduction of bilirubin in these conditions and because the bilirubin glucuronides, although much more water-soluble than unconjugated bilirubin, are nevertheless not sufficiently soluble to remain in solution when bile is concentrated in the gall bladder.

Choosing a group of chemical tests for the investigation of liver disease

Most clinical chemistry departments carry out, as a routine, a limited range of blood tests that may as a group serve as pointers to the presence of liver disease. It is misleading to call these tests 'liver function tests' since every one of them may be abnormal in various conditions other than liver disease. It is only when the results of two or more biochemically and physiologically independent tests in the group become abnormal that patterns of abnormality begin to emerge which, taken together with other evidence (clinical, side-room tests, etc), point with increasing certainty to the presence of liver disease.

For the investigation of overt or suspected liver disease, we advise students and doctors to become familiar with the patterns of abnormalities observed with a limited range of chemical tests on blood, such as the ones listed as routine tests in Table 9.4, and with a limited number of further tests that may provide useful additional information in each category, in special circumstances such as have been described in this chapter.

Computer-assisted diagnosis

In assessing the patient with liver disease, the doctor has to weight the value of various findings, placing most reliance on those tests which have been shown to be good at discriminating one kind of liver disease from another in the past. The problem can be approached mathematically, if the probability with which each indicant occurs in

TABLE 9.4. Chemical tests on plasma for the investigation of liver disease.

Property to be assessed	Routine tests	Special tests
Protein synthesis	[Albumin]	[Immunoglobulins]
Hepatic anion transport	[Bilirubin]	[Bile acids]
Hepatocellular integrity	ALT *or* AST activity	
Presence of cholestasis	Alkaline phosphatase *or* GGT activity	Alkaline phosphatase isoenzymes, Lipoprotein X

Notes

(1) Many laboratories measure plasma alkaline phosphatase *and* GGT activity routinely as part of their group of chemical investigations.

(2) Serum required for lipoprotein X measurements.

each disease is known, from study of a sufficiently large and wide-ranging 'bank' of patients in all of whom the correct diagnosis had eventually been established. Indicants include clinical features and results of radiological and laboratory investigations.

Chemical tests are particularly convenient to use as indicants in this mathematical approach to diagnosis. Blood and urine are easy to obtain, and the results of measurements are quantiative, precise and free from observer bias. However, even the more effective chemical indicants, such as the activities of the aminotransferases and alkaline phosphatase in plasma, behave atypically in a significant minority of patients with liver disease. Difficulty also arises in deciding how to weight measurements which, although seemingly different, may to some extent be assessing the same property, e.g. the ability to synthesize plasma proteins.

FURTHER READING

BOUCHIER, I. A. D. (1983). Biochemistry of gallstone formation. *Clinics in Gastroenterology*, **12,** 25–48.

Editorial (1981). Pathogenesis of Gilbert's syndrome. *European Journal of Clinical Investigation,* **11,** 417–418.

Leading article (1978). Hereditary jaundice. *Lancet,* **2,** 926–927.

Leading article (1979). Diagnosis of cholestasis. *British Medical Journal,* **1,** 1232.

Leading article (1980). Screening tests for alcoholism? *Lancet,* **2,** 1117–1118.

Leading article (1981). Diagnosis of ascites. *British Medical Journal,* **282,** 1499.

Leading article (1982). Serum bile acids in hepatobiliary disease. *Lancet,* **2,** 1136–1138.

MARSHALL, W.J. (1978). Enzyme induction by drugs. *Annals of Clinical Biochemistry,* **15,** 55–64.

REICHEN, J. and PAUMGARTNER, G. (1980). Excretory function of the liver. In *International Review of Physiology,* **21,** pp. 103–150. Ed. N.B. Javitt. Baltimore: University Park Press.

SAUNDERS, J. B. (1983). Alcoholic liver disease in the 1980s. *British Medical Journal*, **287,** 1819–1821.

SHER, P.P. (1977). Diagnostic effectiveness of biochemical liver-function tests, as evaluated by discriminant function analysis. *Clinical Chemistry*, **23,** 627–630.

SHERLOCK, S. (1981). *Diseases of the Liver and Biliary System,* 6th Edition. Oxford: Blackwell Scientific Publications.

STERN, R.B., KNILL-JONES, R.P. and WILLIAMS, R. (1974). Clinician versus computer in the choice of 11 differential diagnoses of jaundice based on formalized data. *Methods of Information in Medicine,* **13,** 79–82.

TRIGER, D.R. (1979). Physiological functions of the liver. *British Journal of Hospital Medicine,* **22,** 424–432.

WARNES, T.W. (1982). A new look at cirrhosis. *Journal of the Royal College of Physicians of London,* **16,** 23–32.

Chapter 10

Gastrointestinal Tract Disease

Carbohydrates and proteins are for the most part broken down to their component monosaccharides and amino acids by the processes of digestion, before being absorbed into the portal blood or the intestinal lymphatics. The other principal dietary constituents, the fats, are emulsified and partially digested, rendered water-soluble and then absorbed.

Mastication helps to break up the food and mix it with saliva, so as to lubricate it prior to swallowing. Salivary amylase begins the digestion of polysaccharides, but the enzyme is inactivated by gastric HCl. Softening and mixing of food continue in the stomach, where the parietal cells secrete HCl and intrinsic factor, and the zymogen cells secrete the pro-enzyme pepsinogen; mucus is secreted by the cells of the pyloric antrum. However, the main processes of digestion occur in the small intestine, and absorption of the products of digestion is mainly from the intestinal lumen.

The brush border of the intestine has both digestive and transport functions. It contains enzymes capable of splitting small molecules, and specific transport mechanisms whereby some constituents of food and of bile can be transferred, against a concentration gradient, into the mucosal cells.

DIGESTION AND ABSORPTION OF CARBOHYDRATES

The main digestible carbohydrates in the diet are the polysaccharide starch (60–70% of total carbohydrate intake) and the disaccharides sucrose (20–30%) and lactose (about 5%).

Digestion of starch occurs in the intestinal lumen, mainly through the action of pancreatic amylase. This splits α-1, 4-glucosidic links in starch to yield maltose (a disaccharide), maltotriose (a trisaccharide) and α-limit dextrins (a mixture of glucose polymers). Further digestion of these products, and the hydrolysis of sucrose and lactose, occurs on the brush border of the mucosa of the small intestine. Isomaltase, which acts also as an α-limit dextrinase, and the disaccharidases are attached to the brush border. The end-products of carbohydrate digestion, the monosaccharides, are absorbed by the mucosal cells and transported to the portal blood. Malabsorption of carbohydrate may occur in:

(1) Generalized intestinal disease, affecting absorption.
(2) Pancreatic disease, causing amylase deficiency.
(3) Intestinal disaccharidase deficiency.
(4) Defects of monosaccharide transport.

DIGESTION AND ABSORPTION OF PROTEINS

Digestion of proteins by pepsin begins in the stomach and continues in the lumen of the small intestine under the action of trypsin and chymotrypsin from the pancreas. These three endopeptidases are all secreted as inactive precursors (e.g. pepsinogen). The pancreas also secretes exopeptidases which split off terminal amino acids from proteins and peptides. Thus, the products of luminal digestion of proteins are oligopeptides (two to six amino acids) and free amino acids.

Some oligopeptides pass without further breakdown directly into the portal circulation. Others are further digested to their constituent amino acids in the cells of the intestinal mucosa, either in the brush border or in the cytoplasm, under the action of a variety of peptidases. Absorption of amino acids across mucosal cells into the circulation is mostly carried out by a number of energy-requiring active processes, each specific for one group of amino acids (basic, acidic, neutral); there is a fourth mechanism for the imino acids (proline and hydroxyproline). Malabsorption of amino acids may occur in:

(1) Generalized intestinal disease, affecting absorption.
(2) Pancreatic disease, causing deficiency of the pancreatic peptidases.
(3) Specific transport defects (e.g. cystinuria, Hartnup desease, p. 98).

DIGESTION AND ABSORPTION OF LIPIDS

These processes can conveniently be considered as occurring in three stages, digestion, solubilization and absorption.

Dietary triglycerides are normally partly hydrolysed by pancreatic lipase to yield a mixture of fatty acids, monoglycerides and diglycerides. Lipase activity occurs most effectively at an oil–water interface, and adequate lipolysis requires the production of a fine emulsion from dietary fat. Emulsification is achieved by a combination of the mechanical effects of intestinal motility and the detergent-like action of bile acids. These various processes together make up the digestive stage.

Diglycerides, monoglycerides, free fatty acids and other dietary lipids (cholesterol, phospholipids, fat-soluble vitamins, etc.) are largely insoluble in water. Prior to absorption, these lipids are rendered water-soluble by incorporation into soluble aggregates of bile acid molecules, called micelles. This is the solubilization stage.

The stage of absorption of lipids is a poorly understood process. In it, the products of digestion of dietary triglycerides are transferred from the mixed micelles into the intestinal mucosal cells. This transfer takes place mainly in the jejunum.

After entry into the mucosal cells, triglycerides are resynthesized and partly incorporated into chylomicrons together with cholesterol, phospholipid and a small amount of protein. The chylomicrons pass into the lymphatics draining the intestinal wall, and enter the systemic circulation via the thoracic duct. The rest of the triglycerides are incorporated into VLDL in the small intestine and enter the bloodstream (p. 247). Some short-chain fatty acids escape re-esterification and pass directly into the portal blood where they bind to albumin and are transported to the liver. Abnormalities of lipid absorption may occur in:

(1) Generalized intestinal disease, affecting absorption.
(2) Pancreatic disease, causing lipase deficiency.
(3) Decreased lipase activity due to high intestinal [H^+].
(4) Deficiency of bile acids.
(5) Abetalipoproteinaemia (p. 260).

Peptides of gastrointestinal origin

Several peptides with hormonal or local effects have been identified as arising from the gastrointestinal tract. Their structures have mostly been established, and their cells of origin located (Table 10.1). Several of these have been found not only in cells and nerve fibres widely distributed in the gastrointestinal tract but also in the central nervous system. Some gastrointestinal peptides have endocrine effects, acting on organs at a distance. Others appear to act locally, by diffusion to their target organs through the extracellular space; these are called paracrine effects.

The functional relationships of these peptides are not fully understood, and the diagnostic value of measuring their concentrations in plasma is in most cases still unkown. However, the Zollinger–Ellison syndrome (p. 198), the Verner–Morrison syndrome

TABLE 10.1. Examples of gastrointestinal peptides.

Peptide	Location	Probable functions
Gastrin	Gastric antrum and duodenum	Stimulates gastric H^+ production, and trophic to gastric mucosa
Secretin	Duodenum and jejunum	Stimulates water and HCO_3^- secretion from pancreas
Cholecystokinin (CCK)*	Duodenum and jejunum	Stimulates enzyme secretion from pancreas, and gall bladder contraction
Glucose-dependent insulinotrophic peptide (GIP)	Duodenum and jejunum	Post-prandial release of insulin
Motilin	Duodenum and jejunum	Stimulates intestinal motor activity
Pancreatic polypeptide	Pancreas	Inhibits enzyme secretion from pancreas, and relaxes gall bladder
Gut glucagon-like immunoreactivity	Ileum and colon	Increases small intestinal mucosal growth and slows transit
Vasoactive intestinal polypeptide (VIP)*	All areas	Secretomotor, vasodilation and relaxation of smooth muscle

*Examples of peptides found both in the gastrointestinal tract and in the CNS

(p. 210) and coeliac disease (p. 441) are three conditions in which abnormal secretion of gastrointestinal peptides will be discussed.

GASTRIC FUNCTION

Two main mechanisms of gastric stimulation are recognized, the cephalic or nervous phase mediated via the vagus nerves, and the gastric or chemical phase in which the stomach is stimulated more directly by the hormone gastrin. Both mechanisms can be tested.

Gastrin is a small polypeptide (18 amino acids), released into the bloodstream from specialized cells in the antral and duodenal mucosa. It is a potent stimulus to gastric acid production. The stimulus for release of gastrin is reduced gastric $[H^+]$. Certain foods, e.g. proteins and protein extracts, are particularly effective physiological stimulants.

Widely used tests of gastric acid-producing function measure the acid secreted in response to maximal chemical stimulation (pentagastrin test) or to nervous stimulation (insulin-hypoglycaemia test).

PENTAGASTRIN TEST

This measures maximal gastric acid production. Pentagastrin is a pentapeptide with the four C-terminal amino acids identical to those of gastrin. It has an effect on acid production similar to gastrin.

To perform the test, the patient should be fasting. A radio-opaque nasogastric tube is passed into the stomach, the position of the tube checked radiographically, and the resting juice (the juice present in the stomach of the fasting patient) aspirated. The basal output of gastric juice (the basal juice) is assessed by collecting the fluid output for the next 60 min with the patient resting. Pentagastrin (6 μg/kg body weight) is then injected subcutaneously and the 'post-pentagastrin secretion' collected for 60 min as four separate 15 min specimens.

The volume and pH of each sample are measured and the acid content determined by titration with NaOH to pH 7.0. The 'peak acid output' is usually reported, as the rate of acid secretion (expressed as mmol/h) derived from the two adjacent 15 min specimins that together show maximal output. The presence of bile or blood in any of the samples should be noted.

Interpretation of pentagastrin tests

The response in healthy subjects to the administration of pentagastrin is variable. Only a limited amount of information, therefore, can be derived from the results of the pentagastrin test. However, the test is much more reproducible and less unpleasant than its predecessors (e.g. augmented histamine test).

Resting juice is normally less than 50 mL, with a low total acid content. An increased volume of resting juice (more than 100 mL) may indicate impaired gastric emptying due to pyloric stenosis. The basal secretion of acid is usually less than 5 mmol/h. Following pentagastrin, the peak acid output is usually less than 45 mmol/h in males and less than 35 mmol/h in females. It is difficult to define the lower reference value, but acid is usually detectable.

Achlorhydria is found in pernicious anaemia, except the very rare juvenile form, and in many cases of gastric carcinoma. Achlorhydria is usually defined as an inability to secrete gastric acid so that the pH of gastric juice produced in response to pentagastrin stimulation does not fall below pH 7.0.

There tends to be an increased output of acid in patients with duodenal ulcer, and a decreased output in patients with gastric ulcer. However, these findings are not sufficiently constant to be of diagnostic value. A high rate of basal acid secretion is found in the Zollinger–Ellison syndrome.

Tests rendered obsolete by the pentagastrin test include fractional test meals and the augmented histamine test. Pentagastrin tests have also largely replaced 'tubeless' gastric analysis which, though simple in theory, has proved unreliable in practice.

INSULIN–HYPOGLYCAEMIA TEST

This test is widely used for investigating patients with peptic ulcer before and shortly after treatment by vagotomy, to determine whether vagal section has been adequate. Hypoglycaemia normally stimulates a vagus-mediated output of gastric acid; this output is abolished if the vagi are completely severed.

Intubation is performed as in the pentagastrin test. Resting juice is aspirated and 'basal juice' collected for 60 min. Insulin (0.2 units/kg body weight) is then injected subcutaneously, and gastric juice collected for 2 h. Plasma [glucose] is measured before the injection of insulin and 15, 30, 45 and 60 min after the injection.

The patient must be kept under observation throughout the test. Glucose solutions should be immediately available, for oral or intravenous administration as necessary, if severe hypoglycaemic symptoms develop.

A different insulin–hypoglycaemia test, in which the dose and route of administration differ from the gastric function test, is used to investigate the HPA axis (p. 362).

Interpretation of insulin-hypoglycaemia tests

No significance can be attached to the results unless hyopoglycaemia occurs. This is usually defined as a plasma [glucose] below 2.2 mmol/L (40 mg/100 mL) in at least one of the blood specimens. If the $[H^+]$ in any of the specimens increases by more than 20 mmol/L over the basal level, it can be assumed that vagal section has been incomplete. Also, if the acid output in any four consecutive 15 min specimens totals 10 mmol or more, this again indicates that vagotomy is incomplete and that there is a strong chance that ulceration will recur.

PLASMA GASTRIN

Plasma [gastrin] reflects the rate of gastrin secretion by the pyloric antrum. It is reduced in the fasting state and increased when gastric H^+ is neutralized by the presence of food or HCO_3.

In diseases causing hyperacidity (e.g. duodenal ulcer), plasma [gastrin] is reduced, except in the case of the Zollinger–Ellison syndrome. Conversely, in achlorhydria or hypochlorhydria due to pernicious anaemia or gastric ulcer, plasma [gastrin] is increased unless atrophic gastritis has destroyed the gastrin-producing cells. Although these findings can be explained in terms of normal physiological responses to alterations in gastric acid production, in duodenal ulcer there is some evidence that gastrin-producing cells contain increased amounts of gastrin.

Measurement of plasma [gastrin] is important for the diagnosis of the Zollinger–Ellison syndrome.

ZOLLINGER–ELLISON SYNDROME

In the original description severe, often multiple and recurrent peptic ulcers had developed in patients who had both excessive production of acid by the stomach and non-insulin-secreting tumours of the pancreas.

The syndrome is due to neoplasia of either pancreatic gastrin-producing cells or gastric gastrin-producing cells. The Zollinger–Ellison syndrome may occur as part of the multiple endocrine neoplasia (MEN) or 'pluriglandular' syndromes (p. 291). In all these conditions the production of gastrin appears to be autonomous, and the normal reciprocal relationship between gastrin secretion and gastric acidity is lost.

The initial diagnosis of Zollinger–Ellison syndrome is made by finding high overnight and basal rates of H^+ secretion by the stomach; these often approach the rates observed following stimulation with pentagastrin. If gastric aspiration is performed overnight, more than one litre of fluid is usually obtained from these patients, containing at least 100 mmol/L of HCl. The diagnosis is confirmed by finding inappropriately high plasma [gastrin] in blood samples collected from fasting patients.

Diarrhoea and evidence of malabsorption, especially malabsorption of fat, are often present. The steatorrhoea is thought to be due to the high $[H^+]$ in the intestinal lumen, caused by the grossly excessive gastric production of acid, inhibiting the action of pancreatic lipase.

PANCREATIC DISEASE

Secretion of pancreatic juice is controlled by several factors. For instance, food in the stomach, products of hydrolysis of proteins and triglycerides present in the duodenum (but not the presence of the proteins or triglycerides themselves), H^+ in contact with the intestinal mucosa, and Ca^{2+} in the small intestinal juice all lead to the stimulation of pancreatic secretion. The principal hormonal mechanisms responsible for stimulating the intestinal phase of secretion of pancreatic juice are:

(1) The release of cholecystokinin-pancreozymin (CCK–PZ). This is the most important known stimulus. It mainly evokes secretion of fluid that is rich in enzymes.

(2) The release of secretin. This is probably also of physiological importance. It stimulates the secretion of fluid that is rich in HCO_3^-.

Pancreatic juice is an alkaline fluid containing electrolytes (Table 4.4) as well as digestive enzymes, some in the form of their precursors or pro-enzymes. The principal enzymes are the pro-enzymes of the proteases (trypsin, chymotrypsin and the carboxypeptidases), amylase and the lipolytic enzymes (lipase and co-lipase).

Acute pancreatitis

There is acute inflammation of the pancreas. There is a predominantly oedematous form and a more serious, haemorrhagic form in which tissue necrosis is more severe. Acute pancreatitis is commonly associated with gallstones or alcoholism; vascular and infective causes have also been recognized. Confirmation of the clinical diagnosis mainly depends on plasma amylase activity measurements.

PLASMA AMYLASE

The activity of this enzyme is usually increased in acute pancreatitis, but not always. Maximum values of more than five times the upper reference value are found in about 50% of cases, and usually occur on the first or second day of the illness. Such high values are sometimes considered pathognomonic of acute pancreatitis, but this is not correct. Similarly high values sometimes occur in the afferent loop syndrome, mesenteric infarction and acute biliary tract disease. Smaller increases in plasma amylase activity may occur in almost any acute abdominal condition.

In patients with acute pancreatitis, plasma amylase activity usually returns to normal within 3–5 days.

Lipase and trypsin activities are also increased in patients with acute pancreatitis, and in other conditions where plasma amylase activity is increased. Measurement of these enzymes is technically more difficult, and these tests offer no advantages.

URINE AMYLASE

In most conditions where plasma amylase activity is increased, urinary amylase activity also rises (p. 151). However, in acute pancreatitis, this test offers no advantages over plasma amylase measurements. The diagnostic value of measuring urinary amylase activity in acute pancreatitis has not been improved by measuring the amylase:creatinine clearance ratio; the promise of early reports has not been confirmed. Measurements of urinary amylase activity have not gained widespread acceptance, except as part of the investigation of macro-amylasaemia.

PLASMA CALCIUM

Hypocalcaemia often occurs in the more severe cases, sometimes not until after a few days. It is probably due to the formation of insoluble calcium salts of fatty acids in areas of fat necrosis caused by excessive lipolysis.

METHAEMALBUMINAEMIA

This sometimes develops in severe cases of acute pancreatitis, probably as a result of proteolytic breakdown of haemoglobin. Both haem and methaem (oxidized haem) are formed. Haem is metabolized normally to bilirubin. Methaem combines with plasma albumin to form methaemalbumin. In healthy individuals there is at most only a trace of methaemalbumin detectable in plasma.

Chronic pancreatitis

Failure to secrete adequate amounts of digestive enzymes may not occur until pancreatic disease is advanced, but may then give rise to malabsorption, especially steatorrhoea. Various chemical methods, many of them time-consuming and lacking in sensitivity until the disease is advanced, have been used for the investigation of these patients.

Assessment of pancreatic secretion, following duodenal intubation, continues to be performed but alternative, non-invasive procedures seem likely to be preferred in time. Both types of test will be described.

STIMULATION TESTS

Stimulation of pancreatic secretion can be achieved directly, by injecting secretin together with or followed by cholecystokinin–pancreozymin (CCK–PZ). Some patients are sensitive to CCK–PZ and caerulein has been used as an alternative stimulus. This is a decapeptide with CCK–PZ–like activity; it occurs naturally, in amphibian skin, and has been synthesized. Indirect stimulation depends on the pancreatic response to a meal.

Secretin/CCK–PZ test

This test can be performed in various ways. The following account indicates principles that are generally applicable.

The patient fasts overnight and a double lumen radio-opaque tube is passed in the morning under radiological control. One opening is positioned for aspiration of gastric secretions, and the other close to the opening of the pancreatic duct so that it can collect duodenal contents from the second part of the duodenum. Continuous suction is applied to both tubes, the gastric juice being aspirated to prevent it contaminating the pancreatic juice. Specimens of duodenal contents are collected and preserved on ice, according to the following schedule:

(1) Basal specimens are collected for 2 periods of 10 min.
(2) After intravenous secretin, 6 more 10 min collections are made.
(3) After intravenous CCK–PZ, 2 more 10 min collections are made.

The 10 specimens are kept cool and taken to the laboratory immediately the test ends. The volume, [HCO_3^-], and trypsin or amylase (or lipase) activity are measured on each specimen. Trypsin and amylase are technically easier to measure than lipase.

Healthy subjects show a wide variation in the pattern of results, partly due to difficulties in securing complete collections of specimens. The post-secretin specimens usually show a maximal hourly secretory rate of at least 2.0 mL/kg body weight and the [HCO_3^-] normally rises above 75 mmol/L. The results for enzyme activity measurements depend so much on the method used by the laboratory that it is necessary to consult locally to find out what represents a normal response, i.e. how great an increase in enzyme activity should follow CCK–PZ injection.

Abnormal results are obtained in most cases of chronic pancreatitis, enzymic activity and [HCO_3^-] tending to fall before there is any obvious reduction in the volume of juice. Results are also abnormal in some cases of pancreatic carcinoma; although all three variables (enzymic activity, [HCO_3^-] and volume) may be affected, a low volume of juice is a particularly marked feature of pancreatic carcinoma (Table 10.4, p. 212).

Lundh test

This involves indirect stimulation of pancreatic secretion by a meal containing carbohydrate, protein and fat. The duodenal contents are collected for 2 h and the activity of trypsin or amylase or lipase measured. Low activity indicates pancreatic exocrine insufficiency.

Serum immunoreactive trypsin (IRT)

Both the secretin/CCK–PZ test and the Lundh test can be used as stimuli to elicit a response in serum enzyme activity. Tests based on amylase, lipase and trypsin measurements have been described; of these, serum [IRT] has so far proved the most promising.

Patients with chronic pancreatitis frequently have a low fasting serum [IRT], and the test has been suggested as a screening investigation for this condition. If the fasting value is normal and duct obstruction is suspected, serial measurement of serum [IRT] at 30 min intervals after a Lundh test meal may provoke a marked increase in concentration.

ANALYSIS OF PANCREATIC JUICE

Endoscopic cannulation of the pancreatic duct is now an established method of investigation, allowing collection of pancreatic juice uncontaminated by either bile or intestinal secretion.

Lactoferrin, an iron-containing glycoprotein secreted by the pancreas, can be measured in pancreatic juice. Its concentration has been reported to be higher than normal in patients with chronic pancreatitis, and normal in patients with pancreatic carcinoma. Similar results have been obtained for measurements of [lactoferrin] in duodenal juice.

BT PABA/^{14}C-PABA TEST

This is an indirect test of pancreatic function. It depends on the fact that pancreatic chymotrypsin specifically hydrolyses the synthetic peptide *N*-benzoyl-L-tyrosyl-*p*-aminobenzoic acid (BT PABA). This releases PABA which is then absorbed from the intestine, partly metabolized by the liver and partly excreted in the urine. In order to correct for these extra-pancreatic factors, the oral dose (0.5 g) of BT PABA is given together with 5 μCi ^{14}C-PABA, as a drink containing 25 g casein.

Urine is collected for 6 h, urinary PABA and ^{14}C content measured, and both results expressed as a percentage of the oral dose. The two percentages are then used to calculate the ratio, the PABA/^{14}C excretion index. Normally this index is 1.0 (SD 0.08). In patients with chronic pancreatitis, hydrolysis of BT PABA is impaired and the excretion index is low.

The results of this test are said to correlate well with the CCK–PZ and Lundh tests. It has the advantage of being non-invasive, and suitable for the investigation of outpatients. However, the BT PABA/^{14}C-PABA test is still essentially a research procedure. It is contraindicated in children, in pregnancy and in patients with renal failure.

INTESTINAL ABSORPTION

The xylose absorption test is the most frequently performed carbohydrate test of the absorptive function of the intestine; other tests include the oral glucose tolerance test and tests related to intestinal disaccharidase activity. Chemical tests of protein absorption are performed much less frequently. The mainstay of chemical tests of fat absorption continues to be faecal fat estimation.

Carbohydrate absorption

XYLOSE ABSORPTION TEST

D-xylose, a pentose, is normally absorbed rapidly from the small intestine and excreted in the urine. Xylose is absorbed by a different mechanism from most dietary carbohydrates. It is partly metabolized in the body, but it can nevertheless be used satisfactorily to test the intestine's ability to absorb monosaccharides. A 5 g dose is now widely accepted as adequate for testing intestinal absorptive capacity. Urinary excretion of xylose is usually measured.

An oral dose of D-xylose (5 g) is given to the patient after fasting overnight. The bladder is emptied immediately afterwards, this first specimen of urine being discarded. Urine passed during the next 5 h is collected. An adequate flow of urine is ensured by giving at least 500 mL water during the early part of the test. Normally, in healthy individuals, more than 2 g xylose is excreted in 5 h.

Some investigators attempt to refine the test by subdividing the urine collection into two periods, 0–2 h and 2–5 h. The xylose content of these specimens is determined and a ratio, which should be greater than one, calculated from the two results. The inaccuracy of urine collections made over such short periods casts doubt on the usefulness of this modification.

Impaired absorption and excretion of xylose is often observed in patients with disease of the small intestine. Low values may also be found when there is bacterial colonization of the small intestine (since bacteria may metabolize xylose) and in renal disease due to impaired excretion of xylose. If renal function is impaired, an alternative test based on blood xylose measurements can be used.

In patients over 50 years, xylose excretion may be low in the absence of intestinal malabsorption. These abnormal results in the xylose absorption test are due to the progressive loss of renal function that tends to occur with increasing age.

The test can be used to monitor the response to therapy, e.g. the response of patients with coeliac disease to a gluten-free diet.

ORAL GLUCOSE TOLERANCE TEST

This test may help to differentiate pancreatic from other causes of malabsorption. Malabsorption caused by chronic pancreatitis is often found to have a 'diabetic' type of response (p. 225) whereas a 'flat' response is more common in other forms of malabsorption.

DISACCHARIDE TOLERANCE TESTS

Disaccharidase deficiency may be exhibited as intolerance to one or more of the disaccharides, lactose, maltose or sucrose. The defect may be congenital or acquired. These tests are performed to determine whether there is impairment of absorption of ingested disaccharides, and to help define whether any impairment is due to intestinal disaccharidase deficiency (Table 10.2).

The patient fasts overnight, and a specimen of blood is collected for measurement of plasma [glucose] before the test is started. An oral dose of 50 g disaccharide (usually lactose or sucrose) is then given, and plasma [glucose] measured at 30 min intervals for the next 2 h.

In healthy individuals, plasma [glucose] should increase by at least 1.1 mmol/L (20 mg/100mL). In disaccharidase deficiency, the rise is

TABLE 10.2. Intestinal disaccharidases.

Enzyme	Substrate	Products
Maltase	Maltose	Glucose
Sucrase	Sucrose	Glucose and fructose
Lactase	Lactose	Glucose and galactose
Isomaltase	(1) α-limit dextrins	Short chain polymers of glucose, plus maltose and glucose
	(2) Isomaltose	Maltose and glucose

usually less than 1.1 mmol/L. To eliminate the possibility that generalized mucosal disease is present, the test should be repeated using a mixture containing 25 g of each of the monosaccharides that together make up the relevant disaccharide (Table 10.2).

In *disaccharide intolerance* measurement of faecal $[H^+]$ may assist in diagnosis since the pH of faeces may be less than 5.5 due to bacterial metabolism of unabsorbed carbohydrate. Also, the presence of reducing substances in the faeces can be detected by side-room tests normally used for testing urine specimens (p. 32). *Disaccharidase activity* can be measured in specimens of small intestinal mucosa obtained by peroral biopsy; this is the most reliable way of specifically diagnosing small intestinal disaccharidase deficiency.

Protein absorption

Intestinal protein loss is usually detected by the parenteral administration of radioisotopically labelled macromolecules such as $[^{125}I]$- or $[^{131}I]$-labelled proteins, or labelled polyvinylpyrrolidone; faecal radioactive iodine is then measured. Alternatively, faecal radioactivity is measured following *in vivo* labelling of plasma proteins with ^{51}Cr. Faecal nitrogen excretion is very rarely measured.

Amino acid transport

Certain specific disorders affect both intestinal and renal epithelial transport. In Hartnup disease (p. 98) there is impaired transport of neutral amino acids, and deficiency of some essential amino acids (especially trytophan) may occur. In cystinuria (p. 98) the basic amino acids and cystine are affected; however, there is no associated nutritional defect despite the fact that lysine is an essential amino acid. These disorders are investigated by examining the pattern of amino acids excreted in the urine.

Fat absorption

FAECAL FAT

In the simplest form of the test, all faecal specimens passed over a 5-day period are collected and sent to the laboratory. The specimens are combined and total fat content estimated. No attempt is made to subdivide the measurement into unhydrolysed and hydrolysed fat. Two important drawbacks of the tests are (1) the difficulty, even under the best conditions, of collecting complete 5-day specimens, and (2) the inherently unpleasant nature of the test.

Before starting this test, the patient should have been on a normal mixed diet containing between 50 and 100 g fat/day for at least 3 days. Provided fat intake does not exceed 100 g/day, the excretion of fat by healthy individuals is up to 18 mmol/24 h (up to 5 g/24 h).

To improve the accuracy of faecal collections, patients should ideally be admitted to a metabolic ward and all faeces collected between the time of appearance in the faeces of non-absorbable dye markers such as carmine. At the start of the test, the first dose of marker is given and a second dose of marker 5 days later; the first specimen in which the marker appears is included in the collection.

Faecal fat excretion may be increased in patients with pancreatic disease and in patients with intestinal malabsorption. Sometimes the output is very high (e.g. 100–150 mmol/24 h, 25–40 g/24 h). In pancreatic disease, faecal fat excretion is only increased when pancreatic function has fallen to less than 10% of normal.

Tests based on the separate estimation of unhydrolysed triglyceride (unsplit fat) and hydrolysed triglyceride (split fat) in the faeces have proved unreliable as a means of differentiating between pancreatic and non-pancreatic causes of fat malabsorption. This is because intestinal bacteria can also hydrolyse triglycerides.

Triglyceride 'breath test'

This is one example of tests that have been devised so as to overcome the difficulties and unpleasantness of collecting faeces over several days. Following digestion and absorption of an oral dose of [^{14}C]-triglyceride (the marker is in the fatty acid component), part of the fatty acid is metabolized to $^{14}CO_2$ which is then excreted in expired air. A high $^{14}CO_2$ excretion is associated with normal fat absorption, whereas $^{14}CO_2$ excretion is low in patients with fat malabsorption. This test is convenient and rapid but, because it is technically difficult, it has not gained wide acceptance.

Bile acids and lipid metabolism

Primary bile acids are formed in the liver, conjugated with glycine or taurine, and secreted into bile. Together with the phospholipid lecithin, they play an important part in preventing cholesterol from precipitating out in the bile ducts and gallbladder. In the small intestine, they are essential for the normal processes of fat digestion and absorption, although some absorption does occur even in their absence. Bile acids are mostly reabsorbed in the terminal ileum by a specific active process, transported to the liver and re-excreted into the bile, thus completing an enterohepatic circulation.

During absorption of a meal, the bile acids must be present in the upper small intestine in concentrations sufficient to allow the formation of micelles, to be effective in promoting the action of pancreatic lipase and co-lipase, and to render water-soluble the products of lipolysis. Insufficient concentrations of bile acids give rise to malabsorption of fat, and this can occur in the following circumstances:

(1) Insufficient synthesis of bile acids in the liver, e.g. hepatic cirrhosis.
(2) Obstruction to the outflow of bile, e.g. gallstones.
(3) Interruption of the enterohepatic circulation:
  (a) Failure of bile acid absorption from the intestine e.g. ileal disease or resection.
  (b) Failure to clear bile acids from portal blood and secrete them rapidly into bile, e.g. cholestasis associated with cirrhosis.
(4) Abnormal bacterial colonization of the upper small intestine. Some colonizing bacteria split the bile acid conjugates, reducing their effective concentration at the site of fat absorption. This is a functional abnormality called the 'stagnant gut' syndrome.

Chemical tests of bile acid metabolism are designed to differentiate between conditions causing reduced [bile acids] in the lumen of the upper jejunum and conditions in which bile acids are functionally abnormal. These tests are not yet widely available on a routine basis.

Bile acids in jejunal juice

These may be measured after giving the fasting patient a meal containing fat. Specimens are obtained at 15 min intervals for 2–5 h. In normal subjects jejunal [bile acids] reaches a minimum of 5 mmol/L in at least one sample. This concentration is not attained in patients with malabsorption of fat due to bile acid deficiency.

Unconjugated bile acids are normally absent from jejunal juice but may be detected in juice collected from patients with bacterial colonization of the upper small intestine.

BACTERIAL COLONIZATION OF THE SMALL INTESTINE

The small intestine is usually virtually sterile. However, when there is stasis (e.g. blind loop or stricture) or a fistula from the colon or, occasionally, when immune mechanisms are impaired, anaerobic bacteria colonize the small intestine. When this occurs, it often causes malabsorption of fat, due at least partly to deconjugation of bile acid conjugates by the bacteria. Vitamin B_{12} deficiency may also develop, due to consumption of the vitamin by the bacteria.

Diagnosis of bacterial colonization is essentially based on microbiological procedures. However, several non-invasive chemical tests have been devised as a means of suggesting the presence of bacterial colonization.

Bile acid 'breath test'

Radioactive glycocholic acid is given to the patient in the fasting state; the glycine portion of the molecule contains radioactive carbon. Normally, most of the glycocholic acid conjugate is absorbed. Any $[^{14}C]$-glycine that is split off is either metabolized by the intestinal bacteria with the formation of $^{14}CO_2$ or, following absorption, metabolized similarly by the patient. Samples of end-tidal expired air are collected for $^{14}CO_2$ analysis, at hourly intervals for 6 h after taking the glycocholic acid. Normally, there is very little $^{14}CO_2$ in the first three samples.

When abnormal numbers of bacteria are present in the small intestine ('stagnant gut' syndrome), large amounts of $^{14}CO_2$ are detected in the first or second breath samples, or both, following administration of radioactive glycocholic acid. Bacterial splitting of bile acid conjugates is also increased in patients who fail to absorb them normally from the terminal ileum. In these patients, the radioactive glycocholic acid passes into the colon where large numbers of bacteria are normally present; increased amounts of $^{14}CO_2$ may sometimes be found in the early breath samples in these patients also.

To distinguish between abnormal ileal function and 'stagnant gut' syndrome as the cause of these findings, faeces are collected for 24 h after the dose of glycocholic acid and faecal radioactivity measured. This is normal in patients with 'stagnant gut' syndrome, but increased if ileal disease has caused failure of bile acid reabsorption.

Other tests

A test using [^{14}C]-xylose in place of [^{14}C]-glycocholic acid has been claimed to be more sensitive and specific. The anaerobic bacteria metabolize the xylose with the production of $^{14}CO_2$. This appears in the breath earlier, and in larger amount, than $^{14}CO_2$ formed as a result of the normal processes of xylose absorption and partial metabolism in the liver.

Urinary indican is increased in patients with bacterial colonization of the small intestine. The bacteria in the intestine convert dietary tryptophan into indican; this is absorbed and then excreted by the kidney.

Absorption of other dietary constituents

VITAMINS

Most water-soluble vitamins are readily absorbed, mainly in the upper small intestine. Malabsorption as a cause of deficiency of the water-soluble vitamins is uncommon unless there is moderately severe general disease of the intestinal mucosa. Vitamin B_{12} is the main exception, its absorption requiring both the secretion of intrinsic factor by gastric parietal cells, and specific ileal receptors for attachment of the complex between intrinsic factor and vitamin B_{12}.

Absorption of vitamin B_{12} may be assessed by the *Schilling test.* After administering radioactive vitamin B_{12} by mouth, urinary radioactivity is measured. The test depends on the fact that the radioactive vitamin B_{12} which has been absorbed will be reversibly bound to receptors in the body, and the radioactive vitamin B_{12} can then be largely released from these receptors by giving a large intramuscular 'flushing' dose of non-radioactive vitamin B_{12}. The radioactive vitamin B_{12} displaced from receptors in this way is then excreted in urine. If defective absorption is demonstrated, the test can be repeated but with intrinsic factor being given at the same time as the radioactive vitamin B_{12}. Impaired absorption may be observed in patients with (1) intrinsic factor deficiency due to pernicious anaemia or gastrectomy, (2) bacterial colonization of the small intestine (the 'stagnant gut' syndrome) or (3) ileal disease or following ileal resection.

Fat-soluble vitamins (A, D, E and K) share absorption mechanisms with other dietary lipids. Malabsorption of fat-soluble vitamins, which is most commonly manifest as vitamin D deficiency, thus occurs in conditions causing fat malabsorption.

WATER AND INORGANIC CONSTITUENTS

About 8 L of intestinal secretions are produced each day (Table 4.4) and must be largely reabsorbed or deficiency states rapidly develop. Reabsorption takes place mainly in the jejunum and ileum, but also in the colon. Acute and severe disturbances may occur in patients following operations, especially operations on the gastrointestinal tract, and losses of K^+ often become severe.

Non-surgical intestinal causes of electrolyte imbalance include severe diarrhoea and cholera, in which there is a defect of Na^+ reabsorption in the jejunum. In the *Verner–Morrison syndrome,* there is severe watery diarrhoea and hypokalaemia, associated with hypersecretion of vasoactive intestinal peptide (VIP); in these patients, fasting plasma [VIP] is high, whereas it is normal or low in patients with diarrhoea due to other causes.

Absorption of calcium and magnesium is discussed in Chapter 14, and iron in Chapter 15. Deficiencies of these ions can occur because of intestinal disorders.

DISORDERS CAUSING INTESTINAL MALABSORPTION

By grouping intestinal malabsorptive disease into the following categories, the planning and interpretation of tests is simplified:

(1) *Pancreatic disease* may cause malabsorption of protein, fat or carbohydrate, due to deficiency of digestive enzymes.

(2) *Biliary disease* may cause malabsorption of fat and fat-soluble vitamins.

(3) *Disease of the intestinal mucosa* may affect digestion or transport, or both, of many dietary constituents, bile acids etc. The effects may be general, or relatively specific (e.g. cystinuria, Hartnup disease).

(4) *Bacterial colonization of the small intestine* may cause a functional deficiency of bile acids, interfere with absorption of vitamin B_{12}, fats and proteins, and possibly have other effects.

Excessive losses into the intestinal lumen may also produce effects similar to those of intestinal malabsorption. Examples include protein-losing gastroenteropathy (p. 116) and carcinoid syndrome (p. 454).

Chemical tests in diagnosis of malabsorption

Chemical investigations may help to establish that a state of malabsorption exists, and to allocate the malabsorption process into the appropriate category. In certain circumstances, specific chemical tests may be available to define the cause precisely. Usually, diagnosis depends on a combination of evidence.

The tests detailed in Table 10.3 are widely used to determine whether a state of malabsorption exists, and they help to place the cause into the appropriate category. Non-invasive chemical tests that may confirm or refine the diagnosis include the following:

(1) Zollinger–Ellison syndrome: plasma [gastrin].
(2) Chronic pancreatitis: BT PABA/^{14}C-PABA test.
(3) Disaccharidase deficiency: appropriate sugar tolerance tests (e.g. lactose, and a galactose-glucose mixture).
(4) Amino acid transport defects (e.g. Hartnup disease): urine chromatography.
(5) Verner–Morrison syndrome: plasma [VIP].
(6) 'Stagnant gut' syndrome: bile acid 'breath test'.

The special investigations listed above are all indirect or non-invasive. Examples of direct methods of chemical investigation include:

(1) Disaccharidase deficiency: enzyme activity in intestinal biopsy specimens.
(2) Pancreatic disease: secretin/CCK–PZ test (Table 10.4).

Several other chemical abnormalities may occur in association with intestinal malabsorption. These may also require investigation and appropriate treatment. Examples of these other abnormalities include:

Defects in calcium absorption. Rickets or osteomalacia may be

TABLE 10.3. Initial chemical investigations in intestinal malabsorption.

Cause	Faecal fat excretion	Xylose absorption	Vitamin B_{12} absorption
Chronic pancreatitis	↑	N	N
Carcinoma of pancreas	↑ or N	N	N
Biliary obstruction	↑	N	Not indicated
Gluten enteropathy	↑	↓	↓
Small bowel resection			
(1) Jejunum	N	N or ↓	N
(2) Terminal ileum	↑	N	↓
'Stagnant gut' syndrome	↑	N or ↓	↓

N = within the range of reference values
↑ = increased
↓ = decreased

TABLE 10.4. Secretin/CCK–PZ test.

	Chronic pancreatitis	Pancreatic carcinoma*
Volume	N or ↓	↓ or ↓ ↓
HCO_3^- output	↓	N or ↓
Enzyme activity	↓	N or ↓
[Lactoferrin]	↑	N

*Carcinoma of the tail of the pancreas may give normal results. Carcinoma of the head of the pancreas may obstruct the common bile duct, and the secretin/CCK–PZ test is not indicated under these circumstances.

present. Plasma [calcium] and [phosphate] may both be reduced and alkaline phosphatase activity increased.

Malabsorption of iron occasionally leads to haematological disorders. Mixed deficiencies of vitamin B_{12}, folate and iron may occur. Investigations include haematological indices, measurement of plasma [iron] and testing faeces for occult blood.

Malabsorption of vitamin K may occur and give rise to coagulation defects. The prothrombin time may be prolonged.

Malabsorption of protein. Reduction in plasma [albumin] most often results, but hypogammaglobulinaemia may be a prominent feature.

Malabsorption giving rise to malnutrition may have other effects, especially the adverse effects of malnutrition on the liver.

FURTHER READING

ADRIAN, T.E. and BLOOM, S.R. (1982). Gut hormones and related neuropeptides. In *Recent Advances in Endocrinology and Metabolism,* **2,** pp. 17–46. Ed. J.L.H. O'Riordan. Edinburgh: Churchill Livingstone.

BLOOM, S.R. and POLAK, J.M. (1981). *Gut Hormones,* 2nd Edition. Edinburgh: Churchill Livingstone.

BRAGANZA, J.M. (1982). Does your patient have pancreatic disease? *Journal of the Royal College of Physicians of London,* **16,** 13–22.

DAWSON, A.M., editor (1970). Intestinal absorption and its derangements. *Journal of Clinical Pathology,* **24,** Supplement (Royal College of Pathologists), **5.**

GOWENLOCK, A.H. (1977). Tests of exocrine pancreatic function. *Annals of Clinical Biochemistry,* **14,** 61–89.

LANKISCH, P.G. (1982). Exocrine pancreatic function tests. *Gut,* **23,** 777–798.

Leading article (1982). Diagnosis of chronic pancreatitis. *Lancet,* **1,** 719–720.

POLAK, J.M. and BLOOM, S.R. (1983). Regulatory peptides: key factors in the control of bodily functions. *British Medical Journal,* **286,** 1461–1466.

WEST, P.G., LEVIN, G.E., GRIFFIN, G.E. and MAXWELL, J.D. (1981). Comparison of simple screening tests for fat malabsorption. *British Medical Journal,* **282,** 1501–1504.

WILSON, F.A. and DIETSCHY, J.M. (1971). Differential diagnostic approach to clinical problems of malabsorption. *Gastroenterology,* **61,** 911–930.

Chapter 11
Disorders of Carbohydrate Metabolism

The principal sources of carbohydrate in the diet are the polysaccharides, starch and glycogen, which are based on glucose units linked by α-glucosidic links, and the disaccharides sucrose and lactose. These carbohydrates are a major source of energy intake. Digestion and absorption of carbohydrates are discussed in Chapter 10. After absorption, the constituent monosaccharides (Table 10.2) are transported to the liver.

The liver plays a key role in carbohydrate metabolism and in maintaining plasma [glucose]. Glycogen is stored mainly in the liver and gluconeogenesis takes place there. It is also the principal site for conversion of fructose and galactose into glucose, via the formation of intermediates in the glycolytic pathway. Inborn errors of carbohydrate metabolism affecting glycogen storage and monosaccharide interconversions are discussed in this chapter.

The hormones mainly concerned with glucose metabolism and with plasma glucose homeostasis are insulin, glucagon, growth hormone, adrenaline and cortisol. Of these, insulin has the most marked effect in man, and is the only hormone with a plasma glucose-lowering effect. The other hormones all have actions that tend, in general, to antagonize insulin. The organ mainly affected by the actions of these hormones on carbohydrate metabolism is the liver, but muscle, kidney and adipose tissue are also important target sites for hormonal action.

INSULIN

This is a peptide hormone synthesized by the β-cells of the pancreas. Initially, a single polypeptide chain is formed, *pro-insulin,* having the following structure:

A chain–connecting peptide–B chain

The pro-insulin molecule folds and the A and B chains become linked by disulphide bridges. Subsequently, the active hormone insulin (which consists of the A and B chains, linked by disulphide bridges) is formed when the connecting peptide (C-peptide) is split off.

The granules in the β-cells of the islets contain insulin and C-peptide and both are released, in equimolar amounts, when insulin secretion occurs in response to hyperglycaemia. Since C-peptide has a longer half-life in plasma than insulin, and is much more stable than insulin in blood specimens after collection, plasma [C-peptide] is sometimes measured as an index of endogenous insulin secretion. Furthermore, plasma [C-peptide] is a valid index of continuing endogenous insulin secretion even when exogenous insulin is being given for treatment, since the various therapeutic insulin preparations contain little or no C-peptide.

The principal effects of insulin are (1) to decrease the output of glucose from the liver, due to inhibition of gluconeogenesis and glycogen breakdown, and (2) to increase the uptake of glucose by insulin-sensitive tissues such as liver and skeletal muscle. Synthesis of protein, glycogen and lipid are also stimulated. These effects on glucose uptake and on intermediary metabolism collectively tend to return the plasma [glucose] to the fasting level.

The effects of insulin on fat metabolism, decreasing the rate of FFA release from adipose tissue into plasma, are described later (p. 216).

GLUCOSE-DEPENDENT INSULINOTROPHIC PEPTIDE (GIP)

This is one of the gastrointestinal peptides (Table 10.1). It is produced by mucosal cells of the upper jejunum, and plasma [GIP] increases after meals containing carbohydrates or fats. The principal action of GIP is to stimulate the release of insulin from the pancreas, thereby enhancing the action of insulin. GIP is probably the most important factor in the entero–insular axis, the mechanism that accounts for the larger release of insulin that occurs in response to an oral glucose load, as compared with the same dose of glucose given intravenously.

The earlier name for GIP was gastric inhibitory peptide (which gave rise to the same acronym). However, the ability of GIP to inhibit gastric secretion is only demonstrable at non-physiological dose levels. It also stimulates secretion of small intestinal juice, but it is uncertain whether this further action of GIP is of physiological importance.

GLUCAGON

This is a polypeptide synthesized by the cells of the pancreatic islets. It is secreted in response to (1) hypoglycaemia, (2) increased plasma [amino acids], (3) some of the gastrointestinal peptides (e.g. GIP in non-physiological dosage) and (4) catecholamines released locally, or from the adrenal medulla.

Glucagon in pharmacological doses stimulates the breakdown of liver glycogen by activating phosphorylase kinase, and thereby increases the breakdown of protein and fat. There is, however, little evidence to suggest that glucagon is of much importance in man.

GROWTH HORMONE (GH)

This anterior pituitary hormone (p. 389) inhibits glucose uptake by the tissues, and inhibits the synthesis of fat from carbohydrate. It causes the release of FFA from adipose tissue.

Plasma [GH] normally varies inversely with plasma [glucose].

ADRENALINE

This catecholamine activates phosphorylase kinase in liver and phosphorylase *b* kinase in muscle, thereby promoting the breakdown of glycogen to glucose. It also increases the breakdown of adipose tissue triglycerides.

Adrenaline release from the adrenal medulla tends to produce increased plasma [glucose] and [FFA].

CORTISOL

This is the principal glucocorticoid produced by the adrenal cortex in man (p. 354). Its actions include the stimulation of gluconeogenesis and the inhibition of glucose metabolism in peripheral tissues. Plasma [glucose] increases as a result.

Plasma glucose homeostasis

Plasma [glucose] is normally maintained within fairly narrow limits. After a carbohydrate-containing meal, glucose is transported in the portal blood to the liver, which takes up about 60% of the glucose load. The immediate effects of a rise in plasma [glucose] are:

(1) Increased entry of glucose into the liver and brain.
(2) Release of insulin.

These effects tend to return plasma [glucose] to the fasting level.

After a period of fasting, plasma [glucose] falls sufficiently to inhibit insulin release and to stimulate production of glucagon. These and other subsidiary hormonal effects stimulate the liver and kidney to produce glucose and stimulate the release of FFA from adipose tissue.

Liver and kidney are the only tissues that add significant amounts of glucose to the blood in response to hypoglycaemia; most tissues remove glucose. This is due to the presence of glucose-6-phosphatase in liver and kidney, but not in other tissues.

The regulatory changes that occur in glucose homeostasis also affect indirectly the metabolism of other substances, the most important of these being triglycerides and FFA, 'ketone bodies' and lactate.

TRIGLYCERIDES AND FFA

In adipose tissue, triglyceride formation and breakdown both go on continuously, there being a fine balance between the two processes. Adipose tissue is unable to re-use glycerol formed as a result of hydrolysis of triglycerides; instead, it requires newly synthesized glycerol phosphate. Glucose is the source of the substrate for the formation of glycerol phosphate, which is needed for the re-esterification of FFA in the formation of triglycerides.

Most of the FFA in plasma are derived from adipose tissue triglycerides, only a small contribution coming from triglycerides hydrolyzed by lipoprotein lipase (p. 247) but not taken up locally. The rate of release of fatty acids from triglycerides in adipose tissue is determined by the activity of *mobilizing lipase* (or hormone-sensitive lipase). This enzyme is activated by the cellular adenyl cyclase system and is, therefore, enhanced by catecholamines, growth hormone and glucocorticoids. Catecholamine-induced release of FFA occurs during muscular exercise and other forms of stress. On the other hand, glucose and insulin both inhibit FFA release. Lack of circulating insulin leads to (1) failure of entry of glucose into adipose tissue and hence a shortage of glycerol phosphate, and (2) stimulation of mobilizing lipase. As a result of these effects of insulin lack, the amount of FFA in adipose tissue and in plasma is increased.

Fatty acids in plasma are transported bound to albumin. There is rapid turnover of plasma FFA; they have a biological half-life of 2–3 min. Much of the FFA is oxidized in tissues and used as a source of energy, yielding up to 50% of the body's energy requirements. However, a substantial proportion of plasma FFA is taken up by the liver. Hepatic uptake of FFA is directly related to plasma [FFA]. In the liver, FFA may be:

(1) used for resynthesis of triglyceride, incorporated into VLDL and secreted into plasma. The triglycerides are then transported back to adipose tissue.

(2) stored in the liver, as triglyceride.
(3) partly oxidized, to 'ketone bodies' (see below).
(4) oxidized completely, to CO_2 and water.

'KETONE BODIES'

Fatty acids are metabolized in the liver to their coenzyme A derivatives (acyl CoA). These are subsequently oxidized, in the mitochondria, to acetyl CoA. In diabetic ketoacidosis, acetyl CoA is largely diverted to form 'ketone bodies'. Two molecules of acetyl CoA condense to form acetoacetyl CoA, from which acetoacetate is derived by hydrolysis.

Acetoacetate may be reduced to 3-hydroxybutyrate (β-hydroxybutyrate) or decarboxylated with the formation of acetone and CO_2. Acetoacetate, 3-hydroxybutyrate and acetone are collectively described as 'ketone bodies' although 3-hydroxybutyrate is not in fact a ketone. They are most commonly found in the bloodstream in excessive amounts in uncontrolled diabetes and in starvation.

The ratio between 3-hydroxybutyrate and acetoacetate varies. When there is a relative excess of NADH (e.g. when there is increased formation of lactate, see below), 3-hydroxybutyrate predominates. This may be of practical importance as side-room methods for ketone bodies detect acetoacetate but not 3-hydroxybutyrate.

LACTATE

This is a normal end-product of tissue glycolysis. Lactate formation occurs when there is insufficient oxygen to allow complete oxidation of carbohydrate to CO_2 and water. Poor tissue perfusion, or a sudden increase in tissue metabolism of the type seen in intensive muscular exercise, is associated with an increased rate of formation of lactate. This is released into the bloodstream, carried to the liver and used for gluconeogenesis or converted to other metabolites.

In liver, and other tissues, the interconversion of lactate and pyruvate is catalyzed by lactate dehydrogenase:

$$\text{Lactate} + \text{NAD}^+ \rightleftharpoons \text{Pyruvate} + \text{NADH} + \text{H}^+$$

The normal tissue [lactate] : [pyruvate] ratio is about 10 : 1. In anoxia this ratio rises, since tissue [NADH] rises and the equilibrium position shifts to the left. The plasma [lactate] : [pyruvate] ratio is similar to the ratio in the tissues. Increase in plasma [lactate] occurs during severe exercise as a consequence of the rapid rate of glycolysis. Under certain circumstances the plasma [lactate] will be increased in association with an increased blood [H^+]; in these cases the pathological state of lactic acidosis is considered to be present.

Lactic acidosis

Lactate itself is not an acid, and addition of lactate anions to blood does not cause acidosis. The cause of the acidosis associated with increased plasma [lactate], and of the condition termed lactic acidosis, is the net overproduction of H^+ that occurs when cellular energy requirements are met by anaerobic metabolism. There is no net overproduction of H^+ when glucose is metabolized aerobically to CO_2 and water.

In lactic acidosis the patient is ill and often has clinical signs of poor tissue perfusion or arterial oxygen desaturation. It may occur in septicaemic shock, in liver disease due to the inability of the liver to metabolize lactate normally, in diabetes mellitus and in uraemia. Acute lactic acidosis carries a poor prognosis, presumably because of the severity of the primary disease. Chronic lactic acidosis sometimes occurs in the glycogen storage diseases, and rarely in some other conditions.

Glucose measurements in blood and plasma

Plasma [glucose] is usually measured by enzymic methods that employ glucose oxidase or hexokinase, both of which are enzymes with a high degree of specificity for glucose. However, some laboratories still use reductive methods, which depend on the fact that glucose is the reducing substance normally present in largest quantity in blood.

Reductive methods are not specific for glucose. They measure other reducing sugars (e.g. galactose) and certain other reducing substances (e.g. creatinine) present in blood. The results of glucose analyses performed by reductive methods tend, therefore, to be higher than results obtained by enzymic techniques. The extent of the difference depends upon the choice of reductive method. With some of the earlier methods, differences were often as much as 1.0 mmol/L (approximately 20 mg/100 mL). With improved reductive methods, however, results are on average usually only 0.3–0.6 mmol/L higher than results for enzymic measurements. Although these differences may not seem large, they assume considerable clinical significance where the glucose concentration falls below 2.5 mmol/L (45 mg/100mL). Both enzymic and reductive methods can be used satisfactorily for the diagnosis and control of diabetes mellitus, but enzymic methods are essential for the investigation and diagnosis of hypoglycaemia.

In this book the term plasma [glucose] is used when the measurement depends on a method specific for glucose, whereas plasma [sugar] or blood [sugar], depending on the nature of the specimen analysed, is used when referring to a method that measures reducing substances.

BLOOD OR PLASMA GLUCOSE?

Many laboratories now measure [glucose] in plasma although some still use whole blood. Plasma is to be preferred, as the measurements yield more reliable results.

The red cell membrane is freely permeable to glucose. However, plasma [glucose] tends to be slightly higher than whole blood [glucose] because red cells contain less 'free water' than plasma; glucose is dissolved in the free water. It is not possible to quote a conversion factor that can be used satisfactorily to calculate plasma [glucose] from blood [glucose] measurements, and *vice versa.*

CAPILLARY OR VENOUS BLOOD?

At normal plasma [glucose], there is little difference between results obtained on capillary and venous blood. However, at hyperglycaemic levels, capillary plasma [glucose] may be significantly higher than venous plasma [glucose]. This is important in the interpretation of glucose tolerance tests.

IS PRESERVATION OF BLOOD SPECIMENS NEEDED?

The interpretation of glucose measurements depends both on the nature of the blood specimen and on the method of assay. Whatever procedure is adopted, it is essential to preserve the glucose in the specimen following collection and prior to analysis, unless the analysis is to be performed almost immediately.

Sodium fluoride, to inhibit glycolysis, and potassium oxalate (to act as anticoagulant) have usually been added to the tubes when collecting specimens for laboratory analysis. The fluoride stabilizes the [glucose] for several hours, and allows specimens to be sent considerable distances to a central laboratory, or to be collected by a general practitioner at evening surgery for analysis the following day.

Analyses that are to be performed using one of the side-room methods, i.e. using enzyme-impregnated 'sticks', need to be carried out immediately on blood specimens that do not contain preservative. Lithium heparin is a satisfactory anticoagulant for these specimens.

SIDE-ROOM MEASUREMENTS

Methods based on the glucose oxidase reaction are available for quantitative determination of blood [glucose] (p. 34). Some of these use small hand-held meters, and others depend on visual comparison between the colour developed by applying blood to the enzyme-impregnated 'stick' and a series of colours on a concentration-based colour chart. Many diabetic patients now use these methods in their homes, for controlling insulin dosage on the basis of blood [glucose].

DIABETES MELLITUS

This may be defined as a state of chronic hyperglycaemia, usually accompanied by glycosuria. It is due to a metabolic disorder in which the effectiveness of insulin in, or the delivery of insulin to, the tissues is reduced. This defect causes biochemical abnormalities that are expressed as a wide range of clinical presentations, ranging from asymptomatic patients with relatively mild biochemical abnormalities to patients admitted to hospital with severe metabolic decompensation of rapid onset that has led to coma.

The diagnosis of diabetes mellitus may first be suggested by the results of side-room tests for glucose carried out, for instance, as part of a full clinical examination of a patient, or as one component of an assessment for life insurance or of a health-screening programme. Sensitive, qualitative urine-testing strips specific for glucose include those marketed by Ames Co. (e.g. Clinistix) and the Boehringer Corporation (e.g. BM-Test Glucose). Semi-quantitative tests for reducing substances in urine include Clinitest (Ames Co.).

Clinically, patients with diabetes mellitus may be classified as having idiopathic (primary or essential) or secondary diabetes. There are also other groups of patients in whom abnormalities of glucose tolerance may be detected, but who do not fit into either of these categories. Oral glucose tolerance tests play an important part in the recognition and categorization of patients.

Idiopathic diabetes

This may be caused by (1) factors in the blood which antagonize or inhibit the action of insulin, (2) production of an abnormal form of insulin, or (3) inability of the pancreas to produce sufficient insulin from the earliest stages of the disorder. It is divided into Types 1 and 2.

Insulin-dependent diabetes (Type 1)

This usually presents acutely in young, non-obese subjects, but it can occur at any age. In general, insulin is required for treatment and ketosis is liable to occur.

Islet cell antibodies that react specifically with the β-cells have been demonstrated in over 90% of patients with newly diagnosed Type 1 diabetes. Over the next few years, these antibodies disappear from the serum of most Type 1 diabetics.

Islet cell antibodies have been demonstrated in the serum of some patients several years before the clinical and chemical features of diabetes develop. Individuals with certain human leucocyte antigens (HLA) have also been shown to carry a particularly high risk of developing Type 1 diabetes.

The importance of these immunological findings lies, at present, in their application to relatives of diabetic patients. If a sibling has islet cell antibodies, or HLA characteristics that are identical with the patient's, the sibling is potentially diabetic, i.e. has an increased probability of developing impaired glucose tolerance or frank diabetes mellitus.

Non-insulin-dependent diabetes (Type 2)

This Type may be further subdivided into (a) non-obese and (b) obese categories. It usually presents less acutely than Type 1, and in older (over 40 years old) obese subjects; it used to be called 'maturity onset' diabetes. Rarely, Type 2 diabetes is recognized in young patients. In general, insulin is not required to prevent ketosis as these patients are relatively resistant to the development of ketosis, but it may be needed for correction of abnormalities of blood [glucose].

There appears to be no association between Type 2 diabetes and either the HLA system or the development of autoimmunity. However, there is a strong genetic element to the disorder. For instance, if one identical twin develops Type 2 diabetes, there is a distinct probability that the other twin will develop it.

Secondary diabetes

This occurs as a consequence of other diseases, either pancreatic or endocrine. With pancreatic diabetes, the secretion of insulin is reduced due to pancreatitis, haemochromatosis or resection of the pancreas. In diabetes secondary to other endocrine disorders, ineffective insulin action is caused by abnormal secretion of hormones with 'diabetogenic' activity; these forms of secondary diabetes occur in acromegaly, phaeochromocytoma, Cushing's syndrome, and occasionally in thyrotoxicosis.

Secondary diabetes may also be caused by a number of drugs that adversely affect carbohydrate metabolism, giving rise first to impaired glucose tolerance and later, in some cases, frank diabetes. These include thiazide diuretics and β-blockers, phenytoin, oestrogen-containing oral contraceptives and corticosteroid therapy.

Several genetic syndromes are rare causes of secondary diabetes. Down's syndrome, Turner's syndrome, Type I glycogen storage disease, and insulin receptor abnormalities are examples of these.

Other abnormalities of glucose tolerance

There are several other categories of patients in whom abnormalities of glucose tolerance may be detected, sometimes only of a temporary nature, or who are able to be recognized as having an increased risk of developing diabetes mellitus. These categories are:

(1) Impaired glucose tolerance.
(2) Gestational diabetes mellitus.
(3) Previous abnormality of glucose tolerance.
(4) Potential abnormality of glucose tolerance.

The recognition of individuals that fit into these various categories depends considerably on the results of oral glucose tolerance tests, and detailed descriptions of these categories will be given later.

Glucose tolerance tests

These measure changes in plasma or blood [glucose] after an oral or an intravenous glucose load. Their main value lies in establishing the diagnosis of diabetes mellitus or impaired glucose tolerance at a stage when the metabolic abnormality is mild.

Following an oral dose of glucose, the plasma [glucose] alters depending on (1) the rate of absorption, (2) the volume of distribution and (3) the rate at which glucose leaves the blood. The last factor mainly depends on the action of insulin.

Intravenous glucose tolerance tests have been used in the past, but are very rarely used nowadays. By eliminating effects due to the variable rate of absorption of glucose, the intravenous test becomes more dependent on insulin action but it has two disadvantages. It is unphysiological, as the insulin response to a rising plasma [glucose] depends partly on the [glucose] in the splanchnic bed, mediated through GIP release. Secondly, thrombophlebitis may be caused by injecting solutions containing high [glucose].

Other glucose tolerance tests now considered obsolete include the cortisone glucose tolerance test and similar provocative tests (e.g. prednisone, prednisolone).

ORAL GLUCOSE TOLERANCE TEST (OGTT)

It is important to recognize that oral glucose tolerance tests are *not* required in every case when the diagnosis of diabetes mellitus is suspected.

If a patient has symptoms of diabetes, the finding of a random plasma [glucose] greater than 11 mmol/L or a fasting plasma [glucose] greater than 8 mmol/L *confirms the diagnosis, and a glucose tolerance test is unnecessary.* Also, according to the WHO Expert Committee on Diabetes Mellitus, a random plasma [glucose] less than 8 mmol/L, or a fasting plasma [glucose] less than 6 mmol/L excludes the diagnosis in these patients.

In all patients, whether or not they have symptoms of diabetes, if the measurements described in the previous paragraph give equivocal results, an oral glucose tolerance test should be performed. Various precautions are necessary in preparing for, and in performing, the test.

Before the test

The patient should have been on an unrestricted diet containing at least 150 g carbohydrate/day for at least 3 days, and should not have indulged in unaccustomed amounts of exercise.

Drugs such as corticosteroids and diuretics may impair glucose tolerance; if the patient has been receiving them, they should be stopped before the test, if possible.

The test should not be performed on patients who are suffering from the effects of trauma or who are recovering from a serious illness. It should also be delayed if the patient has an intercurrent infection.

The patient must not smoke or eat either before or during the test, nor drink anything other than as specified below.

Performing the test

The test is usually performed after an overnight fast, although a fast of 4–5 h may be sufficient. The patient is allowed to drink water during the fast, and may have a cup of unsweetened tea before the test; this helps to reduce any tendency to nausea that might otherwise be caused by the oral glucose drink.

A standard dose of 75 g anhydrous glucose dissolved in 250–350 mL of water, lemon-flavoured or chilled (or both) to avoid nausea, is given by mouth. Smaller amounts of anhydrous glucose (1.75 g/kg body weight) should be given to children, to a maximum dose of 75 g.

During the test the patient should be sitting up or, if lying down, should be lying over on the right side so as to ensure rapid emptying of the stomach. The patient must not lie over on the left side.

Blood specimens are collected before giving the glucose load, and thereafter at 30 min intervals for 2 h. In cases of suspected reactive hypoglycaemia, it may be advisable to prolong the blood collections until 6 h after the dose of glucose.

Urine specimens should be collected before the test and at 1 and 2 h, and tested for glucose and reducing substances in the side-room. Although the results for these urine measurements are not included among the diagnostic criteria for diabetes mellitus, urine needs to be tested if patients with renal glycosuria are to be recognized.

DIAGNOSTIC CRITERIA FOR DIABETICS, BASED ON OGTT RESULTS

Patients with symptoms can be diagnosed using the criteria summarized in Table 11.1. The data all depend on measurements of [glucose] made by a specific enzymic method. If measurements are made using a reductive method rather than an enzymic method, all results may be up to 1.1 mmol/L (20 mg/100 mL) higher than the values given in Table 11.1.

Patients without symptoms can be diagnosed as having diabetes mellitus if, in addition to the criteria for patients with symptoms given in Table 11.1, they fulfil *one* of the following two requirements:

(A) At least one of the intermediate (30, 60 or 90 min) blood specimens has a [glucose] greater than the level specified in Table 11.1.

(B) On repeat testing, either the fasting or the 2-h post-75 g glucose load plasma [glucose] is abnormal.

It should be emphasized that there is day-to-day variation in the results of oral glucose tolerance tests, especially in patients who have only a slight impairment of glucose tolerance. Also, it should be reiterated that the selection of the criteria for normality is somewhat arbitrary, and that those described here were not agreed internationally until 1980.

OTHER ABNORMALITIES OF GLUCOSE TOLERANCE

Four categories of individuals, all having an increased risk of developing diabetes mellitus, were listed above. In addition, there are certain other responses to an oral glucose load that need to be mentioned.

Impaired glucose tolerance

These are individuals who fulfil the diagnostic criteria summarized in

TABLE 11.1. Oral glucose tolerance test.

Oral load, 75 g anhydrous glucose. Analyses performed by an enzymic method specific for glucose. Data given for 3 different types of blood specimen (After the Second Report of the WHO Expert Committee on Diabetes Mellitus, 1980).

DIAGNOSTIC CRITERIA FOR PATIENTS WITH SYMPTOMS

Diabetes mellitus	*Fasting specimen*		*2 hours after 75 g glucose*
Venous plasma	≥8.0 mmol/L	*and/or*	≥11.0 mmol/L
Capillary blood	≥7.0 mmol/L	*and/or*	≥11.0 mmol/L
Venous blood	≥7.0 mmol/L	*and/or*	≥10.0 mmol/L
Impaired glucose tolerance			
Venous plasma	<8.0 mmol/L	*and*	≥8.0 but <11.0 mmol/L
Capillary blood	<7.0 mmol/L	*and*	≥8.0 but <11.0 mmol/L
Venous blood	<7.0 mmol/L	*and*	≥7.0 but <10.0 mmol/L

DIAGNOSTIC CRITERIA FOR PATIENTS WITHOUT SYMPTOMS

For a diagnosis of diabetes mellitus or impaired glucose tolerance to be made in asymptomatic patients, *in addition to the above criteria,* at least one of the intermediate (30, 60 or 90 min) specimens in the OGTT must have:

(1) a venous plasma or capillary blood [glucose] ≥11.0 mmol/L,
or
(2) a venous blood [glucose] ≥10.0 mmol/L.

Note

Conversion factor: 1.0 mmol glucose/L = 18 mg glucose/100 mL.

Table 11.1. Some of these go on to develop overt diabetes mellitus. Others do not, and in them the abnormal response to an oral glucose load may even revert to normal. It should again be noted that different criteria have to be met, depending on whether the patient is symptomatic or asymptomatic.

Gestational diabetes mellitus

This term is used to describe both the impaired glucose tolerance and diabetes mellitus that may develop during pregnancy. In the majority of cases, the oral glucose tolerance test reverts to normal after the pregnancy, but about 50% of cases go on to develop diabetes mellitus within 7 years.

The diagnosis of diabetes mellitus is important in pregnancy since the incidence of maternal and fetal complications appears to be related to the adequacy of control. Glycosuria detected at routine antenatal testing may suggest the diagnosis of diabetes, but this finding may have no significance since the renal threshold for glucose tends to be lowered in pregnancy.

If investigating a pregnant woman for suspected diabetes mellitus, the normal criteria for assessing OGTT (Table 11.1) should be applied. Mild degrees of abnormality should be reassessed not less than 6 weeks after delivery. Even normal women tend to show slightly impaired glucose tolerance in pregnancy.

Previous abnormality of glucose tolerance

These are individuals who may be asymptomatic and have a normal response to an OGTT, but who are known to have previously had an abnormal OGTT, usually under a stress condition such as pregnancy.

Potential abnormality of glucose tolerance

The family history and obstetric history may be important in recognizing these individuals. They include individuals who have an identical twin that is diabetic, mothers who have previously given birth to a child weighing 4.5 kg (10 lb) or more, and individuals found to have islet cell antibodies in plasma.

Flat response to OGTT

This describes the response when the plasma [glucose] fails to rise significantly after a 75 g oral glucose load. If the fasting plasma [glucose] was 5.0 mmol/L, the 30 and 60 min or 60 and 90 min specimens, or all three specimens, would usually have plasma [glucose] of 7.0 mmol/L or more.

A flat response is occasionally due to malabsorptive disease of the small intestine, but is much more often caused by incorrect positioning of the patient during the test, resulting in delayed gastric emptying. A flat curve is also sometimes seen in hypopituitarism and in adrenal hyposecretory states.

'Lag storage' response to OGTT

This is the term used to describe the pattern of results in which there is a sharp rise in plasma [glucose] with the peak values appearing early and sometimes exceeding 11 mmol/L. There is also a tendency for the 2-h value to be much lower than the fasting plasma [glucose]. There may be glycosuria.

This pattern is due to rapid absorption of glucose from the gut. Tissues, mainly the liver, are unable to attain a sufficiently rapid uptake to match the rapid absorption from the intestine. This 'lag storage' is not related to any abnormality of insulin action, and the tendency to hypoglycaemia after 2 h may represent an 'overshoot' phenomenon or 'reactive' hypoglycaemia.

This pattern of response may be seen after gastric surgery (e.g. gastroenterostomy), in patients with severe liver disease and sometimes in apparently healthy individuals.

Renal glycosuria

This term is applied to patients who exhibit glycosuria at some point in the oral glucose tolerance test although the plasma [glucose] remains below 10 mmol/L. It is due to a lowered renal threshold for glucose and is without pathological significance.

Glycosylated haemoglobin

Haemoglobin (Hb) normally consists mostly of two components, A and a small amount of A_2 (p. 321). These are the forms found in reticulocytes when released from the bone marrow. Thereafter, in the course of the normal 120-day life-span of the red cell, a small percentage of the Hb becomes glycosylated, i.e. covalently bonded to glucose, glucose phosphate, etc. These complexes have different electrophoretic mobilities from Hb and are collectively known as

HbA_1. Between 5% and 10% of circulating Hb is normally in the form of HbA_1; the 120-day old cells contain 10–15% HbA_1. Glycosylation of haemoglobin has very little effect on its oxygen-carrying properties.

Several demonstrably different glycosylated derivatives contribute to HbA_1, and are designated A_{1a1}, A_{1a2}, A_{1b}, A_{1c}, etc. The principal glycosylated haemoglobin complex is HbA_{1c}, and it normally forms about 5% of circulating Hb. Methods have been developed for measuring [total HbA_1] and [HbA_{1c}]; for technical reasons [total HBA_1] is the determination most often performed, results being expressed as a percentage of [total Hb].

HbA_1 in diabetic patients

Glycosylated Hb tends to be increased in diabetes because the percentage of HbA_1 in blood is related to the average values of plasma [glucose] over the previous 1–2 months, and these tend to be higher in diabetics than in normal individuals. The extent of the elevation of HbA_1 indicates the overall average degree of plasma glucose control, and in poorly controlled diabetics may rise as high as 30%. In many ways, HbA_1 provides a better index of diabetic control then plasma [glucose], since it is little affected by short-term fluctuations in plasma [glucose].

In diabetic clinics, HbA_1 is mostly used to assess control of insulin-dependent diabetics. In these patients, the plasma [glucose] determined at the time of the clinic attendance can only give limited information and may not represent the overall closeness of control at other times. The results of HbA_1 measurements can be used to motivate patients to achieve better control of their diabetes, since the results recorded at one clinic visit indicate the success they have been having in managing their disease hitherto, and they can be given a lower HbA_1 value as a target which they can aim to achieve in the future.

Many proteins are slowly glycosylated in the course of their life-span, in a manner similar to Hb, and it has been suggested that glycosylation of structural proteins in arterial walls and elsewhere might be responsible for some of the long-term sequelae of diabetes.

It is unlikely that HbA_1 measurements will ever displace plasma or blood [glucose] measurements as the mainstay of day-to-day control of diabetic patients, especially in view of the much improved and widely available techniques for home monitoring of glucose control. Technically, HbA_1 measurements are more time-consuming and more expensive than glucose measurements.

Complications of diabetes mellitus

Diabetic coma is the term used to denote the condition of patients who develop metabolic decompensation that requires treatment with insulin and electrolyte solutions, irrespective of their state of consciousness. It is a serious medical emergency which, despite modern advances in treatment, still carries an average mortality of about 7% *per episode.*

Diabetic coma is most often seen in diabetics who develop an intercurrent infection, or suffer trauma or unusual physical or mental stress. The consequences of metabolic decompensation are primarily due to lack of effective insulin action. The major abnormalities result from hyperglycaemia or ketoacidosis, or both. Lactic acidosis and pre-renal uraemia may also be present.

Hyperglycaemia

This causes extracellular hyperosmolality, which leads to intracellular dehydration. The osmotic diuresis causes loss of water, Na^+, K^+, calcium and other inorganic constituents, and leads to a fall in circulating blood volume. Vomiting may exacerbate all these effects.

Ketoacidosis

This is due to insulin deficiency and results from increased mobilization of FFA from adipose tissue. The insulin deficiency is accompanied by raised plasma concentrations of so-called 'diabetogenic' hormones (adrenaline, cortisol, growth hormone and glucagon).

The FFA are carried to the liver and are partly converted to acetyl CoA. There is overproduction of acetyl CoA and its disposal depends on the activity of various metabolic pathways. Entry of acetyl CoA into the tricarboxylic acid cycle is not depressed as a rule, but the cycle is unable to handle the excessive amounts of acetyl CoA formed. Fatty acid synthesis by the liver is impaired, so acetyl CoA molecules condense to form acetoacetic acid, from which 3-hydroxybutyric acid and acetone are then derived.

Acetoacetic acid and 3-hydroxybutyric acid give rise to a metabolic acidosis as the liver and other tissues cannot, in general, completely metabolize the increased amounts being formed. The acidosis is partly compensated by hyperventilation with reduction in P_{CO_2}, which causes the plasma $[HCO_3^-]$, already reduced by the metabolic acidosis, to fall even further (Tables 5.2b and 5.3). The acidosis causes H^+ to move into cells and K^+ to move out; increased plasma [K^+] often results.

Other features may be present. If tissue perfusion is affected, by extreme dehydration (hyperosmolality) or the factors which precipitated the original metabolic decompensation (e.g. myocardial infarction), tissue anoxia may lead on to lactic acidosis.

DIAGNOSIS AND TREATMENT OF DIABETIC COMA

Diagnosis is usually made initially on the basis of the history, clinical examination and side-room testing of urine and blood for glucose and ketone bodies. Laboratory-based tests on blood are needed to evaluate the nature of the metabolic decompensation and severity of the condition more precisely, and to monitor progress during treatment. However, it is rarely necessary, and indeed may be positively dangerous, to wait for laboratory results before starting treatment. Further measures should be based upon regular clinical and biochemical assessment. The following types of clinical syndrome are recognized and may require different forms of treatment:

(1) *Hyperglycaemia with ketoacidosis,* the commonest combination.

(2) *Ketoacidosis without hyperglycaemia.* This usually occurs in young patients with unstable diabetes in whom insulin therapy has continued in the absence of the normal intake of carbohydrate.

(3) *Hyperglycaemia without ketoacidosis.* Some patients, more commonly elderly, do not develop ketoacidosis despite very high plasma [glucose] and severe dehydration; plasma [Na^+] may be very high. These patients tend to be sensitive to insulin.

(4) *Hyperglycaemia with lactic acidosis.* This is rare. It carries a high mortality rate, and it is important to distinguish it from the much commoner ketotic form. Patients with lactic acidosis require treatment primarily of the underlying cause as well as replacement of fluid losses, $NaHCO_3$ and oxygen. They may be very sensitive to insulin.

INITIAL LABORATORY ASSESSMENT OF DIABETIC COMA

Plasma [glucose], [urea], [Na^+], [K^+] and [total CO_2] are usually measured, on a sample of venous blood.

Plasma [Na^+] may be normal or low initially, except in hyperglycaemia without ketoacidosis, when plasma [Na^+] and osmolality are both high. Plasma [K^+] may be increased or normal, but occasionally it is decreased. Plasma [total CO_2] is nearly always reduced, often being less than 5 mmol/L in severe cases. Plasma [urea] is usually increased in the presence of significant dehydration.

Additional, more definitive assessment of the acid–base status can be obtained from measurement of arterial blood [H^+], $P\text{CO}_2$, [HCO_3^-] and $P\text{O}_2$. Results indicating a severe degree of metabolic acidosis with compensatory reduction in $P\text{CO}_2$ are often found. A lowered $P\text{O}_2$ is an indication for oxygen therapy.

Plasma [acetoacetate] can be estimated semi-quantitatively with side-room test materials (p. 34). Quantitative measurements of [ketone bodies] are still largely research procedures.

Treatment of the hyperglycaemic types of diabetic metabolic decompensation aims to correct fluid and electrolyte deficits by fluid replacement, and the metabolic abnormality by injection of appropriate amounts of insulin. Knowledge of the fluid and electrolyte deficits likely to be present helps in planning appropriate therapy. It has been estimated that there may be deficits of 5–10 L of water, 500 mmol of Na^+, 250–800 mmol of K^+ and 300–500 mmol of base (e.g. HCO_3^-) in cases with severe acidosis.

FURTHER LABORATORY ASSESSMENT DURING THERAPY

Timing of repeat analyses depends on the clinician's experience, on the severity and nature of the diabetic coma, and on the method of treatment (e.g. continuous intravenous infusion of small amounts of insulin). For most of these patients, it is advisable to repeat some of the initially requested analyses, particularly plasma [K^+], after one hour and thereafter at longer intervals (e.g. 2h) if treatment is progressing satisfactorily. The following investigations are likely to be the most valuable for monitoring treatment:

(1) *Plasma [glucose].* Insulin therapy is usually based on the response shown by the plasma [glucose].

(2) *Plasma [K^+].* In many centres potassium replacement is started, intravenously for most patients, within a few minutes of the initial dose of insulin, in the knowledge that (1) the patient has almost certainly developed a large K^+ deficit, (2) insulin will rapidly cause K^+ to enter the cells from the ECF, and (3) serious hypokalaemia is likely to develop fairly quickly in the absence of early corrective action.

(3) *Plasma [total CO_2],* especially where this is being used as an index of the acid–base disturbance. Reassessment of the arterial blood $[H^+]$, $P\text{CO}_2$ and $[HCO_3^-]$ may be indicated in some cases.

It generally proves possible to manage patients with blood analyses performed less often than every 2 h once the plasma [glucose] has fallen satisfactorily, the plasma $[K^+]$ is fairly stable and the plasma [total CO_2] has risen. As convalescence proceeds, dependence on laboratory examinations becomes less as 'side-room' tests regain their usefulness.

Metabolic changes during therapy

Insulin causes plasma [glucose] to fall by facilitating glucose entry into cells, and by reducing the output of glucose from the liver. Potassium also returns to the cells and a rapid fall in plasma $[K^+]$ may result. Lipolysis is inhibited.

Fluid replacement is usually given initially as so-called 'physiological saline' (154 mmol/L NaCl); K^+ and other electrolytes may be added to the infusion fluid once insulin begins to exert its effects. In severe acidosis, HCO_3^- may be infused, but this can be dangerous as rapid correction of acidosis augments the K^+ influx into cells and can precipitate dangerous hypokalaemia. Infusion of HCO_3^- may also cause CSF $[H^+]$ to rise rather than fall, due to the more rapid diffusion of CO_2 than HCO_3^- into the CSF. It is not known whether this rise in CSF $[H^+]$ is harmful to brain function.

In patients with hyperglycaemic coma without ketoacidosis, 'half-strength saline' (77 mmol/L) may be used for fluid replacement initially, because of the patient's hyperosmolality and high plasma $[Na^+]$.

SIDE-ROOM TESTS IN LONG-TERM MANAGEMENT

The use of side-room tests by patients, for measuring blood or urine glucose, is discussed elsewhere (p. 32). These tests should also be performed at clinic visits.

Long-standing diabetes mellitus may give rise to proteinuria, and side-room tests may provide the first indication of the development of diabetic nephropathy.

Ketonuria, which may herald the onset of diabetic ketoacidosis, can also be detected in the side-room. Testing for ketone bodies in urine (and blood) with side-room tests helps in the differential diagnosis of coma.

HYPOGLYCAEMIA

The plasma [glucose] at which symptoms of hypoglycaemia appear is extremely variable, and is often related more to the rate of fall of plasma [glucose] than to the absolute value observed. Arbitrarily, a plasma [glucose] below 2.2 mmol/L (40 mg/100 mL), measured by a glucose-specific (i.e. enzymic) method, is taken as the biochemical definition of hypoglycaemia. This definition is open to criticism, particularly when considering plasma [glucose] 2–4 h after meals, since some reduction of plasma [glucose] occurs during this period when the insulin secreted in response to glucose absorption may produce large arteriovenous differences in plasma [glucose]. However, this definition of hypoglycaemia has been widely accepted since, even in the post-prandial state, the venous plasma [glucose] seldom falls below 2.2 mmol/L.

It would be logical to classify hypoglycaemia into conditions in which release of glucose by the liver is defective, and conditions in which tissue glucose uptake is increased. It is, however, more convenient to distinguish the hypoglycaemia that occurs in response to fasting (Table 11.2) from hypoglycaemia due to some other stimulus (Table 11.3).

It will be apparent, from the range of conditions listed in Tables 11.2 and 11.3, that the finding of hypoglycaemia is, by itself, never diagnostic. Instead, it has to be related to other information concerning the patient's illness, and may be merely one of the first in a series of investigations. It is not possible to review all these here.

TABLE 11.2. Fasting hypoglycaemia. (Modified from Marks and Rose, 1981)

Organ affected	Examples of pathological conditions
Pancreas	Insulinoma (benign, malignant, microadenomatosis)
	Pancreatic tumours as part of MEN syndromes
	Pancreatitis
	Hyperinsulinism of childhood (nesidioblastosis)
Other endocrine glands	Pituitary insufficiency
	Adrenocortical insufficiency
	Hypothyroidism
Liver	Starvation and malnutrition
	Hepatocellular insufficiency
	Glycogen storage disease, Type I
	Primary hepatic carcinoma
Other non-pancreatic tumours	Mesenchymal tumours
	Adrenal tumours
Kidney	End-stage renal failure

TABLE 11.3. Stimulative (reactive) hypoglycaemia (Modified from Marks and Rose, 1981)

Causal group	Examples of causes
Drugs and poisons	Therapeutic hypoglycaemic agents (e.g. insulin)
	Alcohol
	Liver poisons (e.g. chloroform, phosphorus)
	Toadstools (e.g. Amanita phalloides)
Essential reactive hypoglycaemia	Idiopathic
	Alcohol-induced
	Post-gastrectomy
Inborn errors of metabolism	Galactosaemia
	Hereditary fructose intolerance

Although Tables 11.2 and 11.3 subdivide the causes of hypoglycaemia into two main categories, fasting and stimulative (reactive), it should be noted that patients who experience fasting hypoglycaemia are prone also to develop stimulative hypoglycaemia. On the other hand, most patients categorized as having stimulative hypoglycaemia never develop fasting hypoglycaemia.

Insulinoma

These are usually small, solitary, benign adenomas of the pancreatic islets that secrete inappropriate amounts of insulin. Occasionally, multiple pancreatic adenomas may be associated with adenomas in other endocrine organs as part of the multiple endocrine neoplasia (MEN) syndromes (p. 291).

The symptoms may be bizarre or atypical and laboratory investigations have a major place in diagnosis. Most patients develop symptomatic hypoglycaemia after a fast of 24–36 h. In a few, the fast must be continued for up to 72 h before hypoglycaemia develops. Rarely, hypoglycaemia does not develop at all. Plasma [glucose] is measured by an enzymic method and relief of symptoms following glucose administration is observed.

A specimen should be collected for determining plasma [insulin] when the patient has hypoglycaemic symptoms, and must be carefully stored (at −20°C) prior to analysis; plasma [insulin] is measured if hypoglycaemia is confirmed. The diagnostic finding, in patients suspected of having an insulinoma, is a *fasting plasma [insulin] that is inappropriately high,* when considered in relation to the low plasma [glucose].

It can be difficult to demonstrate fasting hypoglycaemia

satisfactorily in some patients. In these, it may still be possible to obtain support for a diagnosis of insulinoma by measuring plasma [C-peptide] during an infusion of exogenous insulin sufficient to induce hypoglycaemia. Exogenous insulin contains little or no C-peptide and failure to detect a fall in plasma [C-peptide] shows that endogenous insulin release is not switched off, as it should be, in response to hypoglycaemia. This finding is strongly suggestive of insulinoma.

Provocative tests involving injection of tolbutamide or glucagon, or oral administration of L-leucine, no longer have a place in the diagnosis of insulinoma.

Stimulative hypoglycaemia

Accidental or deliberate overdose of insulin, giving rise to hypoglycaemia, can be distinguished from insulinoma by measuring both plasma [insulin] and plasma [C-peptide].

Demonstration of a plasma [glucose] below 2.2 mmol/L within 6 h (usually 2–4 h) after a meal, or after an oral load of 75 g glucose, is an important feature in the recognition of essential reactive hypoglycaemia. This is probably due to an exaggeration of the normal insulin response to carbohydrate ingestion.

Neonatal and childhood hypoglycaemia

NEONATAL HYPOGLYCAEMIA

This is arbitrarily defined as a blood [glucose] less than 1.1 mmol/L (20 mg/100 mL) in an underweight (small for gestational age) infant, and less than 1.6 mmol/L in an infant of normal weight. Blood [glucose] may fall as low as 0.5 mmol/L in the neonate without symptoms of hypoglycaemia developing, but such low concentrations can only be tolerated for short periods. The causes of transient neonatal hypoglycaemia have been categorized into 4 types:

(1) Early transitional adaptive hypoglycaemia, as may occur within a few hours of birth in infants of diabetic mothers.

(2) Secondary hypoglycaemia, which may be due to CNS disease or be caused by asphyxia, respiratory distress, infection, etc.

(3) 'Classical' hypoglycaemia due to intra-uterine malnutrition (small-for-dates infants); this type develops hypoglycaemia within 72 h of birth.

(4) Recurrent hypoglycaemia of infancy and childhood; this type usually first develops hypoglycaemia more than 72 h after birth, when it may be due to one of many causes (see below).

RECURRENT HYPOGLYCAEMIA OF INFANCY AND CHILDHOOD

This may be due to any of the causes listed in Tables 11.2 and 11.3. Other possible causes not listed in the Tables include several inborn errors of amino acid metabolism (e.g. phenylketonuria, maple syrup urine disease) and Reye's syndrome. One of the most important causes is hyperinsulinism of childhood (nesidioblastosis).

Functional hyperinsulinism of infancy usually develops before the infant is 6 months old. It is characterized by severe hypoglycaemia, usually accompanied by inappropriately high plasma [insulin] and [C-peptide]. About 50% of these infants show an excessive insulinaemic and hypoglycaemic response to L-leucine. There are diffuse lesions of the pancreas, with a great increase in the mass of islet cells.

Functional hyperinsulinism presenting in children over the age of one is usually due to an islet cell adenoma.

Nutritional causes of hypoglycaemia in this age-group can be classified as ketotic hypoglycaemia or as hypoglycaemia secondary to malnutrition.

Ketotic hypoglycaemia is associated with fasting and usually responds rapidly to administration of carbohydrate. Rarely, it is due to one of the glycogen storage diseases or other inborn errors of metabolism, and very occasionally due to adrenal or pituitary insufficiency.

Malnutrition, severe enough to cause hypoglycaemia, is encountered in children with kwashiorkor (protein–energy malnutrition) or in starvation, including gross parental neglect.

INHERITED METABOLIC DISORDERS

GLYCOGEN STORAGE DISEASES

Glycogen is a polysaccharide composed of straight chains of glucose molecules linked by α-1, 4-glucosidic links, and with branching points connecting these chains by α-1, 6-glucosidic links. It can be synthesized by most tissues and stored (the main stores are in liver and muscle) for use later by that tissue in metabolism. However, only the liver and kidney are capable of hydrolysing glucose-6-phosphate, produced from glycogen, at a rate sufficient to yield glucose for release into the blood.

Glycogen synthesis and breakdown both involve several enzymes, and the pathways are shown in Fig. 11.1. The storage diseases are rare inborn errors of carbohydrate metabolism due to deficiency or reduced activity of one or more of the many enzymes involved (Table 11.4). The classification adopted in Table 11.4 will need to be modified in time, but we think it too early to do so. There are some who consider that a different number (Cori Type) should be assigned to each different enzymic defect. On this basis, another classification already recognizes twelve different Types, with Types II and IX being further subdivided on the basis of age of presentation (IIa and IIb) or the nature of the inheritance, whether autosomal or sex-linked recessive (IXa and IXb).

The common feature in this complex group of rare inherited conditions is an abnormality in the storage of glycogen, usually in increased amount and sometimes with an abnormal structure. As secondary features, there may be hypoglycaemia, abnormalities in

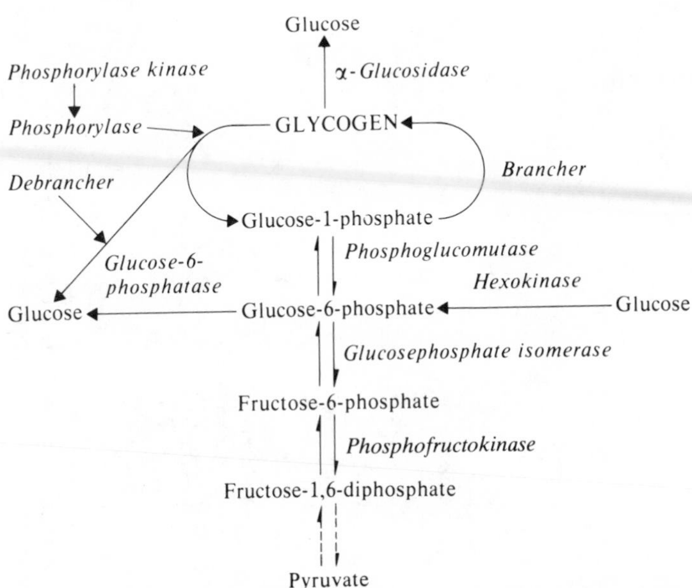

FIG. 11.1. *Glycogen synthesis and breakdown*, showing the sites of enzymic defects in various forms of glycogen storage disease. Table 11.4 relates the sites of the enzyme deficiences to the Cori Types, and indicates the tissues mainly affected.

TABLE 11.4. Glycogen storage diseases (After Ryman, 1980)

Cori Type	Disease		Principal storage sites	Enzyme deficiency	Glycogen structure
I	von Gierke's		Liver, kidney, intestine	Glucose-6-phosphatase	Normal
II	Pompe's		Liver, heart, muscle	α-D-Glucosidase	Normal
III	Cori's limit dextrinosis	(A)	Liver, muscle	Amylo-1, 6-glucosidase (debrancher) and/or	Abnormal
		(B)	Liver	4-α-Glucanotransferase	Abnormal
IV	Anderson's amylopectinosis		Liver	1-4-α-Glucan branching enzyme (brancher)	Abnormal
V	McArdle's		Muscle	Phosphorylase	Normal
VI	Hers'	(A)	Liver, muscle	Phosphorylase kinase	Normal
		(B)	Liver	Phosphorylase	Normal
VII	Tarui's		Muscle	Phosphofructokinase	Normal

Note Other classifications were discuissed at the Symposium on glycogen diseases from which this Table is derived. However, the above set of subdivisions continued to attract support, and it would seem premature to adopt one of the newer classifications.

blood lipids, hyperuricaemia or lactic acidosis, and abnormal responses to various tolerance tests (glucagon, galactose, etc.) have been described. These secondary features, and other indirect forms of testing, are no substitute for tissue enzyme assays; these sometimes need to be supplemented by carbohydrate structural studies. Some of the types have been recognized *in utero,* by culture of amniotic fluid cells, where there has previously been an index case, and this is worthwhile in the case of the lethal Types II and IV.

Von Gierke's disease (Type I) is the best known of the glycogen storage diseases, but the least uncommon is Type VI(A), using the numbering adopted in Table 11.4. Fasting hypoglycaemia may be very marked in Type I.

GALACTOSAEMIA

Two different enzymic defects have been described. Both cause inability to metabolize galactose normally, due to deficiency of galactose-1-phosphate uridylyltransferase or of galactokinase (Fig. 11.2). With the former, galactose-1-phosphate accumulates in the tissues. In both disorders, galactose is partly reduced to the corresponding alcohol.

Deficiency of the transferase enzyme is associated with the more severe clinical manifestations. In the neonatal period nausea and vomiting, often with hypoglycaemia, usually appear. The clinical manifestations in the newborn should lead to consideration of the diagnosis; this depends greatly on chemical investigations. If galactose-containing foods are ingested, the plasma [galactose] rises

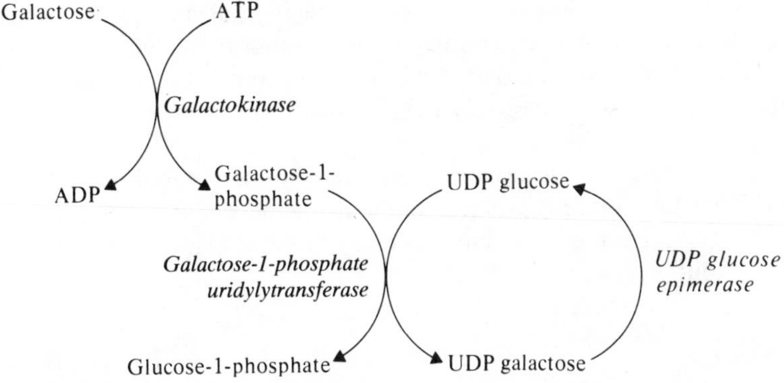

FIG. 11.2. *Conversion of galactose to glucose-1-phosphate*. Galactosaemia may be caused by deficiency of the transferase enzyme or of galactokinase.

and galactose can be detected in the urine. Side-room testing of urine gives positive results for reducing substances and negative results for glucose. The nature of the reducing substance can be determined by urine chromatography. Hepatic damage, cataract and mental retardation develop in time, in the absence of treatment.

Galactokinase deficiency may not give rise to symptoms in the neonatal period, and may escape detection until cataracts develop. Side-room testing of urine and measurement of plasma [galactose] again show that there is an abnormality of galactose metabolism, further proof being obtained by urine chromatography.

The diagnosis is established by measuring the activity of galactose-1-phosphate uridylyltransferase in erythrocytes or lymphocytes, and galactokinase in erythrocytes. These measurements are needed to differentiate between the two types of enzymic defect.

Galactosaemia is rare. Fortunately the more common and severe defect usually manifests itself sufficiently early in infancy, and before irreversible damage has occurred. Diagnosis is nearly always made following clinical presentation, except in the case of infants born subsequently into a family where an index case has been identified, for whom investigations should be specially performed shortly after birth. Galactosaemia does not need to be specially sought as part of a screening programme.

MISCELLANEOUS CAUSES OF GLYCOSURIA

Positive results with side-room tests for reducing substances in urine (e.g. Clinitest, Ames Co.) are most often due to the presence of glucose. However, they may be due to the urine specimen being very concentrated (creatinine is a weak reducing substance) or to the presence of one of several reducing sugars other than glucose.

Lactosuria is of no pathological significance. Lactose may be detected in the urine in pregnancy and while lactation continues after delivery.

Essential pentosuria is a benign inborn error of metabolism in which the sugar L-xylulose is excreted in the urine in excess, due to a defect in NADP-linked xylitol dehydrogenase, one of the enzymes of the glucuronic acid oxidation pathway.

Fructosuria

There are two inborn errors of metabolism that give rise to fructosuria, one of them potentially a serious condition and the other benign.

Hereditary fructose intolerance is due to deficiency of fructose-1-phosphate aldolase, which causes intracellular accumulation of fructose-1-phosphate. Vomiting and hypoglycaemia occur after ingestion of fructose-containing foods, usually the disaccharide sucrose. The age of presentation depends on feeding patterns and on the severity of the defect; most patients develop a strong aversion to sucrose. Damage to the liver and kidney may be followed by death unless fructose-containing foods are removed from the diet.

Investigation can be by means of either an oral or an intravenous fructose tolerance test. Patients with the deficiency show marked and prolonged falls in plasma [glucose] and plasma [phosphate] following fructose administration, as well as fructosuria.

Essential fructosuria, a benign condition, is caused by fructokinase deficiency. Its importance lies in the need to distinguish it from other causes of glycosuria.

FURTHER READING

BONSER, A.M. and GARCIA-WEBB, P. (1981). C-peptide measurement and its clinical usefulness. *Annals of Clinical Biochemistry,* **18,** 200–206.

COHEN, R.D. and WOODS, H.F. (1976). *Lactic Acidosis.* Oxford: Blackwell Scientific Publications.

Leading article (1980). Impaired glucose tolerance and diabetes – WHO criteria. *British Medical Journal,* **2,** 1512–1513.

MARKS, V. (1981). Blood sugar levels. *British Journal of Hospital Medicine,* **25,** 76–78.

MARKS, V. and ROSE, F.C. (1981). *Hypoglycaemia,* 2nd Edition. Oxford: Blackwell Scientific Publications.

RYMAN, B.E. (1980). Recent advances and problems in the glycogen storage diseases. In *Inherited Disorders of Carbohydrate Metabolism,* pp. 289–295. Ed. D. Burman, J.B. Holton and C.A. Pennock. Lancaster: MTP Press.

SCHADE, D.S., EATON, R.P., ALBERTI, K.G.M.M. and JOHNSTON, D.G. (1981). *Diabetic Coma.* Albuquerque: University of New Mexico Press.

TURNER, R.C. and WILLIAMSON, D.H. (1982). Control of metabolism and the alterations in diabetes. In *Recent Advances in Endocrinology and Metabolism,* **2,** pp. 73–97. Ed. J.L.H. O'Riordan. Edinburgh: Churchill Livingstone.

WHO Expert Committee on Diabetes Mellitus, Second report (1980). *WHO Technical Report Series,* **646.** Geneva: World Health Organisation.

Chapter 12
Disorders of Plasma Lipids

Four main classes of lipid can be recognized, from a metabolic standpoint. These are cholesterol and its esters, triglycerides, phospholipids and fatty acids. The principal functions of lipids are to act as energy stores and to serve as important structural components of cells. To fulfil these functions, lipids have to be transported in plasma from one tissue to another, from the intestine or the liver to other tissues such as muscle or fat, or from the other tissues to the liver.

There are complex mechanisms that control the release of lipids from the tissues into plasma, and the uptake of lipids by the tissues from plasma. Abnormalities of these mechanisms may be associated with the development of disease, particularly ischaemic heart disease.

The main lipid classes in plasma

Cholesterol is absorbed from the diet in varying amounts; usually between 40% and 70% of dietary cholesterol is absorbed, the absolute percentage tending to vary inversely with the cholesterol intake. All cells, except erythrocytes, can synthesize cholesterol from acetate but normally only the liver and the small intestine produce significant amounts; hepatic synthesis is controlled, at least in part, by the amount of cholesterol in the diet.

Cholesterol from the liver and small intestine is transported in plasma, about 75% esterified with fatty acids and the rest unesterified. It is taken up from plasma by the tissues. It forms a major structural component of cell membranes.

Cholesterol is the substrate from which the steroid hormones are formed (p. 378), and a small amount is converted to 7-dehydrocholesterol or pro-vitamin D_3 (p.264). Quantitatively, its main route of metabolism is to bile acids, which are secreted into bile as conjugates with glycine or taurine. Unesterified cholesterol is also secreted into bile, and both the bile acids and biliary cholesterol undergo an enterohepatic circulation (p. 171); this process is normally only about 95% efficient for bile acids, and has a variable but generally lower efficiency for cholesterol, depending upon the dietary cholesterol intake. Some loss of cholesterol and bile acids occurs daily in the faeces, the losses being made good by dietary intake and by synthesis.

Triglycerides are present in most diets in large amounts. They are broken down in the small intestine to a mixture of monoglycerides, fatty acids and glycerol. These products of digestion are absorbed from the jejunum, and the monoglycerides and fatty acids are mostly resynthesized into triglycerides in the mucosal cells. The triglycerides are then transported into the intestinal lymph, and thence to the systemic circulation. Some of the shorter chain fatty acids escape the triglyceride resynthesis pathway and pass partly into the portal circulation and partly into the intestinal lymph.

Liver and adipose tissue are also major sites of triglyceride synthesis, those formed in the liver normally being secreted into plasma. Triglycerides synthesized in adipose tissue are either stored locally or else reconverted to fatty acids and glycerol prior to re-entry into the circulation. They are important storage forms of energy.

Phospholipids have a more complex structure than the triglycerides. One of the three fatty acid residues is replaced by a nitrogen-containing compound, linked to the glycerol via a phosphate residue. The nitrogenous compound in cephalin is ethanolamine, in lecithin it is choline, in sphingomyelin it is sphingosine, etc.

Phospholipids are mainly synthesized in the mucosa of the small intestine and in the liver. They circulate in the plasma and are important constituents of all cells, especially the cells of the nervous system. They do not form part of the depot fats.

Fatty acids in the body are mostly straight-chain monocarboxylic acids with an even number of carbon atoms, between 12 and 24. They may be saturated or unsaturated, the latter having one or more double bonds. Although mammals can synthesize most fatty acids, certain polyunsaturated acids (called the essential fatty acids) cannot be synthesized and are therefore necessary dietary constituents.

Most tissues can use fatty acids in place of glucose as readily available sources of energy; muscles use fatty acids preferentially. One of the essential fatty acids, linoleic acid, is converted in the body to arachidonic acid, the major substrate for synthesis of the prostaglandins.

LIPID TRANSPORT

In plasma, fatty acids (1) are present as esters, as constituents of triglycerides, phospholipids or cholesterol esters, or (2) may be unesterified. The unesterified or free fatty acids (FFA) are transported in plasma bound to albumin.

Apart from FFA, plasma lipids are carried in the form of complex particles of varying size and composition, all of which contain both lipid and protein. The protein component and the more hydrophilic of the lipids (e.g. phospholipids) are usually present on the surface of lipoprotein particles, whereas the hydrophobic lipids (e.g. cholesterol esters, triglycerides) are carried in the 'core'. This structure makes it possible for these complex particles to transport water-insoluble lipids in plasma.

Plasma lipoproteins

The lipid-protein particles described in the previous paragraph are called lipoproteins. Their protein components, the apolipoproteins, are a complex family of polypeptides, separable into four main groups (Apo A, B, C and E) and two minor groups (Apo D and Apo F). Each group of apolipoproteins serves a distinct functional role in plasma lipid metabolism (see below).

The lipid/protein complexes that together comprise the plasma lipoproteins can be separated into five main classes, defined according to their behaviour in the ultracentrifuge. Details of the flotation characteristics of lipoproteins are described in more specialized texts. Other, more readily determined differences between the lipoprotein classes include their diameter, density, electrophoretic mobility, and lipid and apolipoprotein composition (Table 12.1). The five classes are:

(1) *Chylomicrons*. These are large particles which consist mainly of triglycerides. They have the lowest density of the lipoprotein classes, and contain very little protein. They are formed in the intestinal mucosa and reach the systemic circulation via the thoracic duct.

TABLE 12.1 Properties

Lipoprotein class	Diameter (nm)	Density (kg/L)	Electrophoretic mobility
Chylomicrons	100–500	<0.95	stay at origin
VLDL	30–80	<1.006	pre-β
IDL	25–30	1.006–1.019	β
LDL	20–35	1.019–1.063	β
HDL	5–10	>1.063	α_1

(2) *Very low density lipoproteins (VLDL).* These are moderately large particles in which triglycerides are again the main lipid component. They are mainly formed in the liver but to a lesser extent by the intestinal mucosa, and are secreted into plasma from these two sites.
(3) *Intermediate density lipoproteins (IDL).* These arise from VLDL, by catabolism within the circulation that results in the removal of some triglyceride and apolipoprotein from VLDL, leaving IDL or 'remnant particles'.
(4) *Low density lipoproteins (LDL).* These are cholesterol-rich particles formed in the liver from IDL, by the removal of more triglyceride and apolipoprotein, before being secreted into plasma.
(5) *High density lipoproteins (HDL).* These are the smallest of the lipoprotein particles and the most dense. They contain a large amount of protein, and approximately equal amounts of cholesterol and phospholipid, but very little triglyceride.

Certain general principles can be derived from this summary. The large particles have a low density, a high content of triglycerides and a low content of protein. As the particles become smaller, their relative concentration of triglyceride falls while the relative concentrations of both cholesterol and protein rise; these changes are accompanied by an increase in density. This description represents an oversimplification, and the following points should be emphasized:

(1) *Plasma lipids exist in a dynamic state.* There is an interchange of lipids between (a) different lipoprotein particles and (b) plasma lipoproteins and tissues.
(2) *The apolipoproteins exist in a dynamic state.* There is an interchange of apolipoproteins between different lipoprotein

of lipoprotein classes.

Lipoprotein composition (approximate %)				Apolipoprotein
Triglyceride	Cholesterol	Phospholipid	Protein	composition*
90	5	5	1	C, B, E, (A)
65	20	10	5	C, B, E, (A)
35	40	15	10	B, (C, E, A)
10	50	20	20	B
5	35	35	25	A, C, E, (B)

*The main apolipoprotein component is listed first; trace components are shown in brackets.

particles. These interchanges can often be related to specific functions of the lipoprotein (see below)

(3) *The lipoproteins in plasma represent a continuum* with respect to size, lipid composition and other properties. Within each class, there may be considerable heterogeneity.

THE APOLIPOPROTEINS

The protein components of the lipoproteins (Apo A etc.) differ widely both in structure and function. Some of the groups are divisible into subgroups (e.g. Apo A–I, Apo A–II etc.). By sensitive and specific analytical techniques, the various apolipoproteins can be detected in all the lipoprotein classes; only the main components are listed in Table 12.1. The roles of the apolipoproteins are:

(1) *Physical.* They promote the solubility of lipids in plasma and enhance the stability of the lipoprotein particles.
(2) *Regulatory.* They have a role in controlling lipid metabolism, e.g. the rate of tissue uptake of cholesterol or triglyceride.

Some of the features of the main apolipoprotein groups will be briefly considered.

Apo A. This group of proteins is synthesized in the liver and the intestine; its two main components are Apo A–I and Apo A–II. Apo A is mainly present in HDL, but to a lesser extent in VLDL, IDL and chylomicrons. Apo A–I is a potent stimulator of the enzyme lecithin cholesterol acyl transferase (LCAT), as discussed below.

Apo B. This is the main protein constituent of LDL, and is present also in the precursors of LDL, i.e. in VLDL and IDL, as well as chylomicrons. It appears that Apo B is essential for the incorporation of triglycerides into lipoproteins, in the liver and intestine. Apo B also plays a part in the uptake of cholesterol into tissues; most cells possess 'surface pits' that contain specific LDL receptors which 'recognize' Apo B.

Apo C. This is a family of 3 proteins (Apo C–I, Apo C–II and Apo C–III), which are synthesized in the liver and incorporated into HDL. In plasma, Apo C proteins transfer to chylomicrons and VLDL. While present on the surface of these particles, Apo C–II exerts its function, as an activator of lipoprotein lipase in tissue capillaries; the Apo C proteins are then transferred back to HDL. These apolipoproteins can also enhance LCAT activity *in vitro*.

Apo E. This family of proteins is mainly synthesized in the liver.

Lipoprotein classes that contain significant amounts of Apo E (Table 12.1) bind strongly to the same receptor sites as bind LDL – Apo B. The function of Apo E is not definitely known, but it may be involved in hepatic uptake of chylomicron remnants and IDL (VLDL remnants), since the liver appears to have Apo E-specific receptors.

Enzymes involved in lipid transport

Four enzymes, important in plasma lipid transport, need to be described. They are lecithin cholesterol acyl transferase (LCAT), lipoprotein lipase, hepatic lipase and mobilizing lipase (alternatively called hormone-sensitive lipase).

LCAT transfers an acyl group (fatty acid residue) from lecithin to cholesterol, forming a cholesterol ester. In plasma, this reaction probably takes place exclusively on HDL, and may be modified by Apo A–I. In familial LCAT deficiency, a very rare inherited metabolic disorder, there is inability to esterify cholesterol and the lipoproteins in plasma are grossly abnormal; lipoprotein X is present (p. 183).

Lipoprotein lipase splits triglycerides to glycerol and FFA. The enzyme is attached to the capillary endothelium in adipose tissue, muscle and other tissues. It exerts its action on triglycerides present in chylomicrons and VLDL. The activity of lipoprotein lipase increases after a meal, partly as a result of activation by Apo C–II, present on the surface of triglyceride-bearing lipoproteins.

Hepatic lipase has an action similar to lipoprotein lipase, but it is not activated in the manner just described. Hepatic lipase seems to be more effective in hydrolysing triglycerides in smaller particles such as VLDL remnants or IDL than the triglycerides contained in chylomicrons.

Mobilizing lipase, present in adipose tissue cells, controls the rate of release of FFA from adipose tissue into plasma. It is activated, through the adenyl cyclase system, by catecholamines, growth hormone and glucocorticoids (e.g. cortisol); it is inhibited by glucose and by insulin.

METABOLISM OF PLASMA LIPOPROTEINS

CHYLOMICRONS

These are formed in the intestinal mucosa, in response to fat-containing meals; the Apo A and Apo B components of the chylomicrons are also formed in the intestinal mucosa. After entering the bloodstream, the chylomicron particles transfer Apo A to HDL and acquire Apo C from HDL. The Apo C–II then activates lipoprotein lipase, and triglycerides are progressively removed from the chylomicrons; this occurs mainly in adipose tissue, where there is the greatest lipoprotein lipase activity.

As triglycerides are removed from the hydrophobic core of the chylomicron particles, the surface area of the chylomicron decreases and the more hydrophilic surface components (Apo C, unesterified cholesterol and phospholipid) transfer to HDL. The triglyceride-poor *chylomicron remnants,* now consisting mainly of cholesterol esters and Apo B (and Apo E), are taken up by the liver where they are catabolized within the hepatic lysosomes.

The net effect, of the metabolic processes just described, is (1) the transfer of dietary triglycerides to the tissues, mainly adipose tissue, and (2) the transfer of cholesterol from the intestine (i.e. both dietary cholesterol and cholesterol excreted in bile) to the liver.

VLDL AND IDL METABOLISM

Most VLDL is secreted into plasma by the hepatocytes ('endogenous' VLDL), but some originates from the intestine ('exogenous' VLDL). Hepatic synthesis of VLDL is increased whenever there is increased hepatic triglyceride synthesis, e.g. when there is increased transport of FFA to the liver.

Nascent VLDL (i.e. VLDL when first produced) consists mainly of triglyceride and some unesterified cholesterol, with Apo B and lesser amounts of Apo E. Apo C is acquired from HDL, and possibly also from VLDL remnants (see below) in plasma.

As with chylomicrons, triglycerides are removed from the VLDL 'core' as a result of the action of lipoprotein lipase, activated by Apo C–II; this occurs in adipose and other tissues (Fig. 12.1). The surface components of VLDL, which are mainly Apo C, unesterified cholesterol and phospholipid, are transferred progressively to HDL as the VLDL particles lose triglycerides from their core and become smaller. After a considerable proportion of the triglyceride has been removed from VLDL, the now smaller particle is known as a 'VLDL remnant', or IDL.

IDL are normally present in the circulation only in low concentration, since they are either rapidly converted to LDL or removed from the circulation. It seems that the liver plays the major role in both these processes, possibly through recognition of IDL–Apo E.

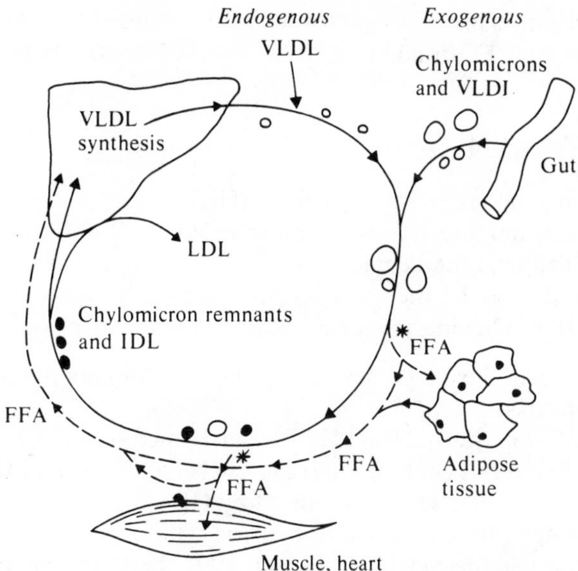

FIG. 12.1. *Diagrammatic representation of triglyceride transport*. O, chylomicrons; ○, VLDL; ●, chylomicron remnant particles. Points of action of lipoprotein lipase, *. Dotted lines used to demonstrate movements of FFA.

LDL METABOLISM

In man, probably all LDL arises normally from VLDL metabolism; there is a one-for-one relationship between the particles, i.e. each LDL particle derives from a single VLDL particle. The LDL particles are rich in cholesterol esters, which have largely replaced the triglycerides that were present in the core of the VLDL particles; the cholesterol esters are probably derived from HDL. Apo B is the only apolipoprotein in LDL particles.

The peripheral tissues remove LDL from the circulation. In the first stage, LDL becomes bound to specific receptors on 'coated pits' on the surface of cells in the peripheral tissues. These receptors recognize LDL–Apo B, and have a high affinity for it. The entire LDL particle is then incorporated into the cell, by invagination of the cell membrane. Inside the cell, the particle fuses with lysosomes; Apo B is then broken down and the cholesterol esters are hydrolysed, thereby making unesterified cholesterol available to the cell. Unesterified cholesterol also regulates:

(1) The rate of synthesis of cholesterol in the cell.
(2) The number of LDL receptors on the cell surface.

The receptors in peripheral tissues that recognise LDL–Apo B also recognise Apo E. In addition, cells possess low-affinity lipoprotein receptors.

HDL METABOLISM

This heterogeneous group of particles (HDL_2, HDL_3, etc.) is formed in the liver and the intestinal mucosa. When first formed, HDL particles mainly contain unesterified cholesterol and phospholipid, Apo C and Apo E; the particles are secreted as flat discs, called nascent HDL. On entering the circulation, the following occur:

(1) HDL acquires Apo A–1, its major apolipoprotein, from chylomicrons.
(2) Apo C is transferred from HDL to chylomicrons and VLDL.
(3) LCAT converts cholesterol to cholesterol esters, within HDL.
(4) More free cholesterol is acquired by HDL particles, from other lipoproteins in plasma and from tissue cells.
(5) The discoid particles of nascent HDL become spherical and mature, with a core consisting of cholesterol esters.

The ester cholesterol, formed on HDL, moves from HDL to other lipoprotein particles (VLDL and LDL) by equilibration; further free cholesterol can then be taken up and esterified in HDL. The delivery of cholesterol, as cholesterol esters, to the peripheral tissues is effected by LDL (Fig. 12.2).

The means whereby cholesterol is returned from the periphery to the liver are not so clear, although HDL certainly has a central role. There is little doubt that free cholesterol can be removed from tissues by HDL, but it seems unlikely that more than a small proportion of such cholesterol is delivered to the liver directly by HDL. Most of the cholesterol esters formed on HDL after uptake of cholesterol from tissues probably transfer to VLDL, and some of these are probably taken up by the liver during the conversion of VLDL to LDL. The remaining LDL–cholesterol is taken up by both liver and other tissues.

INVESTIGATION OF PLASMA LIPID ABNORMALITIES

Plasma [lipids] vary in health with age, sex, diet and other factors. The importance of these variations must be appreciated before attributing changes to disease. Oral contraceptives reduce some of the chemical differences between the sexes (p. 401).

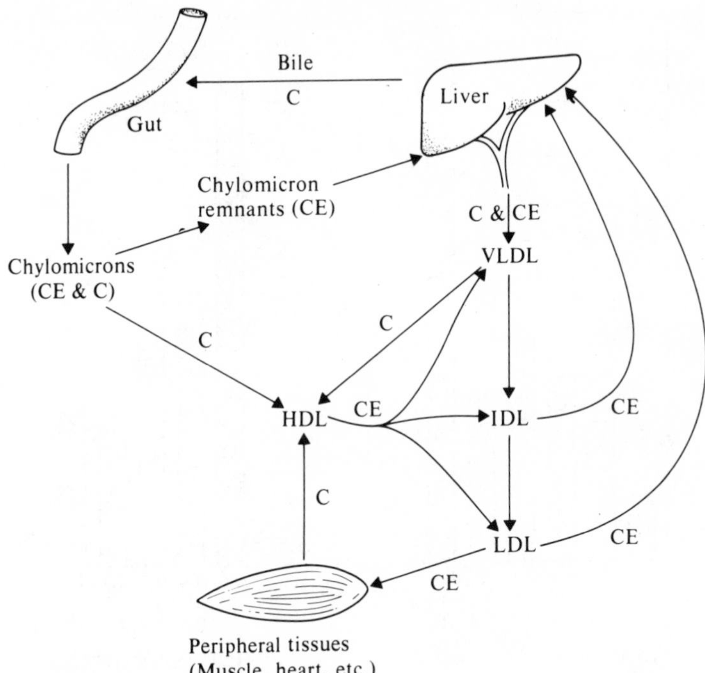

FIG. 12.2. *Diagrammatic representation of the movement of cholesterol (C) and cholesterol esters (CE)* between the lipoproteins in plasma and various tissues, and *vice versa*, and between the individual lipoproteins. *The arrows depicting the transfer of cholesterol to HDL, and of cholesterol esters from HDL, are shown as unidirectional, for simplicity; in fact, they are bidirectional equilibria.*

The figure does not show that cholesterol esterification, catalysed by LCAT, takes place on HDL, nor does it show that the conversion of IDL to LDL takes place in the liver.

Plasma [cholesterol] varies considerably between different ethnic groups, low plasma [cholesterol] occurring in races which are less industrialized and which obtain most of their energy from dietary carbohydrate; such races show no age-related change in plasma [cholesterol]. On the other hand, in most industrialized countries, the average plasma [cholesterol] tends to increase by about 0.2 mmol/L per decade (about 10 mg/100 mL). This begs the question whether to relate plasma cholesterol results in an individual to the age of the individual, since available evidence suggests that the age-related changes in average plasma [cholesterol] are due to the cumulative effects of 'unhealthy' Westernized diets.

Plasma [cholesterol] is correlated with the incidence of ischaemic

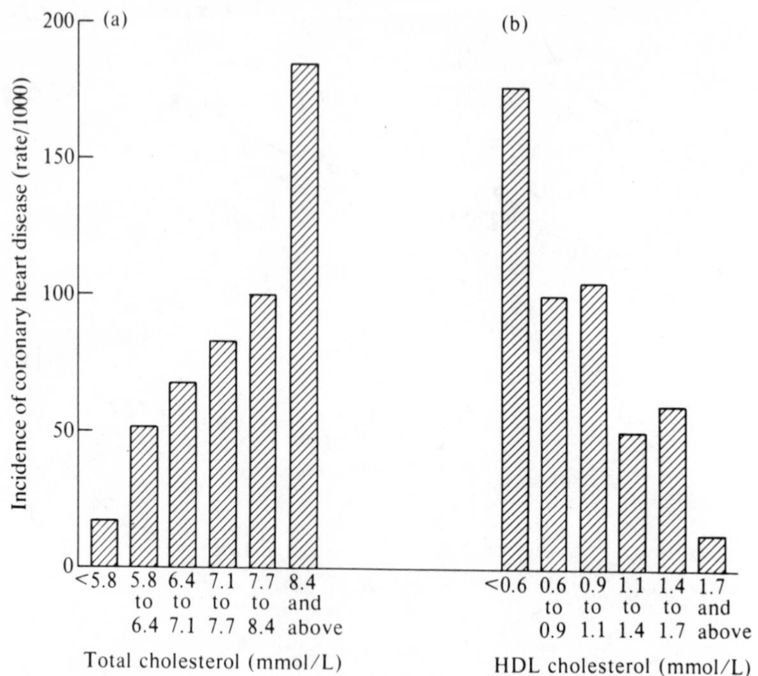

FIG. 12.3. *Variations in the incidence of coronary heart disease* with the concentration in plasma of (a) total cholesterol (data from Westlund and Nicolaysen, 1972) and (b) HDL-cholesterol (data from Gordon *et al.*, 1977).

heart disease (Fig. 12.3), but not necessarily as cause and effect. There is no clear cut-off between normal values and increased risk, so it would seem reasonable to interpret all results for plasma [cholesterol] on the basis of reference ranges determined on young adults. There does, however, appear to be a particularly high risk, and a rapidly rising incidence, of ischaemic heart disease if plasma [cholesterol] exceeds 6.0 mmol/L (235 mg/100 mL).

Plasma [triglycerides] also show variations with age, sex, race and especially with diet. There is also a very considerable within-individual variation, which makes interpretation of a single result difficult.

Full characterization of plasma lipids in every patient is impracticable, in view of the complexities already outlined. Most laboratories measure plasma [triglycerides] and [cholesterol]. They may also undertake tests to determine which of the lipoprotein classes is mainly affected by disease.

It is very important to standardize the conditions under which

TABLE 12.2. The hyperlipoproteinaemias (WHO classification).

WHO Type	Appearance of plasma*	Chylomicrons	VLDL	LDL
I	Cream layer on clear plasma	↑	N or ↑	N or ↓
IIa	Normal	N	N	↑
IIb	Clear or slightly turbid	N	↑	↑
III	Clear or slightly turbid	N or ↑	↑ **	↓
IV	Clear or turbid	N	↑	N or ↓
V	Cream layer on turbid plasma	↑	↑	N or ↓

Key to symbols: N = normal; ↑ = increased; ↓ = reduced
* After storage at 4°C, undisturbed for 12–18 hours (p. 258)
** Floating ß-lipoprotein. This is an abnormal lipoprotein (p. 255) which has the density of VLDL.

specimens for plasma lipid investigations are collected. We specify these in detail later (p. 258).

The primary hyperlipoproteinaemias

These conditions have been classified in several ways. One classification, based principally on the nature of the plasma abnormalities, was proposed by Fredrickson, Levy and Lees and later adopted by the WHO (Table 12.2).

The WHO terminology has been widely used, and indeed is still used to describe patterns of lipoprotein abnormality. However, more is now known about the nature of the defects in some of the genetic hyperlipoproteinaemias, and about their modes of inheritance. The description given here is based on present knowledge of the genetic abnormalities (Table 12.3) and other factors.

Increased plasma lipid concentrations may be:

(1) Due to purely genetic factors, discussed here.
(2) Due to genetic and environmental factors considered above.
(3) Secondary to other diseases (p. 256).

FAMILIAL HYPERCHOLESTEROLAEMIA (WHO TYPES IIa, IIb)

This abnormality is transmitted by an autosomal dominant gene. Homozygotes commonly develop flat cutaneous and tendon

TABLE 12.3. The primary

	Lipoprotein class mainly affected
Familial hypercholesterolaemia	LDL
Familial hypertriglyceridaemia	VLDL (and chylomicrons)
Familial combined hyperlipidaemia	LDL and/or VLDL
Remnant hyperlipoproteinaemia	IDL and chylomicron remnants
Lipoprotein lipase or Apo C–II deficiency	Chylomicrons and VLDL

xanthomas, and usually die from ischaemic heart disease in early adult life. Heterozygotes are less severely affected, and may sometimes be free from heart disease; more often they exhibit tendon xanthomas and are liable to die from ischaemic heart disease before the age of 50.

The characteristic findings are severe hypercholesterolaemia and increased plasma [LDL], but little or no increase in plasma [triglycerides]. The disorder is probably due to a complete lack of high affinity LDL–Apo B receptors in peripheral tissues, in the homozygotes, and to a reduction in the number of these receptors to 50% of normal in heterozygotes.

FAMILIAL HYPERTRIGLYCERIDAEMIA (WHO TYPES IV, V)

This abnormality is transmitted by an autosomal dominant gene. It is usually asymptomatic, although patients may develop eruptive xanthomas and acute pancreatitis may occur; ischaemic heart disease is not a feature.

The characteristic findings are increased fasting plasma [triglycerides], often to very high values; there is increased plasma [VLDL], but plasma [LDL] is either normal or reduced. Plasma [cholesterol] is often increased, although not markedly, and there is chylomicronaemia in more severe cases (Type V phenotype).

The cause of this disorder is not known for certain, but a defect in hepatic VLDL production leading to the formation of plasma VLDL particles that are larger than normal, with a high triglyceride content, has been suggested.

FAMILIAL COMBINED HYPERLIPIDAEMIA

This disorder is difficult to classify and the method of inheritance is unclear. Even in the same family, the gene does not always express itself in the same way. The incidence of ischaemic heart disease is 3–4 times the incidence in the general population. There may be:

hyperlipoproteinaemias (genetic classification).

Plasma concentrations		Corresponding WHO Types
Cholesterol	Triglycerides	
↑↑	N (or ↑)	IIa, IIb
↑ or N	↑↑	IV, V
↑ or N	↑ or N	IIa, IIb, IV
↑	↑	III
↑ or N	↑↑	I, V

Key to symbols: N = normal; ↑ = increased; ↑↑ = greatly increased

(1) Increase in plasma [LDL] only (Type IIa phenotype).
(2) Increase in plasma [VLDL] only (Type IV phenotype).
(3) Increase in plasma [VLDL] and [LDL] (Type IIb phenotype).

The plasma [cholesterol] and plasma [triglycerides] show increases depending on the phenotypic expression of the disorder. Although the cause is unknown, it would seem as if the production of VLDL and LDL rather than their catabolism is at fault.

REMNANT HYPERLIPOPROTEINAEMIA (WHO Type III)

This is an uncommon disorder, in which the clinical features may include both flat and tuberous cutaneous xanthomas; there is a greatly increased risk of ischaemic heart disease. The disorder responds well to therapy, so diagnosis is important. The lipoproteins show unusual features.

There is an increase in cholesterol-rich but otherwise VLDL-like particles, which are probably chylomicron remnants and VLDL-remnants. These abnormal particles behave atypically on electrophoresis, giving rise to a broad β–band. They also behave atypically on ultracentrifugation, where they are described as 'floating β–lipoproteins'. Both plasma [cholesterol] and [triglycerides] are increased, and plasma [LDL] is decreased.

The cause of remnant hyperlipoproteinaemia is uncertain, but the conversion of VLDL, which contains excess Apo E, to LDL is impaired. This may be due to an abnormality in Apo E, since deficiency of Apo E_3 has been demonstrated; this may affect hepatic recognition and catabolism of VLDL particles. However, additional factors must be involved since deficiency of Apo E_3 is not uncommon, and it is not always accompanied by remnant hyperlipoproteinaemia.

LIPOPROTEIN LIPASE DEFICIENCY (WHO Type I)

This is a very rare disorder, transmitted as an autosomal recessive characteristic. There may be eruptive xanthomas. Ischaemic heart disease is not a feature.

There is marked chylomicronaemia, and consequently greatly increased plasma [triglycerides]. The primary abnormality is deficiency of either lipoprotein lipase or of its activating apolipoprotein, Apo C–II; the two causes have identical effects.

OTHER INHERITED DEFECTS

Hyper–α–lipoproteinaemia occurs as an inherited abnormality, giving rise to increased plasma [HDL] and mildly increased plasma [cholesterol]. These patients have a *reduced* incidence of ischaemic heart disease. Similar abnormalities may also occur as a result of alcoholism.

Secondary hyperlipidaemia

Probably less than 20% of cases of hyperlipidaemia are secondary to other diseases. Patterns of abnormality tend to vary, even within a single disease; plasma [cholesterol] or [triglycerides], or both, may be affected.

Hypercholesterolaemia is commonly a feature, often marked, of hypothyroidism and the nephrotic syndrome; in these two disorders, there is increased plasma [LDL]. Hypercholesterolaemia also occurs in cholestatic jaundice, but in this condition there is accumulation of abnormal discoid particles rich in phospholipid and unesterified cholesterol; lipoprotein X (p. 183) is detectable on electrophoresis.

Coronary artery disease tends to develop in those patients with secondary hyperlipidaemia who have increased plasma [LDL] and who have had the underlying disease for a considerable time.

Hypertriglyceridaemia, secondary to other disease, is most commonly due to diabetes mellitus or to alcoholism. It may also occur in chronic renal disease and in patients on oestrogen therapy, including women taking oestrogen-containing contraceptives.

The effects of alcohol on plasma lipids are complex. Regular drinking of small amounts of alcohol increases plasma [HDL] without affecting the other lipoprotein particles. On the other hand, some heavy drinkers develop hypertriglyceridaemia due to increased plasma [VLDL]; this is possibly the result of increased direction of FFA metabolism into triglyceride synthesis by the liver.

The hyperlipidaemia secondary to diabetes mellitus is also complex. Increased plasma [VLDL] is the usual finding, but plasma [LDL] is often increased also and plasma [HDL] reduced.

Hyperlipidaemia and arterial disease

The incidence of myocardial infarction and other types of arterial disease is greater in individuals and races with high plasma [cholesterol] (Fig. 12.3). This is especially the case for those familial disorders in which there is increased plasma [LDL–cholesterol]; in many of these patients, coronary artery disease develops prematurely. The incidence of arterial disease may also be greater in individuals with increased plasma [triglycerides], of the familial combined hyperlipidaemic type.

More detailed analysis of some of the surveys on which these statements are based has shown that plasma [LDL–cholesterol] is *positively* correlated with a high incidence of ischaemic vascular disease. On the other hand, in those communities which have (on average) an increased plasma [LDL], plasma [HDL] shows an even stronger *negative* correlation with these disorders. In other words, increased plasma [HDL–cholesterol] appears to exert a protective effect; this may possibly be explained on the basis of the hypothesis that HDL plays an important role in the removal of cholesterol from tissues. This protective effect of HDL appears to be present even when plasma [LDL–cholesterol] is increased.

LIPID ANALYSES IN DIAGNOSIS

The categories of primary hyperlipoproteinaemia discussed above are becoming better defined in terms of aetiology. However, they still have only limited relevance to the treatment of hyperlipidaemia, and precise classification of the disorder may not always be necessary. Many of the measurements that can be performed, and which were required in order to substantiate the descriptions given earlier in the chapter, are not routine procedures nor does there seem any likelihood that they will soon become so.

Specimen collection

It is important to collect specimens for plasma lipid and lipoprotein studies under the appropriate, standardized conditions:

(1) The patient should have been leading a normal life, in terms of diet (including alcohol consumption) and exercise, for at least the previous fortnight.
(2) Blood specimens should be collected after an overnight 10–14 h fast, if triglyceride measurements are to be performed.
(3) The technique of venepuncture should be standardized, especially if it is intended to follow the progress of patients, or monitor the effects of treatment, on a long-term basis. Points to standardize include:
    (a) Ambulant patients should have been sitting for at least 15 min before venepuncture is performed.
    (b) Venous stasis should be minimal.
(4) The anticoagulant recommended by WHO is EDTA, but lithium heparin is widely regarded as acceptable.

Routine investigations

Certain investigations should be performed as a routine, for diagnostic purposes, in all cases of suspected hyperlipidaemia before treatment is started. We recommend the following:

(1) Plasma [cholesterol].
(2) Plasma [triglycerides]; this specimen *must* be collected under fasting conditions.
(3) Inspection of plasma that has been stored at 4°C, undisturbed, for 12–18 h, if the patient has hyperlipidaemia.
(4) Plasma [HDL–cholesterol] may be indicated in some patients.

Measurements of plasma [cholesterol] and [triglycerides] are the essential preliminary investigations. If the results are abnormal, it is advisable to repeat the observations to confirm the abnormality before starting treatment. Reference values for U.K. adults are given in Table 12.4.

Helpful information about the nature of the lipoproteins giving rise to hypertriglyceridaemia can be obtained by the inspection of stored plasma (Table 12.2). Triglyceride-rich particles (chylomicrons and VLDL) scatter light, whereas triglyceride-poor particles do not. Also, chylomicrons float to the surface if the specimen is left *undisturbed* for several hours (usually overnight), but VLDL do not float under these conditions.

Measurement of plasma [HDL–cholesterol] has become routine in some laboratories. It helps to define the risk of ischaemic heart disease, and is needed for the diagnosis of hyper–α–lipoproteinaemia (p. 256), and Tangier disease (p. 260).

TABLE 12.4. Reference values for routine plasma lipid investigations (fasting conditions). After Mount *et al.* (1982).

		mmol/L	mg/100mL
Cholesterol:		3.6–6.7	140–270
Triglycerides:	males	0.7–2.1	65–190
	females	0.6–1.5	55–135
HDL-cholesterol:	males	0.8–1.6	32–64
	females	0.9–1.8	36–72

Specialized investigations

Occasionally, further and more specialized investigations may be indicated, on the basis of clinical features and the results of the preliminary, routine investigations.

Patients with a mixed hyperlipidaemia, in whom both fasting plasma [cholesterol] and fasting plasma [triglycerides] are increased, may require to be investigated by ultracentifuge studies of plasma, carried out in a specialized laboratory. These ultracentrifuge studies may be needed to establish a diagnosis of remnant hyperlipoproteinaemia, by demonstrating 'floating β–lipoproteins'. However, in most of these patients, lipoprotein electrophoresis is usually the only specialized investigation required, to demonstrate a broad β-band.

Measurement of plasma lipoprotein lipase activity, following an intravenous injection of heparin to release the enzyme, may be needed to establish the diagnosis in some patients in whom there is chylomicronaemia.

Caveat

The results of plasma lipid investigations may be seriously misleading if patients are investigated for suspected hyperlipidaemia within 3 months of a serious illness such as a myocardial infarction or a major operation. Stresses of this kind may cause prolonged hypertriglyceridaemia and may also sometimes temporarily reduce plasma [cholesterol]. Even minor illnesses may cause shorter-lived alterations in plasma lipid concentrations, but these may be sufficient to prevent the correct diagnosis being made.

TREATMENT OF HYPERLIPIDAEMIA

It is possible to lower plasma [LDL–cholesterol] by dietary or other means, but little is known as yet about ways of modifying plasma [HDL]. Both primary and secondary prevention trials have shown that reduction in plasma [cholesterol] is accompanied by a small but significant reduction in the overall rate of non-fatal myocardial infarction; no change in the rate of fatal infarction has been demonstrated.

In *patients with mild hyperlipidaemia,* it is usually only necessary to alter the diet. Reduction in fat intake and substitution of much of its saturated fat content by polyunsaturated fats, combined with an increase in vegetable fibre, will usually result in a lowering of plasma [cholesterol] by 10–20%. Since changes of this magnitude can also be produced by varying the method of specimen collection, the importance of standardizing the conditions of venepuncture should be apparent. Reduction of plasma [triglycerides] is more likely to be achieved by restricting the energy intake and limiting the amount of dietary carbohydrate.

In *patients with more severe hyperlipidaemia,* more aggressive treatment is indicated, with drugs, especially in young patients. Cholestyramine is the drug most often used to reduce plasma [cholesterol]; it binds bile acids in the intestine, thereby interrupting their enterohepatic circulation and increasing bile acid excretion in the faeces. Nicotinic acid may be used to treat patients with mixed hyperlipidaemia, as it reduces both plasma [cholesterol] and [triglycerides]. Clofibrate is also effective in mixed hyperlipidaemia, but is probably best restricted to patients with remnant hyperlipoproteinaemia, in whom it is very effective. All these forms of drug treatment may have side-effects.

The hypolipoproteinaemias

Three rare familial diseases require brief mention. Their recognition has helped with the understanding of normal lipoprotein metabolism.

Tangier disease, α–lipoprotein deficiency, is inherited as an autosomal recessive condition. It is due to very low plasma [Apo A–I], caused by this protein having a very high catabolic rate in these patients rather than by a defect in its synthesis. Only traces of HDL are detectable in plasma, and plasma [LDL–cholesterol] is also reduced. Cholesterol esters accumulate in the lymphoreticular system, probably due to excessive phagocytosis of the abnormal chylomicron and VLDL remnants that result from the Apo A–I deficiency.

Abetalipoproteinaemia, another condition with autosomal recessive inheritance, is associated with a complete absence of Apo B. The lipoproteins that normally contain Apo B in significant amounts, i.e. chylomicrons, VLDL, IDL and LDL, are absent from plasma in consequence. Plasma [cholesterol] and [triglycerides] are very low.

Hypobetalipoproteinaemia. This condition has autosomal dominant inheritance and is distinct from the previous condition. Synthesis of Apo B is decreased, but VLDL and LDL, although reduced, are present in plasma. Plasma [cholesterol] is decreased, but not as markedly as in abetalipoproteinaemia.

Secondary hypolipidaemia

Greatly reduced plasma [cholesterol] and [triglycerides] occur in patients with gross protein-energy malnutrition, e.g. kwashiorkor in young children. Low plasma [cholesterol] may also be a feature of severe, long-standing hepatocellular disease or intestinal malabsorptive disease. Hepatic synthesis of proteins is also reduced.

FURTHER READING

Bachorik, P.S. (1982). Collection of blood samples for lipoprotein analysis. *Clinical Chemistry*, **28,** 1375–1378.

Beaumont, J.L., Carlson, L.A., Cooper, G.R., Fejfar, Z., Fredrickson, D.S. and Strasser, T. (1970). Classification of hyperlipidaemias and hyperlipoproteinaemias. *Bulletin of the World Health Organisation,* **43,** 891–915.

Buckley, B.M. and Bold, A.M. (1982). Managing hyperlipidaemias. *British Medical Journal,* **285,** 1293–1294.

Chait, A. (1978). Hyperlipoproteinaemia — an approach to diagnosis and classification. In *Recent Advances in Clinical Biochemistry,* **1,** 73–109. Ed. K.G.M.M. Alberti. Edinburgh: Churchill Livingstone.

Gordon, T., Castelli, W.P., Hjortland, M.C., Kannel, W.B. and Dawber, T.R. (1977). High density lipoprotein as a protective factor against coronary heart disease – The Framingham study. *American Journal of Medicine,* **62,** 707–714.

Lewis, B. (1976). *The Hyperlipidaemias: Clinical and Laboratory Practice.* Oxford: Blackwell Scientific Publications.

Lewis, B. (1983). The lipoproteins: predictors, protectors and pathogens. *British Medical Journal*, **287,** 1161–1164.

Levy, R.I. (1981). Cholesterol, lipoproteins, apoproteins, and heart disease: Present status and future prospects. *Clinical Chemistry,* **27,** 653–662.

Miller, N. E. and Lewis, B., editors (1981). *Lipoproteins, Atherosclerosis and Coronary Heart Disease*. Amsterdam: Elsevier.

Mount, J., Rao, S., Miller, N., Hammett, F., Whiting, C. and Lewis, B. (1982). Laboratory investigation of lipid disorders. *Association of Clinical Pathologists, Broadsheet*, **101.** London: British Medical Association.

Oliver, M.F. (1984). Hypercholesterolaemia and coronary heart disease: an answer. *British Medical Journal*, **288,** 423–424.

Stanbury, J. B., Wyngaarden, J. B., Fredrickson, D. S. Goldstein, J. L. and Brown, M. S., editors (1983). *The Metabolic Basis of Inherited Disease*, 5th Edition, pp. 589–747. New York: McGraw-Hill.

Westlund, K. and Nicolaysen, R. (1972). Ten-year mortality and morbidity related to serum cholesterol. *Scandinavian Journal of Clinical & Laboratory Investigation,* **30,** *Supplement*, **127.**

Chapter 13
Deficiencies of Vitamins and Trace Elements

VITAMIN DEFICIENCIES

Vitamins are all organic compounds which, as originally defined, cannot be synthesized by man and must be provided in the diet. They are essential for the normal processes of metabolism, including growth and maintenance of health. It is now known that man is able to produce part or even all of his requirements for some of the vitamins, e.g. vitamin D from 7-dehydrocholesterol and nicotinic acid from tryptophan.

There are five main groups of causes of a vitamin deficiency state — inadequate diet, impaired absorption, insufficient utilization, increased requirement, and increased rate of excretion. Vitamin deficiency develops in stages:

(1) *Subclinical deficiency,* in which there is depletion of body stores. These stores are normally relatively large in the case of fat-soluble vitamins (e.g. A and D) and vitamin B_{12}, but are small in the case of most of the water-soluble vitamins. The fact that the stores of a vitamin have become depleted may only become evident at times of increased demand. For instance, late neonatal hypocalcaemia (p. 433) tends to occur particularly in infants of mothers who are themselves deficient in vitamin D, and delayed wound healing may occur in patients with mild, previously subclinical, vitamin C deficiency.
(2) *Overt deficiency.* This stage is usually accompanied by other evidence of malnutrition (e.g. protein-energy malnutrition).

The history of the patient's illness, including the dietary and social history, as well as the findings on clinical, radiological (e.g. rickets) and haematological (e.g. anaemia) examination may all contribute to the diagnosis in the later stages of the development of a vitamin deficiency state.

Chemical investigations not only help to confirm the diagnosis of overt disease but may enable the diagnosis to be made at an earlier stage. Several categories of chemical tests can be recognized, only some of which may be applicable for the investigation of suspected deficiency of a particular vitamin:

(1) Direct measurement of [vitamin] in whole blood, plasma, erythrocytes, leucocytes or tissue biopsy specimens.
(2) Direct measurement of the amount of a vitamin or one of its major metabolites in urine.
(3) Measurement in blood or urine of a metabolite that accumulates as a result of a partial or complete blockage of a metabolic pathway that involves an enzyme which requires the vitamin (or derivative) as a cofactor or prosthetic group (e.g. methylmalonic acid excretion in vitamin B_{12} deficiency).
(4) Measurements as in (3), but only when the pathway has been placed under extra load (e.g. tryptophan load test and formiminoglutamic acid excretion in folic acid deficiency).
(5) Saturation tests (e.g. ascorbic acid saturation test in vitamin C deficiency).
(6) Enzyme cofactor saturation tests, in which the activity of the enzyme in a specimen is measured with and without the addition of the enzyme's cofactor or prosthetic group (e.g. erythrocyte transketolase activity, with and without the addition of thiamin pyrophosphate, in vitamin B_1 deficiency); the degree of enhancement of the enzyme's activity is a measure of the severity of the deficiency.

When interpreting the results of plasma measurements, it is important to note that plasma concentrations of vitamins do not necessarily reflect the vitamin status of the body. They generally fall before cellular and tissue levels fall, but are usually insensitive measures of deficiency, i.e. low or undetectable plasma levels can be obtained in the absence of a deficiency state. Recent dietary intake may cause plasma [vitamin] to fluctuate markedly, even in the presence of severe deficiency, but a sustained high plasma [vitamin] will usually exclude a deficiency state for that vitamin.

Measurements of the concentrations of vitamins in erythrocytes, leucocytes or tissue biopsy specimens are, in general, to be preferred to plasma vitamin measurements, since they give a much better indication of the body's vitamin status.

For those vitamins or vitamin metabolites that can be measured in urine, it should be noted that low output figures may be observed despite the presence of adequate reserves in the tissues. However, a high level of urinary excretion means that a deficiency state is improbable.

Deficiency of fat-soluble vitamins

VITAMIN A

This vitamin is present in the diet as retinol or as β–carotene, which is hydrolysed in the intestine to form retinol. After absorption from the intestine, and esterification in the mucosal cells, vitamin A ester is transported in the blood bound to protein (retinol-binding protein or pre-albumin). Specific binding proteins on cell membranes are involved in the uptake of vitamin A from plasma into the tissues. The vitamin is stored in the liver, mainly as its ester.

The active form of vitamin A, 11–cis–retinal, is involved in rod vision and vitamin A deficiency can cause night blindness. Vitamin A is also involved, in an unknown manner, in mucopolysaccharide synthesis; xeroderma and xerophthalmia are effects of vitamin A deficiency related to defects in mucopolysaccharide formation. Low plasma [vitamin A] has been shown to be associated, in man, with an increased risk of developing cancer.

Plasma [vitamin A] may also be decreased in states of severe protein deficiency, due to lack of retinol-binding protein or of pre-albumin, and may increase if the protein deficiency is corrected, e.g. in treated kwashiorkor. Reduced urinary excretion of sulphated mucopolysaccharides has been reported in children with signs of vitamin A deficiency.

The laboratory measurement most frequently carried out is determination of plasma [vitamin A]. If this is less than 0.5 μmol/L (0.14 mg/L or approximately 0.5 I.U./L) the finding supports a diagnosis of vitamin A deficiency. Plasma [vitamin A], however, provides only limited information about the tissue stores.

Vitamin A absorption tests, based on measurements of plasma [vitamin A] before and 5 h after a large oral dose of vitamin A (50 μmol/kg), have been used as a test of intestinal absorptive function, but have not been widely accepted.

VITAMIN D

Cholecalciferol (vitamin D_3) is normally present in the diet. It can also be formed *in vivo* from 7–dehydrocholesterol (pro-vitamin D_3), a minor metabolite of cholesterol, provided that the skin is exposed to UV light. Some foods (e.g. margarine) may be fortified with ergosterol, which is converted by UV light to pro-vitamin D_2, the precursor of ergocalciferol (vitamin D_2). Endogenous synthesis of vitamin D is important; vitamin D deficiency can develop if exposure to sunlight is inadequate, or can occur solely because of inadequate intake, but usually arises as a result of the combined effects of both these reasons.

Vitamin D is stored in the liver, adipose tissue and muscle. These stores may be large enough to delay the onset of vitamin D deficiency for months or even years after vitamin D intake stops. In the body, both vitamin D_2 and D_3 undergo two hydroxylation steps before attaining full physiological activity. These will be described for vitamin D_3 (Fig. 13.1):

(1) *25–hydroxylation, in the liver,* with the production of 25–hydroxycholecalciferol (25–HCC), or calcidiol. This process is regulated by the amount of 25–HCC in the liver. The main circulating form of vitamin D in the plasma is 25–HCC, which circulates bound to a specific transport protein and which is carried to the kidney for further metabolism there.

(2) *1α–hydroxylation of 25–HCC, in the kidney*, with the production of 1 : 25–dihydroxycholecalciferol (1 : 25–DHCC), or calcitriol; this is the most active derivative of vitamin D. The kidney also contains a 24–hydroxylase, which converts 25–HCC to 24 : 25–dihydroxycholecalciferol (24 : 25–DHCC).

The pathway for further metabolism of 25–HCC in the kidney is determined by the body's requirement for calcium. If there is a tendency to hypocalcaemia, metabolism of 25–HCC is directed by parathyroid hormone (PTH) towards the formation of 1 : 25–DHCC, catalysed by 1α–hydroxylase. If plasma [calcium] is normal, or in the presence of hypercalcaemia, metabolism of 25–HCC is directed towards the formation of 24 : 25–DHCC, by 24–hydroxylase. The activity of renal 1 α–hydroxylase is also affected by the local [1 : 25–DHCC], which exerts a negative feedback effect.

Experimentally, many other factors have been shown to influence the hydroxylation pathway of 25–HCC in the kidney. The activity of 1α–hydroxylase is modified by changes in the plasma concentrations of calcium, phosphate, parathyroid hormone, calcitonin, growth hormone and prolactin.

The most active derivative of vitamin D, 1 : 25–DHCC, has hormonal actions. It is an important factor in the absorption of calcium from the intestine and in the deposition and resorption of calcium salts in bone; it may also have direct, PTH-like effects on the renal tubule. The metabolic effects of 24 : 25–DHCC on intestinal absorption appear to be similar to those of 1 : 25–DHCC, but it is much less potent; 24 : 25–DHCC has little or no action on bone.

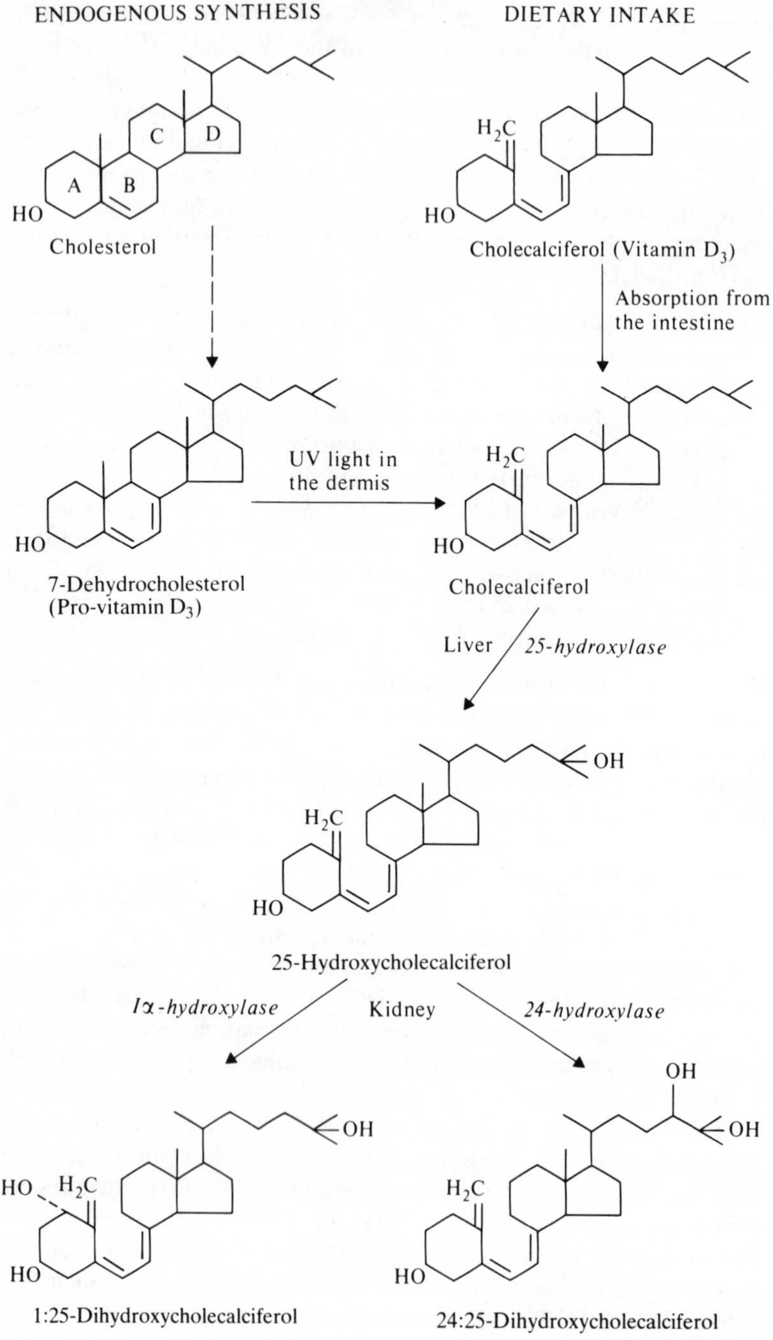

ENDOGENOUS SYNTHESIS

DIETARY INTAKE

A B C D

HO

Cholesterol

H_2C

HO

Cholecalciferol (Vitamin D_3)

Absorption from the intestine

HO

7-Dehydrocholesterol
(Pro-vitamin D_3)

UV light in the dermis

H_2C

HO

Cholecalciferol

Liver *25-hydroxylase*

OH

H_2C

HO

25-Hydroxycholecalciferol

1α-hydroxylase Kidney *24-hydroxylase*

OH

HO H_2C

HO

1:25-Dihydroxycholecalciferol

OH

OH

H_2C

HO

24:25-Dihydroxycholecalciferol

Vitamin D deficiency, when it occurs, tends to do so in conjunction with calcium deficiency; both deficiencies may then be at least partly of dietary origin. However, the calcium deficiency may be entirely due to impaired absorption of calcium from the intestine secondary to vitamin D deficiency. The deficiency of vitamin D may be due to (1) insufficient intake in the diet, (2) failure of absorption (e.g. intestinal malabsorptive disease), (3) inadequate exposure to sunlight, or (4) inadequate conversion of 25–HCC to 1 : 25–DHCC in chronic renal disease (p. 298). Failure to form 25–HCC because of liver disease has not been described.

Until recently, chemical investigations of suspected vitamin D deficiency were restricted to measurements of plasma [calcium], [phosphate] and alkaline phosphatase activity and other indices that only become abnormal in the presence of fairly gross deficiency. The finding of decreased plasma [calcium], often associated with low [phosphate] and increased alkaline phosphatase activity, supports a clinical or radiological diagnosis of rickets or osteomalacia. If there is generalized nutritional deficiency, there may also be decreased plasma [albumin], which tends to exaggerate the degree of hypocalcaemia.

Within the last few years, methods for measuring the hydroxylated derivatives of vitamin D have been developed, for 25–HCC, 1 : 25–DHCC and 24 : 25–DHCC. Of these, plasma [25–HCC] is the most readily available measurement, and it provides a reasonable indication of the overall vitamin D status of the patient, if renal function is normal and renal 1α–hydroxylase activity can therefore be assumed to be normal. There is a seasonal variation in the normal plasma [25–HCC], which can make the interpretation of single measurements difficult.

VITAMIN K

This vitamin is needed for hepatic synthesis of prothrombin and factors VII, IX and X. Vitamin K deficiency is most often the result of treatment with anticoagulants (e.g. phenindione); control of treatment is usually the responsibility of haematology laboratories. Tests performed include the Quick one-stage prothrobin time and Thrombotest (British Drug Houses, Poole, Dorset). These tests are

FIG. 13.1. *Formation of 1:25 dihydroxycholecalciferol* (1:25-DHCC) from dietary vitamin D_3 (the main source) and from cholesterol *in vivo*. By the action of UV light, pro-vitamin D_3 is converted to pre-vitamin D_3 (not shown) in which the B-ring of the steroid skeleton has been opened; pre-vitamin D_3 rearranges spontaneously to give vitamin D_3. Factors that influence the hydroxylation of cholecalciferol and 25-HCC are discussed in Chapter 14.

also important in the investigation and management of jaundiced patients, in whom they provide an index of the liver's ability to synthesize this group of coagulation factors (p. 179).

Deficiency of water-soluble vitamins

THIAMIN (VITAMIN B_1)

As its pyrophosphate, thiamin is one of the cofactors essential for the oxidative decarboxylation of pyruvate and of 2–oxoglutarate. It is also the coenzyme of transketolase, an enzyme concerned with aldehyde group transfers. A clinical diagnosis of thiamin deficiency can be investigated by several chemical tests.

Erythrocyte transketolase

This provides a specific and sensitive index of tissue [thiamin] and is the measurement of choice for investigating possible thiamin deficiency. Enzyme activity is measured in red cell haemolysates before and after addition of thiamin pyrophosphate (TPP). The increase in enzymic activity produced by adding TPP (the TPP effect) gives a direct indication of the degree of thiamin deficiency.

Urinary thiamin

These measurements have been extensively used in nutritional surveys, where they have provided a useful criterion of deficiency. As with all the water-soluble vitamins, however, thiamin excretion is considerably influenced by recent dietary intake and by renal function.

Blood pyruvate

Fasting blood [pyruvate] has been used to provide an index of thiamin deficiency, usually as part of a pyruvate tolerance test in which the pathway of metabolism is stressed by giving the fasting patient a 100 g glucose load. In patients with thiamin deficiency, the fasting blood [pyruvate] or the [pyruvate] in at least one of the post-glucose specimens (or in both types of specimen) is increased. These tests are much less specific for thiamin deficiency than erythrocyte transketolase measurements, and are rarely performed nowadays.

RIBOFLAVIN (VITAMIN B_2)

Deficiency of riboflavin usually occurs as part of a mixed deficiency state involving several vitamins of the B_2 complex, a term used to include all the B–vitamins except thiamin; usually the deficiency state extends to include thiamin (vitamin B_1) as well.

The nucleotides of riboflavin, flavin mononucleotide (FMN) and flavin-adenine dinucleotide (FAD), are the prosthetic groups of many enzymes. The activity of one of these, glutathione reductase, can be measured in haemolysed erythrocytes before and after the addition of FAD (the FAD effect), and this provides a useful test for detecting riboflavin deficiency.

PYRIDOXINE

The term vitamin B_6 includes pyridoxine, pyridoxal, pyridoxamine and their 5–phosphate derivatives. The active form of the vitamin is pyridoxal phosphate (PP). Investigation of suspected pyridoxine deficiency provides a third example of a B–vitamin that can have its status assessed by an enzyme cofactor saturation test (Table 13.1), by determining the effect of adding PP upon ALT or AST activity in erythrocyte haemolysates.

The *tryptophan load test* is another way of investigating pyridoxine deficiency. In this test, the 24–h urinary excretion of xanthurenate or of 3–hydroxykynurenine is measured before and after a loading dose of 2 g L-tryptophan, which places a stress on the main metabolic path (Fig. 13.2), with its pyridoxine-dependent enzymes. Considerably increased amounts of 3–hydroxykynurenine and xanthurenate are excreted following the loading dose of tryptophan in patients with pyridoxine deficiency. The excess formation of xanthurenate in these patients is due to the presence of an alternative metabolic pathway (not dependent on pyridoxine) from 3–hydroxykynurenine.

TABLE 13.1. Enzyme tests on haemolysed erythrocytes for vitamin deficiency states.

Vitamin	Enzyme	Prosthetic group
Thiamin (B_1)	Transketolase	Thiamin pyrophosphate (TPP)
Riboflavin (B_2)	Glutathione reductase	Flavin-adenine dinucleotide (FAD)
Pyridoxine (B_6)	Aminotransferases (ALT and AST)	Pyridoxal phosphate (PP)

NICOTINIC ACID AND NICOTINAMIDE

Nicotinamide is the form in which nicotinic acid is incorporated into nicotinamide-adenine dinucleotide (NAD) and nicotinamide-adenine dinucleotide phosphate (NADP). The body's requirement for nicotinamide is met partly from dietary nicotinic acid and nicotinamide, and partly from metabolism of tryptophan (Fig. 13.2)

Deficiency can be caused by an inadequate intake of nicotinic acid and nicotinamide, or by conditions in which large amounts of tryptophan are metabolized along abnormal pathways (e.g. the

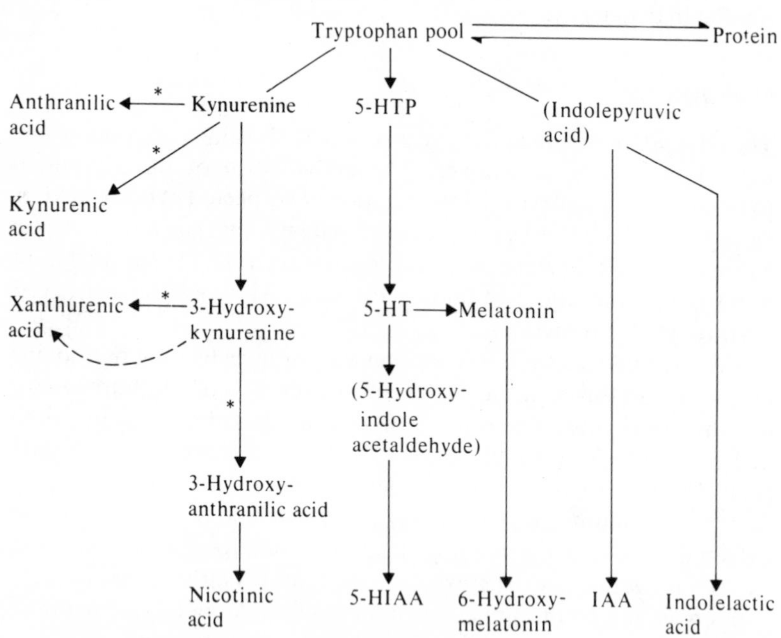

FIG. 13.2. *The metabolism of trytophan*. The main degradation pathway in man is normally via kynurenine and 3-hydroxykynurenine to nicotinic acid. In pyridoxine deficiency, an alternative (normally minor) pathway from 3-hydroxykynurenine to xanthurenate becomes quantitatively important; this pathway (shown by a broken arrow) is not pyridoxine-dependent.

The amount of trytophan metabolized along the 5-HIAA pathway is greatly increased in the carcinoid syndrome. The output of IAA is increased in phenylketonuria and in Hartnup disease.

Key to abbreviations

Arrows marked * all involve enzymes that require pyridoxal phosphate as cofactor.

5-HTP = 5-hydroxytryptophan 5-HIAA = 5-hydroxyindoleacetic acid
5-HT = 5-hydroxytryptamine IAA = indoleacetic acid

carcinoid syndrome, p. 454; Hartnup disease, p. 98). It gives rise to pellagra.

The best chemical methods for detecting deficiency measure the urinary excretion of nicotinic acid metabolites, either N′–methylnicotinamide or 2–pyridone. Alternatively, microbiological assays that measure the active vitamin can be used.

FOLIC ACID

Deficiency of folate is one of the commonest vitamin deficiency states in man; it is one cause of megaloblastic anaemia. Methods of investigation include serum [folate] and erythrocyte [folate]. These tests are usually carried out in haematology departments, by microbiological or competitive protein-binding methods.

Measurement of formiminoglutamate (FIGLU) in urine is essentially a subsidiary investigation in patients with suspected folate deficiency. It is also sometimes used as a screening procedure in patients with malabsorption from the intestine. FIGLU is an intermediate in the degradation of histidine to glutamate, and its further metabolism requires the presence of tetrahydrofolate. If there is folate deficiency, urinary excretion of FIGLU may be increased. However, if the patient is given a dose of 15 g L–histidine by mouth, in a *histidine load test,* this places a stress on the pathway for conversion of histidine via FIGLU to glutamate and FIGLU excretion may then become greatly increased.

VITAMIN B_{12}

Deficiency of vitamin B_{12} also causes megaloblastic anaemia. This is usually diagnosed by haematological examination of blood and bone marrow specimens. Measurements of serum [vitamin B_{12}], by microbiological or competitive protein-binding methods, and investigation of vitamin B_{12} absorption from the intestine before and after the administration of intrinsic factor (p. 209), are tests normally performed by haematology departments.

Examination of gastric secretion for pentagastrin-fast achlorhydria, a feature of Addisonian pernicious anaemia, is discussed elsewhere (p. 196). Another chemical test is measurement of methylmalonic acid (MMA) excretion in urine. Conversion of methylmalonyl-CoA to succinyl-CoA is catalysed by an enzyme which requires vitamin B_{12} as cofactor, and urinary excretion of MMA is increased in vitamin B_{12} deficiency. It is also sometimes increased in hepatic disease.

ASCORBIC ACID (VITAMIN C)

Deficiency of vitamin C causes scurvy, which was recognized as a dietary deficiency disease long before the concept of vitamins was proposed. Frank scurvy rarely occurs nowadays, but its subclinical form is by no means uncommon, especially among elderly people living alone.

The role of ascorbic acid in metabolism is not yet known with certainty. However, it is involved in the hydroxylation of proline and lysine, during the synthesis of collagen. Confirmation of a diagnosis of vitamin C deficiency depends on measurements of [ascorbate] in blood or urine.

Plasma ascorbate

This provides a reasonable index of dietary intake, but is a poor index of tissue stores. Plasma [ascorbate] falls rapidly when the diet is very deficient in vitamin C. While it is possible to say that all patients with scurvy will have plasma [ascorbate] so low as to be undetectable, not all people in whom ascorbate cannot be detected in plasma have scurvy.

Leucocyte ascorbate

These measurements provide a more reliable assessment of tissue stores of ascorbate. In practice, difficulties involved in obtaining leucocytes uncontaminated by other cellular elements mean that the buffy layer, consisting of leucocytes and platelets (and a few erythrocytes) is examined instead.

Leucocytes and platelets take up ascorbate against a concentration gradient, and may retain most of their ascorbate even when plasma [ascorbate] has fallen to undetectable levels. Buffy layer [ascorbate] falls at about the same time as clinical evidence of scurvy appears, and seems to give a good indication of the body's stores of the vitamin. Cellular abnormalities in the components of the buffy coat can cause misleading results unless correcting factors are applied.

Urinary ascorbate

This measurement is only of practical diagnostic value if performed as part of a therapeutic trial, in the *ascorbic acid saturation test.*

When regular daily loading doses of ascorbic acid (10 mg/kg body weight) are given to patients deficient in vitamin C, the tissues preferentially take up large amounts of the vitamin to replenish their stores and little or none is excreted in the urine for several days or even weeks. Patients whose vitamin C intake has been adequate begin to excrete large amounts within 1–2 days. Care has to be taken over the preservation of urine specimens collected in this test as ascorbic acid is unstable.

TRACE ELEMENT DEFICIENCIES

The elements present in the body can be arbitrarily subdivided into groups:

(1) *The major elements:* H, C, N, O, Na, Mg, P, S, Cl, K and Ca. These are either components of vitamins or of compounds discussed elsewhere in this book.
(2) *Essential trace elements:* Cr, Mn, Fe, Co, Cu, Zn, Se, Mo and I. Some of these are discussed in this section. Others (Fe and I) are discussed elsewhere.
(3) *Other elements,* including 5 trace elements (F, Si, V, Ni and Sn) that may be essential. These will not be discussed.

The arbitrary definition of a trace element is one that is present in the body in amounts less than 100 mg/kg.

The essential trace elements (chromium, manganese, iron, cobalt, copper, zinc, selenium, molybdenum and iodine) are all metals, apart from iodine, and all are required for optimum health. Deficiencies of trace elements may cause disease, and deficiency may arise under various circumstances:

(1) *Neonatal period,* especially in premature infants being fed parenterally. Also, infants fed on milk from breast-milk banks may develop trace metal deficiencies since breast milk normally contains fairly high [trace elements] shortly after parturition but these levels fall progressively thereafter.
(2) *Protein-energy malnutrition* (p. 437).
(3) *Synthetic or liquid protein diets,* e.g. children being treated for phenylketonuria.
(4) *Total parenteral nutrition* (TPN, p. 61).
(5) *Dieting,* e.g. patients who have been on very low energy diets for weight reduction, or committed vegans (meat provides one of the main sources of trace elements).
(6) *Excess losses,* due to chronic, severe gastrointestinal disease (e.g. Crohn's disease with fistulae, extensive gut resection, protracted

diarrhoea), or to treatment with chelating agents, or in patients on renal dialysis.

(7) *Inherited disorders* of specific mechanisms for handling the absorption or transport of trace elements. These are all very rare.

In adults, the commonest group of causes liable to give rise to trace element deficiency is excessive losses from the gastrointestinal tract, either postoperative or in patients with chronic, severe gastrointestinal disease, especially when these patients are being managed by total parenteral nutrition.

Single deficiencies of trace elements may develop, but sometimes there are combined deficiencies, and often these accompany deficiencies in the organic or other inorganic components of the body's needs.

Methods of assessing trace element deficiency are inadequate at present, and mainly depend on blood measurements. Although plasma levels are helpful in diagnosis and management, if very low, smaller changes can often be attributed to changes in concentration of the plasma proteins to which trace metals are normally bound. In some cases, measuring the concentration of trace elements in erythrocytes proves helpful, but tissue measurements do not provide a satisfactory alternative. Some of the trace elements require specialized equipment for their determination, this equipment presently being available in only a few laboratories in the U.K.

Diagnosis of trace element deficiency, at present, can usually be only made retrospectively on the basis of (1) a clinical condition likely to give rise to deficiency, occurring in association with clinical symptoms that (2) can be attributed to trace element deficiency, and which (3) respond to treatment with the appropriate trace element or elements.

ZINC

This metal is an essential constituent of several enzyme systems, including those concerned with protein synthesis and nucleic acid synthesis, carbonic anhydrase and ALA dehydratase. The total body content in adults is about 2 g, and the recommended intake of zinc is 15–25 mg/day.

The body does not store zinc to any appreciable extent in any organ. Absorption from the intestine appears to be controlled in a manner similar to iron (p. 305), by sequestration of zinc in enterocytes which are later desquamated. In plasma, zinc is mostly transported bound to albumin, α_2–macroglobulin and transferrin; the reference values are 12–25 μmol/L. Specimens of blood need to be collected before a meal and without venous stasis; plasma [zinc] may fall by as much as 20% after meals.

Zinc deficiency is characterized by dermatitis, diarrhoea, mental disturbance and lethargy. Symptoms and signs rapidly improve with the administration of zinc.

Mild forms of zinc deficiency may be due to any of the causes discussed above, and occur most often in infants or older patients receiving total parenteral nutrition. More marked deficiency occurs in *acrodermatitis enteropathica,* in which there is an inherited defect of zinc absorption that causes low plasma [zinc] and reduced total body content of zinc.

Several much commoner diseases are associated with moderate reductions in plasma [zinc]. These include myocardial infarction, rheumatoid arthritis and chronic liver disease. The reduction in plasma [zinc] appears to be due, in these patients, to reduction in plasma [binding proteins] rather than to primary abnormalities of zinc metabolism. Symptoms of zinc deficiency are not a feature of these conditions.

COPPER

This metal is an essential constituent of cytochrome oxidase, superoxide dismutase and several other important enzyme systems. The total body content in adults is about 100 mg, and the recommended intake is at least 2 mg/day.

Plasma [copper] has a reference range of 12–26 μmol/L, most of this circulating bound to ceruloplasmin and the rest to transferrin. Copper is excreted in bile and urine.

Copper deficiency may develop in any of the conditions listed on p. 273, but is less common than zinc deficiency. It is associated with anaemia that does not respond to treatment with iron, unless copper is also given, and with neutropenia.

Menke's disease is a fatal, sex-linked recessive disorder in which there are cerebral and cerebellar degeneration and abnormalities of connective tissues. Plasma [copper] is very low, and the disease appears to be due to abnormal metabolism, but it does not respond to administration of copper.

FURTHER READING

BARKER, B.M. and BENDER, D.A., editors (1980 and 1982). *Vitamins in Medicine,* volumes 1 and 2. London: Heinemann.

BAYOUMI, R.A. and ROSALKI, S.B. (1976). Evaluation of methods of co-enzyme activation of erythrocyte enzymes for detection of deficiency of vitamins B_1, B_2, and B_6. *Clinical Chemistry,* **22,** 327–335.

CLAYTON, B.E. (1980). Clinical chemistry of trace elements. *Advances in Clinical Chemistry,* **21,** 147–176. New York: Academic Press.

FRASER, D.R. (1981). Biochemical and clinical aspects of vitamin D function. *British Medical Bulletin,* **37,** 37–42.

Leading article (1978). Deficiencies in parenteral nutrition. *British Medical Journal,* **2,** 913–914.

Leading article (1981). Another look at zinc. *ibid,* **1,** 1098–1099.

THURNHAM, D.I. (1981). Red cell enzyme tests of vitamin status: do marginal deficiencies have a physiological significance? *Proceedings of the Nutrition Society,* **40,** 155–163.

WALD, N., IDLE, M., BOREHAM, J. and BAILEY, A. (1980). Low serum-vitamin-A and subsequent risk of cancer. *Lancet,* **2,** 813–815.

Chapter 14

Disorders of Calcium and Magnesium Metabolism

Calcium is the most abundant mineral in the body, there being about 25 mol (1 kg) calcium in a 70 kg man compared with about 3.5 mol (80 g) sodium, 3.0 mol (120 g) potassium and 1.0 mol (24 g) magnesium. In adults calcium intake and output are normally in balance. In infancy and childhood, positive balance normally exists with net retention of calcium, especially at times of active skeletal growth. In old age, and in various disease states, calcium output may exceed input and a state of negative balance then exists.

The normal dietary intake of calcium in the U.K. is about 25 mmol/day (1 g/day); the minimum daily requirement in adults is about 0.5 g. Significant amounts of calcium (over 3 mmol/day) are normally contained in gastrointestinal secretions.

Absorption of calcium occurs from the small intestine and is considerably influenced by hormonal action, described below under homeostatic mechanisms. It may be impaired if the diet contains large quantities of inorganic phosphates, or organic phosphates such as phytate, which is present in cereals. The calcium excreted in faeces is derived partly from the diet and partly from intestinal secretions. The efficiency of calcium absorption tends to decrease with age, and increased dietary intakes may be needed in the elderly. The excretion of calcium in the urine varies considerably with the diet, but the relationship is not a simple linear one.

About 99% of the body's calcium is present in bone. Bone is made up of approximately 40% inorganic material, 20% organic matrix and 40% water, the inorganic material consisting mainly of the complex salt hydroxyapatite, $3\,Ca_3(PO_4)_2, Ca(OH)_2$.

On purely physicochemical considerations, calcium and phosphate should precipitate from the ECF as the solubility product, [calcium] × [phosphate], is exceeded. Deposition of bone salt, however, is not simply a passive physicochemical process. Instead, it seems that inhibitors of the process of calcification, possibly inorganic pyrophosphate, need to be removed. Alkaline phosphatase, which can catalyse the breakdown of inorganic pyrophosphate, may be concerned in this process. Calcification is defective if alkaline phosphatase is deficient, as in hypophosphatasia

(p. 148). Hormonal influences are very important, as discussed below.

Calcium salts in bone have a mechanical role, but are not metabolically inert, there being a constant state of turnover in the skeleton associated with deposition of calcium in sites of bone formation and release at sites of bone resorption. Calcium in bone acts as a reservoir which helps to stabilize the [Ca^{2+}] in the ECF.*

Other important functions, related to the ECF [Ca^{2+}], are the effects of Ca^{2+} on neuromuscular activity, membrane permeability, and the activity of many enzymes. Calcium is contained in many secretions, including milk.

PLASMA CALCIUM

The three forms in which calcium is present in plasma are shown in Table 14.1. These components are in equilibrium with one another, the position of the equilibria being altered by changes in plasma [protein] or [H^+], discussed below, or by complexing or chelating agents such as infusions of EDTA. Whatever the position of the equilibria, it is the plasma [Ca^{2+}] which is the physiologically important component. This fraction directly affects tissue responses (e.g. neuromuscular excitability), and controls the feedback mechanism responsible for PTH secretion.

There is a close physiological regulation of plasma [Ca^{2+}], which is the same as the ECF [Ca^{2+}]. Two hormones are known to be involved in the control process in man, parathyroid hormone (PTH) and 1 : 25-dihydroxycholecalciferol (1 : 25–DHCC); these hormones both act to increase plasma [Ca^{2+}] and hence plasma [calcium]. The importance in

TABLE 14.1. Calcium components in plasma.

Component	Percentage of total plasma calcium
Ionized calcium, Ca^{2+}	50–65
Bound to protein (mainly albumin)	30–45
Complexed with organic ions (e.g. citrate)	5–10

*In this book 'calcium' is used as a composite term (ionized, protein-bound, etc.) whereas Ca^{2+} means that only calcium ions are being considered. The concentration of calcium in a fluid (plasma, urine, etc.) is shown as, for example, plasma [calcium] whereas plasma [Ca^{2+}] refers specifically to the concentration of calcium ions. A similar convention is adopted for magnesium.

man of the calcium-lowering hormones, calcitonin and katacalcin, is less well established. Growth hormone, glucocorticoids (e.g. cortisol), oestrogens, testosterone and the thyroid hormones (T4 and T3) also exert a minor influence.

The body's response to a fall in plasma [Ca^{2+}], in terms of changes in PTH and 1 : 25–DHCC production, are shown in Fig. 14.1.

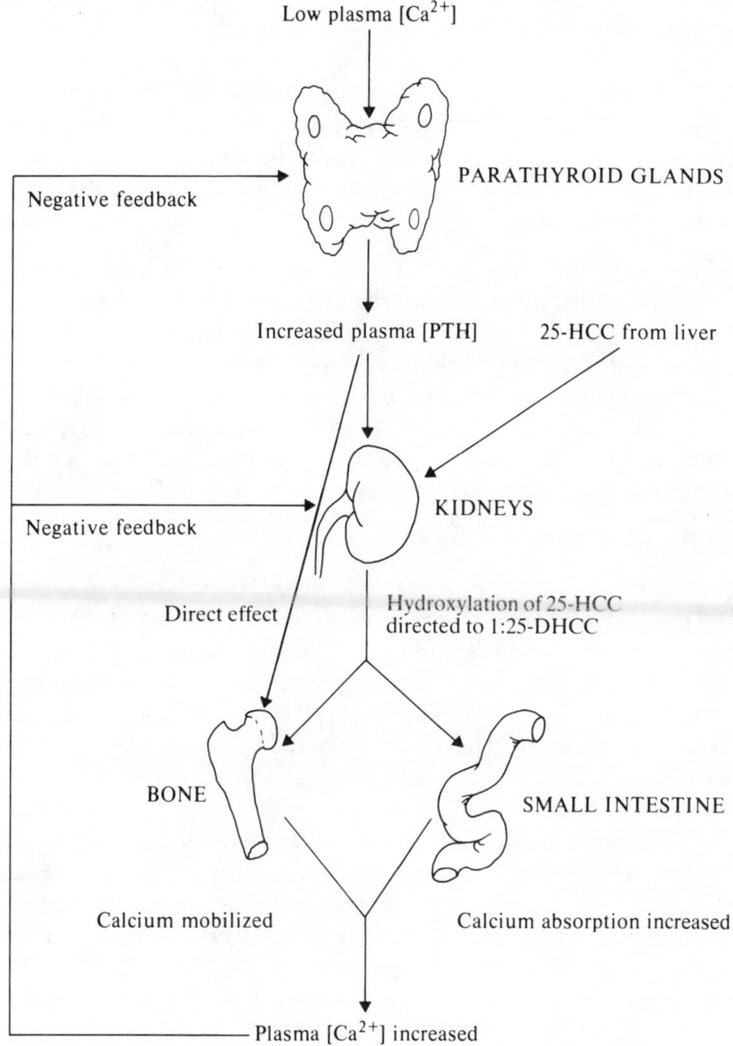

FIG. 14.1. *Calcium homeostasis in man*, showing the main hormonal responses to a fall in plasma [Ca^{2+}].

PARATHYROID HORMONE

Pure PTH is made up of 84 amino acids. The hormone, as first synthesized, contains 115 amino acids (pre-pro-PTH) but, thereafter, in quick succession, first 25 and then a further 6 amino acids are removed giving pro-PTH (90 amino acid residues) and then PTH, respectively. Biological activity lies in the N–terminal 30 amino acids. The secretion of PTH is controlled by the ECF [Ca^{2+}], and the principal function of PTH is to control ECF [Ca^{2+}].

The direct effect of PTH on bone leads to release of Ca^{2+} from bone crystals. It also exerts an important indirect effect, by promoting the production of 1:25–DHCC in the kidney (see below).

Parathyroid hormone has two other important effects on the kidney. It exerts the following direct effects on the proximal tubule, by promoting the release of cyclic AMP:

(1) *Decreases the tubular reabsorption of phosphate,* thereby increasing the amount of phosphate excreted in the urine.
(2) *Reduces the renal clearance of calcium.* However, this effect is more than offset by hypercalcaemia, which causes an increased amount of calcium to be filtered at the glomerulus. Consequently, in the presence of hypercalcaemia due to the effects of PTH, there may be increased renal excretion of calcium despite the fact that PTH has an opposite action directly on the kidney.

DIHYDROXYCHOLECALCIFEROL (DHCC)

The further metabolism of cholecalciferol to DHCC is described on p. 265. The absorption of Ca^{2+} across the small intestinal mucosal cells is promoted by 1 : 25–DHCC, and to a lesser extent by 24 : 25–DHCC. Parathyroid hormone exerts its action on the intestinal absorption of Ca^{2+} indirectly, via the effects of 1 : 25–DHCC. The exact mechanism of DHCC action is imperfectly understood, but involves the synthesis of a calcium-binding protein and facilitated diffusion of calcium across the mucosal cell.

The efficiency of calcium absorption increases if the dietary intake is restricted. The reduced intake tends to cause hypocalcaemia which stimulates PTH release, and this in turn promotes increased formation of 1:25–DHCC by the kidney (Fig. 14.1). Conversely, a high calcium diet directs 25–HCC metabolism in the kidney more towards the formation of 24:25–DHCC (Fig. 13.1), with its much less marked effect on calcium absorption.

The release of Ca^{2+} by osteoclasts is promoted by 1 : 25–DHCC, but 24 : 25–DHCC has little or no effect on bone.

CALCITONIN AND KATACALCIN

Calcitonin (CT) is a peptide containing 32 amino acids, one of two fragments of a larger peptide that both have plasma calcium-lowering actions; the other fragment is called katacalcin. Calcitonin is secreted by the 'C' cells of the thyroid, after cleavage from the precursor protein.

The importance of CT in the physiological regulation of plasma [Ca^{2+}] in man is uncertain. Experimentally its actions have been shown to include the inhibition of bone resorption and the reduction of hypercalcaemia towards normocalcaemia. In pharmacological doses, CT increases the renal excretion of calcium and phosphate, thereby tending to lower plasma [phosphate] as well as plasma [calcium].

Katacalcin (KC) contains 21 amino acids, and is adjacent to CT in the precursor polypeptide. Release of KC and CT occurs in approximately equimolar amounts. Both peptides have a plasma calcium-lowering action, but KC has no effect on plasma [phosphate]. As with CT, the physiological importance of KC is unknown.

Both plasma [CT] and [KC] are considerably increased in patients with medullary carcinoma of the thyroid (p. 457).

Plasma calcium measurements

Plasma [calcium] is normally kept within narrow limits, and the hormonal influences described, collectively serve to control plasma [Ca^{2+}] closely. The reference values for plasma [calcium] are 2.12–2.62 mmol/L (8.5–10.5 mg/100 mL). Calcium in plasma is made up of 3 components (Table 14.1), in equilibrium with one another. The equilibria can be easily disturbed, by physiological or therapeutic means and by pathological causes. The effects of alterations in plasma [albumin] and plasma [H^+] will be described, and the importance of avoiding unnecessary artefacts from these sources, when collecting specimens for plasma calcium measurements, will be stressed.

Plasma albumin

Alterations in plasma [albumin] affect the equilibrium between protein-bound and ionized calcium fractions. Since PTH maintains plasma [Ca^{2+}] constant, alterations in [albumin-bound calcium] are associated with changes in plasma [calcium]. In general, therefore, changes in plasma [calcium] occur in parallel with changes in plasma [albumin] in patients in whom there are disturbances of albumin metabolism (p. 116) but in whom the regulation of calcium metabolism is normal. For instance, in patients with the nephrotic syndrome, plasma [calcium] may be greatly reduced but plasma [Ca^{2+}] remains normal.

Correcting factors have been proposed in an attempt to offset the effects of variation in plasma [albumin] on [calcium], whether due to disease or excessive venous stasis applied at the time of specimen collection. The size of the correction depends on the extent to which plasma [albumin] differs from an arbitrarily selected albumin concentration, e.g. 41 g/L (4.1 g/100 mL). Values have been suggested for correcting factors that differ by as much as 40%. The application of a simple correction factor assumes there is little or no variation between individuals in the avidity of calcium-binding by proteins, a demonstrably incorrect assumption. We prefer not to apply such corrections automatically, and instead recommend:

(1) Standardize conditions of specimen collection. In particular, avoid venous stasis while the blood specimen is being withdrawn.
(2) Measure plasma [albumin] on specimens submitted for plasma [calcium] determinations, and take the results for plasma [albumin] into account when interpreting the plasma [calcium].

Plasma H^+

If plasma [H^+] falls, calcium binds more to albumin and plasma [Ca^{2+}] falls. If plasma [H^+] and [Ca^{2+}] fall rapidly, e.g. as a result of overbreathing, tetany may occur. If the acid–base disturbance develops more slowly, parathyroid activity has time to alter in response to the tendency for plasma [Ca^{2+}] to fall, and plasma [Ca^{2+}] is maintained.

If plasma [H^+] rises, less calcium is bound to protein and plasma [Ca^{2+}] increases. In conditions associated with chronic acidosis, e.g. chronic renal failure, although decreased plasma [calcium] is often observed, plasma [Ca^{2+}] remains normal. However, it may fall acutely if treatment is given to correct the acidosis rapidly (e.g. by infusing $NaHCO_3$), and tetany may then develop.

Where possible, when collecting specimens for plasma calcium measurements, care should be taken to avoid producing acute alterations in plasma [H^+], especially if plasma [Ca^{2+}] is to be determined.

PLASMA IONIZED CALCIUM

It would clearly be preferable to measure plasma [Ca^{2+}] rather than [calcium], and thereby make allowances for possible effects due to abnormalities in plasma [albumin]. However, the measurement of plasma [Ca^{2+}] on all the specimens where, at present, plasma [calcium] is requested would not be practicable.

Specimens for plasma [Ca^{2+}] have to be preserved with as much care as specimens for arterial blood gas analysis (p. 68) or changes in [H^+] will occur after collection and these will alter the equilibria between Ca^{2+} and the other components of calcium in the specimen. Such care is unlikely to be observed regularly, in view of the very large numbers of requests for plasma calcium measurements. Also, there are still technical problems in measuring plasma [Ca^{2+}] accurately, and the reference values for the measurement have not been as precisely established as they have been for plasma [calcium].

Reliable measurements of plasma [Ca^{2+}] are mainly obtained at present only in specialized centres. They are, in general, reserved for the investigation of those few patients in whom the presence of an abnormal [Ca^{2+}] is suspected but in whom difficulty has arisen over the interpretation of other investigations related to calcium metabolism. For instance, plasma [Ca^{2+}] might be of diagnostic importance in a patient in whom a diagnosis of primary hyperparathyroidism seems probable, on clinical and other grounds, but in whom plasma [calcium] is repeatedly found to be normal.

HYPERCALCAEMIA

The causes of hypercalcaemia are summarized in Table 14.2. The two commonest pathological causes are primary hyperparathyroidism and malignant disease. However, the commonest cause of borderline or slight degrees of hypercalcaemia is artefact, i.e. the potentially misleading hypercalcaemia caused by faulty technique (excessive venous stasis) when collecting blood specimens.

In the last 10–15 years, the apparent incidence of hypercalcaemia has increased severalfold. This is mainly due to the greatly increased numbers of measurements of plasma [calcium] being performed on asymptomatic individuals, or on patients whose reason for attendance at hospital was apparently unrelated to a disturbance of calcium metabolism, the measurement then being performed as part of a screening programme (p. 4). In such asymptomatic patients, the finding of hitherto unsuspected hypercalcaemia is due to primary hyperparathyroidism in over 80% of cases.

In a hospital in-patient population, the spectrum of diseases giving

TABLE 14.2. Causes of hypercalcaemia.

Artefact	Excessive venous stasis when collecting blood specimens
Parathyroid disease	Primary and tertiary hyperparathyroidism Multiple endocrine neoplasia (MEN) syndromes Secondary to ectopic PTH production (rare)
Excess intake of vitamin D, or calcium, or both	Self-medication with vitamin D, or overdosage (e.g. in treatment of hypocalcaemia) Milk-alkali syndrome
Bone disease	Carcinoma, with osteolytic secondary deposits Multiple myeloma, leukaemia Reticuloses involving bone Paget's disease (at times of increased activity or after being immobilized) Osteoporosis, if patient immobilized (rare)
Familial conditions	Familial hypocalciuric hypercalcaemia, alone or as part of MEN syndromes (rare)
Drug-induced	Thiazide diuretics
Miscellaneous causes	Sarcoidosis Thyrotoxicosis Adrenal insufficiency Hypercalcaemic periostitis

rise to a finding of hypercalcaemia is different. Malignant disease, considered as a group, is the commonest cause; in these patients, the malignant disease is often overt, or is suspected, and explaining a finding of hypercalcaemia does not present diagnostic difficulty. Hyperparathyroidism, renal disease (tertiary hyperparathyroidism, p. 299) and vitamin D therapy are the next most common causes. Recognition of these conditions, or of the others listed in Table 14.2, as the cause of hypercalcaemia, will usually present little difficulty once the possibility is considered.

Hypercalcaemia often directly damages the kidneys. Acute alterations in plasma [calcium] may cause tubular malfunction (p. 90). More commonly, hypercalcaemia of long duration causes irreversible renal damage due to nephrocalcinosis. The increased clearance of calcium by the kidneys predisposes to renal stone formation.

Primary hyperparathyroidism

This is an important disorder, usually caused by a parathyroid adenoma, less often by multiple adenomas, diffuse hyperplasia, or carcinoma.

Classically, patients present with renal calculi or with metabolic bone disease. However, nowadays, because of the growing practice of screening and the inclusion of measurements of plasma [calcium] in most screening programmes of chemical investigations, the majority of patients are asymptomatic when first diagnosed.

Evidence to support a diagnosis of primary hyperparathyroidism depends to a large extent, especially initially, on finding hypercalcaemia. It may be necessary to perform the measurement of plasma [calcium] on two or more occasions before hypercalcaemia is detected.

Other tests usually performed, and potentially valuable for the initial investigation of suspected primary hyperparathyroidism, are listed in Table 14.3, which also lists further tests that may be performed to substantiate the diagnosis. Of these, the measurement of choice is plasma [PTH], now that reliable methods are available.

PLASMA PHOSPHATE

There is a considerable variation in plasma [phosphate] during the day, especially following meals, and reference values only relate to specimens collected under fasting conditions; these values show considerable variation with age during infancy and childhood. If blood is collected from a fasting patient, low values (below 0.8 mmol/L or 2.5 mg/100 mL, expressed as phosphorus content) are often observed in patients with hypercalcaemia due to primary hyperparathyroidism, at least while renal function is unimpaired. This is a direct consequence of the phosphaturic action of PTH.

The combination of an increased plasma [calcium] and decreased fasting plasma [phosphate] supports a diagnosis of primary hyperparathyroidism. However, it is not possible to place great reliance, by itself, on the finding of a low plasma [phosphate] since it is not a constant feature of primary hyperparathyroidism. Also, even if a low plasma [phosphate] is detected, this finding is relatively non-specific.

ALKALINE PHOSPHATASE

The activity of this enzyme is routinely measured as part of any screening group of investigations for possible abnormalities of calcium metabolism. In patients with hyperparathyroidism, activity may be considerably increased, especially when there is radiological evidence of bone involvement. Alkaline phosphatase activity is increased in

TABLE 14.3. Chemical tests and the diagnosis of primary hyperparathyroidism.

*Initial investigations on plasma**	*Comments*
[Calcium]	If increased, supports the diagnosis
[Albumin]	Needed as a check on plasma [calcium]
[Phosphate], fasting	If decreased, supports the diagnosis
Alkaline phosphatase activity	If increased, supports the diagnosis
[Urea] or [creatinine]	Included as screening tests of renal function, advisable in all patients with suspected disturbances of calcium metabolism
Further investigation of hypercalcaemia	
Plasma [PTH] **	The definitive test. If PTH is detectable in presence of a raised plasma [calcium], and even more so if increased, this confirms the diagnosis.
Other tests that may be necessary	
Plasma [Ca^{2+}]	Needed for the diagnosis of normocalcaemic hyperparathyroidism
TmP/GFR	If reduced, supports the diagnosis. This test is much less essential than plasma [PTH]

* The only investigations required initially are plasma [calcium] and [albumin]. The others listed have been included either because they sometimes provide supporting evidence, or because they are nowadays widely included in a routine group of investigations used in screening for abnormalities of calcium metabolism.

** Plasma [calcium], [albumin] and [phosphate] need to be determined on another portion of the same specimen.

rickets and osteomalacia and in many other conditions giving rise to metabolic bone disease, especially if there is significant osteoblastic activity. It is also frequently increased in malignant disease.

PLASMA TOTAL CO_2

Patients with primary hyperparathyroidism often have a mild metabolic acidosis, whereas other causes of hypercalcaemia are usually associated with a mild metabolic alkalosis. Consequently, in primary hyperparathyroidism, plasma [total CO_2] tends to be reduced and plasma [Cl^-] increased whereas the opposite findings, increased plasma [total CO_2] and decreased plasma [Cl^-], are more usual in hypercalcaemia due to other causes. The mechanism giving rise to the acidosis in primary hyperparathyroidism is not established, but it may be related to a direct effect of PTH on reabsorption of HCO_3^- by the kidney.

PARATHYROID HORMONE

Plasma [PTH] can be measured by radioimmunoassay (RIA), and this test is now available in many laboratories. The techniques initially failed to differentiate satisfactorily between biologically active PTH and some of the inactive fragments, and were lacking in sensitivity. Recently, methods have been introduced that are specific for the N–terminal end of the molecule. This is now the definitive measurement for confirming a diagnosis of primary hyperparathyroidism.

The normal plasma [PTH] ranges from undetectable levels up to approximately 1 μg/L. In patients with hypercalcaemia, because of feed-back suppression of PTH synthesis and release, plasma [PTH] would be expected to be undetectable. However, in patients with primary hyperparathyroidism, plasma [PTH] is not only detectable but is frequently more than 1 μg/L. The finding of an abnormally high plasma [PTH], or even of a detectable amount (i.e. below the upper reference value) that is nevertheless *inappropriately high in the presence of hypercalcaemia,* supports the diagnosis of hyperparathyroidism. However, caution must continue to be exercised when interpreting results for plasma [PTH], since the specificity of the methods presently in use is not the same in all laboratories.

In patients with hypercalcaemia, if plasma [PTH] is undetectable using a reliable method of assay, the diagnosis is most likely to be one of the other conditions listed in Table 14.2.

Localization of parathyroid tumours

If a diagnosis of primary hyperparathyroidism is made, chemical tests to help locate the tumour are not normally required prior to surgery. However, if the surgeon fails to find a parathyroid adenoma or other abnormality (hyperplasia, etc.) in any of the usual sites, specimens can be collected for measurement of plasma [PTH] before completing the operation, if there is good evidence to suggest that a tumour is present somewhere. The specimens are collected from veins draining various possible sites for the tumour, for later examination, and the results may be helpful in planning any further surgical exploration.

URINE EXCRETION STUDIES

Several tests that used to be performed in the investigation of suspected primary hyperparathyroidism are considered nowadays to be largely unnecessary by those who have adopted the combination of plasma [calcium] and [PTH], and who now use these as the key measurements for the diagnosis of primary hyperparathyroidism. However, there is still a limited place for phosphate excretion studies, in the determination of the maximum tubular capacity for reabsorbing phosphate, relative to the GFR (TmP/GFR). Some of the other tests that used to be performed will also be briefly described, as they still have a place in the investigation of other disorders of calcium metabolism.

Renal excretion of phosphate

Many tests based on the renal excretion of phosphate have been described. Of these, the TmP/GFR has proved the most reliable. It is independent of phosphate intake and urine volume and can be determined directly by measuring plasma [phosphate] and [creatinine], and urinary [phosphate] and [creatinine], during a phosphate infusion. Alternatively, it can be derived from a nomogram constructed from data obtained with phosphate infusions, if these same measurements are made as part of a short clearance investigation.

The TmP/GFR is normally 0.75–1.35 mmol/L of glomerular filtrate. It is reduced in hyperparathyroidism and increased in hypoparathyroidism. The test is not as specific a measure of parathyroid activity as is measurement of plasma [PTH].

Urinary calcium

The excretion of calcium in the urine varies considerably with the diet, but the relationship is not a simple linear one. On a normal diet, healthy adults may excrete up to 12 mmol/24 h (500 mg/24 h); this upper limit overlaps with the urinary calcium output observed in some patients who are stone-formers. Even if dietary intake is severely restricted, urinary calcium does not fall below 1.2–2.5 mmol/24 h (50–100 mg/24 h). Most laboratories, when reporting the results of urine calcium measurements, relate these to urinary creatinine, and some prefer to express the results as a calcium : creatinine ratio in an attempt to improve the discriminating power of the test.

In cases of suspected primary hyperparathyroidism, urinary calcium output was sometimes measured after the patient had been on a low calcium diet (less than 3.8 mmol (150 mg) calcium/day) for at least seven days. These tests were time-consuming, lacked specificity and have largely been abandoned.

The maximum tubular reabsorption of calcium (TmCa/GFR) can be determined in a manner similar to TmP/GFR. Reference values are 1.6–2.1 mmol/L of glomerular filtrate. The TmCa/GFR is increased in primary hyperparathyroidism, but it is also increased by the action of various drugs, e.g. lithium, thiazide diuretics. Nevertheless, the test is more specific than the urinary calcium : creatinine clearance ratio.

The excretion of calcium in the urine may fall if renal function becomes impaired, e.g. as a complication of nephrocalcinosis developing in patients with primary hyperparathyroidism.

The main value of urinary calcium measurements is for the investigation of familial hypocalciuric hypercalcaemia (p. 292).

Urinary hydroxyproline

This is an index of collagen turnover, as the imino acids (proline and hydroxyproline) are not re-used in the synthesis of new collagen. It is raised whenever collagen turnover is increased.

Urinary hydroxyproline excretion may be increased in hyperparathyroidism (whether or not there is X-ray evidence of bone disease), Paget's disease, malignant disease (primary or secondary) involving bone, and hyperthyroidism. Results are greatly influenced by the nature of the diet; patients must be put on a special diet beforehand. The test is little used.

Urinary cyclic AMP

Parathyroid hormone exerts its actions on the kidney via adenyl cyclase, and urinary cyclic AMP may be increased in patients with hyperparathyroidism. There is, however, considerable overlap with the cyclic AMP excretion observed in healthy individuals.

Attempts have been made to improve the specificity of the test by measuring cyclic AMP clearance. This tends to be increased in hyperparathyroidism due to the addition of cyclic AMP, produced locally by renal tubule cells, to the cyclic AMP that has been filtered from plasma at the glomerulus. Other modifications include measurement of the changes in cyclic AMP clearance that occur in response to a calcium infusion; the clearance falls if the parathyroid glands are responsive to increases in plasma [Ca^{2+}].

Diagnostic tests based on urinary cyclic AMP measurements are not as specific as determination of plasma [PTH] for confirming the diagnosis of primary hyperparathyroidism.

Urinary cyclic AMP measurements are used in the investigation of patients with pseudohypoparathyroidism (p. 295). If the amount excreted does not increase in response to PTH injection, this confirms the diagnosis.

OBSOLETE TESTS

These include the *hydrocortisone test*, in which the effects of large doses of hydrocortisone (40 mg 8–hourly for 10 days) on plasma [calcium] were investigated, and the *calcium infusion test* in which the effects of the infusion on renal excretion of phosphate were assessed in patients who were normocalcaemic at the time. Various indices, e.g. the phosphate:creatinine clearance ratio and the phosphate excretion index, are also obsolete.

Differential diagnosis of primary hyperparathyroidism as the cause of hypercalcaemia

TERTIARY HYPERPARATHYROIDISM

This term is used to describe those patients who develop adenomas of the parathyroid gland as a complication of previously existing secondary hyperparathyroidism (p. 294). It gives rise to hypercalcaemia and needs to be considered in patients with intestinal malabsorption or severe renal failure, if they develop hypercalcaemia not attributable to treatment with vitamin D or its hydroxylated derivative, 1α–HCC.

The plasma [calcium] is almost always increased in these patients and plasma [PTH] is always increased. Unlike patients with uncomplicated primary hyperparathyroidism, the plasma [phosphate] may be increased, especially if the tertiary hyperparathyroidism develops in a patient with renal failure.

MULTIPLE ENDOCRINE NEOPLASIA (MEN) SYNDROMES

Three types of MEN syndrome have been described, all of them familial. There are adenomas of several endocrine tissues and sometimes hypersecretion of the hormones they produce. Primary hyperparathyroidism is the commonest abnormality in MEN I (MEN 1), and is present in about 20% of patients with MEN IIa (MEN 2) but is not a feature of MEN IIb (MEN 3); the MEN II syndromes are discussed elsewhere (p. 457).

The tissues most commonly affected in the MEN I syndrome are the parathyroid gland, the islet cells of the pancreas (30%) and the anterior pituitary (15%). The commonest islet cell tumour is gastrinoma, but insulinoma and vipoma have also been described. Overproduction of ACTH, growth hormone and prolactin by the pituitary tumours have all been reported in these patients.

ECTOPIC HORMONE PRODUCTION

Most malignant tumours that give rise to hypercalcaemia do so because they have metastasized to bone. Rarely, the tumours secrete PTH, or an active fragment of the hormone. Squamous cell carcinoma of the lung is the primary tumour most often responsible for ectopic PTH production.

Other factors, e.g. a vitamin D-like sterol or the prostaglandins, have been suggested as being responsible for hypercalcaemia in some patients with malignant disease. However, these suggestions have not been substantiated.

OTHER CAUSES OF HYPERCALCAEMIA

The history of the patient's illness, the findings on clinical examination and various investigations, as suggested by the provisional diagnosis, will usually mean that the other conditions listed in Table 14.2 can be recognized. Chemical investigations may contribute to the making of several of these diagnoses, as follows:

(1) *Carcinoma,* with osteolytic secondary deposits. Plasma acid phosphatase, alkaline phosphatase and its isoenzymes, and GGT activities (p. 163).

(2) *Multiple myeloma*. Serum and urine protein electrophoresis (p. 134).

(3) *Vitamin D overdose*. Plasma [25–HCC] (p. 267).

(4) *Thyrotoxicosis*. Thyroid function tests (p. 349).

(5) *Adrenal insufficiency*. Adrenal function tests (p. 367).

Familial hypocalciuric hypercalcaemia (FHH)

This is an uncommon disorder, transmitted by an autosomal dominant gene. It may occur as an isolated abnormality or as part of the MEN I or MEN II synromes.

Symptoms of FHH, if present, are usually mild. However, in a few patients, severe neonatal primary hyperparathyroidism has been described and, in a few other patients, acute pancreatitis. Patients with FHH mostly have lifelong hypercalcaemia unaccompanied by the hypercalciuria that would normally be expected under these circumstances. Plasma [PTH] is not usually suppressed and may even be increased, but the patients are not cured by parathyroidectomy.

This is one of the few conditions in which it is important to measure urine calcium output. The relatively low urinary calcium excretion in FHH is the best way of distinguishing it from primary hyperparathyroidism.

Sarcoidosis

About 10–20% of patients with sarcoidosis may have mild hypercalcaemia, which is frequently intermittent. Much more often, these patients have hypercalciuria. Plasma [phosphate] may be slightly increased but alkaline phosphatase activity, if it is increased, is of hepatic origin, possibly in response to hepatic infiltration (p. 185).

The hypercalcaemia of sarcoidosis usually reverts to normal if cortisol is given in high dosage, but this test is not a reliable means of differentiating the hypercalcaemia of sarcoidosis from the hypercalcaemia of primary hyperparathyroidism.

Hypersensitivity to vitamin D has in the past been suggested as the cause of hypercalcaemia in sarcoidosis. More recently, since methods for measuring the metabolites of cholecalciferol became available, it has been shown that plasma [1 : 25–DHCC] is increased at times when hypercalcaemia is present.

HYPOCALCAEMIA

The main causes of a low plasma [calcium] are listed in Table 14.4. Of these, by far the commonest cause is hypoproteinaemia, and plasma [albumin] should be measured routinely in these patients in order to exclude this explanation before considering the other conditions in Table 14.4. In the developed countries, chronic renal failure is probably the next commonest cause. However, on a worldwide basis, rickets and osteomalacia are probably more common than chronic renal failure, and in these cases the hypocalcaemia is most frequently nutritional in origin. Table 14.5 summarizes the results of various chemical investigations that may be performed in patients with hypocalcaemia; measurement of plasma [PTH] is only required for investigation of hypoparathyroidism and pseudohypoparathyroidism.

Tetany

Hypocalcaemia may be asymptomatic or associated with tetany, the symptom which classically suggests the presence of a low plasma [Ca^{2+}]. It is appropriate to distinguish between (1) low plasma [calcium], unaccompanied by tetany and with a normal plasma [Ca^{2+}], (2) low plasma [calcium] associated with tetany and a low plasma [Ca^{2+}] and (3) tetany occurring in patients with a normal plasma [calcium].

With few exceptions, tetany only occurs if plasma [Ca^{2+}] is low, regardless of whether or not plasma [calcium] is low. The exceptions

TABLE 14.4. Causes of hypocalcaemia.

Hypoproteinaemia	Various causes (p. 116)
Renal disease	Chronic renal failure (p. 106)
Inadequate intake of calcium *and/or* vitamin D	Dietary deficiency (p. 267, 277) Malabsorption, due to gastrointestinal, hepatobiliary or pancreatic disease
Hypoparathyroidism	Idiopathic (rare) or acquired
Target organ resistance	Pseudohypoparathyroidism (rare)
Neonatal hypocalcaemia (p. 433)	
(a) Early (first 3 days of life)	Premature infants, infants with low birth weight, asphyxia at birth Maternal hyperparathyroidism (rare)
(b) Late (within the first 3–4 weeks)	Cow's milk (bottle-fed babies) Maternal vitamin D deficiency
Miscellaneous	Acute pancreatitis (p. 200) Cystinosis (p. 98) Cytotoxic drugs

TABLE 14.5. Chemical investigations

	Plasma phosphate (fasting)	*Plasma PTH*
Chronic renal failure	↑ or N	↑
Deficiency of calcium and vitamin D	↓ or N	N or ↑
Hypoparathyroidism	↑	↓
Pseudohypoparathyroidism	↑	↑
Renal tubular defects	↓	N

N = normal ↑ = increased ↓ = decreased

Note: Plasma [albumin] should be measured so as to help with the initial interpretation of plasma [calcium].

are the tetany caused by low plasma [magnesium], and very infrequently by low plasma [K^+]. Thus, tetany may occur in any of the conditions listed in Table 14.4, but it may also occur whenever there is a rapid fall in plasma [H^+]. Alkalosis may occur, for instance, in patients with hysterical overbreathing or in patients being given an intravenous infusion of $NaHCO_3$, and the acute reduction in plasma [Ca^{2+}] as [H^+] falls accounts for the tetany in these patients.

Secondary hyperparathyroidism

This is a term used to describe conditions in which increased amounts of PTH are secreted in response to some other disorder which has given rise to hypocalcaemia and especially a reduced plasma [Ca^{2+}]. Examples of such disorders include chronic renal failure and long-standing intestinal malabsorptive disease.

There is usually hyperplasia of all four parathyroid glands, in response to the continuing hypocalcaemia, or tendency to hypocalcaemia. Plasma [calcium] may be abnormally low or near the lower end of the reference range, depending on the size of the response in terms of secondary hyperparathyroidism. It is never raised unless tertiary hyperparathyroidism develops (p. 297).

The increase in plasma [PTH] in secondary hyperparathyroidism is appropriate for the correction of the hypocalcaemia, and the response of the parathyroid gland can be suppressed by treating the primary condition where this is possible.

Hypoparathyroidism

There may be a history of an operation on the neck or previous in patients with hypocalcaemia.

Plasma alkaline phosphatase	*Other investigations*
↑ or N	Plasma urea, creatinine, total CO_2 etc.
↑	Plasma 25-HCC, urinary calcium
N	Urinary cyclic AMP
N	Urinary cyclic AMP
↑ or N	Urine amino acid chromatography

treatment with radioactive iodine for thyrotoxicosis. The idiopathic condition is uncommon. There may be symptoms suggesting the diagnosis, principally tetany, and related signs may be elicited on physical examination. Radiological investigation may reveal ectopic calcification, especially in the basal ganglia. Chemical investigations also contribute to the diagnosis.

Hypoparathyroidism of acute onset is nearly always postoperative. The other conditions giving rise to hypoparathyroidism are all chronic.

Idiopathic and acquired hypoparathyroidism

The diagnosis of hypoparathyroidism is supported by the following findings:

(1) *Plasma [calcium].* This is reduced, sometimes markedly, to values as low as 1.5 mmol/L (6.0 mg/100 mL). Plasma [Ca^{2+}] is also much reduced.
(2) *Plasma [phosphate].* This is usually increased, sometimes markedly. The combination of a low plasma [calcium] and high [phosphate] is very suggestive of a diagnosis of hypoparathyroidism.
(3) *Plasma [PTH].* This is greatly reduced, being undetectable by presently available assay methods.

Plasma alkaline phosphatase activity is usually normal in these patients.

Pseudohypoparathyroidism

This is a rare inborn error of metabolism in which there is end-organ failure to respond to parathyroid hormone. The diagnosis may be suggested by the family history and by the presence of characteristic skeletal abnormalities. As in patients with idiopathic and acquired hypoparathyroidism, there is reduced plasma [calcium] and increased

TABLE 14.6. Chemical investigations on plasma

	Calcium	Phosphate (fasting)
Hyperparathyroidism		
Primary	↑ (or N)	↓ or N
Secondary	↓ or N	↑ or N
Tertiary	↑ or N	↑ or N
Rickets & osteomalacia		
Deficient intake	↓ or N	↓ or N
Renal failure	↓ or N	↑ or N
Renal phosphate loss	↓ or N	↓ or N
Osteoporosis	N	N
Paget's disease	N	N

N = normal ↑ = increased ↓ = decreased

plasma [phosphate]. However, patients with pseudohypoparathyroidism have increased plasma [PTH].

RESPONSE TO PTH INFUSION

Further differentiation between hypoparathyroidism and pseudohypoparathyroidism can be made by investigating the response of the renal tubule cells to intravenous bovine PTH. The response is assessed by measuring plasma [cyclic AMP] and urinary cyclic AMP excretion.

In patients with hypoparathyroidism, idiopathic or acquired, there is a normal response to the PTH infusion and both plasma and urinary cyclic AMP increase. However, in patients with pseudohypoparathyroidism, due to end-organ loss of responsiveness, there is little or no change in plasma and urinary cyclic AMP. This test has rendered obsolete the potentially dangerous sodium edetate infusion test.

METABOLIC BONE DISEASES

These are diseases which affect bone because of a disturbance of calcium metabolism. The principal descriptive types are:

(1) *Hyperparathyroidism.* There is excessive resorption of bone, with proliferation of osteoclasts and replacement of bone by fibrous tissue. Bone cysts may form.

(2) *Osteoporosis.* There is loss of matrix and reduction in bone mass. However, the deposition of calcium salts (mineralization) occurs normally.

in patients with metabolic bone disease.

PTH	*Alkaline phosphatase*	*Ca^{2+}*
↑ or detectable	N or ↑	↑ (or N)
↑	↑ or N	N
↑	↑ or N	↑
↑ (or N)	↑	N (or ↓)
↑	↑	N
N	↑	N
N	N	N
N	↑	N

(3) *Rickets and osteomalacia.* There is failure of deposition of calcium salts in new bone. In consequence, there is an increased amount of osteoid or uncalcified matrix.

In many examples of metabolic bone disease, patients show features of two or more of these conditions. It can be difficult to define the pathological process in full, even with the aid of radiological examination, bone biopsy, etc. Results of chemical investigations (Table 14.6) need to be interpreted in relation to all the available evidence.

HYPERPARATHYROIDISM

This has already been extensively discussed. Primary hyperparathyroidism may give rise to renal functional impairment and eventually lead on to renal failure; this is probably due to the raised plasma [calcium], which may result in nephrocalcinosis. Secondary hyperparathyroidism may develop in patients with long-standing intestinal malabsorptive disease or with chronic renal failure, and may progress to tertiary hyperparathyroidism.

It can very occasionally be difficult to decide which was the primary disorder in patients who present with renal failure and hyperparathyroidism. This combination of metabolic abnormalities comprises one form of uraemic osteodystrophy, discussed below.

OSTEOPOROSIS

This is a very common disorder, but it is usually not diagnosed with certainty until there is a marked loss in bone density, revealed by radiological investigation. Results of routine chemical investigations are all normal, as a rule. However, prolonged and careful balance studies may reveal a negative calcium balance, especially if these studies are performed during active stages of the disease.

The diagnosis depends on the finding of skeletal rarefaction in a patient who does not have hyperparathyroidism, osteomalacia, carcinoma, myeloma, etc. and in whom plasma [calcium], [phosphate], alkaline phosphatase activity, etc. are all normal.

RICKETS AND OSTEOMALACIA

Rickets occurs in growing children. Osteomalacia is a term that applies to disease in adults. By definition, rickets implies impaired mineralization of the cartilage in the long bones. Osteomalacia, on the other hand, has a less precise meaning; it should be applied to conditions where there is excess osteoid.

Two principal pathophysiological abnormalities may lead to rickets and osteomalacia. The more common of these is deficiency or impaired action of vitamin D which, as its hydroxylated derivative 1 : 25–DHCC, promotes calcium absorption from the intestine and has a direct action on bone. The less common mechanism is excessive loss of phosphate in the urine, usually due to an inherited or acquired renal tubular disorder. The main causes of rickets and osteomalacia are:

(1) *Lack of dietary vitamin D.* There may also be an accompanying inadequate intake of dietary calcium.
(2) *Ineffective conversion of 7-dehydrocholesterol* (pro-vitamin D_3) to pre-vitamin D_3 by UV light in the dermis (p. 266).
(3) *Intestinal malabsorptive disease,* especially when there is long-standing steatorrhoea.
(4) *Vitamin D resistance,* which may be due to:
  (a) Chronic renal disease. The commonest cause of ineffective conversion of 25–HCC to 1:25–DHCC.
  (b) Inherited deficiency of renal 1α–hydroxylase.
  (c) End-organ receptor abnormality. This is rare.
(5) *Excessive renal phosphate loss,* which may be due to:
  (a) Fanconi syndrome (p. 98). Whatever the cause of the syndrome, this is most often responsible for the excess loss.
  (b) A specific defect in renal phosphate handling (p. 99), that can cause hypophosphataemia.

The results of chemical investigations performed on patients with rickets or osteomalacia depend partly on the underlying cause. As far as any disturbance of calcium metabolism is concerned, as long as there is no evidence of impairment of renal glomerular function or acid–base disturbance, the following findings are commonly observed:

(1) Plasma [calcium] and [phosphate] both tend to be decreased.
(2) Plasma alkaline phosphatase activity is increased. This measurement is of limited diagnostic value in childhood because of the marked physiological variations in activity that occur with age (p. 148).
(3) Urinary calcium excretion is nearly always low or very low.
(4) Plasma [25–HCC] is usually low, but this measurement is not yet widely available and the diagnosis can usually be made without it.

In patients with renal failure, the pattern of abnormalities is different (Table 14.6).

RENAL DISEASE AND METABOLIC BONE DISEASE

Metabolic bone disease frequently develops in patients with chronic renal failure, when it may manifest itself in various ways; the conditions are often collectively called uraemic osteodystrophy. Metabolic bone disease may also develop in patients with renal tubular defects.

Uraemic osteodystrophy

Virtually all patients with chronic renal failure have increased plasma [PTH], and therefore secondary hyperparathyroidism. Some go on to develop tertiary hyperparathyroidism. The types of bone lesion detected radiologically in chronic renal failure include one or more of the following:

(1) Delayed epiphyseal closure, in children and young adults.
(2) Rickets or osteomalacia, depending on the age-group.
(3) Osteitis fibrosa, due to secondary or tertiary hyperparathyroidism.
(4) Osteosclerosis, localized or generalized (rare).

Chemical investigation in these patients usually provides evidence of a marked degree of renal functional impairment, with plasma [urea] that is often grossly increased and markedly reduced creatinine clearance. There is usually a metabolic acidosis. Plasma [calcium] is reduced or at the lower limit of the reference values, but plasma [Ca^{2+}] is normal unless steps are taken to treat the acidosis actively (e.g. by infusing $NaHCO_3$); this can precipitate a severe attack of tetany.

Plasma [phosphate] and [Cl^-] are almost always raised, but plasma alkaline phosphatase activity is too variable to be able to generalize about it. In those patients who have progressed to the development of tertiary hyperparathyroidism, plasma [calcium] is increased.

Patients with uraemic osteodystrophy who develop rickets or osteomalacia are almost always resistant to treatment with vitamin D, since their ability to hydroxylate 25–HCC is impaired. However, they respond to treatment with synthetic 1α–hydroxycholecalciferol 1α–HCC).

Patients being treated on a long-term basis by haemodialysis are liable to develop a complex combination of biochemical disturbances, including the various forms of uraemic osteodystrophy. The nature of the disturbance is considerably influenced by the particular dialysis regime, especially the composition of the dialysis fluid. These complications appear to be less frequently observed with patients maintained by continuous ambulatory peritoneal dialysis.

Chemical investigations play an important part in checking the composition of dialysis fluid and in monitoring the effects of treatment. The principal objectives of long-term intermittent or continuous dialysis are to remove nitrogenous waste products and to maintain acid–base balance, and to minimize the effects of secondary hyperparathyroidism. Plasma [calcium] and [phosphate] need to be measured regularly as part of the control of treament by 1α–HCC or 1 : 25–DHCC, and by phosphate-binders such as aluminium hydroxide. In addition, the aluminium content of dialysis fluid and plasma should be regularly monitored, in view of the possible associations between aluminium toxicity, dialysis osteodystrophy and dialysis dementia.

Patients with renal failure who have been treated by renal transplantation may continue to have hyperparathyroidism, with raised plasma [PTH], for as much as three years after operation. It would seem that the parathyroid glands may not always 'switch off' immediately after a long period of secondary hyperparathyroidism. Sometimes these patients develop marked hypercalcaemia and may require subtotal parathyroidectomy to correct this.

Renal tubular defects

Patients with metabolic bone disease secondary to inherited defects of the renal tubules (e.g. the Fanconi syndrome), in which there is excessive loss of phosphate in the urine, develop bone lesions similar to those seen in rickets and osteomalacia. Depending on the cause, chemical abnormalities may include reduced plasma [phosphate], increased urinary phosphate output, amino aciduria and glycosuria.

Those patients in this group who lose excessive amounts of phosphate in the urine are best treated with dietary phosphate supplements as well as vitamin D. The hydroxylated derivative, 1α–HCC, may be required in some patients.

PAGET'S DISEASE

This is a common disorder of bone, and a common cause of a markedly increased plasma alkaline phosphatase activity. Plasma [calcium] and [phosphate] are usually normal. Some would not include Paget's disease among the metabolic bone diseases because it is often local rather than general in its manifestations.

MAGNESIUM METABOLISM

Magnesium is the second most abundant intracellular cation, only a small proportion of the body's content being in the ECF. Bone contains about 50% of the body's magnesium.

Daily intake of magnesium in the diet is normally about 10 mmol (250 mg). Significant quantities are contained in gastric and biliary secretions. Magnesium can be absorbed from both the small and large intestine, and normally only small amounts of magnesium are excreted in faeces. Excretion occurs mainly in the urine, output being closely related to the dietary intake.

Magnesium homeostasis

Plasma [magnesium] is normally kept within narrow limits, which again implies close homeostatic control. The reference values are 0.7–1.0 mmol/L (1.7–2.4 mg/100 mL). About 35% of the magnesium in plasma is protein-bound.

The factors concerned with control of magnesium metabolism have not been defined. There is no factor known to influence its absorption. However, the recognition of specific magnesium malabsorption syndromes in man has led to the suggestion that the ion is actively transported from the intestine. When the dietary intake is restricted, renal conservation mechanisms are normally so efficient that depletion, if it develops at all, only comes on very slowly.

MAGNESIUM DEFICIENCY

This may develop for several reasons (Table 14.7), but usually as part of a deficiency state involving other electrolytes (e.g. Na^+, K^+). Symptoms and signs attributable to magnesium depletion usually go unrecognized until the other components of the depleted state have been partially or wholly corrected.

Tetany may develop as a result of magnesium deficiency, but the cause is often not recognized until it has been shown that the tetany fails to respond to administration of calcium salts. Other symptoms include muscular weakness.

The existence of magnesium deficiency states should be suspected in patients who have hypokalaemia or hypocalcaemia that is failing to respond to treatment. Plasma [magnesium] should be measured.

Plasma magnesium

These estimations may not be helpful, at least initially, in the diagnosis of magnesium deficiency. Marked alterations in the body's content can occur with little or no change detectable in plasma [magnesium]. In this respect, magnesium is very similar to the other major intracellular cation, potassium, since there may be reduced plasma [magnesium] when the intracellular content is normal. Conversely, the plasma

TABLE 14.7. Magnesium deficiency.

Cause	Examples
Abnormal losses	
Gastrointestinal tract	Prolonged aspiration; diarrhoea; malabsorptive disease; fistula; jejuno-ileal bypass; small bowel resection
Urinary tract	
(a) *Renal disease*	Renal tubular acidosis; chronic pyelonephritis; hydronephrosis
(b) *Extra-renal disorders*	(i) Factors modifying renal function (e.g. aldosteronism, primary and secondary; diuretic therapy; osmotic diuresis)
	(ii) Diseases affecting magnesium transfer from cells or bone (e.g. hyperparathyroidism, primary and tertiary; disorders giving rise to ketoacidosis)
Lactation	If lactation is excessive
Reduced intake	If severe and prolonged (e.g. kwashiorkor, marasmus)
Mixed aetiology	Chronic alcoholism; hepatic cirrhosis

[magnesium] may be within the range of reference values although a marked state of intracellular depletion exists. Nevertheless, plasma [magnesium] should be measured before patients in whom a diagnosis of magnesium deficiency is suspected start to be treated with magnesium salts.

In magnesium deficiency plasma [magnesium] may be as low as 0.5 mmol/L, but in many patients the concentration is normal.

Other tests for magnesium deficiency

These include erythrocyte [magnesium], 24-h urinary excretion of magnesium, a magnesium loading test, and muscle biopsy for tissue [magnesium].

Erythrocyte [magnesium] is reduced when a deficiency state exists, and urinary excretion falls to less than 1 mmol/day unless the cause of the deficiency is abnormal loss of magnesium due to renal disease. Even when a diagnosis of magnesium deficiency seems probable on clinical grounds, if the plasma [magnesium], erythrocyte [magnesium] and 24-h urinary output of magnesium are all normal, this renders the diagnosis of magnesium deficiency very unlikely.

Patients who have normal renal function can be further investigated for possible magnesium deficiency by means of a loading test. An intravenous infusion of 40 mmol magnesium sulphate is given over 60–90 min, and urine collected for 48 h from the start of the infusion. In the presence of normal renal function, at least 35 mmol magnesium is excreted in the urine during the period of collection. In states of magnesium deficiency, however, much of the dose is either retained in the body or lost by another route that is giving rise to the state of deficiency, usually the gastrointestinal tract; the urinary excretion may then be as low as 25 mmol magnesium, or even less.

MAGNESIUM EXCESS

There may be no symptoms unless plasma [magnesium] exceeds 2.5 mmol/L. Nausea, vomiting, weakness and impaired consciousness may then develop. However, these symptoms are not necessarily caused solely by the hypermagnesaemia. The main causes of raised plasma [magnesium] are:

(1) Acute renal failure.
(2) Advanced stages of chronic renal failure.
(3) Haemodialysis against hard water, which has a high [magnesium].
(4) Intravenous injection of magnesium salts.

(5) Adrenocortical hypofunction (usually only a slight increase occurs).

Hypermagnesaemia is most often due to severe impairment of renal function. Its presence is readily confirmed by measuring plasma [magnesium].

FURTHER READING

DELUCA, H.F. (1975). The kidney as an endocrine organ involved in the function of vitamin D. *American Journal of Medicine,* **58,** 39–47.

FISKEN, R.A., HEATH, D.A., SOMERS, S. and BOLD, A.M. (1981). Hypercalcaemia in hospital patients. *Lancet,* **1,** 202–207.

GOODWIN, F.J. (1982). Hypercalcaemia. *British Journal of Hospital Medicine,* **28,** 50–58.

HEATH, D.A. and MARX, S.J., editors (1982). *Calcium disorders.* London: Butterworth.

HILLYARD, C.J., MYERS, C., ABEYASEKERA, G., STEVENSON, J.C., CRAIG, R.K. and MACINTYRE, I. (1983). Katacalcin: a new plasma calcium-lowering hormone. *Lancet,* **1,** 846–848.

Leading article (1977). Dietary calcium. *British Medical Journal,* **2,** 1105–1106.

Leading article (1978). Dialysis osteodystrophy. *Lancet,* **2,** 451–452.

Leading article (1979). Serum-calcium. *ibid.,* **1,** 858–859.

Leading article (1981). Asian rickets in Britain. *ibid.,* **2,** 402.

MACINTYRE, I. (1983). The physiological actions of calcitonin. *Triangle,* **22,** 69–74.

MCKENNA, M.J., FREANEY, R., CASEY, O.M., TOWERS, R.P. and MULDOWNEY, F.P. (1983). Osteomalacia and osteoporosis: evaluation of a diagnostic index. *Journal of Clinical Pathology,* **36,** 245–252.

MARX, S.J., SPIEGEL, A.M., LEVINE, M.A. *et al.* (1982). Familial hypocalciuric hypercalcaemia. *New England Journal of Medicine,* **307,** 416–426.

PATERSON, C.R. (1976). Hypocalcaemia: differential diagnosis and investigation. *Annals of Clinical Biochemistry,* **13,** 578–584.

RUSSELL, R.G.G. (1976). Regulation of calcium metabolism. *ibid.,* **13,** 518–539.

VERNON, W.B. and WACKER, W.E.C. (1978). Magnesium metabolism. In *Recent Advances in Clinical Biochemistry,* **1,** 39–71. Ed. K.G.M.M. Alberti. Edinburgh: Churchill Livingstone.

WALTON, R.J. and BIJVOET, O.L.M. (1975). Nomogram for derivation of renal threshold phosphate concentration. *Lancet,* **2,** 309–310.

WILLS, M.R. and SAVORY, J. (1983). Aluminium poisoning: dialysis encephalopathy, osteomalacia and anaemia. *ibid,* **2,** 29–34.

WOODHEAD, J.S. (1983). Regulation of whole body calcium in man. In *Biochemical Aspects of Human Disease,* pp. 217–246. Ed. R.S. Elkeles and A.S. Tavill. Oxford: Blackwell Scientific Publications.

Chapter 15

Iron, Porphyrin and Haemoglobin Metabolism

Haemoglobin (Hb), myoglobin and cytochromes are all important members of the group of haem-proteins. Their prosthetic group, haem, is a complex of Fe^{2+} with protoporphyrin IXα (Fig. 15.1, p. 311). Disorders of iron, poryphyrin and haemoglobin metabolism have been grouped together in this chapter because of their interrelationship, through the haem-proteins, but the individual groups of disorders have little in common with one another.

IRON METABOLISM

The body contains about 70 mmol (4 g) of iron*; about 70% of this is in Hb and up to 25% in iron stores, mainly in the liver, spleen and bone marrow. The remaining 5–10% is mostly contained in other haem-proteins. Only about 0.1% of total body iron is present in plasma.

Iron balance is regulated by alterations in intestinal absorption of iron. There is only a limited capacity to increase or decrease the rate of loss of iron.

DIETARY IRON AND IRON ABSORPTION

The normal intake of iron is about 0.2–0.4 mmol/day (10–20 mg/day), mostly as meat, grain products and vegetables. Normally about 5–10% of this is absorbed. The rate of absorption is controlled by complex physiological factors, and is influenced by the contents of the diet and by the nature of gastrointestinal secretions, as follows:

Physiological regulatory mechanisms

Two factors *control* iron absorption, by means that are still not fully understood:

(1) *State of iron stores* in the body. Absorption is increased in states of iron deficiency, and decreased when there is iron overload.

*In this book, 'iron' is used to refer to all forms of iron, organic and inorganic. The symbols Fe^{2+} and Fe^{3+} refer specifically to the inorganic ions.

(2) *Rate of erythropoiesis.* In conditions associated with increased rates of erythropoiesis, absorption may be increased even though the iron stores are adequate or overloaded (e.g. in thalassaemia).

Local factors in the gastrointestinal tract

The following factors *influence* the absorption of iron:

(1) *The chemical state of the iron.* Iron in the diet does not usually become available for absorption unless released during digestion. Release of iron depends, at least partly, on gastric acid production; Fe^{2+} is more readily absorbed than Fe^{3+}, and the presence of H^+ helps to keep iron in the Fe^{2+} form. The iron in haem, present in meat products, can be absorbed while still contained in the haem molecule.
(2) *Complex formation.* Substances that tend to form complexes with iron affect absorption. Ascorbate (vitamin C) forms a soluble complex and thus facilitates absorption. Substances which form insoluble complexes (e.g. phytate) inhibit absorption.

IRON TRANSPORT, STORAGE AND UTILIZATION

After being taken up by the intestinal mucosa, iron may be transported across the mucosa directly to the plasma. Alternatively, it may be incorporated into the iron-storage protein, ferritin, present in the mucosal cell. Iron stored in mucosal ferritin is later lost from the body when the mucosal cell sloughs into the intestinal lumen.

In the plasma, iron is transported mainly combined with transferrin, which binds two Fe^{3+} ions reversibly. The binding sites on transferrin are normally about 30–50% saturated; the total iron circulating bound to transferrin is normally about 50–70 μmol (3–4 mg).

In the bone marrow and in certain other sites, e.g. the liver, iron is chelated with protoporphyrin IXα to form haem, which is incorporated into the various haem-proteins. Any excess of iron is stored as either ferritin or haemosiderin.

Ferritin is a soluble protein made up from an iron-free protein, apoferritin, which consists of a number of polypeptide subunits that encircle a central core in which is contained a variable amount of iron. Haemosiderin is a less well-defined protein, probably formed by condensation of several molecules of ferritin, which are then partially catabolized.

Iron released by the breakdown of Hb, at the end of the erythrocyte's life, is efficiently conserved and later re-used.

IRON EXCRETION AND SOURCES OF LOSS

Iron excreted in the faeces is principally exogenous, dietary iron that has not been absorbed.

In males there is an average loss of endogenous iron of about 20 μmol/day (1 mg/day) in cells desquamated from the skin and the intestinal mucosa. In females there are additional sources of loss, due to menstruation and pregnancy. The urine contains negligible amounts of iron. To a limited extent, intestinal losses of endogenous iron can be varied according to the state of the body's iron stores, being approximately doubled in iron overload and halved in states of iron depletion.

Investigation of disorders of iron metabolism

Haematological investigations are of primary importance in evaluating many disorders of iron metabolism. Only the biochemical tests are considered here.

Plasma iron

There are considerable day-to-day fluctuations in plasma [iron], even in health. Plasma [iron] may vary, in the same healthy person, from half to double its initial value over a few days. Much of this variation is apparently random but some specific associations are known:

(1) *Diurnal variation.* This is quite marked, higher values occurring in the morning.
(2) *Menstrual cycle.* Variation occurs with the stage of the cycle, low values occurring just before and during the menstrual period.
(3) *Pregnancy and oral contraceptives.* These may both be associated with a rise in plasma [iron]. However, in pregnancy a low plasma [iron] is more commonly observed, due to concurrent iron deficiency.

In most patients in whom the investigation is requested, plasma [iron] does not provide useful diagnostic information, mainly because of these sources of variation.

Changes observed in iron deficiency and in iron overload tend to occur relatively late in the progress of the disorder. Plasma [iron] only gives an indication of available iron stores when these are grossly abnormal. Plasma [iron] tends to fall in patients with chronic inflammatory disease, such as rheumatoid arthritis, and in patients with neoplastic disease (Table 15.1).

Plasma total iron-binding capacity (TIBC)

This is a measure of the amount of iron with which the plasma sample can combine. Normally, almost all the binding capacity is due to transferrin. Plasma TIBC shows smaller physiological variations than plasma [iron]. In pregnancy and in women taking oestrogen-containing oral contraceptives, plasma TIBC increases.

Plasma TIBC rises in iron deficiency, but often tends to be low in patients with iron overload (Table 15.1). In these conditions, it alters in the opposite direction to the plasma [iron].

In protein-losing states, infections, neoplasms and after trauma, plasma TIBC tends to fall. However, these changes are of little diagnostic significance.

TABLE 15.1. Plasma iron and total iron-binding capacity.

	Plasma [iron]	Plasma TIBC	Saturation of TIBC
Reference values			
Males μmol/L	14–32	45–72	30–40%
μg/100 mL	80–180	250–400	30–40%
Females μmol/L	10–28	45–72	30–40%
μg/100 mL	55–160	250–400	30–40%
Physiological changes			
Pre-menstrual	↓	N	↓
Contraceptive pill	↑	↑	N
Pregnancy	↑*	↑	N
Disease states			
Iron deficiency	↓	↑	Below 30%
Iron overload	↑	N or ↓	Up to 100%
Infection, neoplasms	↓	↓	Normal
Hypoplastic anaemia	↑	N or ↓	Over 40%
Hypoproteinaemia	N	↓	Over 40%

N = normal, ↑ = increased, ↓ = decreased

*Plasma [iron] more commonly reduced in pregnancy due to associated iron deficiency.

Plasma ferritin

As already noted, plasma [iron] is only a crude index of iron status and is only useful in diagnosis when gross abnormalities of iron metabolism are present. Instead, it has been shown that plasma [ferritin] correlates directly with the state of the body's iron stores, regardless of whether these are decreased, normal or increased. It should be noted, however, that in patients with liver disease, leukaemia and certain other neoplastic disorders, plasma [ferritin] does *not* correlate with iron stores, since ferritin is released from the diseased tissue.

Plasma [ferritin] is a more difficult assay than plasma [iron]. However, the test is becoming more widely available, and there are several circumstances where its measurement yields information which cannot readily be obtained by other means:

(1) When a patient is suspected of having a mild degree of iron deficiency or iron overload.
(2) In patients with anaemia due to chronic infection; in these patients, hypochromic anaemia may be due to the primary condition or to associated iron deficiency.
(3) In patients with iron overload, to monitor progress and response to treatment either with chelating agents or with venesection.

IRON DEFICIENCY

This is a common condition, often first suspected on the basis of routine haematological investigation. The main causes are:

(1) *Deficient intake.*
(2) *Impaired absorption* (e.g. intestinal malabsorptive disease, abdominal surgery).
(3) *Excessive loss* (e.g. menstrual loss, gastrointestinal bleeding).

Tests for the presence of occult blood in faeces (p. 33) should be performed when the source of blood loss is not apparent. Plasma [iron], TIBC and [ferritin] may need to be measured where the diagnosis is not straightforward, e.g. if other causes of hypochromic anaemia might be present.

In patients who develop iron deficiency, the sequence of events, in terms of development of abnormal test results, is as follows:

(1) Plasma [ferritin] falls.
(2) Plasma TIBC increases.
(3) Plasma [iron] falls.
(4) Anaemia becomes evident.

Plasma [iron] has been advocated as a screening test for detecting unsuspected iron deficiency in the ambulant apparently healthy population. Unfortunately, this screening application overlooks the fact that measurements of plasma [iron] are of only limited value in detecting mild states of iron deficiency; the test lacks sensitivity.

IRON OVERLOAD

This is much less common than iron deficiency. Diagnosis is not usually difficult, once the possibility has been considered. Raised plasma [iron] and fully saturated TIBC (Table 15.1) are characteristic of iron overload, whatever the cause. The following are the causes:

(1) *Increased intake and absorption:*
  (a) Acute, (b) chronic.
(2) *Parenteral administration* of iron, blood transfusions.
(3) *Idiopathic haemochromatosis.*

Abnormally large amounts of iron are occasionally ingested and absorbed and can then give rise to a state of severe iron intoxication. This may occur acutely in children who have accidentally ingested iron tablets. A more chronic state of overload can occur when the diet contains excess absorbable iron, as with the Bantu, who tend to have an acid-containing diet and to cook their food in iron pots. Under these conditions, there is a slow and progressive build-up of iron deposits in the body, and liver disease may develop.

Iron overload may develop in patients receiving repeated blood transfusions (e.g. for aplastic anaemia) or prolonged courses of parenteral iron therapy. The excess iron is usually deposited mainly in the liver and spleen; hepatic fibrosis may be present but is often not severe. Prognosis is more likely to be related to the amount of myocardial damage caused by relatively small deposits of iron in the heart.

Idiopathic haemochromatosis

This is a rare hereditary disorder that is probably due to an increased absorption of dietary iron consequent upon a defect in the mechanism controlling absorption from the intestine. Excessive deposits of iron build up as haemosiderin in several tissues. The deposits in the liver cause cirrhosis, while pancreatic deposits lead to damage and fibrosis with the development of diabetes mellitus. Myocardial damage may also occur.

Idiopathic haemochromatosis can be recognized, at the preclinical stage, in affected members of a family in which an index case has occurred. The haemochromatosis gene is closely linked with the genes that determine HLA antigens. In families at risk, apparently unaffected members with a susceptible HLA genotype should be kept under observation with regular measurements of plasma [iron], TIBC and [ferritin].

PORPHYRIN METABOLISM

The porphyrins are tetrapyrroles (Fig. 15.1), some of which are

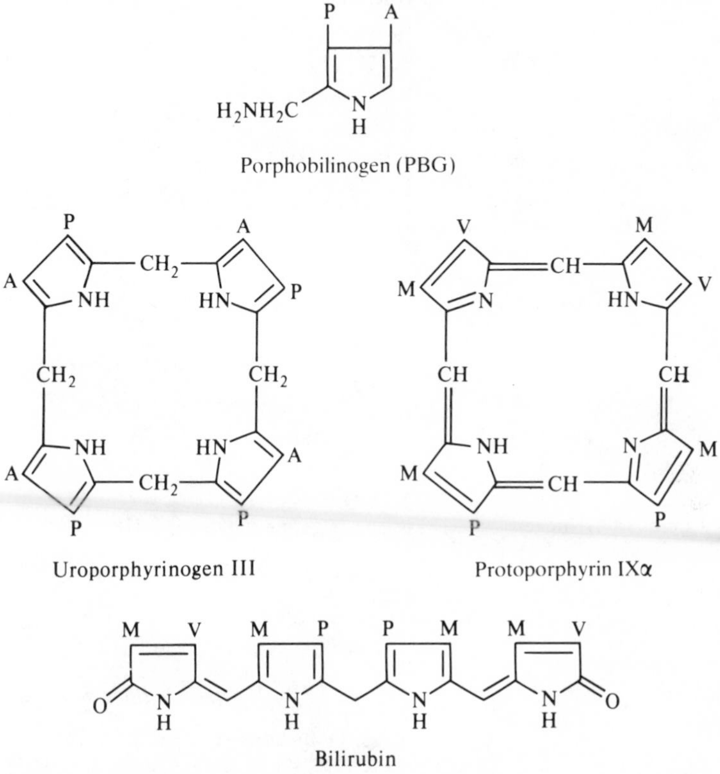

FIG. 15.1. The formulae of porphobilinogen and some of the compounds that derive from it. Uroporphyrinogen, formed from four molecules of PBG, can have four isomers depending on the arrangement of the acetic (A) and propionic acid (P) side-chains, but only series I and III isomers occur naturally. Decarboxylation of the four acetic acid side-chains to methyl (M) gives rise to coproporphyrinogen, and oxidative decarboxylation of two of the propionic acid side-chains to vinyl (V) gives rise to protoporphyrinogen. Of the fifteen possible isomers of protoporphyrinogen, only the series IX isomer occurs naturally. This is oxidized to protoporphyrin IX**a**, after which iron is incorporated to give haem. Bilirubin arises mainly from the breakdown of haem.

intermediates in the formation of haem. Most body cells can synthesize haem but bone marrow and liver are the most active, and the sites most commonly affected by disorders of porphyrin metabolism. To understand the use of chemical tests in investigating porphyrias, the stages in the synthesis of haem (Fig. 15.2) need to be reviewed:

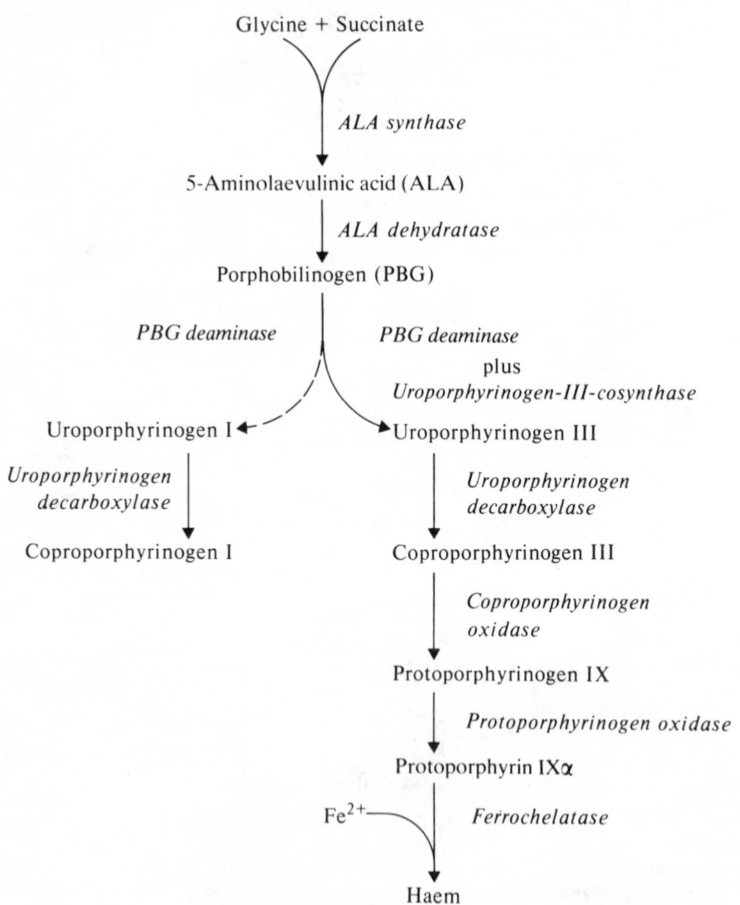

FIG. 15.2. *The enzymic steps in the synthesis of haem*. PGB deaminase (uroporphyrinogen-I-synthase), acting by itself, converts PBG to hydroxymethylbilane, a linear tetrapyrrole which can cyclize spontaneously to yield uroporphyrinogen I. However, in the presence of uroporphyrinogen-III-cosynthase, the reaction is directed mainly along the series III pathway.

(1) 5-Aminolaevulinic acid (ALA) synthase catalyses the condensation of glycine and succinate, to form ALA. This is the rate-limiting step in haem biosynthesis.
(2) ALA-dehydratase catalyses the condensation of two molecules of ALA to form porphobilinogen (PBG).
(3) The formation of uroporphyrinogen III from PBG is catalysed by two enzymes acting in succession, PGB deaminase and uroporphyrinogen-III-cosynthase. If the cosynthase is absent, series I porphyrin isomers are formed, not series III. The tetrapyrrole isomers are named according to the order of substitution of the side-chains in the eight positions on the tetrapyrrole nucleus where H atoms can be replaced (Fig. 15.1).
(4) Successive modifications of the side-chains in uroporphyrinogen III by uroporphyrinogen decarboxylase and coproporphyrinogen oxidase, followed by oxidation of the methane bridges to methene bridges, by protoporphyrinogen oxidase, lead to the formation of protoporphyrin IXα. Coproporphyrinogen oxidase does not act on coproporphyrinogen I.
(5) Ferrochelatase catalyses the final incorporation of iron into protoporphyrin IXα to form haem.

All the porphyrinogens (uro-, copro-, proto-isomers) are unstable and are readily oxidized. Any of these compounds that are excreted in urine or faeces are present there as their corresponding porphyrins.

The porphyrias

There is overproduction of porphyrins or their precursors. Most of the disorders are at least partly genetically determined. Different stages in haem synthesis are affected in each disease, causing a variety of clinical and biochemical manifestations, as follows:

(1) *Photosensitivity* is often a feature of those porphyrias in which there are excessive amounts of uro-, copro-, or protoporphyrin in tissues. In practice, photosensitivity may be present in all the porphyrias except the acute intermittent type. The symptom is explicable in terms of the strong absorption by porphyrins, present in the upper epidermis, of light of 400 nm wavelength. This causes the porphyrin molecule to become activated, and it has been suggested that the consequent release of free radicals causes local lysosomal damage.
(2) *Acute abdominal and neurological symptoms* tend to occur in those

porphyrias in which there is overproduction of ALA and PBG, both of which probably have direct neurotoxic effects. This overproduction can be attributed to the combined effects of increased ALA synthase activity, which occurs in all porphyrias, and decreased PBG deaminase activity, which occurs only in acute porphyrias. In those porphyrias where acute symptoms do not occur, there is usually an increase in PBG deaminase activity.

In all the porphyrias, the underlying defect is a partial deficiency of one of the enzymes in the pathway leading to the synthesis of haem (Fig. 15.2). The porphyrias can be subdivided initially into two groups, depending on whether erythrocyte [porphyrins] is (a) increased or (b) normal. In the second group, the main site of porphyrin overproduction is the liver, and these are collectively described as the hepatic porphyrias.

ERYTHROCYTE [PORPHYRINS] INCREASED

Congenital erythropoietic porphyria

This very rare, recessively inherited disorder presents in infancy. Exposed areas of skin (hands, head) are affected by severe blistering due to photosensitivity. The teeth stain pink and there may be splenomegaly and haemolytic anaemia. It may be due to deficiency of uroporphyrinogen-III-cosynthase.

There is overproduction of series I porphyrins, and the erythrocytes and their precursors contain excessive amounts of these. The large amounts of uroporphyrin I (and to a lesser extent coproporphyrin I) excreted in urine cause it to become a pink to red colour. Faecal coproporphyrin I (and uroporphyrin I) excretion is increased.

Protoporphyria

This is a rare disorder, dominantly inherited. It usually presents with photosensitivity; blistering does not occur, however, as this disease is less severe than congenital erythropoietic porphyria. Protoporphyria is associated with a partial deficiency of ferrochelatase.

The diagnosis can be established by measuring red cell [protoporphyrin]; the main biochemical abnormality is increased red cell [protoporphyrin]. Urinary porphyrins are normal but faecal protoporphyrin excretion may be increased.

ERYTHROCYTE [PORPHYRINS] NORMAL

The hepatic porphyrias are more common and more important clinically than the two conditions described above. They are largely genetically determined, but environmental factors may precipitate clinical manifestations. The hepatic porphyrias may mimic a number of other diseases. Four forms are usually distinguished:

(1) Acute intermittent porphyria (Swedish type).
(2) Variegate porphyria (South African type).
(3) Hereditary coproporphyria.
(4) Porphyria cutanea tarda (Symptomatic cutaneous porphyria).

Acute intermittent porphyria

This usually takes the form of a series of acute episodes of severe abdominal pain. Intermittent neuromuscular and psychiatric disturbances are frequent, but skin manifestations do not occur. Attacks may be precipitated by drugs (e.g. barbiturates, alcohol), changes of hormonal balance (e.g. pregnancy, oral contraceptives, steroid treatment) or by infections.

During acute attacks, large amounts of ALA and PBG are always present in the urine. Uro- and coproporphyrin are also excreted in excess, but much of this increase may be due to spontaneous condensation of PBG in the bladder. Most of these abnormalities can be detected between attacks but are then less marked. It is particularly important to test the urine for PBG in patients with symptoms that could be attributable to acute porphyria. Faecal excretion of porphyrins is usually normal.

The basic defect is a deficiency of PBG deaminase. However, there is also a deficiency of steroid 5α-reductase in the liver; this allows certain steroids known to be potent inducers of ALA synthase to accumulate. The steroid reductase defect could explain the known relationship between acute attacks of porphyria and hormonal imbalance.

Variegate porphyria

This disorder may present acutely with symptoms similar to those of acute intermittent porphyria. Alternatively, photosensitivity may be the most marked feature. The two types of symptom may co-exist.

Faeces contain excess porphyrins, protoporphyrin IX and coproporphyrin III, both during and between attacks, but urinary coproporphyrin III excretion may only be increased during an attack. Abnormal porphyrin-peptide complexes are also excreted in faeces in excess. Urinary ALA and PBG output are often only increased during acute attacks.

The basic biochemical defect is not known with certainty, but is probably due to deficiency of protoporphyrinogen oxidase.

Hereditary coproporphyria

This is a rare disease in which the usual features are acute attacks of the type seen in acute intermittent porphyria. Excess coproporphyrin III is present in faeces and urine, but urinary porphyrin excretion may be normal between attacks. Excess ALA and PBG are excreted in urine, but only during acute attacks. The disorder is probably due to a deficiency of coproporphyrinogen oxidase.

Porphyria cutanea tarda (PCT)

This is the commonest type of porphyria. Photosensitivity is the principal symptom, and bullous lesions develop on areas of skin exposed to sunlight. The disorder is usually due to a combination of two factors:

(1) A genetic deficiency of uroporphyrinogen decarboxylase.
(2) Acquired hepatic disease.

Alcoholic liver disease is almost always present in patients with PCT, in Europe and North America and other 'Western' countries, but occasionally hepatic tumours may cause the symptom complex. Haemosiderosis is associated with the disorder, especially if alcoholism is common, as among the Bantu. Toxic damage to the liver may also give rise to PCT, the best-known example having occurred in Turkey, in 1957, where a combination (occasionally fatal) of photosensitivity, hirsutism and liver disease was caused by eating bread made from wheat where the seed wheat had been treated with the fungicide hexachlorobenzene.

The pattern of porphyrin excretion in PCT is complex. There is excessive excretion of uroporphyrin I and III in urine, and both urine and faeces contain 7-carboxyl-substituted porphyrins.

CHEMICAL INVESTIGATION OF THE PORPHYRIAS

Precise diagnosis of abnormalities of haem biosynthesis requires

TABLE 15.2. *The hepatic porphyrias*, i.e. porphyrias with normal erythrocyte [porphyrins] (Modified from Elder, 1983)

	Urinary excretion			*Faecal*	*Enzyme defect*
	ALA	*PBG*	*Porphyrins*	*Porphyrins*	
Acute intermittent porphyria					
During an attack	++	+++	PBG polymers ++	N	Uroporphyrinogen-
During remission	N	+	±	N	I-synthase
Variegate porphyria					
During an attack	+	++	Copro III +	Proto IX, Copro III and X-porphyrin +	Protoporphyrinogen oxidase?
During remission	N	N	±	As above	
Hereditary coproporphyria					
During an attack	+	++	Copro III +	Copro III ++	Coproporphyrinogen oxidase
During remission	N	N	±	±	
Porphyria cutanea tarda	N	N	Uro I and III, 7-carboxyl-porphyrins etc ++	Acetate-substituted porphyrins +	Uroporphyrinogen decarboxylase

Key to abbreviations:
N = normal; ± = normal or slight increase; +, + + and + + + = increasing degrees of abnormality
Copro = coproporphyrin Uro = uroporphyrin Proto = protoporphyrin

specific identification of the stage affected, by measuring the appropriate enzyme activities. This is not practicable on a routine basis. Instead, the diagnosis can nearly always be made on the basis of the following qualitative or semi-quantitative tests:

(1) Total urinary porphyrins.
(2) Total faecal porphyrins.
(3) Total erythrocyte porphyrins.
(4) Urinary excretion of PBG.
(5) Urinary excretion of ALA (mainly of value in lead poisoning).

In terms of routine diagnosis, the importance of testing urine for the presence of PBG, in the side-room (p. 31) or in the laboratory, must be reiterated. If a patient with acute porphyria is not recognised, it is all too possible that the administration of a barbiturate anaesthetic will have serious consequences.

The results of investigations of disorders of porphyrin metabolism are summarized, depending on whether erythrocyte [porphyrins] are normal (Table 15.2) or increased (Table 15.3). The value of urinary ALA and PBG measurements, as preliminary investigations especially of the hepatic porphyrias, is evident.

Porphyrinuria

This term is used to describe various conditions in which there is a disturbance of porphyrin metabolism but in which the clinical features are not due to this disturbance. In some of these haem synthesis is normal, but in some conditions that give rise to porphyrinuria haem synthesis is abnormal.

The commonest cause of porphyrinuria is impaired excretion of

TABLE 15.3. Diseases in which erythrocyte [porphyrins]

	Urinary excretion			
	ALA	*PBG*	*Porphyrins*	
Congenital erythropoietic porphyria	N	N	Uro I Copro I	++ +
Protoporphyria	N	N	N	
Lead poisoning	++	N	Copro III	++

Key to abbreviations: See Table 15.2

porphyrins in bile. This may be due to hepatocellular disease, or to any of the causes of cholestasis (p. 185). It may also occur with oestrogen therapy, including oestrogen-containing oral contraceptives (p. 400) and in pregnancy. In all these conditions, haem synthesis is normal.

Porphyrinuria, secondary to abnormalities of haem synthesis, may occur as a result of lead poisoning or as one of the manifestations of ethanol ingestion.

LEAD POISONING

There are similarities, in both the clinical and metabolic features, between lead poisoning and the porphyrias. Lead inhibits many enzyme systems. In the haem synthetic pathway (Fig. 15.2), it inhibits ALA dehydratase and, to a lesser extent, coproporphyrinogen oxidase and ferrochelatase. Two types of diagnostic problem need to be considered:

(1) *Patients* with symptoms or signs that may be explicable in terms of a diagnosis of lead poisoning. If the diagnosis is correct, biochemical findings are usually grossly abnormal.
(2) *Occupational hazard,* where healthy members of the population may be exposed to increased risk of poisoning (e.g. lead smelters). Screening of those at risk by sensitive tests is required.

The following tests may be abnormal in patients and in people affected by lead poisoning:

(1) *Tests demonstrating the presence of excess lead:*
  (a) Blood [lead].
  (b) Urinary lead excretion after giving EDTA, calcium salt.

are increased (Modified from Elder, 1983).

Faecal Porphyrins		*Erythrocyte Porphyrins*		*Enzyme defect*
Copro I	+++	Copro I	+++	Uroporphyrinogen-III-cosynthase?
Uro I	±	Uro I	+++	
Proto IX	±	Proto IX	+++	Ferrochelatase
N		Zinc protoporphyrin	++	ALA dehydratase, coproporphyrinogen oxidase and ferrochelatase

(2) *Tests demonstrating toxic effects of lead:*

(a) ALA dehydratase activity in erythrocytes.
(b) Urinary ALA excretion.
(c) Erythrocyte [protoporphyrin].
(d) Urinary coproporphyrin (rarely used as too non-specific).

Blood [lead] is a useful measurement, but it has several drawbacks as the sole measure of either lead poisoning or overexposure. Although blood [lead] correlates significantly with other biological indices of lead poisoning, there are variations in individual susceptibility to a given blood [lead]. Partly for this reason, it has proved difficult to establish 'acceptable' blood levels in a healthy population. Also the measurement of blood [lead] is technically demanding.

Chelatable lead excretion is a measure of the amount of lead excreted in the 24-h period after a 60 min intravenous infusion of 1 g Ca EDTA. The test is an excellent index of lead overexposure. However, it is rather tedious and is therefore not suitable for screening purposes.

ALA dehydratase activity, measured in erythrocyte haemolysates, is probably the most sensitive index of exposure to lead. However, the enzyme activity is also subject to genetic variation. In addition, there are differences in methodology between laboratories which can render interpretation difficult.

Urinary ALA excretion is another sensitive index of lead overexposure and is especially suitable for screening workers exposed to lead. Timed urine collections are required, and this represents a potential source of inaccuracy.

Erythrocyte [protoporphyrin] is the most practicable measure of overexposure to lead. The protoporphyrin, in lead poisoning, is present as a chelate with zinc. There are simple screening methods available, as well as more complex and specific techniques. This test is preferred to urinary ALA excretion, when screening children for suspected exposure to lead, since timed urine collections in children are often unreliable.

ETHANOL

One of the effects of ethanol ingestion over a prolonged period is increased excretion of coproporphyrin III in urine. This is probably due to increased hepatic production of coproporphyrin III, but the cause of the coproporphyrinuria has not been definitely established.

HAEMOGLOBIN AND ITS DERIVATIVES

Haemoglobin consists of a protein part, globin, associated with four haem molecules. Globin possesses two pairs of polypeptide chains, each chain being linked through a histidine residue with one haem molecule. There are four different types of polypeptide chain (α, β, γ, δ) which make up the globin component of the naturally occurring haemoglobins, giving rise to the combinations shown in Table 15.4.

About 200 human haemoglobins have been described. The description that follows has had to be very selective. Inherited abnormalities of globin synthesis may be of two types:

(1) *The haemoglobinopathies* in which there is a structural abnormality of one of the globin chains.
(2) *The thalassaemias,* in which there is deficient production of either the α- or the β-chain.

Although these abnormalities have a biochemical basis, the present discussion will be brief as their routine investigation is most often undertaken in haematology departments.

The haemoglobinopathies

These result from (1) the substitution of an amino acid in, or (2) the deletion of one or more amino acids from, or (3) the addition of extra amino acids to, one of the peptide chains.

Different functional effects may arise from the consequent abnormalities. These include changes in (1) the solubility of Hb, (2) the stability of Hb (and consequently of the erythrocytes), (3) the O_2-carrying capacity of the blood and (4) the rate of Hb production. Sometimes, however, no abnormality of function can be detected.

Altered solubility of haemoglobin

This gives rise to the commonest haemoglobinopathy, *sickle cell*

TABLE 15.4. The naturally occurring haemoglobins.

Designation	Occurrence	Globin structure
Hb-A	About 97–98% of normal adult Hb	$\alpha_2\beta_2$
Hb-A_2	About 2–3% of normal adult Hb	$\alpha_2\delta_2$
Hb-F	Fetal Hb; present in the circulation until the age of 3–6 months.	$\alpha_2\gamma_2$

disease. This disease presents at 3–6 months, when the synthesis of γ-chains is normally replaced by the synthesis of β-chains.

Abnormal β-chains are produced. There is a single substitution, of valine for glutamic acid, at the sixth position from the N-terminal end. The resulting haemoglobin, Hb-S, differs from Hb-A in its electrical charge properties. The reduced (deoxygenated) form of Hb-S tends to aggregate and to come out of solution. When this happens, it leads to the formation of the characteristic sickle cells. These have increased mechanical fragility and become haemolysed.

In the homozygous condition, no normal β-chains are produced; Hb-S predominates, accompanied by small amounts of Hb-F and Hb-A_2. Heterozygotes are usually symptom-free; their Hb is mostly Hb-A but they have about 20–40% Hb-S. Other haemoglobinopathies are much rarer than the sickle cell group of diseases.

Altered stability of haemoglobin

Erythrocytes containing haemoglobins that have altered stability lyse more readily than normal erythrocytes, giving rise to haemolytic anaemia. The various haemoglobinopathies associated with the formation of unstable haemoglobins together make up a rare group of causes of congenital non-spherocytic haemolytic anaemia.

Altered oxygen-carrying capacity of the blood

Several rare Hb variants have been discovered in some of which the affinity of the Hb for O_2 is increased while in others it is reduced.

Increased affinity for O_2 results in a decreased delivery of O_2 to the tissues. This gives rise to polycythaemia, if the abnormal Hb is stable.

Reduced affinity for O_2 in some cases (e.g. Hb-Milwaukee) results from the tendency of the Fe^{2+} in the porphyrin rings of these abnormal haemoglobins to oxidize to Fe^{3+}; methaemoglobinaemia results.

Deficient production of haemoglobin

Some haemoglobinopathies are associated with a hypochromic anaemia, caused by an impaired rate of synthesis of the abnormal Hb.

INVESTIGATION OF THE HAEMOGLOBINOPATHIES

The abnormal haemoglobins may be compared with Hb-A. Demonstrable changes include:

(1) Altered mobility on electrophoresis.

(2) Ease of denaturation in the presence of alkali.
(3) Changes in the oxygen dissociation curve.
(4) Alterations in peptide chain structure. This is the definitive method of characterizing the mutational change.

The thalassaemias

This is a group of hereditary anaemias caused by deficient production of either the α- or the β-chains. The diseases may present with different grades of severity, mainly due to there being several different genetic forms of each disease. Heterozygotes (thalassaemia trait) are usually symptom-free unless subjected to stress (e.g. severe infection).

Classical β-chain thalassaemia is common. In this condition, β-chain production is reduced or absent. Production of Hb-A_2 and Hb-F is increased; neither of these contains β-chains.

In α-chain thalassaemia, α-chain synthesis is reduced or absent. The tetramer γ_4 (Hb-Barts) is produced by homozygotes for one form of the disease, and a stillborn fetus results. A less severe form is associated with increased levels of the tetramer β_4(Hb-H).

Some patients are heterozygous for a thalassaemic gene on the one hand, and for a structurally abnormal Hb gene (e.g. Hb-S) on the other. Depending on the particular combination of abnormal genes, a spectrum of disorders with varying degrees of severity results.

INVESTIGATION OF THE THALASSAEMIAS

The diagnosis is suggested by a combination of the clinical and haematological findings, and confirmed by haemoglobin electrophoresis and alkali-denaturation studies. Electrophoresis can be used to detect Hb-A_2, Hb-H, Hb-Barts and other Hb variants; Hb-F is measured by a test based on its resistance to denaturation by alkali.

In homozygous β-chain thalassaemia, a high percentage of Hb-F is present, but the amount of Hb-A_2 is more variable. In heterozygous β-chain thalassaemia, the percentage of Hb-A_2 is increased, but Hb-F is usually not increased. Patients with β-chain thalassaemia will also show the biochemical features of iron overload, especially if they have been frequently transfused.

Abnormal derivatives of haemoglobin

METHAEMOGLOBIN

This is oxidized Hb, the Fe^{2+} normally present in haem being replaced by Fe^{3+}; the ability to act as an O_2 carrier is lost.

The normal erythrocyte contains small amounts of methaemoglobin, formed by spontaneous oxidation of Hb. Methaemoglobin is normally reconverted to Hb by reducing systems in the red cells, the most important of these being NADH-methaemoglobin reductase.

Excess methaemoglobin may be present in the blood because of increased production or diminished ability to convert it back to Hb. If there is more than 20 g/L (2 g/100 mL) of methaemoglobin, cyanosis develops. Both genetically determined and acquired conditions can cause methaemoglobinaemia; the acquired group is much the commoner. Haemolysis sometimes occurs in cases of methaemoglobinaemia, and methaemoglobin then appears in the urine, giving it a brownish colour.

Genetic causes of methaemoglobinaemia include the following groups of conditions:

(1) A group of haemoglobinopathies, collectively called Hb-M.
(2) A group having a deficiency of NADH-methaemoglobin reductase.

Patients with Hb-M are heterozygous for a gene that codes for a structurally abnormal Hb having an amino acid substitution in the vicinity of the haem portion of the molecule that stabilizes the haem in the Fe^{3+} form. About 40% of the Hb is present as methaemoglobin. Treatment with reducing agents, e.g. methylene blue or ascorbic acid, is ineffective in reducing the methaemoglobin to Hb.

Deficiency of NADH-methaemoglobin reductase leads to an accumulation of methaemoglobin that usually amounts to 20–50% of the total Hb. Treatment with ascorbic acid or methylene blue is effective in reducing the methaemoglobin in these patients.

Acquired methaemoglobinaemia usually arises following the ingestion of large amounts of drugs, e.g. phenacetin or the sulphonamides; excess of nitrites or certain oxidizing agents present in the diet may also cause methaemoglobinaemia. Acquired methaemoglobinaemia can be reversed by treatment with methylene blue or ascorbic acid.

SULPHAEMOGLOBIN

This is formed when Hb is acted upon by the same substances as cause acquired methaemoglobinaemia, if they act in the presence of sulphur-containing compounds. Hydrogen sulphide, which may arise from bacterial action in the intestine, is an example of these.

Sulphaemoglobin and methaemoglobin are often present at the same time in these patients.

Sulphaemoglobin cannot act as an O_2 carrier nor be converted back to Hb. Because of its spectroscopic characteristics, patients with even a mild degree of sulphaemoglobinaemia are cyanosed.

CARBOXYHAEMOGLOBIN (COHb)

Carbon monoxide combines at the same position in the Hb molecule as O_2, but with an affinity about 200 times greater than oxygen. As a result, even small quantities of CO in the inspired air cause the formation of relatively large amounts of COHb, with a corresponding reduction in the O_2-carrying capacity of the blood. This is due not only to the blocking effect of CO on O_2-binding sites but also to the fact that there is a shift to the left of the HbO_2 dissociation curve when only one of the 4 binding sites on Hb is occupied by CO.

Small amounts of COHb (up to 10%) may be present under 'normal' conditions in city dwellers. Concentrations above 40% usually result in unconsciousness, and may be fatal.

HAEMATIN

This is a protein-free Fe^{3+} complex of protoporphyrin. It may be released from the erythrocytes in patients with methaemoglobinaemia, or may be formed following intravascular haemolysis. In the plasma, haematin combines with albumin to form *methaemalbumin*, thereby making the plasma brown in colour.

Methaemalbuminaemia sometimes occurs in patients with acute pancreatitis (p. 200).

TESTS FOR ABNORMAL DERIVATIVES OF HAEMOGLOBIN

It is possible to identify abnormal derivatives of Hb by means of their characteristic absorption spectra, and it is sometimes possible to measure them quantitatively. For spectroscopic examination and quantitative determination of abnormal derivatives of Hb, the blood specimen should be collected with an anticoagulant (e.g. lithium heparin). Most effort has been devoted to developing methods for measuring COHb; its concentration in blood can be determined accurately.

FURTHER READING

AISEN, P. and LISTOWSKY, I. (1980). Iron transport and storage proteins. *Annual Reviews of Biochemistry*, **49**, 357–393.

ELDER, G.H. (1980). The porphyrias: clinical chemistry, diagnosis and methodology. *Clinics in Haematology*, **9**, 371–398.

ELDER, G.H. (1983). Disorders of haem synthesis. In *Biochemical Aspects of Human Disease*, pp. 367–400. Ed. R.S. Elkeles and A.S. Tavill. Oxford: Blackwell Scientific Publications.

GRANICK, J.L., SASSA, S. and KAPPAS, A. (1978). Some biochemical and clinical aspects of lead intoxication. *Advances in Clinical Chemistry*, **20**, 288–339. New York: Academic Press.

KAPPAS, A., SASSA, A. and ANDERSON, K.E. (1983). The porphyrias. In *The Metabolic Basis of Inherited Disease*, 5th Edition, pp. 1301–1384. Ed. J.B. Stanbury, J.B. Wyngaarden, D.S. Fredrickson, J.L. Goldstein and M.S. Brown. New York: McGraw-Hill.

MOORE, M.R. (1983). Laboratory investigation of disturbances of porphyrin metabolism. *Association of Clinical Pathologists Broadsheet*, **109**, London: British Medical Association.

MUNRO, H.N. and LINDER, M.C. (1978). Ferritin: structure, biosynthesis and role in iron metabolism. *Physiological Reviews*, **58**, 317–396.

WALDRON, H.A. (1979). Lead in the environment. *Journal of the Royal Society of Medicine*, **72**, 753–755.

WINSLOW, R.M. and ANDERSON, W.F. (1983). The haemoglobinopathies. In *The Metabolic Basis of Inherited Disease*, 5th Edition, pp. 1666–1710. Ed. J.B. Stanbury, J.B. Wyngaarden, D.S. Fredrickson, J.L. Goldstein and M.S. Brown. New York: McGraw-Hill.

Chapter 16

Disorders of Purine Metabolism

Adenine and guanine are nitrogenous bases, present in nucleic acids and in a large number of lower mol. mass compounds such as ATP and NAD. They may be obtained from the diet or synthesized in the body. Most cells can synthesize purines, the product of the synthetic process being a nucleotide (base-ribose-phosphate), not the free base (Fig. 16.1). Regulation of nucleotide synthesis probably occurs through the reaction for which 5-phosphoribosyl-1-pyrophosphate (PRPP) is the substrate. The enzyme that catalyses this reaction, PRPP-amidotransferase, is inhibited by 5′-nucleotides, and the amount of purine synthesis may be controlled by the concentration of adenosine 5′-phosphate (AMP) and guanosine 5′-phosphate (GMP) within the cell. The nucleotides are incorporated into nucleic acids.

Nucleotides derived from pyrimidines (cytosine, thymine, uracil) will not be discussed as derangements of their metabolism are very rare.

Nucleic acids are broken down by a variety of enzymes initially to nucleotides, which mix with the intracellular nucleotide pool or are

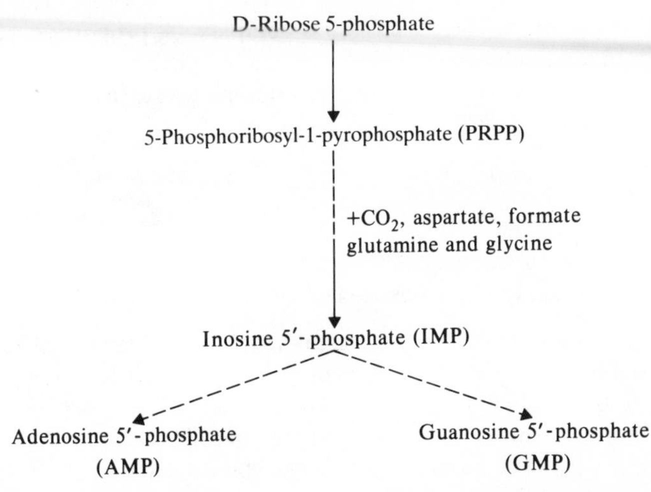

FIG. 16.1. Simplified representation of the synthetic pathway leading to the formation of purine nucleotides.

further degraded to a nucleoside (base-ribose) and then the free base. It used to be thought that most of the free bases so formed, mainly hypoxanthine and guanine, were thereafter converted to urate and excreted. However, an alternative 'salvage pathway' has since been found to be important (Fig. 16.2).

Much of the hypoxanthine formed by the breakdown of AMP and IMP is converted back to IMP, and some of the guanine is converted back to GMP, by reactions catalysed by hypoxanthine phosphoribosyl transferase (better known as hypoxanthine-guanine phosphoribosyl transferase or HGPRT). Some of the adenine is reconverted to AMP, by the action of adenine phosphoribosyl transferase (APRT). Both AMP and GMP can be formed from IMP and these 'salvaged' nucleotides can be used for nucleic acid synthesis. In these ways (Fig. 16.2), purines are retained in the body by processes which involve much less energy expenditure than would be needed for *de novo* synthesis (Fig. 16.1) of the whole of the body's requirements for purine nucleotides.

URATE POOL

Urate formed by metabolism of purines, together with urate formed from dietary nucleic acids, makes up the body's pool, which amounts to approximately 6 mmol (1 g) in a 70 kg adult; about 50% of the urate pool turns over each day.

Urinary excretion of urate is related to the level of renal function. The mechanism, discussed below, is complex. A proportion of the urate pool, up to about 30%, is eliminated by other pathways, mainly via the intestine; output via these other pathways is increased in patients with renal disease.

Plasma urate

There is a wide variation in plasma [urate], even in health. Much of this can be attributed to physiological factors, which include:

(1) *Sex.* Plasma [urate] tends to be higher in males than females. The upper limit of reference values in males is about 0.42 mmol/L (7.0 mg/100 mL); in females it is about 0.36 mmol/L (6.0 mg/100 mL).
(2) *Obesity.* Plasma [urate] tends to be higher in the obese.
(3) *Social class.* The more affluent social classes tend to have a higher plasma [urate].
(4) *Diet.* Plasma [urate] rises in individuals taking a high protein diet,

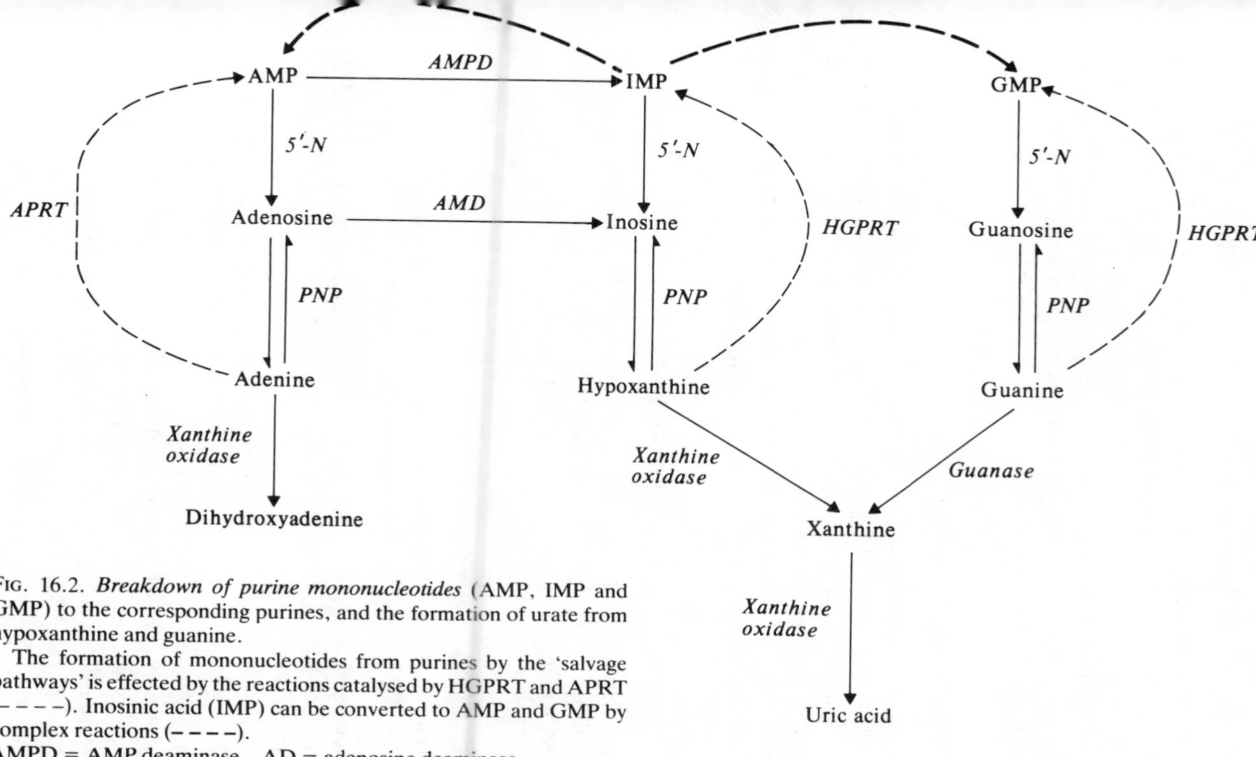

FIG. 16.2. *Breakdown of purine mononucleotides* (AMP, IMP and GMP) to the corresponding purines, and the formation of urate from hypoxanthine and guanine.

The formation of mononucleotides from purines by the 'salvage pathways' is effected by the reactions catalysed by HGPRT and APRT (– – – –). Inosinic acid (IMP) can be converted to AMP and GMP by complex reactions (– – – –).

AMPD = AMP deaminase AD = adenosine deaminase
5′-N = 5′-nucleotidase PNP = purine nucleoside phosphorylase
APRT = adenine phosphoribosyl transferase
HGPRT = hypoxanthine-guanine phosphoribosyl transferase

i.e. a diet which is also rich in nucleic acids, and in those who have a high alcohol consumption.

(5) *Genetic factors*. These are undoubtedly important.

The method of analysis used by the laboratory can have a slight but significant effect on the urate concentrations reported. Measurements that are based on uricase tend to be slightly lower (up to 0.06 mmol/L or 1 mg/100 mL) than measurements made by less specific colorimetric techniques. Several drugs or their metabolites interfere with the colorimetric methods of analysis.

Solubility of urate

Solutions of monosodium urate become supersaturated when the concentration exceeds 0.42 mmol/L (7.0 mg/100 mL). However, the relationship between the presence and severity of hyperuricaemia and the development of symptoms, either arthritis or renal calculi, is more complex than consideration simply of solubility data might suggest. This is discussed again later, under primary gout.

Hyperuricaemia

Increased plasma [urate] may be attributable to a number of mechanisms. Viewed in the simplest terms, however, there must either be overproduction or defective elimination of urate.

Overproduction of urate

Metabolic abnormalities have only been defined in a small number of diseases. The following abnormalities may occur:

(1) Increased activity of the rate-limiting enzyme, PRPP-amidotransferase, or increased concentration of its substrate, PRPP.
(2) Increased activity of those pathways of nucleotide metabolism which lead to urate formation, as compared to those which lead to the formation of nucleic acids.
(3) Increased rate of nucleic acid breakdown, as occurs whenever there is increased rate of cell turnover or destruction.
(4) Decreased activity of the salvage pathway due to absence or deficiency of HGPRT.
(5) Increased activity of xanthine oxidase. When this occurs, it is probably a secondary rather than a primary effect.

Renal excretion of urate

Urate excretion is a complex process. Except for a small fraction bound to plasma proteins, urate is completely filtered at the glomerulus, but this is then mostly reabsorbed in the proximal tubule. Distal tubular secretion of urate is normally responsible for generating the urate present in urine. These processes may all be affected by disease or drugs:

(1) *Glomerular filtration rate.* When the GFR becomes reduced, urate retention occurs. However, the reduction in GFR must be fairly severe before this gives rise to marked hyperuricaemia.
(2) *Proximal tubular reabsorption.* Most uricosuric drugs act by decreasing proximal tubular reabsorption of urate. This is the major effect, for example, of probenecid and of *high* doses of salicylates.
(3) *Distal tubular secretion.* Lactic acid and 3-hydroxybutyric acid apparently compete with urate for this excretory pathway, and any condition giving rise to lactic acidosis or ketosis tends to be associated with hyperuricaemia. Some drugs, e.g. chlorothiazide and *low* doses of salicylates, affect distal tubular secretion by inhibiting urate secretion.

It should be noted that salicylates and many other uricosuric agents have opposite effects on proximal and distal renal tubular handling of urate. With low doses of salicylates, the inhibitory effect on distal tubular secretion is the important action, tending to cause hyperuricaemia, whereas with high doses of salicylates the reduction in proximal tubular reabsorption is the dominant effect and there is increased urate excretion in the urine.

Gout

PRIMARY GOUT

This condition is characterized by recurrent attacks of arthritis. The distribution of the affected joints is an important feature in diagnosis. Men suffer from primary gout much more frequently than women.

Attacks of acute gouty arthritis only occur in about 30% of men who have plasma [urate] greater than 0.42 mmol/L. Conversely, attacks of acute gouty arthritis occur occasionally when plasma [urate] is less than 0.42 mmol/L. This reflects the fact that the occurrence of gout is not related to plasma [urate] in a simple manner. However, individuals with plasma [urate] greater than 0.60 mmol/L (10 mg/100 mL) are very liable to develop acute gouty arthritis, although the factors that precipitate an acute attack remain uncertain.

Gouty arthritis is due to the intra-articular formation of crystals of monosodium urate. The crystals are phagocytosed by leucocytes and then cause damage to lysosomal membranes within the leucocytes. Lysosomal contents are released and damage both the leucocytes and surrounding tissues, causing the acute joint manifestations of gout.

Patients with gout often show deposition of urate as tophi in soft tissues. About 10% of patients also develop renal stones, mainly composed of uric acid. However, stone formation is again not a matter simply of crystallization from a supersaturated solution.

The reason for the onset of acute attacks of gout is frequently unclear. A sharp rise in plasma [urate] is often not demonstrable, nor is plasma [urate] the sole predictive factor in determining whether acute attacks are likely to occur.

The metabolic defect in most patients with primary gout is unknown but almost certainly a number of distinct abnormalities may be responsible. A minority of patients show clear evidence of overproduction of urate and have markedly increased urinary urate output. Most patients appear to have a combined defect, with evidence of (1) overproduction of urate and (2) impaired renal excretion, reflected in an inappropriately low urinary urate output if the raised plasma [urate] is taken into consideration. In a few cases, partial deficiency of HGPRT has been demonstrated, but less severe than the deficiency in the very rare Lesch-Nyhan syndrome (see below).

Diagnosis of primary gout

Diagnosis is often made on clinical grounds. An increased plasma [urate] and a satisfactory response to treatment in many cases provide sufficient confirmation of the diagnosis. Unfortunately, the interpretation of results for plasma [urate] is not straightforward, as discussed above. It is only possible to make the following generalizations:

(1) A high plasma [urate] makes the diagnosis of gout probable.
(2) A consistently low plasma [urate] excludes the diagnosis.

For the definitive diagnosis of gout, it may be necessary to aspirate joint fluid collected during an acute attack. The fluid is examined microscopically and the finding of needle-shaped urate crystals, which show negative birefringence, establishes the diagnosis. This finding clearly differentiates gout from *pseudogout*, or *pyrophosphate arthropathy,* since pyrophosphate crystals do not show these characteristics. It should be emphasized that consideration of clinical features (e.g. distribution of joints affected, etc.) means that aspiration of joint fluid is not required as a routine. Instead, it seems likely that measurement of plasma [urate] will retain its key role in the investigation of primary gout for a considerable time.

Treatment of primary gout

In an acute attack, anti-inflammatory drugs (e.g. indomethacin) are usually employed but long-term treatment aims to achieve reduction in plasma [urate]. Factors known to increase the plasma [urate], e.g. high protein diet, alcohol and certain drugs should be avoided. Weight reduction, uricosuric drugs (e.g. probenecid), and inhibition of urate synthesis by means of allopurinol may be required. In patients in whom renal stones seem likely to form, a high fluid intake and an alkaline urine reduce the likelihood of urate stone formation.

Allopurinol, an isomer of hypoxanthine, inhibits xanthine oxidase, with the result that there is a fall in both the plasma [urate] and the urinary output of urate. There is increased urinary excretion of xanthine and hypoxanthine, but this is usually less than the corresponding fall in urinary urate, presumably due to reconversion of hypoxanthine to IMP via the salvage pathway and further conversion of IMP to the other purine nucleotides (Fig. 16.2). Allopurinol also reduces the synthesis of purines.

SECONDARY GOUT

Hyperuricaemia may occur as a complication of a number of disorders all of which affect either urate production or excretion, or both. These conditions, although commonly causing hyperuricaemia, are only rarely associated with the joint manifestations of gout. They will be briefly considered:

Myeloproliferative disorders. Almost all these disorders may be associated with hyperuricaemia, but polycythaemia rubra vera is probably the disease most commonly associated with signs of gout. Hyperuricaemia is due to increased cell turnover.

Cytotoxic drug therapy. Patients treated with cytotoxic drugs have an increased rate of cell destruction causing hyperuricaemia and increased excretion of urate in the urine. Renal stones may occur.

Psoriasis. The hyperuricaemia is thought to be due to increased rate of cell turnover in the skin.

Hypercatabolic states and starvation. Two mechanisms may be responsible for hyperuricaemia. There may be an increased rate of cell destruction, or the conditions may be associated with lactic acidosis, which causes diminished distal tubular secretion of urate.

Chronic renal disease. The plasma [urate] rises in uraemia, due to reduced GFR. However, clinical gout is very unusual.

Diuretic therapy. Most effective diuretics, e.g. chlorothiazide and frusemide, cause hyperuricaemia, probably by reducing distal tubular secretion of urate.

Inborn errors of metabolism. Some of these disorders are associated with lactic acidosis, which causes urate retention. The most striking example occurs in Type I glycogen storage disease (p. 237).

Hypertension. This is associated with raised plasma [urate] much more commonly than would be expected by chance. This may be explicable partly in terms of the association of both gout and hypertension with obesity, and partly in terms of the urate-retaining effects of some anti-hypertensive drugs.

Other inherited disorders of purine metabolism

The conditions to be discussed are all rare. Some of them give rise to primary gout as part of their complex of symptoms and signs. The points of action of the various enzymes are shown in Fig. 16.2.

The *Lesch-Nyhan syndrome* is a very rare inherited condition which gives rise to one form of primary gout. It is of biochemical interest because of the light its investigation has thrown on the normal pathways of purine metabolism. The disease usually presents in early childhood with choreoathetosis, mental retardation and self-mutilation. Plasma [urate] is increased and urate stones may be present in the renal tract. The activity of HGPRT is greatly diminished or undetectable. In the absence of HGPRT, the salvage pathway is inoperative and purines cannot be reconverted to nucleotides; instead, they are converted to urate.

Partial HGPRT deficiency causes a severe form of primary gout, usually presenting in early adult life. Both plasma [urate] and urinary excretion of urate are increased.

APRT deficiency is associated with renal stones formed from dihydroxyadenine, normally a minor metabolite of adenine. Hyperuricaemia is not present and gout is not a feature. Allopurinol prevents the formation of renal stones in APRT deficiency since it inhibits xanthine oxidase.

Xanthine oxidase deficiency is a rare disorder that results in a very low plasma [urate] and an increased urinary excretion of xanthine and hypoxanthine. It may be asymptomatic, but some patients develop xanthine stones in the urinary tract.

Adenosine deaminase deficiency and *purine nucleoside phosphorylase deficiency* are both rare disorders that are accompanied by immunodeficiency, particularly affecting T lymphocytes.

Hypouricaemia

This is not an important chemical abnormality, in itself. It may occur as one feature of a variety of disorders:

(1) Decreased production of urate:
- **(a)** Inherited defect of purine synthesis, e.g. xanthine oxidase deficiency.
- **(b)** Acquired disease, e.g. severe liver disease.

(2) Increased excretion of urate:
- **(a)** Renal tubular defects, e.g. Fanconi syndrome (p. 98).
- **(b)** Treatment with uricosuric drugs (e.g. probenecid).

Increased urinary excretion is the more common cause of mild hypouricaemia, and is then most often due to drug treatment.

FURTHER READING

BALIS, M.E. (1976). Uric acid metabolism in man. *Advances in Clinical Chemistry*, **18**, 213–246. New York: Academic Press.

CAMERON, J.S. and SIMMONDS, H.A. (1981). Uric acid, gout and the kidney. *Journal of Clinical Pathology*, **34**, 1245–1254.

HUSKISSON, E.C. and BALME, H.W. (1972). Pseudopodagra. *Lancet*, **2**, 269–272.

Leading article (1972). Diagnosis of gout. *British Medical Journal*, **4**, 1–2.

McKERAN, R.O. and WATTS, R.W.E. (1978). Purine metabolism and cell physiology. In *Recent Advances in Endocrinology and Metabolism*, **1**, 219–252. Ed. J.L.H. O'Riordan. Edinburgh: Churchill Livingstone.

WYNGAARDEN, J.B. and KELLEY, W.N. (1983). Gout. In *The Metabolic Basis of Inherited Disease*, 5th Edition, pp. 1043–1114. Ed. J.B. Stanbury, J.B. Wyngaarden, D.S. Fredrickson, J.L. Goldstein and M.S. Brown. New York: McGraw-Hill.

Chapter 17

Abnormalities of Thyroid Function

The diagnosis of thyroid dysfunction is often straightforward. In difficult cases, however, laboratory tests prove particularly valuable, but their results must always be reviewed in the light of clinical findings. There are several pitfalls in the interpretation of laboratory tests of thyroid function.

FORMATION AND RELEASE OF THYROID HORMONES

Iodine in the diet is mainly inorganic iodide. This is absorbed from the intestine, transported in plasma and removed from the circulation by the thyroid gland and kidneys, and to a lesser extent by other organs. The thyroid normally contains about 90% of the iodine in the body; iodide is used by the thyroid in the synthesis of thyroxine (T4) and tri-iodothyronine (T3). Four stages in the formation and release of thyroid hormones (Fig. 17.1) can be recognized, all stimulated by thyroid-stimulating hormone (TSH). The stages are briefly described here:

(1) *Trapping of iodide* from the plasma by the thyroid gland. This stage may be blocked by perchlorate and thiocyanate.
(2) *Oxidation of iodide* to iodine. This stage may be blocked by methimazole, thiourea and related compounds.
(3) *Incorporation of iodine* into tyrosyl residues of thyroglobulin. Mono- and di-iodotyrosines are formed first. Later, T3 and T4 (Fig. 17.2) are formed by coupling of iodotyrosyl residues in the thyroglobulin molecule. Iodine incorporation can be blocked by thiouracil and related compounds, and by sulphonamides. Thyroglobulin normally remains mostly within the thyroid.
(4) *Splitting off of T3 and T4* from thyroglobulin and subsequent release of thyroid hormones into the circulation.

TRANSPORT OF T4 AND T3

Both T4 and T3 are present in plasma, partly as free hormone and partly protein-bound, the two forms being in equilibrium with one another. The protein-binding of T4 and T3 is so strong that only about

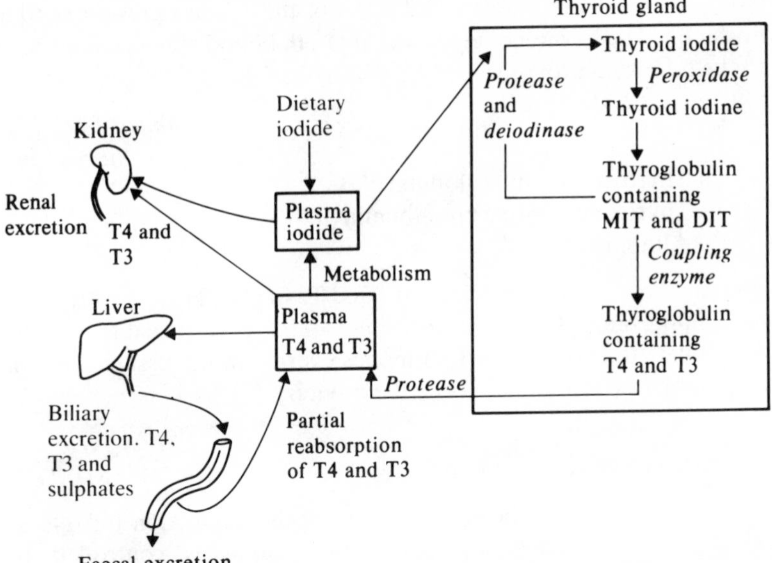

FIG. 17.1. Simplified diagram to show the formation and metabolism of the thyroid hormones, T3 and T4. Most of the T3 is formed from T4 in extra-thyroidal tissues.

Key to abbreviations

MIT = mono-iodotyrosine	T4 = thyroxine
DIT = di-iodotyrosine	T3 = tri-iodothyronine

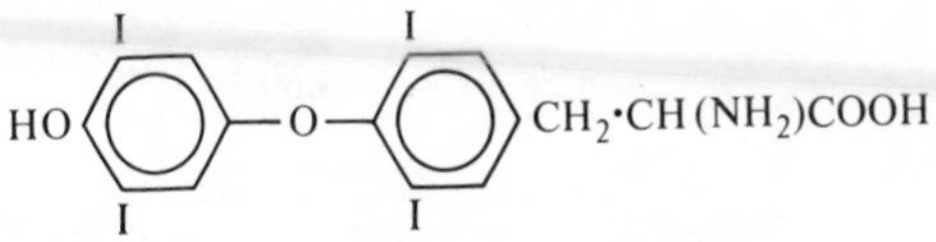

HO–(ring, I at 3 and 5)–O–(ring, I at 3′ and 5′)–$CH_2 \cdot CH(NH_2)COOH$

Thyroxine or 3:5, 3′:5′ tetra-iodothyronine (T4)

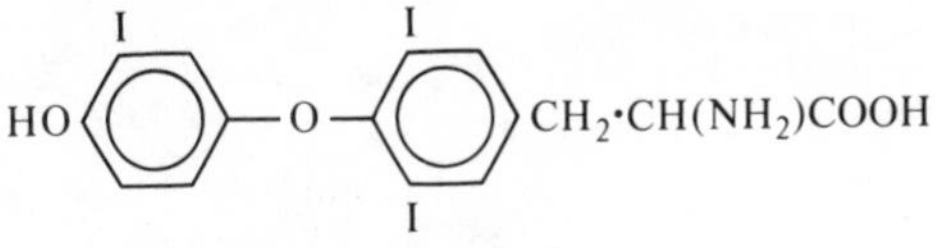

HO–(ring, I at 3′)–O–(ring, I at 3 and 5)–$CH_2 \cdot CH(NH_2)COOH$

3:5,3′ tri-iodothyronine (T3)

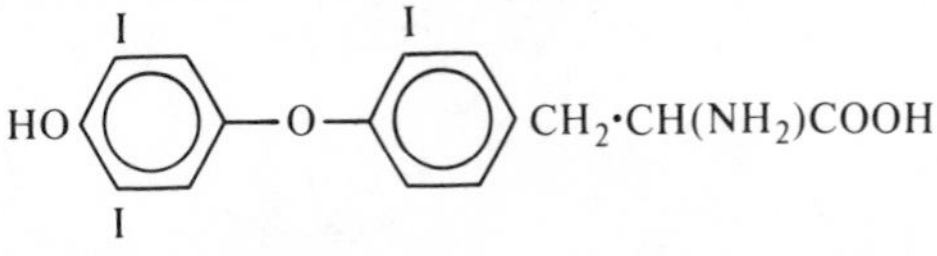

HO–(ring, I at 3′ and 5′)–O–(ring, I at 3)–$CH_2 \cdot CH(NH_2)COOH$

3,3′:5′ tri-iodothyronine (reverse T3)

FIG. 17.2. Formulae of thyroid hormones and of reverse T3.

0.05% of the T4 in plasma, and 0.5% of the T3, are present as free hormone. The hormones in plasma are both bound almost completely to the following proteins:

	Amount of T4 or T3 bound to the protein
Thyroxine-binding globulin (TBG)	75%
Thyroxine-binding pre-albumin (TBPA)	15%
Albumin	10%

The small concentrations of free T4 and T3 in plasma are nevertheless very important. They are diffusible and are responsible for the metabolic effects of thyroid hormones exerted in the tissues, and for regulating the output of TSH by the pituitary.

PRODUCTION, RELEASE AND METABOLISM OF T4 AND T3

The output of TSH by the pituitary controls the production and release of thyroid hormones. In turn, TSH output is controlled by thyrotrophin-releasing hormone (TRH), a tripeptide (pyroglutamyl-histidyl-prolinamide) secreted by the hypothalamus. In health, the main control is by an important feedback mechanism whereby both plasma [free T4] and [free T3] inhibit the output of TSH by the pituitary, and possibly also the output of TRH by the hypothalamus (Fig. 17.3).

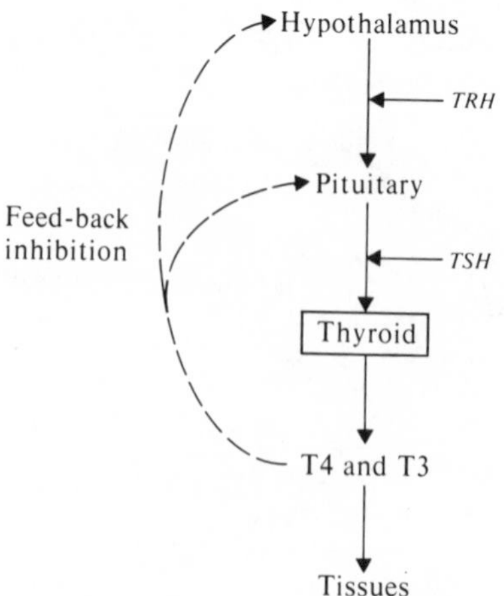

FIG. 17.3. The hypothalamic-pituitary-thyroid axis, and its regulation by feedback inhibition.

Most of the T3 present in plasma is derived from T4, by mono-deiodination in tissues other than the thyroid; only a small percentage of the T3 in plasma is formed in and released from the thyroid. It is possible that T3 is the only metabolically active hormone. The biological effects of thyroid hormones include (1) increased synthesis of cellular protein and (2) increased rate of mitochondrial respiration.

Some T4 is mono-deiodinated in the tissues to reverse T3 (Fig. 17.2), an isomer of T3 that is devoid of hormonal activity.

The further metabolism of T4, T3 and reverse T3 is by deiodination and by oxidative deamination of their alanine side-chains. Iodine released by deiodination is mostly taken up by the thyroid and re-used. The deiodinated and deaminated metabolites are conjugated, mainly in the liver, with the formation of sulphate and glucuronide derivatives.

About 10% of the daily production of T4 and T3 is excreted in bile, partly conjugated as sulphate esters but mainly in the free form. On reaching the intestine, these conjugates are partly hydrolysed; the unconjugated T4 and T3 are then reabsorbed from the intestine and re-enter the pool of circulating thyroid hormones.

Small amounts of unmetabolized T3 and T4 are excreted in urine.

INVESTIGATION OF DISORDERED THYROID FUNCTION

The main contributions of chemical measurements to the investigation of thyroid disease are in:

(1) helping to subdivide patients into the categories of euthyroid, hyperthyroid and hypothyroid states.
(2) monitoring the effects of treatment.

Special chemical investigations, and other tests, are needed for the full elucidation of complex diagnostic problems such as those posed by the rare inborn errors affecting iodine metabolism. The investigations discussed here are grouped under the following headings:

(1) Total thyroxine.
(2) Total tri-iodothyronine.
(3) Free thyroid hormones in plasma.
(4) Tests of hypothalamic-pituitary-thyroid function.
(5) *In vivo* radioactive uptake tests.
(6) Thyroid auto-antibodies.
(7) Miscellaneous tests.

Total thyroxine

Plasma [total T4] can be measured by radioimmunoassay and by enzyme-mediated immunoassay techniques (EMIT). However, alterations in plasma concentrations of thyroxine-binding proteins (TBP) can cause both misleadingly high and misleadingly low results for measurements of plasma [total T4].

The relationship between the amount of T4 bound to thyroxine-binding proteins (TBP-T4), and the amount of free TBP or unoccupied T4-binding sites on proteins (U-TBP) in plasma, is described by the equilibrium:

$$\text{Free T4} + \text{U-TBP} \rightleftharpoons \text{TBP-T4}$$

Because of the strong affinity of TBP for T4, over 99.9% of plasma T4 is in the form of TBP-T4, and the determination of plasma [total T4] effectively measures plasma [TBP-T4].

In health, it is plasma [free T4] rather than plasma [total T4] that is maintained constant by physiological control mechanisms. This means that, *in the euthyroid individual,* alterations in the concentrations in plasma of the thyroxine-binding proteins will be accompanied by corresponding changes in plasma [total T4]. Thus, if plasma [TBP] falls, plasma [total T4] also falls and, if plasma [TBP] rises, plasma [total T4] also rises although, in each case, plasma [free T4] remains unchanged.

Increases in plasma [TBP] may be due, for example, to pregnancy or to oestrogen therapy, and plasma [total T4] is increased in these conditions. On the other hand, low values for plasma [total T4] are encountered in most hypoproteinaemic states, e.g. chronic liver disease or the nephrotic syndrome. Plasma [total T4] may also be abnormally low in euthyroid patients treated with drugs that effectively lower plasma [TBP] by occupying T4-binding sites on U-TBP, e.g. salicylates.

Patients treated with androgens or with corticosteroids, or with some of the anticonvulsants (e.g. phenytoin), may have spuriously low plasma [total T4], probably due mainly to an increased rate of peripheral metabolism of T4. Also, treatment with lithium and with phenylbutazone both inhibit the secretion of thyroid hormones, again giving rise to low plasma [total T4]. Patients with chronic renal failure tend to have low plasma [total T4] for a variety of rather complex reasons.

In all the examples mentioned in the two previous paragraphs, if the patients are euthyroid, plasma [free T4] is normal whereas plasma [total T4] may frequently be abnormal. In practice, the commonest sources of potentially misleading results for plasma [total T4] occur in:

(1) *Pregnancy or oestrogen therapy,* including women taking oestrogen-containing oral contraceptives. On average, plasma [total T4] is 30–40% higher in pregnant than in non-pregnant women.
(2) *Ill patients*, who often have low plasma [total T4], for a variety of complex reasons that include their having low plasma [TBP]. These abnormalities are particularly common in the elderly.

A rare cause of sometimes greatly increased plasma [total T4], in euthyroid patients, is hereditary TBG excess. The condition is benign, but needs to be recognized if unnecessary treatment is to be avoided.

In *hyperthyroidism,* plasma [total T4] is usually increased to more than 150 nmol/L (11.7 μg/100 mL), although a few patients with clinical thyrotoxicosis have normal values. These patients can usually be shown to have raised plasma [total T3], as discussed below.

In *hypothyroidism* plasma [total T4] is usually low, with values less than 70 nmol/L (5.5 μg/100 mL). It provides a reliable test for hypothyroidism, but is less sensitive than plasma [TSH] for detecting early cases.

Protein-bound iodine (PBI)

Very few laboratories still measure plasma [PBI], because the measurements are frequently affected seriously by iodine contamination, e.g. iodine-containing drugs including many proprietary cough mixtures, X-ray contrast media. Plasma [PBI] provides a measure of plasma [total T4], but has been largely replaced by T4 measurements, which are not subject to these common potential sources of interference.

Total tri-iodothyronine

Plasma [total T3] is usually measured by radioimmunoassay. Using suitable antisera, there is little cross-reaction between T3 and T4. Changes in plasma [TBP] alter plasma total [T3] in the same manner as those described for plasma [total T4].

The main value of plasma [total T3] measurements is in the diagnosis of hyperthyroidism. In most cases of thyrotoxicosis, both plasma [total T3] and [total T4] are raised, but in a small percentage of clinically thyrotoxic patients plasma [total T4] and [free T4] are both normal whereas both plasma [total T3] and [free T3] are increased. This condition is called *T3-thyrotoxicosis*, or T3-toxicosis, and T3 measurements are required for its recognition.

Plasma [total T3] is usually low in hypothyroidism, but it appears to be a less sensitive test for detecting hypothyroidism than plasma [total T4], possibly due to a relative increase in the proportion of thyroid hormone secreted as T3 by the hypoactive thyroid.

Low plasma [total T3] is often present in patients in the absence of thyroid disease, especially in the old and in the severely ill. It may also be observed in certain acute illnesses such as myocardial infarction and after surgery. In most of these cases there is impaired conversion of T4 to T3 in the peripheral tissues, reverse T3 being formed instead but, in some of these patients, the low plasma [total T3] is due to a fall in plasma [TBP].

Free thyroid hormones in plasma

Measurements of free thyroid hormone concentration in plasma provide more accurate, i.e. sensitive and specific, *in vitro* tests of thyroid status than plasma [total T4] or plasma [total T3], since they indicate the amounts of thyroid hormones available to tissue cells. Several techniques have been developed for their measurement:

(1) Free T4, measured by equilibrium dialysis or by commercially available methods (e.g. Amerlex Free T4 RIA kit*).
(2) Free thyroxine index (FTI).
(3) Total T4: TBG ratio.
(4) Free T3, measured by the Amerlex Free T3 RIA kit*.

Free T4

The equilibrium dialysis technique is the reference method but is too time-consuming for routine use. More recently, the Amerlex Free T4 RIA kit became available and this is now widely used. Because free T4 can now be readily measured, few laboratories still determine the FTI or the total T4 : TBG ratio.

Results for plasma [free T4] have, in general, been found to provide a clinically more reliable index of thyroid status (euthyroid, hypothyroid or hyperthyroid) than plasma [total T4]. In particular, the free T4 methods are not affected by alterations in plasma [TBG], such as occur in pregnancy and in women taking oestrogen-containing oral contraceptives.

*Amersham International Ltd., Amersham, U.K.

Interpretation of results obtained by the Amerlex Free T4 RIA kit is influenced by reductions in plasma [albumin], and possibly by increased plasma [FFA] as well as by other factors. The subject is too complex to be able to be adequately summarized here, but it is important to note that those euthyroid patients who have low plasma [albumin] are likely also to have low plasma [free T4], if this is measured using the Amerlex Free T4 RIA kit. Such sources of interference should be reduced as the methods are improved.

With the presently available kit method, about 20% of women in the third trimester of pregnancy tend to have low plasma [free T4], but without any pathological significance necessarily attaching to this finding; reference values appropriate to the stage of pregnancy should be used.

Free thyroxine index

This provides an indirect measure of plasma [free T4], by determining (1) plasma [total T4] and (2) plasma [U-TBP] and expressing the result as:

$$[\text{Free T4}] = \text{K.}\ \frac{[\text{total T4}]}{[\text{U-TBP}]}$$

This equation is derived from the equilibrium reaction discussed on p. 340; the ratio [total T4]:[U-TBP] is known as the free thyroxine index.

A direct measurement of [U-TBP] can be obtained by a *serum uptake test,* in which radioactive T3 is added to the patient's serum. The amount of the added T3 that becomes bound to protein is a measure of U-TBP.

Similar information is obtained from a *resin uptake test,* which is performed in the same way as a serum uptake test initially, but then the amount of the added T3 that is *not* taken up by the proteins in the patient's specimen is measured. To perform this measurement, the unbound radioactive T3 is adsorbed on to an ion-exchange resin or other suitable adsorbent and the radioactivity taken up by the resin is counted.

Both the serum uptake test and the resin uptake test provide a measure of plasma [U-TBP]. This can then be used to 'correct' the result of the plasma [total T4] measurement for variations in plasma [TBP].

The free thyroxine index has been shown to correlate well with results for plasma [free T4] in patients with mild abnormalities (increase or decrease) of plasma [TBP], as may occur in pregnancy or mild hypoproteinaemia, whatever the cause (p. 116). However, it does not correct adequately for TBP abnormalities in elderly sick patients, nor in familial TBG abnormalities, nor in patients with dyshormonogenesis. Thyroidal dyshormonogenesis is caused by a group of rare enzyme defects in the synthesis of T4, in which considerable amounts of mono- and di-iodotyrosines may be released into the blood by the thyroid.

Total T4 : TBG ratio

In this method, plasma [TBG] is measured by radioimmunoassay. The ratio of the two measurements, plasma [total T4] : plasma [TBG], has been found to be more satisfactory than the free thyroxine index as an indication of thyroid status. However, this test has never gained wide acceptance since, like the more long-standing free thyroxine index, it requires two measurements whereas plasma [free T4] only requires one. Also, by only measuring TBG, allowance cannot be made for changes in concentration of the other T4-binding proteins, TBPA and albumin.

Free T3

The Amerlex Free T3 RIA kit became available in 1983. Preliminary evidence suggests that this will provide a valuable alternative to previously available measurements of thyroid hormones in plasma, especially for distinguishing between hyperthyroidism and the euthyroid state.

Tests of hypothalamic-pituitary-thyroid function

THYROID-STIMULATING HORMONE (TSH)

Plasma [TSH] can be measured by radioimmunoassay, and is now a widely available investigation. The methods presently available cannot detect very low concentrations of TSH, but stimulation tests that depend on the use of synthetic TRH have greatly extended the value of plasma TSH measurements.

In *hypothyroidism,* measurement of plasma [TSH] is particularly valuable. Plasma [T4] and [T3], and the *in vivo* radioactive uptake tests, are usually normal in the early stages of the disease. On the other hand, plasma [TSH] is almost invariably increased in hypofunction due to primary thyroid disease, as feedback inhibition of the pituitary has been removed (Fig. 17.3). Later in the progress of the disease, plasma [T4] and [T3] fall, but plasma [TSH] remains raised.

Much more rarely, hypothyroidism is secondary to hypofunction of the pituitary. In these patients, plasma [TSH] is low in fully developed cases, but may be normal earlier on in the disease. If it is suspected that there is only a minor impairment of pituitary function, and the plasma [TSH] is normal, a test of pituitary reserve function can be performed (p. 394). However, primary hypothyroidism and hypothyroidism secondary to pituitary disease can usually be readily distinguished clinically.

In most cases of *hyperthyroidism,* plasma [TSH] is decreased, due to feedback inhibition of the pituitary. However, measurement of plasma [TSH] in thyrotoxicosis is as yet of little diagnostic value, due to the inability of presently available methods to measure abnormally low concentrations of TSH satisfactorily.

DYNAMIC TESTS: THE TRH TEST

This test measures the response of the pituitary to TRH. It provides a sensitive measure of hypothalamic-pituitary-thyroid function.

To perform the test, 200 μg TRH is injected intravenously. Blood samples for measuring plasma [TSH] are collected before (basal) and 20 min after the injection. In normal individuals, plasma [TSH] rises to more than 2 mU/L above the basal level after 20 min.

In *hyperthyroidism,* plasma [TSH] always fails to respond to TRH, due to inhibition of pituitary TSH release by raised concentrations of thyroid hormones in the plasma. A normal response to the TRH test rules out the presence of mild or subclinical hyperthyroidism.

Failure to respond to TRH may occur in (1) patients with non-toxic multinodular goitre, (2) some patients with ophthalmic Graves' disease, (3) patients who have been treated for hyperthyroidism and rendered euthyroid within the previous four months, and (4) a few patients with various non-thyroidal conditions (e.g. chronic renal failure, Cushing's syndrome).

In *primary hypothyroidism,* basal plasma [TSH] is high and, following TRH injection, rises still further and remains high much longer. For instance, if a further blood specimen is collected 60 min after TRH injection, plasma [TSH] is usually still very high in these patients whereas in normal individuals it is usually returning towards basal levels by this time.

In patients with hypothyroidism secondary to pituitary disease, basal plasma [TSH] may be low, and remains low after TRH injection or only shows an impaired response.

In general, the TRH test is of limited value in patients with suspected hypothyroidism because:

(1) *In primary hypothyroidism,* the test is unnecessary since basal plasma [TSH] is increased, and this is sufficient to confirm the diagnosis, when taken in conjunction with clinical features and the results of plasma [T4] measurements.
(2) *In secondary hypothyroidism,* other indices of pituitary function, e.g. plasma growth hormone response to hypoglycaemia or basal plasma [gonadotrophins], become abnormal before the TSH response to TRH becomes abnormal.

The main value of the TRH test is in the diagnosis of hyperthyroidism, since this diagnosis can be excluded if there is an increase in plasma [TSH] in response to TRH injection.

In vivo radioactive uptake tests

These tests measure the uptake of an oral dose of radioactive iodine or an intravenous injection of technetium 99m (^{99m}Tc), by direct counting over the thyroid.

The uptake of ^{131}I (half-life, 8 days), measured 4 h after the oral dose, was for many years the standard test. Additional measurements of ^{131}I uptake at 24 and 48 h were also sometimes made, especially in patients with suspected hypothyroidism. There is now a trend towards the use of *early uptake tests* in which counting over the thyroid is performed less than 30 min after the intravenous administration of a radioactive material. In the U.K., these tests make use of ^{99m}Tc (half-life, 6 h); ^{132}I (half-life, 140 min) is also suitable, but is not available in the U.K. The uptake of radioactivity by the thyroid is measured after 20 min. The advantages of these short half-life isotopes include the smallness of the radiation dose and the ability to repeat tests of uptake within a short period. However, even the short half-life isotopes cannot be used for investigations carried out during pregnancy or in childhood.

In *hyperthyroidism,* the rate and extent of thyroid uptake of radioactivity and subsequent discharge is increased in over 90% of cases. In *hypothyroidism,* iodine uptake is both reduced and much slower than normal, but radioactive iodine uptake tests are of little value in hypothyroidism since there is considerable overlap with normal.

Radioactive iodine uptake tests are likely to be affected by the renal clearance of iodide and the diminished clearance of iodide by the kidney in renal disease makes interpretation of the tests difficult. Interpretation is also influenced by the size of the extra-thyroidal iodide pool and by recent medication with thyroid hormones or with anti-thyroid drugs.

In iodine deficiency, the size of the extra-thyroidal iodide pool is decreased, so the specific activity of the administered ^{131}I or ^{132}I in the body is increased and the thyroid uptake appears high. Conversely, following increased dietary intake of iodine or treatment with iodine-containing drugs, the uptake of radioactivity by the thyroid is low.

Thyroid hormone medication suppresses thyroid activity and therefore depresses the uptake of ^{131}I, ^{132}I and ^{99m}Tc. Treatment with anti-thyroid drugs may depress their uptake by direct action on the gland.

The appropriate *in vivo* radioactive uptake tests still have a limited place in the diagnosis and management of thyroid disease, but have otherwise been largely replaced by the chemical tests already described. The uses of these tests are:

(1) To help calculate the dose of radioactivity for patients being treated for hyperthyroidism. Technetium is unsuitable for this purpose.

(2) To monitor thyroid function in patients being treated with anti-thyroid drugs.

(3) In association with 'scanning' of the thyroid, to determine whether:

- **(a)** a thyroid nodule is 'hot' or 'cold'.
- **(b)** extra-thyroidal functioning thyroid tissue is present.

Thyroid auto-antibodies

Several types of antibody to thyroid tissue have been detected in serum, usually from patients with thyroid disease. The importance of measuring these antibodies lies mainly in the further investigation of patients in whom thyroid function has been assessed by other means as they help demonstrate the presence of auto-immune disorders.

Complement-fixing antibodies specific for thyroid tissue. These are present in serum in over 80% of patients with Hashimoto's disease and may be detected by complement fixation or fluorescence tests.

Antibodies to thyroglobulin can be detected in most cases of early or incipient hypothyroidism, but long-standing cases yield abnormal results less frequently. About 80% of hyperthyroid patients also have antibodies to thyroglobulin in the serum. Thyroglobulin antibodies may be found in a small proportion of healthy individuals, the incidence being higher in relatives of patients with hyperthyroidism.

Antibodies to a second colloid antigen have been reported in all forms of auto-immune thyroiditis. They may also help in the diagnosis of de Quervain's thyroiditis.

Thyroid-stimulating immunoglobulins

The serum of patients with Graves' disease contains IgG antibodies directed against the TSH receptors in the thyroid. Cells to which these antibodies are bound are directly stimulated to produce thyroxine; these antibodies are called thyroid-stimulating immunoglobulins (TSI).

In addition to TSI, other immunoglobulins have been recognized that stimulate thyroid growth, but not hormone production. They are called thyroid growth immunoglobulins (TGI).

Miscellaneous tests

There are several tests which may be affected by increased or decreased thyroid hormone activity, and which require brief mention.

Basal metabolic rate (BMR) is raised in hyperthyroidism and reduced in hypothyroidism. The test is now of historical interest and is very rarely performed.

Glucose tolerance tests are sometimes abnormal in patients with hyperthyroidism, who may show a diabetic type of response to an oral glucose load (p. 225).

Plasma calcium. Hyperthyroidism is an occasional cause of increased plasma [calcium] and increased plasma alkaline phosphatase activity. It is one of the rarer causes of metabolic bone disease.

Plasma LDL-cholesterol concentration is often markedly increased in patients with hypothyroidism. It is also sometimes decreased in patients with hyperthyroidism.

REVERSE T3

Some patients with no evidence of thyroid disease may have normal plasma [T4] but low plasma [T3]. This combination of findings has been reported after surgical operations, in patients with acute infections and in obese patients on energy-restricted diets. It would appear that one effect of these various conditions is an alteration in the nature of mono-deiodination of T4 in the periphery, with decreased formation of T3 and increased formation of reverse T3.

Reverse T3 can be measured by specific radioimmunoassay techniques. Its determination can help explain the finding of low plasma [T3] in patients with non-thyroidal illness, but this test is not available routinely.

SELECTIVE USE OF THYROID FUNCTION TESTS

It is not usually necessary to carry out more than a limited number of tests in the investigation of a case of suspected thyroid disease. Selection of the most appropriate initial tests will depend on the clinical findings, including the age and sex of the patient. If the clinical index of suspicion is low, or if thyroid function assessment is being performed routinely as part of an admission screen (p. 501), the choice of tests may be different and should probably be more restricted than if the index of suspicion is high. All patients in whom a diagnosis of thyroid dysfunction is made should have it confirmed by appropriate laboratory tests before treatment is instituted, including even those patients in whom the diagnosis appears clinically obvious. It is important to accept that, unless there are clinical grounds for suspecting that a patient has thyroid disease, 'routine' tests of thyroid function should be postponed, if the patient has some intercurrent illness, until the patient has recovered from its effects.

DIAGNOSIS OF HYPERTHYROIDISM

The diagnosis can generally be made and confirmed on the basis of the following programme, the laboratory tests being carried out in the order suggested:

(1) *Clinical findings*. The clinical features are described in medical textbooks. It is possible to attribute numerical values to various symptoms and signs by means of a diagnostic index, negative values being given to findings which are unlikely to occur in thyrotoxicosis; a score is thus obtained for each patient. The likely thyroid status, or diagnostic index, can be determined from the final score.

(2) *The initial set of chemical investigations* should consist of *one* of the following combinations of tests:
   (a) *Plasma [free T4]* and *[free T3], or*
   (b) *Plasma [total T4]* and *[total T3]* supplemented, where clearly appropriate (i.e. pregnancy, women taking oestrogen-containing oral contraceptives), by measurement of the *free thyroxine index* (serum uptake test or resin uptake test required) *or* the *total T4 : TBG ratio* (plasma [TBG] needs to be determined).
(3) *TRH test.* This is indicated in all patients in whom a *provisional* diagnosis of hyperthyroidism has been made, since a normal TSH response to TRH will then exclude the diagnosis of hyperthyroidism (i.e. will show that the provisional diagnosis was incorrect).

This combination of tests is sufficient for the investigation of most patients with suspected hyperthyroidism. However, in a small number of patients, mostly those who are over sixty and some of whom have atrial fibrillation, plasma [total T4] and [total T3] are normal and yet they (1) show no response to TRH and (2) respond clinically to anti-thyroid treatment. For these patients, a TRH test is required if the clinical diagnosis is to be confirmed by laboratory investigations.

Screening for hyperthyroidism

Many geriatric assessment units include measurement of plasma [T4] in their routine of initial intensive screening of patients, the screen usually involving a range of chemical tests. The T4 measurements have proved valuable in that they have revealed the presence of hitherto unsuspected hyperthyroidism in a small percentage (1-2%) of the patients admitted for assessment. However, the value of these measurements has been greater in the detection of unsuspected hypothyroidism among geriatric patients.

DIAGNOSIS OF HYPOTHYROIDISM

The diagnosis can generally be made and confirmed on the basis of the following programme, the laboratory tests again being carried out in the order suggested:

(1) *Clinical findings,* with calculation of the diagnostic index.
(2) *The initial set of chemical investigations* should consist of:
   (a) *Plasma [free T4]* and *basal [plasma TSH], or*

(b) *Plasma [total T4]* supplemented, where clearly appropriate (i.e. pregnancy, women taking oestrogen-containing oral contraceptives), by measurement of the *free thyroxine index or* the *total T4 : TBG ratio.* In addition, *basal plasma [TSH]* should be measured.

Basal plasma [TSH] is a sensitive test for the diagnosis of early primary hypothyroidism. The results of TSH measurements also help to distinguish between primary and secondary hypothyroidism. The TRH test is not indicated.

Screening for hypothyroidism

Hypothyroidism can be difficult to diagnose in the neonatal period. Many countries now include blood [TSH], or less frequently blood [T4], as one of the measurements made in recent extensions to the well-established programmes of screening for phenylketonuria (pp. 427, 511).

As indicated above, many geriatric assessment units include measurement of plasma [T4] as part of their screening programmes of chemical tests, and these have proved valuable in the detection of hitherto unsuspected hypothyroidism in a small percentage (2-6%) of these geriatric patients.

ADDITIONAL DIAGNOSTIC CHEMICAL TESTS

The main role of chemical measurements in the initial investigation of thyroid dysfunction is in helping to classify patients as euthyroid, hypothyroid or hyperthyroid. The tests required for this purpose have been summarized above. More precise recognition of the specific pathological cause of thyroid dysfunction may require additional information.

Further tests may include histological examination of thyroid tissue or thyroid antibody studies. For investigation of the rare inherited metabolic disorders of thyroid hormone synthesis, measurement of iodinated tyrosines or enzymes involved in the synthesis of thyroxine may be required.

Some carcinomas of the thyroid synthesize and secrete thyroglobulin. In these patients, measurement of serum [thyroglobulin] may be of value in monitoring the progress of the disease, and in assessing the response to treatment. Medullary carcinoma of the thyroid is discussed elsewhere (p. 457).

MANAGEMENT OF HYPERTHYROIDISM

Plasma [T4] provides the best index of progress of the untreated disease. It is also the best index of the adequacy or otherwise of antithyroid drug treatment, since the therapeutic objective is to maintain plasma [total T4] at about 100 nmol/L (8 μg/100 mL).

After radioactive iodine treatment, the likelihood of eventually developing hypothyroidism is very high, and long-term follow-up with periodic measurements of plasma [T4] is essential. Plasma [TSH] is also valuable; it is likely to be raised for a considerable time before hypothyroidism develops.

If plasma [TSH] is increased, this indicates that the patient is likely to develop hypothyroidism and is a clear pointer to the need for more frequent and regular follow-up. Replacement treatment with thyroxine should be instituted on the basis of clinical assessment and the finding of a low plasma [T4] rather than on the detection of increased plasma [TSH].

Patients treated by sub-total thyroidectomy are much more likely to remain free from further disorders of thyroid function. There may, however, be temporary disturbances of thyroid function tests in the early postoperative period, and it is advisable to wait six months before deciding whether replacement treatment is needed. This decision should be based on clinical assessment supplemented by measurements of both plasma [T4] and [TSH].

MANAGEMENT OF HYPOTHYROIDISM

Most patients have their dose of thyroxine determined by a combination of clinical assessment and measurement of plasma [T4]; plasma [total T4] should be maintained at about 100 nmol/L.

The objective of replacement therapy in patients with primary hypothyroidism is to attain a T4 concentration in plasma that is sufficient to suppress plasma [TSH] to normal. Measurement of plasma [TSH], therefore, should be performed periodically as part of the follow-up assessment. It is particularly useful in the management of patients suspected of not taking their thyroxine regularly, but who take their tablets shortly before attending the follow-up clinic. In these patients, poor compliance is indicated by finding that plasma [T4] is normal but plasma [TSH] is increased.

FURTHER READING

BURR, W.A., RAMSDEN, D.B. and HOFFENBERG, R. (1980). Hereditary abnormalities of thyroxine binding globulin concentration. *Quarterly Journal of Medicine, New Series*, **49**, 295–313.

DONIACH, D., CUDWORTH, A.G., KHOURY, E.L. and BOTTAZZO, G.F. (1982). Autoimmunity and the HLA-system in endocrine diseases. In *Recent Advances in Endocrinology and Metabolism*, **2**, pp. 99–132. Ed. J.L.H. O'Riordan. Edinburgh: Churchill Livingstone.

International Symposium (1982). *Free T4–The way ahead in thyroid diagnosis*. Oxford: The Medicine Publishing Foundation.

KENDALL-TAYLOR, P. (1978). Thyroid function and disease. In *Recent Advances in Endocrinology and Metabolism*, **1**, 37–59. Ed. J.L.H. O'Riordan. Edinburgh: Churchill Livingstone.

Leading article (1972). The computer and thyroid disease. *British Medical Journal*, **1**, 457–458.

Leading article (1977). Problems with serum-thyroxine. *Lancet*, **2**, 74–75.

Leading article (1978). Thyroid disease and pregnancy. *British Medical Journal*, **2**, 977–978.

Leading article (1979). Screening for congenital hypothyroidism. *Lancet*, **2**, 678–679.

Leading article (1981). Thyroid autoimmune disease: a broad spectrum. *ibid.*, **1**, 874–875.

Leading article (1981). Screening for thyroid disease. *ibid.*, **2**, 128–130.

Leading article (1982). Hyperthyroxinaemia: does it mean hyperthyroidism? *ibid.*, **1**, 1286.

Leading article (1983). Thyroid function tests — progress and problems. *ibid.*, **1**, 164–165.

STANBURY, J.B. and DUMONT, J.E. (1983). Familial goiter and related disorders. In *The Metabolic Basis of Inherited Disease*, 5th Edition, pp. 231–269. Ed. J.B. Stanbury, J.B. Wyngaarden, D.S. Fredrickson, J.L. Goldstein and M.S. Brown. New York: McGraw-Hill.

TUTTLEBEE, J.W. (1982). Further experience with free thyroxine assays with particular reference to pregnancy. *Annals of Clinical Biochemistry*, **19**, 374–378.

Chapter 18

Steroid Hormones

The adrenal cortex secretes a range of steroid hormones, all formed in the gland from cholesterol. On the basis of their principal activities, these hormones can be classified as glucocorticoids, mineralocorticoids, androgens, oestrogens and progestogens. Three zones can be recognized in the adrenal cortex. Of these, the zona fasciculata and zona reticularis function as a single unit in the synthesis of glucocorticoids. The third zone, the zona glomerulosa, is the site of synthesis of aldosterone.

In the circulation, the glucocorticoids are mainly bound (about 90%) to proteins; the most important quantitatively is cortisol-binding globulin (CBG), sometimes called transcortin. No specific aldosterone-binding protein comparable to CBG has been demonstrated, and a higher proportion of aldosterone circulates in the unbound form.

Cortisol and aldosterone are removed from plasma by the liver and thereafter converted to a number of metabolically inactive compounds. These are excreted in the urine mainly as conjugated metabolites (e.g. glucosiduronates, often called glucuronides). Most of these metabolites retain the side-chain and the hydroxyl group at the 17-position. Small amounts of cortisol are excreted unchanged in the urine and in saliva.

Separate mechanisms control the formation and release of the glucocorticoids and of aldosterone.

THE GLUCOCORTICOIDS

Three control mechanisms are concerned with regulating the secretion of glucocorticoids by the adrenals:

(1) *Negative feedback control,* responsible for controlling the release from the hypothalamus of corticotrophin-releasing factor (CRF); CRF stimulates the release of adrenocorticotrophic hormone (ACTH) from the anterior pituitary into the blood. Finally, ACTH stimulates the secretion of glucocorticoids, of which cortisol is quantitatively the most important in man. Increased plasma [cortisol] suppresses secretion of CRF, while a fall in plasma [cortisol] leads to increased secretion. Synthetic steroid hormones with glucocorticoid activity (e.g. prednisolone, dexamethasone) also influence the secretion of endogenous glucocorticoids by reducing the secretion of CRF.

(2) *The response to stress* is also mediated by CRF and ACTH. It is initiated by centres in the CNS higher than the hypothalamus and results in a sudden large increase in hormone secretion; the feedback mechanism is temporarily overriden. Many stimuli, including surgical operations and emotional stress, can elicit this response. A standardized stimulus for testing the response to stress is provided by the insulin-hypoglycaemia test (p. 362).

(3) *The nychthemeral rhythm* of plasma [cortisol] is related to the rhythm of an individual's sleeping-waking cycle (Fig. 18.1). The pathway for its control also depends on the CRF and ACTH mechanism. It is apparently not due to rhythmic changes in the sensitivity of the adrenals to the effects of ACTH.

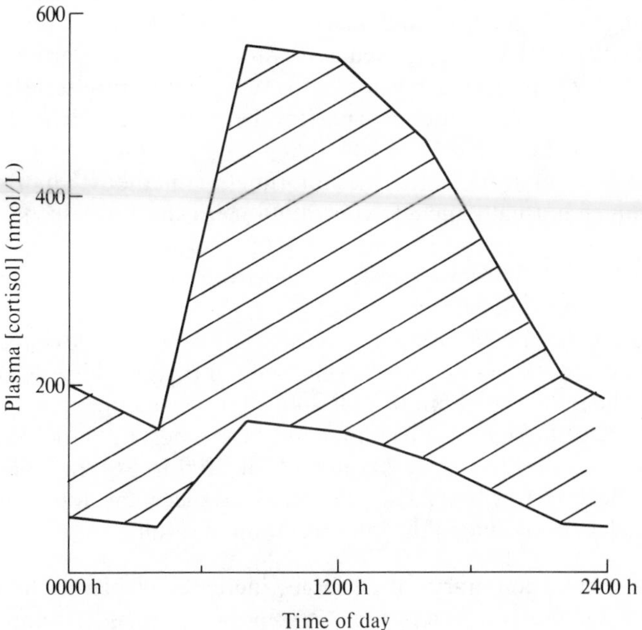

FIG. 18.1. Nychthemeral rhythm of plasma [cortisol]. The hatched area represents values in the reference range.

The immediate effects of stimulation of the adrenals by ACTH are to increase the output of cortisol, and of adrenal androgens and oestrogens; these changes are due to release of stored (presynthesized) hormones and are of only brief duration. The other immediate effect of ACTH is to stimulate new synthesis leading especially to increased secretion of cortisol. The long-term effects of stimulation by ACTH are an increase in the sensitivity of the gland to ACTH and hypertrophy of the adrenal cortex.

The tests of glucocorticoid metabolism are used in the first place to support a clinical diagnosis of hypofunction or hyperfunction of the adrenal cortex. They help to classify the level of activity as normal, reduced or increased. Further tests may provide information about the nature of the disease process.

Glucocorticoid production: basal measurements

PLASMA CORTISOL

This is usually measured nowadays by radioimmunoassay (RIA); most clinical chemistry laboratories use RIA techniques because of their high degree of specificity and relative freedom from interference. The specificity of RIA methods tends to vary with the antiserum that is used, but when using a specific antiserum significant interference is only likely to be encountered in patients being treated with prednisone or prednisolone.

A few laboratories still use fluorimetric methods; fluorimetric techniques normally include contributions to the fluorescence from certain other steroid hormones (e.g. corticosterone, synthetic glucocorticoids), from cholesterol and from drugs such as librium, spironolactone and tetracyclines.

Specimens for measurement of plasma [cortisol] must be collected at standard times because of the nychthemeral rhythm (Fig. 18.1). The times widely adopted are 0800 h and 2200 h since the most reliable reference values have been defined for these times, i.e. 160–565 nmol/L at 0800 h and less than 205 nmol/L at 2200 h. Several important points need to be observed when collecting specimens for measuring plasma [cortisol], and in the interpretation of results:

(1) *Anxiety*. Temporary, often large, increases in plasma [cortisol] may be observed as a response to emotional stress. It is important to allay any anxiety felt by the patient before collecting samples. Using an indwelling venous catheter can help in this respect.

(2) *Venous stasis* must be avoided or misleadingly high results may be obtained because cortisol in plasma is mainly bound to protein.
(3) *Storage*. After collection, blood specimens are best kept at 4°C; they must *not* be frozen. The need to store specimens prior to analysis arises frequently, especially for samples collected at 2200 h. If the analysis is to be delayed more than 12 h, plasma should be separated immediately and then frozen.
(4) *Cortisol-binding globulin*. Changes in plasma [CBG] may be associated with marked alterations in plasma [cortisol] without necessarily implying any alteration in adrenal function. Plasma [CBG] is increased in pregnancy, in patients receiving oestrogen treatment, and in women taking oestrogen-containing oral contraceptives; in all these states plasma [cortisol] is increased. Conversely, in hypoproteinaemic states (e.g. nephrotic syndrome), plasma [CBG] and [cortisol] may both be decreased. Recently, measurement of salivary [cortisol] has been recommended, as it would seem to give an indication of plasma [unbound cortisol], but these measurements are not yet widely available.

One of the most important features, as a pointer to early recognition of overproduction of glucocorticoids, is *loss of the nychthemeral rhythm* of plasma [cortisol]. In hypersecretory states, early in their development, the 0800 h value for plasma [cortisol] may still be normal (i.e. less than 565 nmol/L) but the 2200 h value increased much above 205 nmol/L (the upper reference value for this time).

PLASMA ACTH

This measurement can only be performed reliably in a few specialized laboratories. Moreover, since ACTH is unstable, the very detailed instructions that these laboratories issue for collecting, preserving and transporting the specimens must be followed.

There is a nychthemeral rhythm in the secretion of ACTH, and specimens for plasma ACTH measurements should be collected at specified times, for which reference values are established. These are 0800 h and 2200 h, or closely equivalent times. Stress should be avoided when collecting specimens. Measurement of plasma [ACTH] provides one of the definitive ways of differentiating between primary and secondary adrenal hypofunction.

Lipotrophin (LPH) is secreted by the anterior pituitary in response to the same stimuli as cause release of ACTH; there are, therefore, parallel changes in the plasma concentrations of these two compounds. However, since LPH is more stable than ACTH, measurement of plasma [LPH] has some advantages over measurement of plasma [ACTH].

URINARY GLUCOCORTICOIDS AND METABOLITES

Urinary free cortisol. Small amounts of unmetabolized cortisol are excreted in urine, and can be measured by specific RIA methods. Urinary free cortisol excretion is related to (1) the output of cortisol by the adrenal cortex during the period of urine collection, and (2) the plasma [free (unbound) cortisol], i.e. the level of active hormone to which tissues are exposed. The technique is relatively simple and easy. It has largely replaced the measurements of urinary steroid metabolites because, in patients with suspected adrenal hyperfunction, there is very little overlap with this test between patients with Cushing's syndrome and patients with other diagnoses.

Urinary 17-hydroxycorticosteroids (17-OHCS) include approximately the same group of compounds as urinary 17-oxogenic steroids (17-OGS). They include in their measurement the majority of the metabolites of cortisol, but are not entirely specific since they also measure other compounds (e.g. metabolites of 11-deoxycortisol and of 17α-hydroxyprogesterone). Many drugs (e.g. metabolites of antihypertensive agents, cardiac glycosides and tranquillizers) interfere with the determinations. Both urinary 17-OHCS and 17-OGS measurements are now virtually obsolete.

Urinary 17-oxosteroids (17-OS) comprise a group of metabolites formed mainly from androgens; these may be adrenal or testicular (or ovarian) in origin. About 5–10% of cortisol is metabolized to 17-OS. Measurement of urinary 17-OS is another virtually obsolete test, since specific RIA methods are now available for many of the individual androgens.

INVESTIGATION OF DISORDERS OF GLUCOCORTICOID PRODUCTION

Whether a patient is being investigated for suspected adrenocortical hypofunction or hyperfunction, it is important to plan the selection and performance of the tests so as to make the most economical use of the time of the patient, the clinician conducting the investigation, and the laboratory that is performing the measurements. To this end, the following order of questions offers the basis for a logical scheme:

(1) *Screening tests.* Is a disorder of adrenocortical function (hypofunction or hyperfunction) a likely diagnostic possibility?
(2) *Confirmatory tests.* How can the provisional diagnosis be confirmed or excluded?
(3) *Determining the cause:* if adrenocortical dysfunction is confirmed:
  (a) *What is the site* of the pathological lesion (i.e. is it in the adrenal cortex, the pituitary or elsewhere)?
  (b) *What is the nature* of the pathological lesion?

The need for screening tests

The suspicion that adrenocortical disease may be present arises quite commonly in hospital practice whereas the incidence is very low, if drug-induced (i.e. iatrogenic, usually due to steroid treatment) causes are discounted. Effective screening tests are therefore required, in order to answer the first question quickly and reliably.

The screening test should be sensitive, i.e. it should detect all patients with adrenocortical disease. However, it does not need to be specific, since further tests will be performed in order to answer the second question.

Screening tests that can be performed on an outpatient basis are to be preferred. More detailed investigations can, thereafter, be carried out on an inpatient basis on the small minority of patients in whom the results of screening tests are abnormal.

Suspected adrenocortical hyperfunction

The logical plan of investigation (Fig. 18.2) will depend, to some extent, on the investigations available in the local laboratory. It must be accepted, for instance, that measurements of plasma [ACTH], although potentially very important, are specialized investigations. For their reliable performance, requests for plasma ACTH measurement usually have to be referred elsewhere, and results may not be available for a considerable time.

SCREENING TESTS

The initial screening investigations should consist of either a dexamethasone (or betamethasone) screening test or measurement of the urinary free cortisol : creatinine ratio. Both tests can be performed on an outpatient basis.

Dexamethasone screening test. The patient is instructed to take a 2 mg tablet of dexamethasone (or betamethasone) at 2200 h, the night before visiting the outpatient clinic. A blood specimen for measuring plasma [cortisol] is collected the following morning, about 0900 h.

At 0900 h, plasma [cortisol] is normally in the range 160–565 nmol/L. In normal subjects, ingestion of 2 mg dexamethasone (or betamethasone) results in the suppression of plasma [cortisol] to 150 nmol/L or less. The principal value of the dexamethasone screening test is to help exclude the diagnosis of Cushing's syndrome. Patients with simple obesity respond to this 2 mg dose of dexamethasone, but in patients with Cushing's syndrome plasma [cortisol] may remain well above 300 nmol/L.

Urinary free cortisol:creatinine ratio and *24-h urinary cortisol excretion* are both increased in patients with Cushing's syndrome. In practice, the collection of a complete 24-h urine specimen on an outpatient basis is inconvenient and often inaccurate. By determining

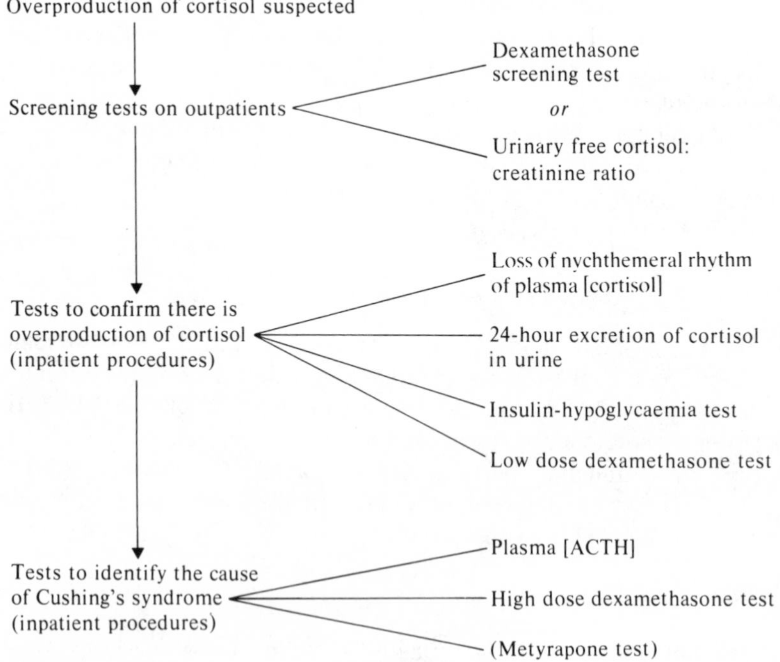

FIG. 18.2. Stepwise scheme for the investigation, by means of hormone measurements, of patients in whom adrenocortical hyperfunction (Cushing's syndrome) is thought to be the possible diagnosis.

the urinary free cortisol:creatinine ratio, the need for a 24-h urine specimen is avoided; the urine specimen for this screening test can be obtained at the outpatient clinic.

Whichever screening test is selected, if the result is normal, there is usually no need of further investigation for the presence of adrenocortical hyperfunction, unless there is a strong clinical suspicion that this is the diagnosis. However, if the result of the screening test is abnormal, the patient will then need to undergo further investigation (1) to confirm the diagnosis and (2) to determine the site of the lesion.

These screening tests usually serve to distinguish simple, non-endocrine obesity from obesity due to Cushing's syndrome as the results are nearly always normal in simple obesity. The further investigations to be described in this section require the patient to be admitted to hospital, for their performance under properly controlled conditions.

CONFIRMATION OF CORTISOL OVERPRODUCTION — CUSHING'S SYNDROME

Loss of nychthemeral rhythm

The characteristic finding at an *early* stage of the disease is an increase in plasma [cortisol] in the blood specimen collected at 2200 h, accompanied by loss of the nychthemeral rhythm. In some of these patients there is also an increase in plasma [cortisol] in the 0800 h specimen. At a later stage, abnormally high plasma [cortisol] can be detected throughout the 24 hours.

Urinary free cortisol

The finding of an increased 24-h excretion provides valuable supporting evidence for the clinical diagnosis. Hypersecretion of cortisol continues in these patients throughout the 24-h period, and may be detected at an early stage in the development of the disease by measuring urinary excretion of cortisol. Measurements of plasma [cortisol] may not reveal the presence of disease until later, as they can only provide information about the concentrations obtaining at the times of specimen collection.

Differential diagnosis

In the absence of adrenal disease, the common causes of abnormal results in the tests so far described for investigating suspected adrenocortical hyperfunction are depressive illness (either exogenous or endogenous) and any cause of stress to the patient. The insulin hypoglycaemia test distinguishes these other causes of increased cortisol production from Cushing's syndrome. The low dose dexamethasone test (p. 363) also helps to confirm a diagnosis of Cushing's syndrome.

Stressing the hypothalamic-pituitary-adrenal (HPA) axis

The insulin-hypoglycaemia test provides a convenient and reproducible means of inducing stress, and is the most widely used stress test of HPA integrity. The adrenal response to a rapid reduction in plasma [glucose], induced by an intravenous injection of insulin, is assessed by measuring plasma [cortisol]. The test is contra-indicated in patients with epilepsy or heart disease.

Preparation for the test: The patient fasts overnight, and has nothing to eat or drink (apart from water) on the morning of the test. An indwelling intravenous needle should be inserted at least 30 min before the test, and the patient warned that the test may induce sweating or a feeling of drowsiness.

The dose of insulin, to be given as a single intravenous injection, must be related to the weight of the patient and to the tentative clinical diagnosis. The standard dose of soluble insulin is 0.15 units/kg. Arrangements must be made to ensure that the patient will not be unattended at any time during the test. A syringe and 50% glucose solution must be immediately available so that the test can be terminated rapidly if signs of marked hypoglycaemia appear (e.g. severe and prolonged sweating, loss of consciousness or fits).

Performing the test: At least 30 min after inserting the indwelling needle, specimens of blood are collected for measurement of basal plasma [cortisol] and plasma [glucose]. The insulin is then given via the indwelling needle and further blood specimens are collected 30, 45, 60 and 90 min later for cortisol and glucose measurements. The test is ended by giving the patient a glucose-containing drink and a meal.

It is essential to produce an adequate degree of hypoglycaemia for the test to be valid. An adequate hypoglycaemic stimulus is usually defined as a rapid fall in plasma [glucose] to less than 2.2 mmol/L (40 mg/100 mL). In patients with Cushing's syndrome, if adequate hypoglycaemia is not induced with the standard dose of insulin (0.15 units/kg), the test can be repeated with increasing doses of insulin, incremented in steps of 0.05 units/kg; up to 0.30 units/kg may be required in some patients.

Interpretation of results: The cortisol response to the stimulus of hypoglycaemia depends on normal function of the whole of the HPA axis. Plasma [cortisol] normally reaches its maximum at 60 or 90 min, the maximum level reached being at least 425 nmol/L, with an increment above basal of at least 145 nmol/L.

A high basal plasma [cortisol] may be due to anxiety, or to insufficient time being left between inserting the indwelling needle and starting the test. However, these patients show a normal response to insulin-induced hypoglycaemia. The response is usually normal also in patients with simple obesity, and in patients with depression.

Patients with Cushing's syndrome, whatever the cause, do not respond normally to insulin-induced hypoglycaemia. There is a high basal plasma [cortisol], but little or no increase in plasma [cortisol] despite the production of an adequate degree of hypoglycaemia. The test is very valuable in the diagnosis of Cushing's syndrome.

Low dose dexamethasone test

This test is usually performed as the immediate preliminary to a high dose dexamethasone test. The patient needs to be in hospital throughout, but can remain ambulant and eat a normal diet during the test. The protocol for one of several versions of the test is detailed in Table 18.1. This protocol is easiest to follow if specimens are collected each day at 0800–0900 h, and it is convenient to change the dexamethasone dosage at this time. Each day a blood specimen is required, for measuring plasma [cortisol], and each 24-h collection period ends at this time, if urinary free cortisol (or 17-OHCS) is to be measured.

In normal subjects, plasma [cortisol] is suppressed by low dose

TABLE 18.1. Low dose and high dose dexamethasone tests.

Day of test	Dexamethasone (oral dose)	Plasma cortisol (specimen description)
1 and 2	Nil	Basal
3	0.5 mg 6-hourly	Basal
4	0.5 mg 6-hourly	"Low dose"
5	2 mg 6-hourly	"Low dose"
6	2 mg 6-hourly	"High dose"
7	Nil	"High dose"

Note Specimens are collected at 0800 h–0900 h, and this is the time when the dosage of dexamethasone for the ensuing 24-hour period is started.

dexamethasone (0.5 mg q.i.d.) to less than 125 nmol/L. Failure of suppression at this low dose is generally found in Cushing's syndrome, whatever the cause, and is often observed also in severely depressed patients.

If the response in the low dose dexamethasone test is being assessed on the basis of urinary steroid measurements, comparison is made between the basal urinary output of free cortisol (or 17–OHCS) and the output on Day four. It is considered that suppression of cortisol secretion has occurred if the output falls to less than 50% of the basal level.

DETERMINING THE CAUSE OF CUSHING'S SYNDROME

If the diagnosis of Cushing's syndrome has been established, by finding an impaired cortisol response in the insulin-hypoglycaemia test, or if the diagnosis is supported by lack of response in the low dose dexamethasone test, the results of other investigations may help to determine the cause of the adrenocortical hyperfunction. This may be due to:

(1) Adrenal neoplasm.
(2) Pituitary neoplasm causing adrenal hyperplasia (Cushing's disease).
(3) Non-endocrine tumour, most often a small cell carcinoma of the lung.

Chemical tests may enable this differentiation to be made, especially if results are considered together with the results of other methods of investigation (e.g. radiological).

High dose dexamethasone test

This test usually follows on immediately after the low dose test, i.e. before the results of administering dexamethasone (0.5 q.i.d.) are known. A protocol for the high dose test is given in Table 18.1. If the response is determined by measuring plasma [cortisol], suppression is as defined above for the low dose test, i.e. to less than 125 nmol/L. If the response is being determined on the basis of urinary steroid measurements, the 24-h outputs on Days six and seven are compared with the basal output; suppression is again defined as a fall in output to less than 50% of the basal level.

With high dose dexamethasone (2 mg q.i.d.), suppression of plasma cortisol to less than 50% of the basal concentration *suggests* that the patient has a pituitary-dependent cause for Cushing's syndrome, i.e.

Cushing's disease. Failure of suppression *suggests* that the patient either has a functioning adrenal tumour or a non-endocrine ACTH-secreting tumour.

The responses to high dose dexamethasone, outlined in the previous paragraph, are in fact more variable. The test has not proved to be a reliable means of distinguishing pituitary-dependent Cushing's disease from other causes of Cushing's syndrome. If measurements of plasma [ACTH] are available, these are to be preferred since they make this distinction much more reliably.

Plasma [ACTH]

This should be measured in blood specimens collected both in the morning and the evening (e.g. 0800 h and 2200 h). If an adrenal neoplasm is present, ACTH will not be detectable in plasma. On the other hand, plasma [ACTH] will be increased, particularly in the evening specimen (due to loss of the nychthemeral rhythm), if there is a pituitary or non-endocrine neoplasm causing hypersecretion of cortisol. Very high plasma [ACTH] is a feature of non-endocrine ACTH-secreting neoplasms.

Metyrapone test

Metyrapone is an inhibitor of 11ß-hydroxylase, the enzyme which catalyses the conversion of 11-deoxycortisol to cortisol. Since 11-deoxycortisol does not cause feedback inhibition of the hypothalamus, unlike cortisol, increased amounts of CRF are secreted by the hypothalamus, as its normal response. This in turn stimulates the release of ACTH and the normal response of the adrenal cortex, when 11ß-hydroxylase is inhibited, is to secrete larger amounts of 11-deoxycortisol.

Metyrapone inhibits other hydroxylation reactions in addition to the 11ß-hydroxylase involved in the synthesis of cortisol. The test is contra-indicated in early pregnancy, since metyrapone may interfere with the production of oestrogens. Care is also needed if metyrapone is to be given to patients receiving drugs inactivated by hydroxylation reactions in the liver (e.g. some anticoagulants), since its administration may potentiate the effects of these drugs.

The metyrapone test can be used to help differentiate (1) patients with pituitary-dependent adrenal hyperplasia, in whom there is usually an increased response to administration of this 11ß-hydroxylase inhibitor, from (2) patients with adrenal tumours or non-endocrine

ACTH-secreting tumours, in whom metyrapone administration elicits little or no response.

The response to metyrapone can be assessed by measuring urinary 17-OHCS, since the metabolites of 11-deoxycortisol are among the compounds that contribute to 17-OHCS. Alternatively plasma [11-deoxycortisol] can be measured. However, this test is rarely performed nowadays, since few laboratories still measure 17-OHCS, and the 11-deoxycortisol assay is specialized and of only limited availability.

OTHER CHEMICAL ABNORMALITIES IN ADRENOCORTICAL HYPERFUNCTION

Cortisol and other glucocorticoids have important effects on electrolyte metabolism. There is retention of Na^+ accompanied by water retention, and loss of K^+. These effects are due to the actions of glucocorticoids on the renal tubules; the loss of K^+ may be severe. Plasma [Na^+] usually remains normal, but plasma [K^+] may be reduced and there is sometimes also a metabolic alkalosis. Hypokalaemic alkalosis is a particularly prominent feature of patients with non-endocrine ACTH-secreting tumours.

Patients with adrenocortical hyperfunction may develop steroid-induced diabetes, and have a diabetic type of response to glucose tolerance tests (p. 224).

Patients with pituitary tumours giving rise to Cushing's disease may show signs of disturbed production of other pituitary hormones, able to be confirmed by the appropriate chemical tests.

TABLE 18.2. Hormonal tests for

To confirm the diagnosis	Cushing's disease
Plasma cortisol	↑
Low dose dexamethasone test	Not suppressed
Urinary free cortisol	↑
Nychthemeral rhythm of plasma cortisol	Lost
Insulin-hypoglycaemia test	No response
To differentiate the cause	
Plasma ACTH	↑
High dose dexamethasone test	Suppressed
Metyrapone test (rarely performed)	Increased response

Key to abbreviations ↑ = increased; ↑ ↑ = greatly increased

The results of hormonal tests carried out on patients with suspected adrenocortical hyperfunction are summarized in Table 18.2.

Suspected adrenocortical hypofunction

The logical plan of investigation (Fig. 18.3) again depends, to some extent, on the range of investigations available locally.

Measurements of basal plasma [cortisol] and 24-h urinary excretion of cortisol (or 17-OHCS) in patients with suspected Addison's disease may, by themselves, be of little value since they may all be normal in patients with adrenocortical hypofunction. Instead, it is often necessary to investigate the response of the adrenal cortex to stimulation. Tests that stimulate the gland directly, with ACTH or tetracosactrin, should always be performed before tests that stimulate the gland indirectly by acting on the hypothalamus or the anterior pituitary, such as the insulin-hypoglycaemia test.

TETRACOSACTRIN SCREENING TEST

ACTH is a single polypeptide chain composed of 39 amino acids, with the biological activity lying in the N-terminal sequence of 24 amino acids. Tetracosactrin (most often used in the form of Synacthen) is a synthetic polypeptide with a structure identical to the N-terminal 24 amino acids of ACTH.

Tetracosactrin has a short duration of action, and its use as a stimulus to the adrenals provides a convenient and rapid screening test for the assessment of adrenocortical function. The test can be

adrenocortical hyperfunction.

Adrenal tumour	Non-endocrine ACTH–secreting tumour
↑	↑
Not suppressed	Not suppressed
↑	↑
Lost	Lost
No response	No response
Not detectable	↑ or ↑ ↑
Not suppressed	Not suppressed
No response	No response

performed with the patient on a normal diet, and as an outpatient procedure, as follows:

(1) A blood specimen is collected for measuring basal plasma [cortisol].
(2) Tetracosactrin, 0.25 mg, is given by intramuscular injection.
(3) Further blood specimens for plasma [cortisol] measurements are collected exactly 30 and 45 min after the injection.

Emotional stress can increase plasma [cortisol], so any anxiety that the patient may feel should be allayed as much as possible beforehand.

Reasons that may invalidate the test include treatment with glucocorticoids within 12 h of the tetracosactrin injection, and the taking of oestrogen-containing oral contraceptives.

A normal response is defined as a rise in plasma [cortisol] to at least 425 nmol/L along with an increase of at least 145 nmol/L over the basal concentration. It may be wise to collect more than one pre-test sample, to establish the baseline concentration of plasma [cortisol]. A normal response excludes primary adrenocortical insufficiency, but an impaired or absent response requires further investigation by the 4-day tetracosactrin test.

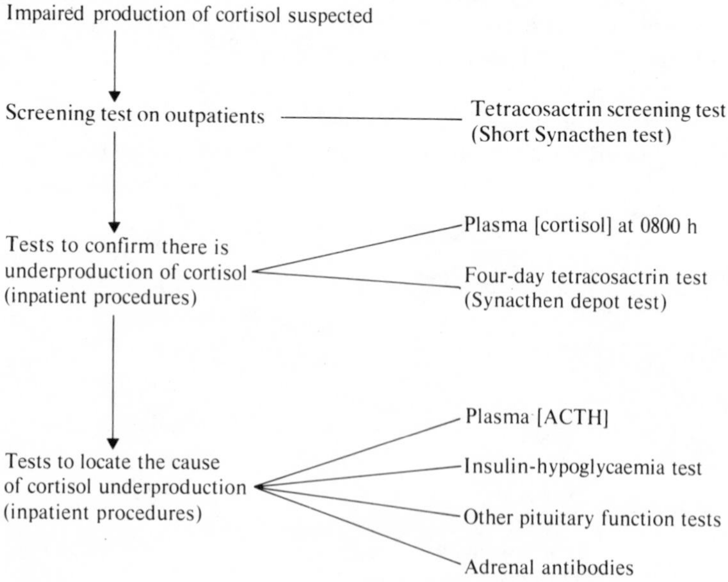

FIG. 18.3. Stepwise scheme for the investigation, by means of hormone measurements, of patients in whom adrenocortical hypofunction (Addison's disease) is suspected.

CONFIRMATION OF CORTISOL UNDERPRODUCTION

If the response to the tetracosactrin screening test is normal, primary adrenocortical hypofunction can be ruled out. Also, adrenocortical hypofunction secondary to hypothalamic or pituitary disease is extremely unlikely; in the prolonged absence of ACTH, the cells of the adrenal cortex would have atrophied.

If the response to the tetracosactrin screening test is impaired or absent, the patient should be admitted to hospital. The plasma [cortisol] at 0800 h should be measured, and a 4-day tetracosactrin test performed.

Plasma cortisol at 0800 h

At an *early* stage of the disease, plasma [cortisol] in blood specimens collected at 0800 h (or at any other time during the day for which reference values have been established) may be low or in the lower part of the reference range. Plasma [cortisol] may still be normal in specimens collected at 2200 h, as the diseased adrenal can still maintain this lower level of output (Fig. 18.1). At a later stage plasma [cortisol] may be abnormally low throughout the 24 hours.

Four-day tetracosactrin test

Tetracosactrin adsorbed on to a zinc phosphate complex (Synacthen depot) is used in this test; it has a longer duration of action than tetracosactrin. The principal diagnostic value of the 4-day test lies in the assessment of adrenocortical function when prolonged stimulation is needed, and the 4-day test is always required if there is a subnormal response to tetracosactrin screening tests.

The patient needs to be admitted to hospital for the 4-day tetracosactrin test, but can remain ambulant and on a normal diet throughout. The procedure is as follows:

(1) *Day 1.* A tetracosactrin screening test is performed, as described above. As soon as the third blood specimen has been collected, 1 mg tetracosactrin zinc phosphate complex is injected intramuscularly.
(2) *Day 2.* A further injection of 1 mg tetracosactrin zinc phosphate complex is given.
(3) *Day 3.* As for Day two.
(4) *Day 4.* A second tetracosactrin screening test is performed, as described above.

If the response on Day one is seriously subnormal, dexamethasone (1 mg daily) can be given as steroid replacement treatment during the rest of the 4-day tetracosactrin test without affecting the interpretation of the test. The response is assessed by comparing the results of the first tetracosactrin screening test, performed on Day one, with the results of the second tetracosactrin screening test, performed on Day four.

A normal response to the 4-day test is defined in the same way as for the tetracosactrin screening test (see above), i.e. in terms of a rise in plasma [cortisol].

A subnormal or absent response on both Day one and Day four indicates the presence of primary adrenocortical insufficiency. A definite but slight increase (i.e. on Day 4), with maximum plasma [cortisol] greater than 550 nmol/L, indicates that the patient has secondary adrenocortical insufficiency. Occasionally, patients with secondary adrenocortical hypofunction fail to produce an adequate response to tetracosactrin after three days of injections, but do so if the test is continued for another two days.

Poor responses to the 4-day tetracosactrin test may occur in patients with hypothyroidism, whether due to thyroid disease itself or secondary to pituitary insufficiency. In patients with hypothyroidism, adrenal function cannot be satisfactorily assessed until the thyroid deficiency has been corrected.

DETERMINING THE CAUSE OF ADRENOCORTICAL HYPOFUNCTION

Plasma ACTH

These measurements are valuable in confirming the presence of minor degrees of primary hypofunction of the adrenal cortex. Results may be very high in these patients, and disproportionately high when related to the corresponding data for plasma [cortisol]. In patients with secondary adrenocortical hypofunction, plasma [ACTH] is almost invariably low.

Insulin-hypoglycaemia test

Stress in various forms normally induces an increased secretion of ACTH and corticosteroids. Patients who have been shown to have adrenal hypofunction, by the results of tests of basal secretion, but in whom the adrenal cortex has been shown to respond to direct stimulation (in the tetracosactrin tests) with an increased output of cortisol, can have the functional integrity of the whole HPA axis tested by means of the insulin-hypoglycaemia test. However, the test is strongly contra-indicated if there is severe unequivocal panhypopituitarism.

Performing the test: In these patients, the initial dose of soluble insulin should be reduced to 0.10 units/kg. Indeed, this dose of insulin should be used initially whenever it seems likely that a patient might be unduly sensitive to insulin. This might be the case in patients with hypothalamic or pituitary hypofunction, or in patients who are severely undernourished (e.g. due to anorexia nervosa). Apart from using a reduced dose of insulin, the test is performed as described above (p. 362).

For the test to be valid, it is again essential to produce an adequate degree of hypoglycaemia. However, in patients with adrenocortical hypofunction, the baseline plasma [glucose] may be low and it may therefore be inappropriate to attempt to produce a rapid fall of plasma [glucose] to hypoglycaemic levels (less than 2.2 mmol/L). Instead, careful observation of the patient for symptoms of hypoglycaemia may be a more satisfactory criterion for assessing the response to insulin.

Interpretation of results: An absent or impaired plasma cortisol response despite the production of an adequate degree of hypoglycaemia may be due to hypothalamic, pituitary or adrenal hypofunction. It is also observed in patients with anorexia nervosa or severe malnutrition due to some other cause.

Failure to respond to the insulin-hypoglycaemia test, combined with an adequate response to the 4-day tetracosactrin test, confirms that there is secondary hypofunction of the adrenal cortex, but fails to distinguish between the pituitary, hypothalamus or 'higher centres' as the site of the cause.

Adrenal antibodies

One of the causes of primary hypofunction of the adrenal cortex is auto-immune disease. In those patients who have idiopathic Addison's disease, adrenal antibodies can nearly always be detected in serum.

OTHER CHEMICAL ABNORMALITIES IN ADRENOCORTICAL HYPOFUNCTION

There are important effects on electrolyte metabolism, but changes in plasma electrolyte concentrations only tend to appear at a relatively late stage in the development of Addison's disease and secondary adrenocortical hypofunction.

There is loss of Na^+ in urine, the volume of ECF tends to fall and there may be hyponatraemia, which can become severe in untreated patients, especially in Addisonian crisis. There is retention of K^+, with diminished urinary excretion and a tendency to hyperkalaemia. Acid–base disturbances are not usually observed unless complications develop, e.g. due to vomiting.

Various special methods of investigation (e.g. radiological studies) can be used to locate the cause of the adrenocortical hypofunction. Chemical tests of pituitary, thyroid and gonadal function may also help localize the disease. Indeed, by the time hypopituitarism is severe enough to be the cause of secondary adrenocortical hypofunction and reduced plasma [cortisol], other pituitary functions will almost certainly be abnormal. It is usual, therefore, to measure basal concentrations in plasma of T4 and TSH, LH and FSH, and to investigate the growth hormone response to insulin-induced hypoglycaemia (p. 394) in addition to the cortisol response, in these patients.

The results of the principal chemical investigations on patients with adrenocortical hypofunction are summarized in Table 18.3.

INVESTIGATIONS IN PATIENTS TREATED WITH STEROIDS

The majority of patients who are receiving long-term treatment with steroids, or who have received such treatment in the past, usually respond quickly to stimulation by tetracosactrin zinc phosphate complex injections.

There are sometimes circumstances in which it is necessary to investigate adrenocortical function while the patient is still receiving steroids. If the patient is being treated with cortisol or cortisone, the

TABLE 18.3. Hormonal tests for

Test	Primary hypoadrenalism
Plasma cortisol	↓ or N
Tetracosactrin tests	
(a) screening test	No response
(b) 4-day test	No response
Plasma ACTH	↑

Key to abbreviations N = normal; ↓ = reduced; ↑ = increased.

drug should be replaced by one of the more powerful synthetic steroids (e.g. dexamethasone or betamethasone), since these steroids do not interfere with measurements of plasma [cortisol] or urinary [free cortisol]. However, prednisone and prednisolone are not suitable for this purpose, as they interfere with the radioimmunoassay of cortisol.

Having satisfactorily stabilized the patient on the low dosage of steroid, baseline values for plasma [cortisol] or urinary [free cortisol] are determined. A '4-day' tetracosactrin test is then performed. However, in these patients, it is advisable to continue daily injections of 1 mg tetracosactrin zinc phosphate complex for two extra days. The response to tetracosactrin may be delayed as a consequence of adrenal suppression due to the long-term steroid therapy.

THE MINERALOCORTICOIDS

The most important steroid with mineralocorticoid activity is aldosterone. Its principal physiological function is to conserve Na^+; it does this mainly by facilitating the exchange of Na^+ for K^+ and H^+ in the distal renal tubule and in other epithelial cells. The mechanisms controlling the output of aldosterone are complex. The renin-angiotensin system (Fig. 18.4) is one of the main factors, but ACTH, K^+ and Na^+ also affect aldosterone secretion:

(1) *Renin* is a proteolytic enzyme produced by the juxtaglomerular apparatus of the kidney. Its release into the circulation is stimulated particularly by a fall in circulating blood volume and loss of Na^+. After its release, renin acts on its substrate and splits off the pharmacologically inactive decapeptide angiotensin I. Angiotensin-converting enzyme (ACE), present in lung and plasma, then converts angiotensin I to the pharmacologically

adrenocortical hypofunction.

Secondary hypoadrenalism	Congenital adrenal hyperplasia
↓ or N	↓ or N
No response	Not indicated
Response	Not indicated
↓ or N	↑

active angiotensin II, by splitting off a dipeptide from the C-terminal end of the molecule. The actions of angiotensin II include stimulation of aldosterone release. Angiotensin II is broken down by an aminopeptidase to angiotensin III, which has little or no pressor activity but which also stimulates aldosterone production. Angiotensin III is then further broken down by angiotensinases.

(2) *ACTH* output influences aldosterone release, but this control mechanism does not seem to be important, except possibly in stress conditions and in congentital adrenal hyperplasia due to 21-hydroxylase deficiency.

(3) *Potassium.* Changes in plasma [K^+] can have marked effects on aldosterone output. In response to increases in plasma [K^+], aldosterone is released and exerts its effect on the distal renal tubule, causing an increased output of K^+ in the urine.

Hyperaldosteronism

Primary hyperaldosteronism, Conn's syndrome, or low renin hyperaldosteronism, is due to an adrenal adenoma in about 80% of

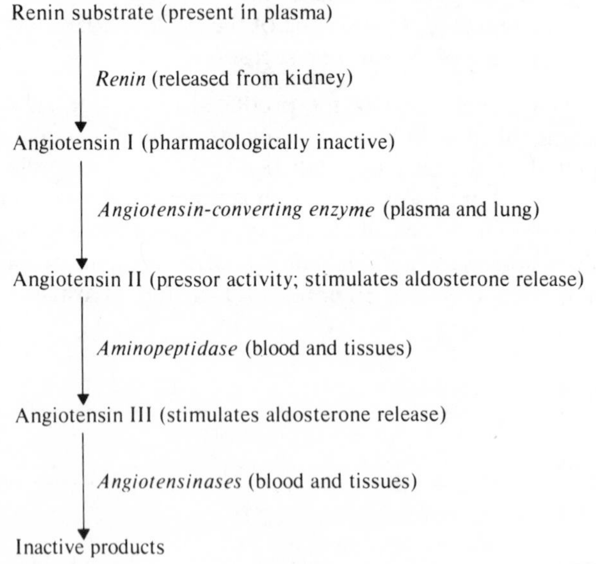

FIG. 18.4. *The renin-angiotensin system.* Only angiotensin II has significant pressor activity. Both angiotensin II and III stimulate aldosterone production. Renin substrate (angiotensinogen) is formed in the liver. Angiotensin I, II and III contain ten, eight and seven amino acids, respectively.

cases. In the remaining 20%, there is diffuse bilateral hyperplasia of the zona glomerulosa (idiopathic adrenal hyperplasia). Adrenal carcinoma is a very rare cause of primary hyperaldosteronism.

Conn's syndrome is a rare cause of hypertension. One of the suggestive, but by no means constant, features is the finding of a reduced plasma [K^+]. The depletion of K^+, consequent upon the renal loss of K^+, gives rise to a number of effects, principally muscle weakness and renal tubular defects. There are, however, many other causes of hypertension occurring in association with a reduced plasma [K^+] and these need to be considered in the differential diagnosis.

Liquorice and carbenoxolone (a synthetic derivative of glycyrrhizic acid, a constituent of liquorice) ingestion may lead to a syndrome similar to primary hyperaldosteronism, by stimulating mineralocorticoid receptors directly.

Secondary hyperaldosteronism occurs much more commonly than primary. It is due to conditions which stimulate the secretion of renin, possibly caused by an abnormality of Na^+ metabolism. The various conditions thus act on the adrenal cortex indirectly, by increasing the output of renin from the juxtaglomerular apparatus.

Secondary hyperaldosteronism occurs in association with a wide range of disorders that do not directly involve the adrenal cortex. These include Na^+ deprivation, K^+ excess, haemorrhage, congestive cardiac failure, the nephrotic syndrome, cirrhosis with ascites and renal artery stenosis. In many of these conditions the diagnosis is clear and there is no need to consider investigation of the renin-angiotensin-aldosterone system. Probably the commonest cause of secondary hyperaldosteronism is diuretic therapy.

INVESTIGATION OF HYPERALDOSTERONISM

The definitive identification of primary hyperaldosteronism requires investigations which are expensive and of limited availability. It is important, therefore, to undertake preliminary studies designed to show a strong likelihood that primary hyperaldosteronism is present. The preliminary studies depend on the fact that aldosterone causes loss of K^+, associated with retention of Na^+ by the kidney. Precautions to observe include ensuring that patients are not taking diuretics, which may lower plasma [K^+], nor any form of liquorice preparation since these have aldosterone-like effects.

Plasma potassium

This is the principal investigation used for initial screening. Ideally, it should be performed either before the patient starts treatment with hypotensive drugs or diuretics, or after the patient has been taken off such treatment for at least one month.

The finding of a reduced plasma [K^+] in a hypertensive patient is an important first step in the diagnostic process. It may be lower than 3.0 mmol/L but values of 3.3 mmol/L and below should be regarded as suspicious. In addition to reduction in plasma [K^+], there may be a slight increase in plasma [Na^+], a mild metabolic alkalosis and urinary K^+ excretion that is inappropriately high, considering that there is hypokalaemia.

It is advisable to use *plasma* for these determinations since any leakage of K^+ from the erythrocytes, as may occur during blood clotting, can raise *serum* [K^+] appreciably and thus delay the diagnosis. The diagnosis might also be made more difficult if the specimen of blood is collected with excessive venous stasis, or after forearm exercise, both of which tend to promote release of K^+ from cells.

Aldosteronism is occasionally associated with an intermittently reduced plasma [K^+]. If the first result is normal but the diagnosis is still considered probable on clinical grounds, measurement of plasma [K^+] should be repeated at intervals of a few weeks. Alternatively, or additionally, these patients can be placed on a high Na^+ intake (150–200 mmol/day) for two weeks, after which plasma [K^+] is measured again. Patients who develop hypokalaemia should be further investigated, as described below.

Special investigations

Patients requiring further investigation (1) will be hypertensive, (2) will have had certain other causes of hypertension excluded (e.g. phaeochromocytoma, coarctation of the aorta), and (3) will have been shown to have hypokalaemia, or to be liable to develop hypokalaemia in the presence of a Na^+ load. Other causes of hypokalaemia (e.g. diuretics, laxatives, liquorice) should have been excluded.

Primary hyperaldosteronism can be distinguished from secondary hyperaldosteronism by measuring *plasma renin activity (PRA)*. Also, plasma [Na^+] tends to be lower in secondary than in primary hyperaldosteronism. The reasons for choosing PRA, as the measurement for investigating the renin-angiotensin system, are:

(1) The rate-limiting factor in the formation of angiotensin II is the renin activity in plasma, except in patients with severe liver disease in whom synthesis of renin substrate may be so reduced as to make plasma [renin substrate] the rate-limiting factor.

(2) Renin has a half-life in plasma of 20 min, whereas the half-life of angiotensin II is only one minute.

If the index of suspicion of aldosteronism is high, on the basis of the preliminary investigations (i.e. plasma [K^+] principally), PRA should be measured:

(1) *If PRA is high,* the patient has secondary hyperaldosteronism and steps should be taken to identify its cause, if not already apparent.
(2) *If PRA is low,* the patient may have primary hyperaldosteronism. Further investigation of these patients requires measurement of plasma [aldosterone] or 24-h urinary excretion of aldosterone, or both; if results for these tests are high, this confirms the diagnosis.

The distinction between aldosterone-producing adenoma, the common cause of Conn's syndrome, and idiopathic adrenal hyperplasia is important since the adenomas are curable by surgery whereas surgery only occasionally cures patients with adrenal hyperplasia. It is possible to distinguish between these conditions pre-operatively, but the subject is too complex to be considered here (See article by Vaughan *et al.*, 1981).

CONGENITAL ADRENAL HYPERPLASIA

Striking anatomical changes take place in the adrenal cortex immediately after birth; these are associated with marked alterations in the pattern of urinary steroid output. There is a period of transition during the first six months of an infant's life, during which the pattern of fetal steroid metabolism changes to the normal childhood pattern, which closely resembles the adult pattern. Soon after delivery, cortisol is synthesized by the infant's adrenal cortex at a rate similar to that found in adults (corrected for the surface area of the body) and the infant adrenal becomes fully capable of responding normally to stress.

Several different inherited enzymic defects affecting the synthesis of cortisol have been identified in congenital adrenal hyperplasia (CAH). These mostly relate to the hydroxylation reactions in the final stages of steroid biosynthesis (Fig. 18.5). However, desmolase defects have also been described affecting earlier stages in the synthetic pathway. In an individual patient with CAH, the defect usually results in deficiency of a single enzyme.

Underproduction of cortisol, in CAH, leads to overproduction of ACTH due to failure of the negative feedback mechanism. Increased ACTH production causes accumulation of compounds proximal to the

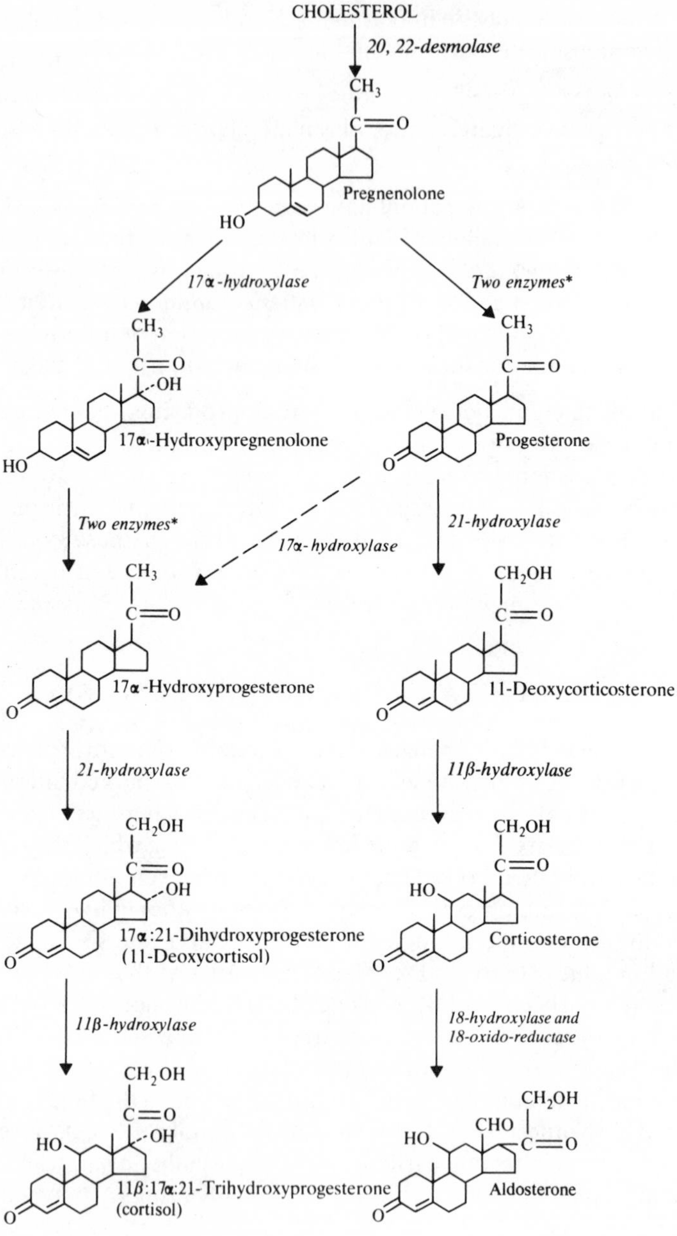

CHOLESTEROL

20, 22-desmolase

CH_3 — C=O

Pregnenolone

HO

17α-hydroxylase

*Two enzymes**

CH_3 — C=O — OH

17α-Hydroxypregnenolone

HO

CH_3 — C=O

Progesterone

O

*Two enzymes**

17α- hydroxylase

21-hydroxylase

CH_3 — C=O

17α-Hydroxyprogesterone

O

CH_2OH — C=O

11-Deoxycorticosterone

O

21-hydroxylase

11β-hydroxylase

CH_2OH — C=O — OH

17α:21-Dihydroxyprogesterone
(11-Deoxycortisol)

O

CH_2OH — C=O

HO

Corticosterone

O

11β-hydroxylase

18-hydroxylase and
18-oxido-reductase

CH_2OH — C=O — OH

HO

11β:17α:21-Trihydroxyprogesterone
(cortisol)

O

CH_2OH

CHO

HO

C=O

Aldosterone

O

***3β-hydroxysteroid dehydrogenase and Δ^5-Δ^4-isomerase**

FIG. 18.5. Pathways in the synthesis of cortisol and aldosterone from cholesterol.

site of the enzymic block, and these may be metabolized to steroid hormones with virilizing effects; sometimes, also, there is hypertension or failure to conserve salt. All the conditions are rare, and the non-virilizing group is extremely rare. Only the commoner conditions will be considered in any detail.

21-hydroxylase defect

This is the commonest defect; it accounts for about 90% of cases. The deficiency leads to failure of cortisol synthesis but to much less marked effects, in most patients, on aldosterone synthesis. There is increased synthesis of 17α-hydroxyprogesterone, the substrate of the deficient enzyme, and large amounts of this are metabolized along normally minor pathways with the formation of pregnanetriol and adrenal androgens, mainly androstenedione (Fig. 18.6).

Diagnosis is usually confirmed by demonstrating greatly increased plasma [17α-hydroxyprogesterone] or excretion of greatly increased amounts of pregnanetriol in the urine, or both.

Since the overproduction of cortisol precursors and their metabolites is due to failure of the feedback control mechanism, treatment with glucocorticoids (e.g. cortisol) rapidly suppresses the

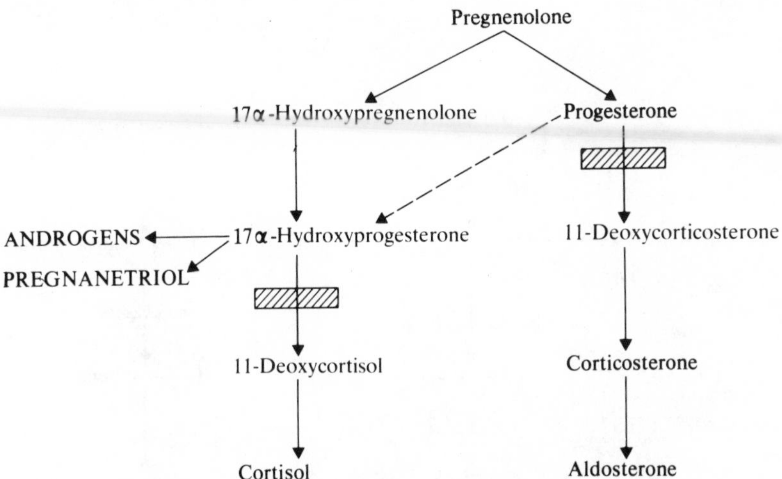

FIG. 18.6. Congenital adrenal hyperplasia, 21-hydroxylase defect. The principal steroids excreted in abnormal amounts are shown in capitals; the main androgen is androstenedione.

[hatched block symbol] indicates the reaction affected in this form of CAH.

excessive output of ACTH. The effectiveness of treatment can be assessed by measuring either plasma [17α-hydroxyprogesterone] or plasma [androstenedione]; these rapidly revert to normal if cortisol dosage is adequate. Long-term treatment must take account of growth-rate and bone-age.

It is possible to diagnose 21-hydroxylase deficiency before birth by examining the pattern of steroids present in amniotic fluid. This investigation is only likely to be performed where there has already been an index case in the family.

11β-hydroxylase defect

This deficiency accounts for about 10% of cases of congenital adrenal hyperplasia. The block in synthesis of cortisol, at 11-deoxycortisol, leads to increased excretion of metabolites of 11-deoxycortisol and, to a lesser extent, metabolites of its precursor, 17 α-hydroxyprogesterone (Fig. 18.7). Antenatal diagnosis is possible, as with 21-hydroxylase deficiency.

EFFECTS OF HYDROXYLASE DEFICIENCY

The severity of the 21-hydroxylase defect influences the clinical

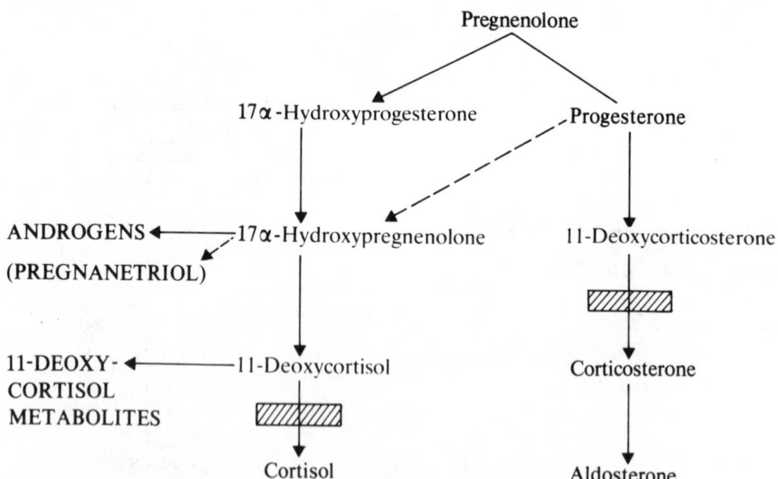

FIG. 18.7. Congenital adrenal hyperplasia, 11 β-hydroxylase defect. The principal steroids excreted in abnormal amounts are shown in capitals.

[hatched box symbol] indicates the reaction affected in this form of CAH.

presentation. If the defect is incomplete, adequate synthesis of cortisol may be maintained in response to the greatly increased output of ACTH. However, marked adrenal hyperplasia results and there is greatly increased synthesis of adrenal androgens since the precursors of cortisol, notably 17α-hydroxyprogesterone, are formed in large amounts and are partly metabolized along normally minor pathways (Fig. 18.6). With more complete degrees of 21-hydroxylase deficiency, the defect in cortisol synthesis is only partly compensated and there is an associated tendency to lose salt, due to the impaired synthesis of 11-deoxycorticosterone and aldosterone.

The 11ß-hydroxylase defect is usually only partial, and the nature of the clinical presentation is again influenced by the severity of the defect. Adequate amounts of cortisol are usually synthesized in response to the increased output of ACTH by the pituitary. The block on the pathway to aldosterone synthesis leads to a build-up of 11-deoxycorticosterone. (Fig. 18.7) and this is responsible for the hypertension which is a feature of 11ß-hydroxylase deficiency.

The 21-hydroxylase and 11ß-hydroxylase defects are both usually incomplete, and adequate amounts of glucocorticoid can be synthesized to meet normal requirements, in response to the increased output of ACTH. However, these patients may be unable to respond rapidly and with an increased output of cortisol under conditions of stress. It is, therefore, important to recognize and to treat partial deficiency of these hydroxylation systems.

MALE SEX HORMONES

The testes produce testosterone, androstenedione and dehydro*epi*androsterone (DHA); these androgens are synthesized in the interstitial (Leydig) cells. Testosterone is produced in the largest amount, and is the most important of the testicular androgens. The adrenal cortex also secretes appreciable amounts of androgens of low potency, particularly DHA.

The development of the Leydig cells and their output of testosterone is influenced by the output of luteinizing hormone (LH) by the pituitary; LH stimulates cholesterol desmolase (Fig. 18.5). The development of the Sertoli cells is controlled by follicle-stimulating hormone (FSH). Stimulation and maintenance of spermatogenesis, by the Sertoli cells, require both FSH and LH; FSH is the initial stimulator and testosterone (controlled by LH) is the maintenance factor.

The pituitary secretion of FSH and LH is controlled by gonadotrophin-releasing hormone (Gn-RH), from the hypothalamus. The control of Gn-RH release and the effects of feedback inhibition by gonadal steroids are complex subjects, beyond the scope of this book.

Testosterone circulates in the plasma mainly bound to plasma proteins, about 65% bound to sex hormone-binding globulin (SHBG) and about 30% to albumin; only about 2% is unbound. In the tissues, testosterone is converted to 5α-dihydrotestosterone (DHT) by 5α-reductase, and the biological activity of testosterone depends on this enzyme. DHT binds more strongly than testosterone to the androgen receptors and is very important in exerting androgen action on protein synthesis.

Testosterone concentrations can be measured in plasma by sensitive and specific methods. Small amounts can be detected in females, more being present in the luteal than the follicular phase of the menstrual cycle.

Testosterone is metabolized in the liver with the formation of inactive conjugates (e.g. glucosiduronates). Some of the testosterone is oxidized to form androstenedione which contributes to the urinary excretion of 17-oxosteroids. Only limited information about gonadal function can be obtained from estimation of the urinary 17-oxosteroids in view of the fact that much of the DHA produced in the body is of adrenal origin; DHA gives rise to most of the steroid metabolites contributing to the 17-oxosteroid measurement.

DISORDERS OF MALE SEX DIFFERENTIATION

Many different conditions have been described, all of them rare. In some the gonads degenerate and in others there is an enzyme defect affecting steroid synthesis. In a third group there is androgen resistance at the end organ, and a fourth group consists of the true hermaphrodites.

In the testicular feminization syndrome, which is probably due to end-organ resistance to androgens, plasma [testosterone] is abnormally high. In many of the other conditions plasma [testosterone] is low, both in childhood and adult life.

HYPOGONADISM IN ADULT MALES

It is convenient to distinguish two forms, hypergonadotrophic and hypogonadotrophic hypogonadism, depending on whether the primary defect is, respectively, in the testes or in the hypothalamic-pituitary region. Both forms lead to infertility. However, the majority of infertile men do not fit into either of these categories, but instead are eugonadotrophic. They have normal (often at the lower end of the reference range) plasma [FSH], [LH] and [testosterone] and oligospermia due to failure of the seminiferous tubules. These infertile, eugonadotrophic males may represent a group that is likely to go on to develop hypogonadotrophic hypogonadism

Hypergonadotrophic hypogonadism

The primary abnormality is in the testes, and plasma [testosterone] is reduced. There may be increased plasma [LH], and this finding indicates that there is failure of function of the Leydig cells, whereas increased plasma [FSH] indicates that the failure of function is in the seminiferous tubules. However, this summary description must be recognized as an oversimplification. This group of conditions includes Klinefelter's syndrome (usually 47XXY).

Hypogonadotrophic hypogonadism

The primary abnormality is in the hypothalamus or the pituitary. There may be selective deficiency of gonadotrophin production, both LH and FSH, or only LH. Alternatively, the deficiency may be part of a generalized failure of pituitary hormone production. There is reduced plasma [testosterone] accompanied by reduced plasma [LH], and usually low plasma [FSH] as well.

The effect of injecting human chorionic gonadotrophin (e.g. daily for several days) on plasma [testosterone] or urinary testosterone output helps to differentiate between primary and secondary hypogonadism. Testicular biopsy is also helpful.

SKIN DISORDERS

Increased plasma [testosterone] has been reported in patients with acne vulgaris.

HIRSUTISM IN FEMALES

Many women complain of hirsutism. If menstruation is normal, hormonal investigations are not indicated. However, patients who have hirsutism together with amenorrhoea or oligomenorrhoea need to be investigated for possible disorders of hormone production (e.g.

Cushing's syndrome, polycystic ovaries), although hormonal abnormalities are only demonstrable in a small proportion of these patients. Plasma [testosterone] should be determined and plasma [SHBG], if available. Plasma [LH] and [FSH] should be measured in patients thought to have polycystic ovaries.

In idiopathic hirsutism, plasma [testosterone] may be normal or moderately increased. It has been shown that the increased plasma [testosterone], observed in some of these patients, is associated with increased testosterone secretion rates.

Plasma [SHBG] may be decreased in patients with idiopathic hirsutism. The effect of this is to increase plasma [free testosterone] and thereby the concentration of DHT to which the tissues are exposed.

Plasma [LH] is increased in patients with polycystic ovaries.

Less common causes of hirsutism include Cushing's syndrome and ovarian tumours. Increased plasma [testosterone] has been observed in female patients with adrenal disorders (tumour, hyperplasia), and in patients with ovarian tumours giving rise to virilization.

FEMALE SEX HORMONES

The principal sex hormones in females are oestradiol-17β and progesterone, normally formed mainly by the ovaries and by the placenta. The adrenal cortex also secretes small amounts of oestrogens and progestogens.

Oestradiol-17β is the most active of the oestrogens synthesized by the ovary. In the plasma, like testosterone, oestradiol is normally mainly bound to SHBG. The concentration of this protein rises in pregnancy and in response to oestrogen therapy, including the oestrogen-containing oral contraceptives.

Oestradiol-17β is partly metabolized to oestrone and oestriol, both of which have significant oestrogenic activity. The oestrogens are conjugated (e.g. as glucosiduronates) prior to excretion in the urine. The main metabolite of progesterone is pregnanediol, which is similarly conjugated before being excreted in the urine.

OESTROGEN MEASUREMENTS

The fetoplacental unit is responsible for the production of large quantities of oestrogens, and the main requirement for oestrogen determinations is in the assessment of fetoplacental function and for the early detection of placental insufficiency (p. 406).

Plasma and urinary oestrogen measurements may be performed as part of the investigation of amenorrhoea and of subfertility in females (p. 416).

Oestrogen-secreting tumours

Granulosa cell tumours of the ovaries secrete large quantities of oestrogens and may lead to precocious sexual development. Occasionally, adrenal and testicular tumours produce large quantities of oestrogens; in male patients these cause feminization. In patients who show evidence of these hormonal disturbances, determination of plasma [oestrogens] or the urinary output of these hormones usually confirms the diagnosis.

PROGESTERONE MEASUREMENTS

Determinations of plasma or salivary [progesterone] have almost completely replaced measurements of urinary pregnanediol. They may be used for the following purposes:

(1) To indicate the *development of a functional corpus luteum,* after spontaneous ovulation or occurring as a result of ovarian stimulation (p. 414).
(2) In the *investigation of recurrent abortion* (p. 405).
(3) In the investigation of *disease of the adrenal cortex.* Excessive formation of progesterone and excretion of pregnanediol occur in a small percentage of adrenal tumours.

In the adrenal cortex, progesterone is a precursor of cortisol and aldosterone and in CAH the normal synthetic pathways are disturbed (Figs. 18.6 and 18.7). In this group of rare diseases the urinary excretion of pregnanetriol, normally a minor metabolite of progesterone, may be greatly increased and metabolism of progesterone is also directed more towards androgen synthesis.

FURTHER READING

ANDERSON, D.C. (1978). Endocrine function of the testis. In *Recent Advances in Endocrinology and Metabolism,* **1,** 111–136. Ed. J.L.H. O'Riordan. Edinburgh: Churchill Livingstone.

BURKE, C.W. (1978). Disorders of cortisol production: diagnostic and therapeutic progress. *ibid.*, **1,** 61–90.

CAPMAN, M.G. (1981). Management of hirsutism. *British Journal of Hospital Medicine,* **26,** 270–272.

Croze, F. and Franchimont, P. (1982). Regulation of secretion and actions of gonadotrophins. In *Recent Advances in Endocrinology and Metabolism,* **2,** 133–156. Ed. J.L.H. O'Riordan. Edinburgh: Churchill Livingstone.

Drury, P.L., Al-Dujaili, E.A.S. and Edwards, C.R.W. (1982). The renin-angiotensin-aldosterone system. *ibid.,* **2,** 157–186.

James, V., editor (1979). *The Adrenal Gland* (Comprehensive Endocrinology). New York: Raven Press.

Leading article (1977). Pituitary-dependent Cushing's disease. *British Medical Journal,* **1,** 1049–1050.

Leading article (1978). Endocrine basis for sexual dysfunction in men. *ibid.,* **2,** 1516–1517.

New, M.I., Dupont, B., Grumbach, K. and Levine, L.S. (1983). Congenital adrenal hyperplasia and related conditions. In *The Metabolic Basis of Inherited Disease*, 5th Edition, pp. 973–1000. Ed. J.B. Stanbury, J.B. Wyngaarden, D.S. Fredrickson, J.L. Goldstein and M.S. Brown. New York: McGraw-Hill.

O'Riordan, J.L.H., Malan, P.G. and Gould, R.P., editors (1982). *Essentials of Endocrinology.* Oxford: Blackwell Scientific Publications.

Rudd, B.T. (1983). Urinary 17-oxogenic and 17-oxosteroids: A case for deletion from the clinical chemistry repertoire. *Annals of Clinical Biochemistry,* **20,** 65–71.

Swales, J.D. (1983). Primary aldosteronism: how hard should we look? *British Medical Journal*, **287,** 702–703.

Vaughan, N.J.A., Jowett, T.P., Slater, J.D.H. *et al* (1981). The diagnosis of primary hyperaldosteronism. *Lancet,* **1,** 120–125.

Vining, R.F., McGinley, R.A., Maksvytis, J.J. and Ho, K.Y. (1983). Salivary cortisol: a better measure of adrenal cortical function than serum cortisol. *Annals of Clinical Biochemistry*, **20,** 329–335.

Chapter 19

Hypothalamic and Pituitary Hormones

THE HYPOTHALAMUS

The hypothalamus secretes several hormones or factors. These control the release of hormones from the anterior pituitary, in some cases by stimulating hormonal release and in others by inhibiting it (Table 19.1).

The hormones produced by the target glands that are controlled by the anterior pituitary may exert a feedback effect on the secretion of the corresponding hypothalamic hormone or factor. For example, plasma [cortisol] influences the output of CRF from the hypothalamus. Alternatively, the main feedback effect may be on the anterior pituitary itself as is the case with plasma [thyroxine], which directly influences the secretion of TSH but which has relatively minor effects on TRH secretion. Release of hypothalamic hormones or factors are also influenced by nervous stimuli from higher CNS centres.

Some of the hypothalamic hormones have been characterized and are now available as synthetic materials. They have been used to test anterior pituitary function, but so far only TRH has gained an established place for this purpose.

TABLE 19.1. Hypothalamic hormones and factors and their principal effects on the anterior pituitary.

Hypothalamic hormone or factor	*Pituitary hormone output stimulated*
Corticotrophin-releasing factor (CRF)	ACTH and LPH
Thyrotrophin-releasing hormone (TRH)	TSH, prolactin and (in males) FSH
Gonadotrophin-releasing hormone (Gn-RH)	LH and FSH
Growth hormone-releasing factor (GH-RF)	GH
	Pituitary hormone ouput suppressed
Prolactin release-inhibiting hormone (PRIH)	Prolactin
Growth hormone release-inhibiting hormone (GH-RIH)	GH, TSH and FSH

THYROTROPHIN–RELEASING HORMONE

This is a tripeptide, pyroglutamyl-histidyl-prolinamide. It controls the secretion of TSH by the anterior pituitary. It also has an independent effect in causing prolactin secretion, but it is doubtful if this latter effect is of major physiological importance; prolactin release is not affected in patients with isolated TSH deficiency.

The main use of TRH is to confirm or exclude a provisional diagnosis of hyperthyroidism (p. 345). In the investigation of hypothyroidism, the TRH test is rarely indicated and then only for the investigation of patients with secondary hypothyroidism (e.g. as a test of reserve capacity to secrete TSH, p. 394).

GONADOTROPHIN–RELEASING HORMONE (Gn-RH)

This hormone is a decapeptide. It exerts its effects mainly by stimulating the synthesis of LH and controlling its release from the anterior pituitary. The effects of Gn-RH on FSH synthesis and release are much less marked.

A Gn-RH test, similar in its design to the TRH test, has been described for use in the investigation of infertility. This test was claimed to be of value in assessing the reserve capacity of the pituitary to secrete LH and FSH. However, its usefulness for this purpose has not been established. Results are difficult to interpret since plasma [LH] and [FSH] do not always alter in parallel in disease, and LH responses to Gn-RH injection often become abnormal before FSH. Furthermore, a normal response to Gn-RH does not exclude the presence of pituitary disease.

ANTERIOR PITUITARY

All the pituitary hormones listed in Table 19.1 can be measured, in plasma or urine as appropriate. These measurements nowadays form part of the standard assessment of hypothalamic and pituitary function.

Hypofunction of the pituitary is most often due to therapeutic action, e.g. removal of a pituitary tumour or destruction of the pituitary by irradiation as part of the treatment of patients with metastatic carcinoma of the breast. If these cases are excluded, pituitary hypofunction is most commonly due to an adenoma of the pituitary or to some other space-occupying lesion close to the pituitary fossa. To begin with, impairment of function will usually only be revealed by the use of stimulation tests. Later, basal concentrations of the pituitary hormones in blood will be affected. In general, the secretion of GH, LH and FSH are affected relatively early in the disease, whereas ACTH and prolactin are not affected until much later.

Less emphasis is placed nowadays on the importance of seeking evidence for multiple disturbances of target gland function, to confirm a diagnosis of anterior pituitary dysfunction. Nevertheless, demonstration that there is an abnormal (i.e. reduced) output of hormones normally produced by at least *two* of the target glands controlled by the anterior pituitary is important evidence pointing to a lesion in the anterior pituitary or hypothalamus. It is, however, important to note that hypofunction of the thyroid may itself influence the response to adrenal function tests (p. 370), so the demonstration of dysfunction of two or more target glands is not necessarily incontrovertible evidence for hypofunction of the pituitary. The demonstration that a target gland is (1) hypofunctional, and (2) able to respond to the administration of the appropriate trophic hormone (ACTH, FSH, LH, TSH) provides more direct support for a diagnosis of anterior pituitary hypofunction.

Hyperfunction of the anterior pituitary is almost always restricted to overproduction of one hormone. Recognition of hypersecretion depends on measurement of the plasma concentration of the appropriate pituitary hormone.

GROWTH HORMONE

This is a polypeptide with mol. mass of about 21 kDa. It is structurally similar to prolactin. Basal plasma [GH] is normally less than 10 μg/L and is often undetectable by present assay methods. Marked increases are mediated through GH-RF in response to stress or exercise. Release of GH is inhibited by GH-RIH (somatostatin). Because of the variable effects of stress and exercise on basal plasma [GH], this measurement by itself has very little diagnostic value. There are no tests of GH activity related to target gland function.

Acromegaly and giantism

These disorders are caused by an adenoma of the anterior pituitary. Giantism results if the disorder occurs before closure of the epiphyses of the long bones; acromegaly results if it occurs after their closure.

Basal plasma [GH] is often very high in these patients, but the concentrations are too variable for accurate diagnosis. In all these patients, the diagnosis should be confirmed by measuring the response of plasma [GH] to an oral glucose tolerance test (p. 223). Normally, plasma [GH] falls to less than 2 mU/L at some time during this test. However, in patients with acromegaly or giantism, plasma [GH] does not fall in response to the stimulus of hyperglycaemia, and may even increase.

Hypopituitarism

In these patients, growth hormone secretion is one of the first functions to be lost and plasma [GH] may be very low or even undetectable. The ability of the pituitary to increase its output of GH in response to stress may then need to be tested. This can best be done as part of an insulin-hypoglycaemia test (p. 362). However, in patients with suspected hypopituitarism, the initial dose of insulin in the insulin-hypoglycaemia test should be reduced, i.e. 0.10 unit/kg body weight (p. 371).

Normally, hypoglycaemia causes a marked increase in plasma [GH], to more than 20 mU/L, but this increase is not observed in severe pituitary hypofunction. A limited response, with plasma [GH] rising but to less than 20 mU/L, may be observed in patients with partial pituitary failure. The insulin-hypoglycaemia test can be used to assess the HPA axis at the same time as investigating GH production (see below).

Deficiency of GH is a rare cause of shortness of stature. Its recognition in children is important as the condition can now be treated, especially if detected early (e.g. when the child is 3–4 years old).

ADRENOCORTICOTROPHIC HORMONE

This is a polypeptide of 39 amino acids (mol. mass 4 500 Da); its biological activity is contained in the 24 amino acids at the N-terminal end (p. 367). Secretion of ACTH is controlled by the hypothalamic regulating factor, CRF, the release of which is subject to feedback inhibition that depends on plasma [cortisol].

It has been reported that ACTH and β-lipotrophin (LPH) are secreted together, and in equimolar amounts, in response to CRF. ACTH (39 amino acids) and LPH (93 amino acids) exist next to each other, at the C-terminal end of a much larger precursor molecule, pro-opiocortin. Separating the LPH from the ACTH component, and separating the ACTH component from the rest of the precursor, are two basic amino acid residues; in both sites these are lysine and arginine. Proteolysis appears to occur simultaneously at two sites, releasing ACTH and LPH into the circulation.

There is a nychthemeral rhythm of ACTH output by the pituitary. This, in turn, is responsible for the nychthemeral rhythm in the secretion of cortisol by the adrenal cortex (Fig. 18.1, p. 355). The output of ACTH is also normally modified rapidly in response to stress. Specimens for plasma ACTH measurement should, therefore, be collected at standard times (e.g. 0800 h and 2200 h) and with the minimum of disturbance to the patient. Close attention to detail in the collection and storage of specimens prior to analysis is essential if reliable results are to be obtained as ACTH is chemically unstable.

Plasma ACTH measurements are helpful in the investigation of adrenocortical hyperfunction (Table 18.2, p. 366). Patients with Cushing's disease (pituitary-dependent adrenal hyperplasia) have high plasma [ACTH], or at least inappropriately high in relation to plasma [cortisol]. Patients with the non-endocrine ACTH-secreting tumours have very high plasma [ACTH] or [ACTH-like material]. Plasma [ACTH] is low or undetectable in patients with autonomous adrenal tumours.

Plasma ACTH measurements can also help to differentiate between primary and secondary adrenocortical hypofunction. Patients with low plasma [cortisol] due to primary hypofunction of the adrenal cortex have high plasma [ACTH]. Patients with secondary adrenal hypofunction, on the other hand, have low or undetectable plasma [ACTH], but the adrenal cortex is able to respond to prolonged stimulation by tetracosactrin (p. 369).

The residual or the reserve capacity of the pituitary to secrete ACTH, e.g. following radiotherapy, can be assessed directly by measuring plasma [ACTH] as part of an insulin-hypoglycaemia test (p. 394). It is important, in these patients, to use a small dose of soluble insulin (0.10 units/kg) and to be ready to terminate the test if signs of hypoglycaemia develop; an impaired ACTH response can be anticipated in patients with hypopituitarism. Instead of measuring plasma [ACTH], the response to the hypoglycaemic stimulus can be measured indirectly by determining plasma [cortisol].

Because of difficulties caused by the instability of ACTH, and by some of the technical features of the RIA method, plasma [LPH] is measured by some laboratories instead of, or in addition to, plasma ACTH; LPH is more stable than ACTH.

THYROID-STIMULATING HORMONE

This is a glycoprotein, mol. mass approximately 28 kDa, secreted in response to TRH. The output of TSH is subject to negative feedback control that depends on plasma [free T4] and [free T3] (Fig. 17.3, p. 338).

Measurement of plasma [TSH] plays an important part in the diagnosis of hypothyroidism. In the common form of hypothyroidism, due to disease of the thyroid gland itself, plasma [TSH] is high. In hypothyroidism secondary to pituitary disease, plasma [TSH] is normal or low, i.e. its concentration is inappropriate in relation to the low plasma [T4].

The TRH test (p. 345) depends on measurements of plasma [TSH]. This test is an important part of the investigation of patients with suspected hyperthyroidism. Because of the lack of sensitivity of the RIA methods presently used for measuring plasma [TSH], differentiation between euthyroidism and hyperthyroidism cannot be made simply by measuring basal plasma [TSH].

LUTEINIZING AND FOLLICLE–STIMULATING HORMONES

Both LH and FSH are secreted in response to the single hypothalamic regulatory hormone Gn-RH. The main effects of Gn-RH are on the synthesis and release of LH.

The pituitary gonadotrophins are glycoproteins, each consisting of two subunits; one of these, the α subunit, is common to FSH, LH and TSH. A complex interplay of factors, as yet incompletely understood but including both oestrogen feedback effects and androgen effects, modulates the output of FSH and LH from the pituitary.

FSH and LH can be determined separately and specifically in plasma and in urine. These measurements are important in the investigation of gonadal failure, as they help differentiate primary failure of the gonads from failure secondary to pituitary dysfunction (pp. 383, 414). In organic pituitary disease, ability to secrete gonadotrophins is one of the earliest functions to be lost.

Low plasma [FSH] and [LH] may be secondary to hypothalamic dysfunction rather than due to pituitary failure. Hypothalamic or pituitary functional impairment may also be sufficient to cause infertility but insufficient to cause significant reductions in plasma [FSH] or [LH].

To test the entire hypothalamic-pituitary-gonadal axis, *clomiphene* may be administered. It acts by blocking the oestrogen receptor sites in the hypothalamus, thereby inhibiting normal negative feedback control by plasma oestrogens. If both the hypothalamus and pituitary are normal, clomiphene administration causes an increase in plasma

[FSH] and [LH], and these stimulate steroid output by the ovaries; all these hormones can be measured. Clomiphene must be used with caution because of the danger of overstimulation of the ovaries.

After the menopause, plasma [FSH] and [LH] both increase. In the child-bearing period, increased plasma [FSH] and [LH] indicate gonadal failure. In gonadal dysgenesis (Klinefelter's syndrome) and in the majority of subfertile males, in whom dysfunction of the seminiferous tubules is the common abnormality, plasma [FSH] is increased. Failure of the Leydig cells is much less common; plasma [LH] is high in these patients.

PROLACTIN

This is a separate pituitary hormone, distinct from growth hormone but having some common structural features. It is responsible for lactation, and has a role in the complex processes controlling gonadal function.

Prolactin secretion increases in response to oestrogens. Plasma [prolactin] is much higher in women of child-bearing age, especially during pregnancy, than in girls before puberty, in post-menopausal women or in males.

Prolactin secretion is normally under tonic inhibitory control, unlike other pituitary hormones, by one or more of the prolactin release-inhibiting hormones (PRIH); the most important of these is probably dopamine. Thyrotrophin-releasing hormone and vasoactive intestinal peptide (VIP), one of the gastrointestinal peptides that is found also in the CNS, have both been shown to stimulate prolactin release.

Before collecting blood for measuring plasma [prolactin], it is important to enquire about the intake of drugs. Many centrally acting drugs (e.g. phenothiazines, methyldopa, imipramine) inhibit the release of PRIH; some (e.g. methyldopa) act by their adrenergic receptor blocking effects. Oestrogens and oestrogen-containing oral contraceptives may also cause raised plasma [prolactin].

There is a circadian rhythm of prolactin secretion. Superimposed on these gradual changes, there are occasional sharp increases in plasma [prolactin]. Increases also occur in response to stress. It can, therefore, be difficult to interpret the results of single measurements of plasma [prolactin].

Prolactin measurements should be performed as part of the investigation of subfertility, whether or not there is galactorrhoea. A significant proportion of subfertile patients (male and female) have been shown to have hyperprolactinaemia; these may respond dramatically to bromocriptine therapy.

Increased plasma [prolactin] occurs in some patients with acromegaly. Hyperprolactinaemia has also been reported in patients with pituitary tumours but in whom no other evidence of hyperfunction could be detected. Some non-endocrine tumours secrete prolactin, and plasma [prolactin] should be measured in any patient with spontaneous inappropriate lactation.

MELANOCYTE–STIMULATING HORMONE

Pro-opiocortin contains within its molecule three components with MSH activity; α-MSH, contained in the ACTH component, β-MSH contained in the LPH component, and γ-MSH. The MSH components are formed by enzymic degradation of pro-opiocortin and its ACTH and LPH components.

The importance of MSH is unknown. Some patients with pituitary dysfunction (e.g. Cushing's disease, Nelson's syndrome) show excessive pigmentation as do some patients with non-endocrine ACTH-secreting neoplasms, and some with adrenocortical hypofunction (Addison's disease). In most of these patients, changes in plasma [ACTH] and plasma [MSH] occur in parallel.

COMBINED TEST OF ANTERIOR PITUITARY RESERVE

The pituitary reserve capacity for secretion of growth hormone, TSH, ACTH and prolactin can all be tested at one time as part of a co-ordinated procedure. This may be indicated for the assessment of residual pituitary function in patients who have been treated by pituitary ablation procedures.

An insulin-hypoglycaemia test is performed using a reduced dose (0.10 unit/kg) of insulin; the stress of hypoglycaemia reveals the pituitary's ability to secrete growth hormone, ACTH and prolactin. At the time of giving the insulin injection, the patient is also given an intravenous injection of 200 μg TRH, and TSH measurements are performed on blood samples collected at the same time as those obtained for growth hormone, ACTH (or cortisol) or LPH, and prolactin determinations. Plasma [glucose] is also measured to confirm that an adequate stress stimulus has been applied.

POSTERIOR PITUITARY

The posterior pituitary is an integral part of the neurohypophysis, and its secretion is directly subject to nervous control. It produces at least two hormones, arginine-vasopressin (AVP), usually still called the antidiuretic hormone (ADH), and oxytocin. Investigation of the functions of the posterior pituitary is mainly based on tests affected by the output of ADH.

ANTIDIURETIC HORMONE

Synthesized in the hypothalamus, the ADH-containing neurosecretory granules migrate to the posterior pituitary. The hormone itself, a decapeptide, is released in response to a rise in plasma osmolality or to a fall in plasma volume.

Overproduction of ADH has been described in association with a large number of conditions including head injury, cerebral tumour, cerebral abscess, pneumonia, pulmonary tuberculosis, acute porphyria and non-endocrine tumours. In some of these, ADH appears to be of pituitary origin, but in others the diseased tissue itself produces ADH or a peptide with similar pharmacological actions.

The syndrome of inappropriate secretion of ADH (SIADH) is defined as the excessive secretion of ADH in the absence of either of the normal major stimuli for ADH secretion, plasma hypertonicity or contraction of ECF. Whatever its cause, this syndrome gives rise to water retention, dilutional hyponatraemia and reduced osmolality of the ECF and ICF. The finding of plasma [Na^+] less than 125 mmol/L, accompanied by a urine osmolality that is inappropriately high, supports the diagnosis. Determination of plasma [ADH] would confirm the diagnosis, but this measurement is not often required, and can only be performed by a few laboratories.

Deficiency of ADH gives rise to diabetes insipidus. It can arise for a number fo reasons, including primary causes (idiopathic, familial) and a number of secondary causes all of which are related to disease or injury in or close to the pituitary. Deficiency of ADH may be the sole hormonal abnormality or there may be disturbances of anterior pituitary hormone production as well in the case of secondary causes of ADH deficiency.

Recognition of hyposecretion of ADH depends on urine concentration tests, i.e. the fluid deprivation test and the DDAVP test (p. 92), and requires careful osmolality studies. Patients with diabetes insipidus due to ADH deficiency do not respond to the fluid deprivation test but injection of DDAVP causes a marked increase in urine osmolality. Patients with nephrogenic diabetes insipidus do not respond in either of the tests. Results obtained in patients with psychogenic diabetes insipidus are very variable, and depend on the extent to which the patients co-operate in the performance of the tests.

Measurements of plasma [ADH] are not usually required for the diagnosis of ADH deficiency.

FURTHER READING

COWDEN, E.A., RATCLIFFE, W.A., BEASTALL, G.H. and RATCLIFFE, J.G. (1979). Laboratory assessment of prolactin status. *Annals of Clinical Biochemistry,* **16,** 113–121.

HALL, R. and GOMEZ-PAN, A. (1976). The hypothalamic regulatory hormones and their clinical applications. *Advances in Clinical Chemistry,* **18,** 173–212. New York: Academic Press.

HOCKADAY, T.D.R. (1983). Assessment of pituitary function. *British Medical Journal,* **287,** 1738–1740.

Leading article (1977). Melanocyte-stimulating hormone? *ibid,* **1,** 533–534.

Leading article (1977). Prolactin updated. *ibid.,* **2,** 846–847.

McGOWAN, G.K. and WALTERS, G. (1976). Hypothalamic and pituitary hormones. *Journal of Clinical Pathology,* **30,** Supplement (Association of Clinical Pathologists), **7.**

MORTIMER, C.H. and BESSER, G.M. (1977). Endocrinology of the hypothalamus. In *Recent Advances in Medicine,* 17th Edition, pp. 441–456. Ed. D.N. Baron. Edinburgh: Churchill Livingstone.

O'RIORDAN, J.L.H., editor (1978). *Recent Advances in Endocrinology and Metabolism.* **1,** Several articles. Edinburgh: Churchill Livingstone.

WILLIAMS, R.H., editor (1981). *Textbook of Endocrinology.* Philadelphia: Saunders.

Also see the reading lists at the end of Chapters 17, 18 and 20.

Chapter 20

Clinical Chemistry in Obstetrics and Gynaecology

The physiological changes that occur during the normal menstrual cycle depend on cyclical variations in the output from the pituitary of follicle-stimulating hormone (FSH) and luteinizing hormone (LH). The release of FSH and LH is influenced by the output of gonadotrophin-releasing hormone (Gn-RH) from the hypothalamus (p. 387). Although Gn-RH appears to control the pituitary's output of both FSH and LH, its different effects (in terms of the amounts secreted at different stages of the menstrual cycle) are thought to be caused by changes in the responsiveness of the pituitary, induced by feedback control effects exerted by gonadal steroids.

The developing Graafian follicles in the ovaries respond to the cyclical stimulus of FSH and LH by secreting two oestrogens, oestradiol and oestrone; these are metabolized to oestriol. After ovulation, the corpus luteum secretes progesterone as well as oestrogens. The cyclical changes in the uterus are determined by the ovarian steroid output, and these changes are further modified if pregnancy occurs.

The changes in plasma concentrations of FSH, LH and the principal gonadal steroids that occur in the normal menstrual cycle (i.e. a cycle unmodified by oral contraceptives) are shown in Fig. 20.1. Reference values for these hormones in females of different ages are given in Table 20.1.

TABLE 20.1. Reference values for plasma concentrations of pituitary gonadotrophins and the principal sex hormones in females (data from London and Shaw, 1978).

	LH (U/L)	FSH (U/L)	Oestradiol-17β (pmol/L)	Progesterone (nmol/L)
Children	1–3	1–3	40–120	<6
Menstruating adults				
Follicular phase	1–10	1–6	40–600	<6
Mid-cycle peak	8–60	4–15	500–1600	4–10
Luteal phase	2–14	1–5	280–1000	>20
Post-menopausal	>15	>20	<150	<6

Chemical changes associated with oral contraceptives

Large numbers of women now take oral contraceptives. These preparations contain either an oestrogen or a progestogen, or both. This section is concerned with certain metabolic effects of oral contraceptives; these effects, it must be emphasized, are not essential for efficient contraception. The metabolic changes caused by oral contraceptives are important because:

(1) Doctors need to know about the ways in which results of chemical tests may be modified by contraceptive treatment.

(2) The possibility has to be examined that the modifications might not regress when treatment is stopped.

A wide range of biochemical changes have been shown to occur as a result of treatment with oral contraceptives. These will be briefly reviewed under the following headings:

(1) Carbohydrate metabolism.
(2) Lipid metabolism.
(3) Plasma proteins.
(4) Plasma hormones.
(5) 'Liver function tests'.
(6) Plasma iron.

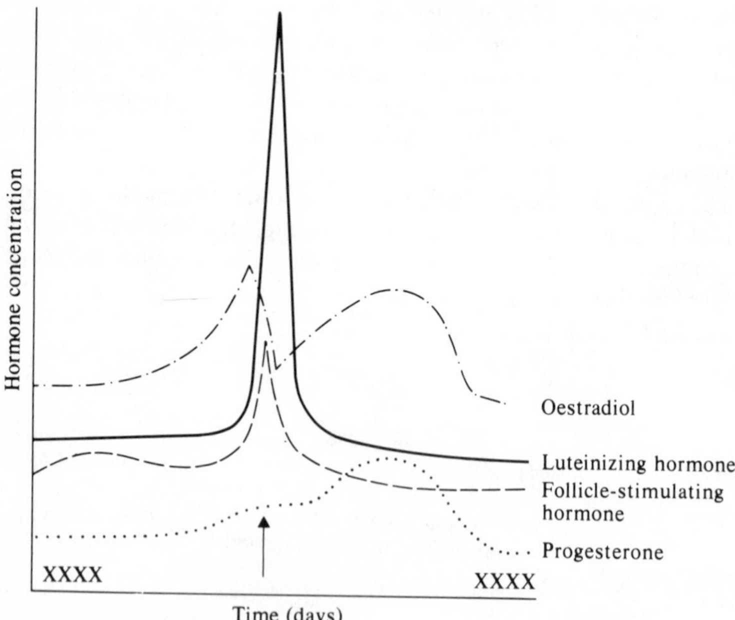

FIG. 20.1. Diagrammatic representation of the cyclical changes in plasma concentrations of gonadotrophins and the principal sex hormones in a normally menstruating female. (Reproduced with permission from London and Shaw, 1978). XXXX = Menses; ↑ = Time of ovulation.

CARBOHYDRATE METABOLISM

About 75% of women starting to take oral contraceptives develop a minor degree of carbohydrate intolerance. In the oral glucose tolerance test, fasting plasma [glucose] usually remains normal but tolerance is impaired. These changes do not progress during prolonged administration, and usually return to normal when the drug is stopped.

The changes are caused by the oestrogen component of oral contraceptives. They are usually insufficient to cause diagnostic problems. If marked abnormalities of glucose tolerance are found, it is advisable to stop the administration of oral contraceptives and then repeat the glucose tolerance test.

LIPID METABOLISM

Oestrogen-containing oral contraceptives tend to cause a small increase in plasma [triglycerides], mainly present in VLDL, but occasionally this increase may be marked. They also tend to cause a rise in plasma [HDL-cholesterol] and a fall in plasma [LDL-cholesterol]; progestogens have the opposite effects.

It is not certain whether these changes in plasma lipid concentrations have any relevance to the increased incidence of thrombotic episodes in women taking oral contraceptives. Alterations in coagulation factors and in the tendency of platelets to aggregate are probably more important.

The changes described above relate to the low-oestrogen preparations which are now used extensively. Higher doses of oestrogens may cause larger increases in plasma [triglycerides]. It has been suggested that the triglyceride changes are due to induction by oestrogens of the hepatic microsomal enzyme that is rate-limiting in the synthesis of triglycerides. The changes are usually no longer detectable a few months after stopping treatment.

Oral contraceptives also cause changes in the composition of bile, mainly an increase in biliary [cholesterol]. Since bile is normally supersaturated with cholesterol at certain times of day, the increased incidence of gall bladder disease in women who are taking oral contraceptives would seem to be simply explained. It is, however, not yet possible to predict the likelihood of gallstones developing in particular individuals.

PLASMA PROTEINS

The concentrations in plasma of many proteins may be altered by taking oral contraceptives. The changes are generally attributable to the oestrogen component and include:

(1) *Albumin.* Plasma [albumin] is often decreased.
(2) *Carrier proteins.* The concentrations of several carrier proteins are increased including sex hormone-binding globulin, transcortin, thyroxine-binding globulin, ceruloplasmin and transferrin. As a result, the plasma concentrations of compounds transported by these proteins also rise.
(3) *Coagulation factors.* The concentrations of fibrinogen, factor VII and Factor X may be increased.
(4) *Apoproteins of the lipoproteins.* The concentrations of these mostly show an increase.

Other changes include an increase in plasma [α_2-macroglobulin] and a decrease in plasma [haptoglobin]. The immunoglobulins are not affected.

PLASMA HORMONES (Other than gonadal steroids)

Increases of as much as 50% in plasma [cortisol], [total T4] and [total T3] often occur in women taking oral contraceptives, in response to the rise in concentration of hormone-binding proteins; occasionally still greater increases are observed. The oestrogen component of the contraceptives appears to be responsible for these effects, by promoting hepatic synthesis of the carrier proteins. The importance of these changes lies in their possible effect upon the interpretation of tests carried out in the investigation of suspected adrenal or thyroid dysfunction. Plasma [free cortisol], however, shows at most only a small increase and the urinary excretion of cortisol is likewise only slightly raised. It thus seems unlikely that the tissues are exposed to significantly increased amounts of hormonally-active cortisol.

Plasma [free T4] and [free T3] are not affected.

'LIVER FUNCTION TESTS'

Oral contraceptives occasionally cause frank jaundice, but minor asymptomatic changes in 'liver function tests' may be detected much more frequently. Three types of effect need to be considered:

(1) *Cholestatic effects.* Up to 40% of women taking oral contraceptives

have slight impairment of hepatic anion transport function, as revealed by measurement of plasma [bile acids] or by the bromsulphthalein or related tests (p. 178). Plasma [bilirubin] is usually not increased but alkaline phosphatase activity may be moderately increased due to induction of the hepatic isoenzyme. Induction of GGT has also been reported.

(2) *Idiosyncratic jaundice.* A small group of women responds to oral contraceptive treatment in a rather specific fashion, by developing overt clinical jaundice of a cholestatic type. Bilirubin is present in the urine, and plasma [bilirubin] and alkaline phosphatase activity are increased. The mechanism of this reaction is not clear. Oestrogens alone (even those occurring naturally) can have some effect on liver function, and some progestogens are structurally related to 17 α-methyl testosterone, well-known for its cholestatic effects. Jaundice has occasionally been reported in women taking a progestogen alone. There is an association between the development of idiosyncratic cholestatic jaundice, when taking oral contraceptives, and the occurrence of jaundice in pregnancy.

(3) *Aminotransferase activity.* Increases in the plasma activities of AST and ALT may occur soon after starting treatment with oral contraceptives. The increases are not marked. Also they are usually short-lived, reverting to normal although the contraceptive continues to be taken, but otherwise becoming normal again in most women soon after stopping the treatment.

PLASMA IRON

Oral contraceptive treatment is associated with increases in plasma [iron] of approximately 5 μmol/L (30 μg/100 mL). This change eliminates the difference between the sexes in plasma [iron]. Plasma TIBC rises by about 20 μmol/L after starting oral contraceptive treatment, and similar changes are found in plasma [transferrin]. Although plasma [iron] and TIBC both increase, it seems that they alter independently of one another. The progestogen component of oral contraceptives is thought to be responsible for both these changes.

Chemical changes occurring normally in pregnancy

Many changes discussed in the section on oral contraceptives are also observed as physiological changes occurring in pregnancy, where they reach their maximum in the third trimester. A few points specifically relating to pregnancy will be added here.

PLASMA VOLUME

During pregnancy, the plasma volume increases, sometimes by as much as 50%. These changes in volume are accompanied by alterations in the concentrations of many plasma constituents. For instance, plasma [urea], [Na^+] and [albumin] all tend to fall, and their reference values differ significantly from those for non-pregnant women, e.g. the upper reference value for plasma [urea] in pregnancy is 6.0 mmol/L (6.6 mmol/L in non-pregnant women). Not all plasma constituents show falls in concentration during pregnancy, however, and some increase; for instance, plasma [triglycerides] increase throughout pregnancy.

CARBOHYDRATE METABOLISM

There are special problems associated with the diagnosis of diabetes mellitus in pregnancy (p. 226).

PLASMA PROTEINS

There is a significant and quite marked fall in plasma [total protein] and [albumin] during pregnancy. These changes occur mostly during the first trimester.

Several proteins characteristic of pregnancy have been described, the best known examples being human placental lactogen (HPL) and pregnancy-specific β_1-glycoprotein (PSβ_1G). The concentrations of these proteins in maternal plasma correlate reasonably well with the stage of gestation, and measurements of both these proteins have been advocated as tests of fetoplacental function (see below).

PLASMA HORMONES

There are large increases in plasma [total T4], due to the increased plasma [thyroxine-binding proteins]. Plasma [free T4], on the other hand, tends to fall during pregnancy, particularly if it is measured by the Amerlex Free T4 RIA kit (p. 342), but usually this is without pathological significance.

Hyperthyroidism is particularly common in women in the 25–50 age-group, and the need to consider the diagnosis occurs quite frequently in pregnant women. It is important that plasma [free T4] rather than [total T4], or plasma [free T3] rather than [total T3], be measured in these patients if potentially misleading results are to be avoided.

There are large increases in plasma [cortisol] due to increased [cortisol-binding globulin]. However, there is also an increase in plasma [free cortisol] and in the 24-h urinary excretion of cortisol, probably due to the production of an ACTH-like substance by the placenta. This may help explain why pregnant women often show intolerance of glucose, and occasionally Cushingoid features.

PLACENTAL ALKALINE PHOSPHATASE

Measurements of alkaline phosphatase activity are of less diagnostic value in the investigation of hepatobiliary disease in pregnancy, especially in the third trimester, than at other times.

In pregnancy, there is an additional isoenzyme of alkaline phosphatase in plasma, originating from the placenta. Total plasma alkaline phosphatase activity is often increased to two, or sometimes three, times the activity present in non-pregnant women. The placental component can be readily measured since it is not destroyed when plasma is heated at 65°C for 10 min. Measurement of heat-stable alkaline phosphatase has been used as a test of placental function (p. 409).

PLASMA IRON

Plasma [iron] normally falls in pregnancy, mainly due to changes in plasma volume, unless iron supplements are taken. However, if pregnant women are treated with iron supplements, the tendency for plasma [iron] to fall is largely prevented. Indeed, if the changes in plasma volume are taken into account, treatment with iron results in an increase in the total *content* of iron in the plasma.

There is an increase in plasma [transferrin], so plasma TIBC increases and the percentage saturation of the iron-binding capacity falls unless iron supplements are taken.

Maternal complications in pregnancy

GLYCOSURIA AND PROTEINURIA

When women attend the ante-natal clinic, urine specimens should be routinely tested for glucose and protein.

If a specific test for glucose is used, detection of glycosuria does not necessarily indicate the presence of diabetes mellitus; the abnormality often disappears after delivery. If a non-specific test for reducing

14

substances is used, a positive result may indicate the presence of lactose rather than glucose; lactosuria has no pathological significance.

Proteinuria, if detected, may be due to urinary tract infection, or the first evidence of toxaemia. Patients with pre-eclampsia may develop impairment of renal function with increase of plasma [urea] as the condition worsens, or as a result of vomiting and dehydration. A plasma [urea] of 7.0 mmol/L (42 mg/100 mL) should be regarded as definitely abnormal, because plasma [urea] is normally reduced in pregnancy due to the increase in plasma volume. Other tests to determine the severity of pre-eclampsia, and to provide an index of prognosis, include plasma [urate] which increases approximately in proportion to the severity of the illness.

JAUNDICE IN PREGNANCY

Intra-hepatic cholestasis is an uncommon feature of pregnancy. It is associated with the occasional occurrence of idiosyncratic jaundice in some women when taking oral contraceptives.

There is retention of bilirubin and bile acids, the retained bilirubin being mainly conjugated; there is also bilirubinuria. The increase in plasma alkaline phosphatase activity observed in these patients is partly hepatic, but is also partly placental.

THE FETUS AND THE PLACENTA

The placenta is the earliest fetal tissue to differentiate. It produces (human) chorionic gonadotrophin (HCG) and (human) placental lactogen (HPL) and several other proteins. Both HCG and HPL are synthesized by the trophoblast. The peak of HCG secretion occurs early in pregnancy, and detection of HCG is widely used as a sensitive test for early recognition of pregnancy. The peak of HPL synthesis occurs much later in pregnancy.

There are considerable structural similarities between HCG and the pituitary gonadotrophins, FSH and LH. The functions of HCG include the maintenance of the corpus luteum and the stimulation of steroid production by the corpus luteum and the placenta. There are also structural similarities between HPL, prolactin and growth hormone.

The placenta contains a range of enzymes involved in the production of steroid hormones; oestrogen synthesis by the placenta begins at six weeks' gestation. The placenta is the source of the large quantities of progesterone produced in pregnancy.

Several reaction sequences involved in steroid metabolism in pregnancy require the participation of both the placenta and the fetus. Some steps in these pathways take place more readily in the fetus and some in the placenta. This results in a complex interplay of fetal and placental functions, with steroids transferring between the placenta and fetus, and *vice versa* (Fig. 20.2). Recognition of this interplay or interdependence has led to the concept of the fetoplacental unit.

Steroid measurements in pregnancy

Early in pregnancy the placenta takes over the production of progesterone from the corpus luteum and urinary pregnanediol excretion rises. Recurrent abortion may be caused by abnormalities of placental function, associated with low plasma and salivary [progesterone] and reduced urinary excretion of pregnanediol. Serial determinations of plasma or salivary [progesterone], or less often nowadays of daily urinary pregnanediol output, can be used to monitor

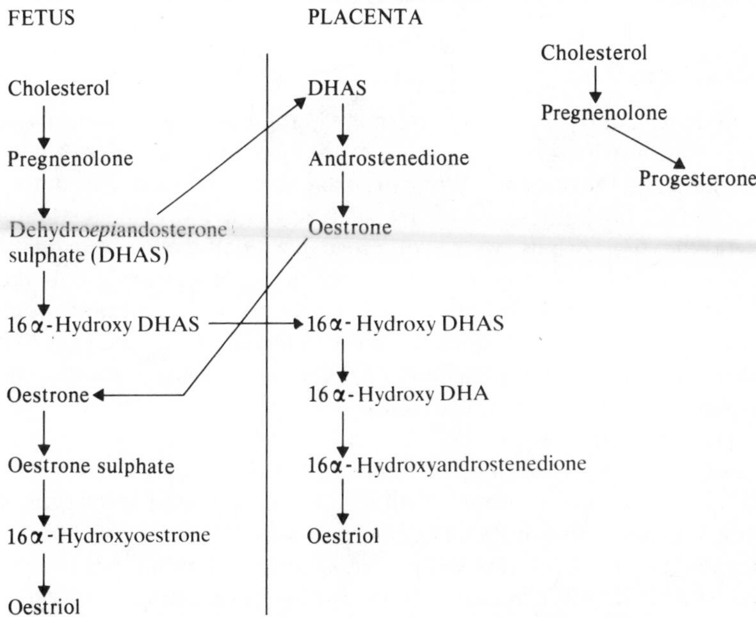

FIG. 20.2. Formation of steroid hormones by the fetus and the placenta (for simplicity, the formation of oestradiol has been omitted). Note that the placenta and the fetus can each form oestrogens, if provided with the appropriate precursors, their complements of enzymes being interdependent.

women with a history of recurrent abortion. The collection of the specimens of saliva is described later (p. 415).

OESTROGENS AND FETOPLACENTAL FUNCTION

Oestriol is the oestrogen produced in greatest amount during pregnancy. Its formation from cholesterol depends on a complex set of interrelated metabolic steps, some dependent on enzymes located in the fetal adrenals and liver, others on enzymes located in the placenta. The placenta can synthesize oestrone and oestradiol if supplied with dehydro*epi*androsterone sulphate from the maternal circulation. Much of the oestrone produced by the placenta is metabolized further, to oestriol, in the fetus (Fig. 20.2)

Oestrogens produced by the fetus and placenta escape into the maternal circulation and are excreted in the urine. During normal pregnancy the maternal excretion of total oestrogens (consisting mainly of oestriol) rises rapidly. The rate of increase is such that the amount excreted doubles approximately every two and a half weeks until eighteen weeks' gestation, after which the rate of rise diminishes.

Urinary oestrogens

The daily rate of excretion, determined either as total oestrogens or better as oestriol, gives a measure of the functional integrity of both the fetus and the placenta. While the excretion rate rises rapidly during pregnancy, it is rather variable from day to day in individual patients.

In the first twenty weeks of pregnancy, daily measurement of urinary oestriol may be of value after threatened abortion. If the fetoplacental unit is no longer viable, rates are low and remain low. Also excretion rates close to the lower reference value indicate that fetoplacental function may be failing, with an increased risk that the pregnancy will not continue to term.

During the third trimester, a decision to terminate a pregnancy may need to be made because of indications that placental function is failing. The value of a single result for 24-h urinary total oestrogens or oestriol excretion is difficult to interpret and, because of day-to-day variability in the excretion rate, *serial* determinations need to be performed. If these reveal a sudden drop in excretion rate, which thereafter continues to fall or at least fails to recover, this information is of considerable value in deciding whether labour should be induced. There is a moderately good correlation between the urinary oestrogen excretion and fetal maturity.

Persistently low urinary oestrogen excretion may be due to fetal death or imminent fetal death, anencephaly, fetal adrenal hypoplasia, and placental sulphatase defect. An iatrogenic cause to be excluded is treatment with steroids.

Urinary oestrogen : creatinine ratio. Difficulties associated with obtaining complete 24-h collections of urine have led to investigation of alternative ways of assessing oestrogen production, as an index of fetoplacental function. The oestrogen : creatinine ratio in a random specimen does not correlate well with total daily oestrogen output, but the ratio determined on an early morning specimen seems to be a reliable means of predicting fetal death, if serial observations are made.

Plasma oestrogens

With the introduction of specific RIA methods, measurement of plasma [oestriol] has tended to replace urinary oestriol or total oestrogen determinations as predictors of risk to the fetus.

The wide range of plasma [oestriol] in healthy pregnant women, and the additional effects of considerable diurnal and day-to-day variation in individuals, mean that plasma [oestriol] needs to be measured *serially,* in order to detect a trend, before basing a clinical decision on the results of these measurements. Serial determinations of plasma [HPL] are the main alternative to oestriol (or total oestrogen) measurements.

Proteins of trophoblastic origin

There are several pregnancy-specific proteins, all of which normally originate in the trophoblast. Examples include HCG, HPL, $PS\beta_1G$ and heat-stable alkaline phosphatase (HSAP).

Chorionic gonadotrophin

This is produced by the syncytiotrophoblast of the placenta, secreted into the maternal circulation and excreted in maternal urine from a very early stage in pregnancy. Measurement of urinary HCG has long been used as a test for pregnancy, peak output occurring between the seventh and tenth weeks; it is also useful in the management of threatened abortion in the first half of pregnancy. There is a second small peak in urinary HCG excretion about the thirty-fourth week, but HCG measurements have no place in monitoring placental function in the second half of pregnancy.

Although described as a 'pregnancy-specific protein', HCG is produced not only in pregnancy, but also by *tumours of the trophoblast,* which can occur in males and females. These tumours include hydatidiform mole and choriocarcinoma, both of which may secrete HCG in very large amounts. A female who is found to be excreting HCG, and who is not pregnant, most frequently has a tumour of the trophoblast while in males testicular teratoma is the commonest source. However, HCG is also a non-specific marker for a number of carcinomas, and HCG or HCG-like proteins are secreted by a minority of non-trophoblstic, non-gonadal neoplasms, especially those of gastrointestinal origin.

Placental lactogen

This is produced by the syncytiotrophoblast from an early stage of pregnancy, the output being proportional to placental mass. It can be detected by about six weeks' gestation and the output rises progressively thereafter until the last few weeks of pregnancy; the high output is then normally sustained until term. In the last few weeks of pregnancy, the placenta may produce about 12 g HPL/day, and the peak plasma concentration is about 8 mg/L. Plasma HPL measurements may be of value in:

(1) *Early pregnancy.* In patients between ten and twenty weeks' gestation, with bleeding due to threatened abortion, *serial* (daily) measurements give a useful indication of the likely outcome of the pregnancy. Patients with a plasma [HPL] that is low for the stage of gestation, and which remains low, tend to abort.

(2) *Late pregnancy,* as an index of placental function at this time. There is a considerable variation in plasma [HPL] from one woman to another, and marked overlap exists between results for healthy women and women who develop complications of pregnancy. As it is difficult to reach a definite conclusion based on a single observation, *serial* (daily) measurements are usually necessary.

The correlation between measurements of plasma [HPL] and oestrogen measurements, e.g. plasma [oestriol] or urinary oestriol excretion, is poor. This could be due to the fact that oestrogen production depends on the functioning of the fetoplacental unit, whereas HPL is purely a placental product. Present practice seems generally to be to perform serial measurements of either plasma [HPL] or plasma [oestriol], on a daily basis, depending on local preference.

Both these tests avoid the inconvenience and potential inaccuracy of measurements that depend on 24-h collections of urine.

Other trophoblastic proteins

Pregnancy-specific β_1*-glycoprotein,* a product of the syncytiotrophoblast, is synthesized in increasing amounts during pregnancy. Plasma [PS β_1G] appears to correlate well with the length of gestation, and concentrations of 200 mg/L have been reported at term. It can be detected in maternal blood as early as fourteen days' gestation. However, technical difficulties mean that the test has not yet been properly evaluated.

Heat-stable alkaline phosphatase (HSAP) is no longer advocated as a test of placental function, since no constant pattern of abnormal results has been described in patients with developing placental insufficiency. However, placental-type (heat-stable) isoenzymes of alkaline phosphatase are produced by some tumours and HSAP measurements are sometimes of value in their investigation (p. 460).

PRENATAL DIAGNOSIS OF FETAL ABNORMALITIES

Chemical tests for this purpose can be subdivided into screening tests, suitable for application to all pregnancies, and tests which are only appropriate for use in at-risk pregnancies.

A clear distinction needs to be drawn between tests which are non-invasive and which carry no risk to mother and fetus, e.g. maternal blood sampling or ultrasound examination, and techniques such as amniocentesis that are invasive.

Amniocentesis carries a risk to the fetus estimated to be between 0.5% and 2% loss. To justify the performance of amniocentesis, therefore, there must be a high index of suspicion that an abnormality is present in the fetus. An appropriate course of action, based on the findings obtained by amniocentesis, must be available in the event that an abnormality is discovered.

INHERITED METABOLIC DISEASE

Amniocentesis can be carried out about the fifteenth week of pregnancy for (1) cytogenetic reasons (e.g. for the detection of Down's syndrome), or (2) the detection of the appropriate *one* of a wide range of rare untreatable metabolic disorders in families where there has already been an index case and the nature of the defect to be sought in the current pregnancy has previously been identified. These studies take time, as they mostly require cells to be cultured from the amniotic fluid before the requisite specialized investigations, principally enzymic activity measurements in the case of chemical tests, can be carried out. Recently, relatively rapid techniques for enzyme analysis on single cells have been described, but these are not generally available.

The metabolic disorders that have been characterized by finding a deficiency in enzymic activity in cultured amniotic fluid cells include a few X-linked disorders, e.g. the Lesch-Nyhan syndrome (p. 334), and a large number of autosomal recessive disorders of which Tay-Sachs disease (G_{M1} gangliosidosis type 1, characterized by a deficiency of hexosaminidase A) is perhaps the best known. The investigations are complex and only performed in a few centres.

It is possible to diagnose congenital adrenal hyperplasia (p. 377) by amniocentesis late in pregnancy, at least when due to deficiency of 21-hydroxylase or 11β-hydroxylase. Amniotic fluid [17α-hydroxyprogesterone] may be measured; this becomes abnormal late in pregnancy when CAH is present. However, there is little to be gained from diagnosing CAH before birth, as the condition is usually recognizable in the neonatal period and is then amenable to steroid replacement treatment.

NEURAL TUBE DEFECTS

The fetal liver begins to produce α_1-fetoprotein (AFP) from the sixth week of gestation onwards, and the highest concentration of this protein in fetal serum is reached at about thirteen weeks, after which it falls progressively until term. If the fetus has an open neural tube defect, i.e. open spina bifida or anencephaly, at sixteen weeks' gestation increased amounts of AFP are present in amniotic fluid in about 95% of cases, and in maternal serum in about 80% of cases.

Screening for neural tube defects

It is possible to use maternal serum (or plasma) AFP (MSAFP) as a screening test for neural tube defects. The test should be carried out at 16–18 weeks' gestation with a view to identifying those women who should be further investigated by ultrasound and then, in appropriate cases, by amniocentesis. If the diagnosis of open neural tube defect is confirmed before the twentieth week, termination of pregnancy can be offered.

It is essential to know the gestational date for the interpretation of MSAFP measurements, since [MSAFP] varies considerably with the length of gestation. The best discrimination between normality and abnormality is obtained at 16–18 weeks' gestation. There is, nevertheless, a considerable overlap between [MSAFP] in normal pregnancies and in women carrying affected fetuses. If an upper reference value at the 95th percentile for the normal population (a value which is about 2.5 times the median value for this population) is selected as the level above which further investigations should be performed, less than 10% of women with 'abnormal' [MSAFP] so defined will, in fact, be carrying affected fetuses. In other words, the MSAFP test can do no more than identify a 'high risk' group of women.

Other causes of high [MSAFP] include incorrect gestational dates, multiple pregnancy and some rare non-neurological fetal abnormalities (e.g. oesophageal or duodenal atresia, exomphalos).

It is possible to miss abnormal fetuses if the length of gestation has been overstated or overestimated. This means that the result for [MSAFP] will be interpreted in relation to data for an inappropriate gestational date, and is likely to appear normal whereas the result might have been recognized as abnormal if related to data for the correct length of gestation. The main reason for missed cases, however, is failure to book in at the antenatal clinic until it is too late for MSAFP screening.

Follow-up examinations

Whether or not [MSAFP] is raised, ultrasound examination and amniocentesis are indicated in women who have previously had a pregnancy affected by a neural tube defect. These women have a higher risk of having another affected fetus.

If an abnormal result for [MSAFP] is reported, it is usual to repeat the measurement and, if the result is again abnormal, to proceed to ultrasound examination. Ultrasound examination detects multiple pregnancy, an important physiological cause of 'abnormal' [MSAFP], and will also often detect neural tube defects. Ultrasound examination helps with placental localization for subsequent amniocentesis, if this is to be performed. Amniocentesis would not be indicated, for instance, if the ultrasound examination had revealed a twin pregnancy, or an anencephalic fetus.

Two tests for neural tube defects are normally carried out, if amniocentesis is performed. These are measurements of amniotic fluid [AFP] and acetylcholinesterase activity.

Amniotic fluid [AFP] is increased in over 95% of cases where there is a neural tube defect. However, false positive results may be obtained when the fluid is blood-stained, or when certain other defects (e.g. exomphalos) are present.

Amniotic fluid acetylcholinesterase activity is increased when the fetus has a neural tube defect, presumably due to leakage of the enzyme (which is relatively specific for the nervous system) through the defect. The amniotic fluid is examined by electrophoresis, a qualitative test which serves to distinguish acetylcholinesterase from the relatively non-specific cholinesterase (p. 153) that is normally present in amniotic fluid. This test is generally considered to be more sensitive and specific than amniotic fluid AFP measurements. It is not affected by contamination of the fluid with maternal blood. However, the activity of this enzyme in amniotic fluid may be increased if there is fetal exomphalos.

Combination of test results.

If the results for all three tests, i.e. [MSAFP] on two occasions, amniotic fluid [AFP] and amniotic fluid acetylcholinesterase, are all abnormal, the diagnosis of neural tube defect can be made with a fairly high degree of confidence. However, certain other uncommon congenital abnormalities (e.g. exomphalos) may yield a similar pattern of abnormal test results.

TESTS FOR FETAL WELL-BEING AND MATURITY

Amniocentesis may be performed late in pregnancy to investigate fetal well-being and to detect potentially serious but treatable conditions. This allows preparations to be made for treatment to be started immediately after birth, or sometimes even before birth. Several different chemical examinations can be carried out on the amniotic fluid:

Creatinine is detectable in amniotic fluid from twenty weeks' gestation onwards; its concentration rises progressively as term approaches. Measurement of amniotic fluid [creatinine] gives an approximate estimate of fetal maturity, to within four weeks. Closer estimation of maturity by this determination is not possible owing to the wide variation in amniotic fluid [creatinine] normally observed.

Lecithin. Fluid within the fetal lung tissues normally contributes to the formation of amniotic fluid. Before its discharge from the alveoli, this fluid is in contact with alveolar epithelial cells; these produce a surface-active material which contains a large amount of phospholipid. Amniotic fluid [lecithin] can be used as an index of fetal lung maturity, and to assess the likelihood than an infant will develop the respiratory distress syndrome. An alternative measurement, the lecithin-sphingomyelin ratio (L:S ratio) in amniotic fluid, is not often performed nowadays.

Bilirubin. Small amounts are normally detectable in amniotic fluid. However, if a fetus is affected by Rhesus haemolytic disease, the concentration rises, often in direct relation to the severity of the haemolytic process. False predictions of the risk to the fetus, based on measurements of amniotic fluid [bilirubin], have been reported in an appreciable percentage of patients. Errors in the assessment of gestational age adversely affect the interpretation; the specimen of amniotic fluid should be examined also for characteristics that vary with the stage of gestation, e.g. the cytological appearance of desquamated cells.

Monitoring the fetus during labour

The most widely available chemical investigation is blood [H^+], measured on samples obtained from the fetal scalp. As an indicator of fetal hypoxia, the recording of [H^+] over 63 nmol/L (pH below 7.20) is a serious finding. These measurements are, however, less direct indicators of fetal oxygenation than fetal blood $P\text{O}_2$ which can be continuously monitored transcutaneously, once the cervix is sufficiently dilated.

The increasing availability in labour wards of equipment for measuring blood [H^+], $P\text{CO}_2$ and, in due course, other indices of fetal well-being places growing emphasis on the reliable performance by ward staff of measurements hitherto regarded as the province of trained laboratory workers. Staff training in the use and maintenance of ward–based equipment, in the recording of results and other essential data needed for retrospective assessment of the reliability of observations, are essential. In many hospitals it will be necessary to depend on laboratory staff for regular checks on the equipment, and to provide a back-up service in the event of a breakdown. It is worth setting up a collaborative quality control programme between the ward and the laboratory.

SUBFERTILITY IN FEMALES

This section will be concerned with endocrine tests of gonadal function, but with neither anatomical nor gametogenic abnormalities. The normal physiology of the menstrual cycle has been briefly outlined (p. 397).

Amenorrhoea may be due to primary (gonadal) or secondary (pituitary) causes. In discussing the place of chemical measurements in the investigation of the hypothalamic-pituitary-gonadal (HPG) axis, the assumption is made that pregnancy has been excluded as the cause of amenorrhoea.

Before embarking on any of the test procedures discussed below, plasma or salivary [progesterone] or the 24-h urinary excretion of pregnanediol should be determined on or close to the twenty-first day of the menstrual cycle in patients complaining of subfertility but who are nevertheless menstruating. These preliminary tests are performed so as to determine whether ovulation is occurring and a functional corpus luteum forming.

BASAL TESTS OF THE HPG AXIS

These include measurements of plasma [oestradiol-17β] and [oestriol], the 24-h urinary excretion of either of these oestrogens or of total urinary oestrogens. If the results are abnormally low, this confirms the presence of gonadal failure, but does not differentiate between primary and secondary failure.

To help differentiate between the ovaries and the pituitary as the site of the disease, plasma [FSH] and [LH], or urinary excretion of FSH and LH, and plasma [prolactin] are the main measurements to be performed next.

Gonadal failure due to disease of the gonads

The ovaries fail to respond to endogenous gonadotrophins. Their failure to produce oestrogens and progesterone means that there is lack of feedback inhibition of the hypothalamus and pituitary. Plasma [FSH] and [LH] are, therefore, increased in these patients.

Gonadal failure due to non-gonadal causes

This may be due to disease of the hypothalamus or the pituitary, or both. Plasma [FSH] and [LH] are low or low-normal in most of these patients; plasma [oestradiol-17β] and [progesterone] are consequentially low.

In the polycystic ovary (Stein-Leventhal) syndrome there is amenorrhoea or oligomenorrhoea, ovulatory failure, hirsutism and sometimes virilism. The primary pathological abnormality is thought to lie in the hypothalamus or the pituitary. Plasma [LH] and [testosterone] may both be increased, especially testosterone, and there is frequently hyperprolactinaemia. Plasma [oestrogens] are consistently low, and there is no evidence of cyclical pattern if measurements are made at weekly intervals.

About 20% of women with secondary amenorrhoea and ovulatory failure have *hyperprolactinaemia.* Some of these patients may have galactorrhoea. Patients with hyperprolactinaemia may respond to treatment with bromocriptine; the response can be monitored by serial measurements of plasma [prolactin].

Detection of ovulation

The measurement of 24-h urinary excretion of oestrogens or of pregnanediol, as a means of detecting the occurrence of ovulation, has been largely or entirely replaced by daily determination of plasma or salivary [progesterone]. These measurements are used to detect ovulation, either occurring spontaneously or as a result of ovarian stimulation (ovulation induction), e.g. by clomiphene.

If these studies are to be based on serial determinations of salivary [progesterone], the patient is given a series of labelled tubes and is instructed to spit into one of these each morning. The date is noted on each label and the tubes are stored in a refrigerator until brought to the laboratory for analysis.

DYNAMIC TESTS OF THE HPG AXIS

Many tests have been devised, including the following:

(1) Gonadotrophin-releasing hormone (Gn-RH) stimulation test.
(2) Clomiphene stimulation test.

These tests will be briefly discussed, but readers are referred to gynaecology textbooks and to original articles for more detailed descriptions of these and other tests.

Gn-RH stimulation test

This test can be used for the further investigation of patients with gonadal failure of non-gonadal origin; plasma [FSH] and [LH] are measured before and 20 and 40 min (or 20 and 60 min) after injection of 100 μg Gn-RH. It is questionable whether this stimulation test adds to the information gained from measurements of basal plasma [FSH] and [LH].

Clomiphene stimulation test

Clomiphene stimulates release of Gn-RH by occupying oestrogen-receptor sites in the hypothalamus, thereby blocking the normal negative feedback effects of oestrogens. Given daily for five days (the recommended dosage varies from 50 to 200 mg), clomiphene provides an alternative means of stimulating the pituitary to produce gonadotrophins.

The response to clomiphene stimulation can be monitored by measuring plasma [FSH] and [LH], but both plasma [oestradiol] or urinary [oestrogens] and plasma or salivary [progesterone] provide better methods of assessment. If a patient fails to respond to clomiphene but responds to stimulation with Gn-RH, this is evidence for the presence of a hypothalamic lesion.

Scheme of endocrine tests for investigating female subfertility

The foregoing account must not be used to justify an uncritical approach to the use of investigations. Instead, the following step-by-step procedure is suggested:

(1) *Plasma or salivary [progesterone]* or 24-h urinary pregnanediol excretion about the twenty-first day, in patients who are still having menstrual cycles. These measurements will help to support the clinical observations (e.g. basal temperature charts), to show whether the patient is ovulating.

(2) *Plasma [oestradiol-17β]*, or 24-h urinary excretion of oestrogens. Low values confirm that there is gonadal failure, primary or secondary.

(3) *Plasma [FSH] and [LH]*. If these are increased, the patient probably has primary ovarian failure. If the results are low or normal, proceed to (4).

(4) *Plasma [prolactin]*. If the result is high, confirm that the patient is not under stress and is not taking oral contraceptives or other drugs. If plasma [prolactin] is low or normal, proceed to (5).

(5) *Dynamic tests*. If the response to the Gn-RH test is subnormal, this suggests that the subfertility is due to pituitary failure, which may be primary or secondary to disease of the hypothalamus. A clomiphene test may then help to determine the level of the lesion.

FURTHER READING

BARSON, A.J., editor. (1981). *Laboratory Investigation of Fetal Disease*. Bristol: Wright.

BROCK, D.J.H. (1982). Early diagnosis of fetal defects. *Current Reviews in Obstetrics and Gynaecology*, **2.** Edinburgh: Churchill Livingstone.

CHARD, T. and KLOPPER, A. (1982). *Placental Function Tests*. Berlin: Springer-Verlag.

DAVISON, J.M. (1983). The kidney in pregnancy: a review. *Journal of the Royal Society of Medicine*, **76,** 485–501.

FERGUSON-SMITH, M.A., editor (1983). Early Prenatal Diagnosis. *British Medical Bulletin*, **39,** 301–404.

HURLEY, R., editor (1976). The pathology of pregnancy. *Journal of Clinical Pathology*, **29,** Supplement (Royal College of Pathologists), **10.**

JEFFCOATE, S.L. (1978). Diagnosis of hyperprolactinaemia. *Lancet*, **2,** 1245–1247.

Leading article (1978). Knowing the oxygenation of the fetus. *British Medical Journal*, **1,** 671.

LONDON, D.R. and SHAW, R.W. (1978). Gynaecological endocrinology. In *Recent Advances in Endocrinology and Metabolism*, **1,** 91–110. Ed. J.L.H. O'Riordan. Edinburgh: Churchill Livingstone.

SANDLER, M. and BILLING, B., editors (1970). The pill. Biochemical consequences. *Journal of Clinical Pathology*, **23,** Supplement (Association of Clinical Pathologists), **3.**

STRIKE, P.W. and SMITH, J. (1982). Neural-tube defect risk assessment for individual pregnancies using alphafetoprotein and acetylcholinesterase test results. *ibid.*, **35,** 1334–1339.

WARREN, M.P. (1973). Metabolic effects of contraceptive steroids. *American Journal of the Medical Sciences*, **265,** 4–21.

WATHEN, N.C., PERRY, L., LILFORD, R.J. and CHARD, T. (1984). Interpretation of single progesterone measurement in diagnosis of anovulation and defective luteal phase: observations on analysis of the normal range. *British Medical Journal*, **288,** 7–9.

Chapter 21

Inherited Metabolic Disorders

Many metabolic diseases are due to an inborn error, and the spectrum of inherited metabolic disorders subdivides into two main groups:

(1) Disorders due to the addition or deletion of chromosomes or parts of chromosomes.

(2) The Mendelian disorders, in which the primary defect is probably in the structure of a gene.

The first group has an incidence of about 6 per 1000 live births, and leads to clinically significant abnormalities (e.g. Down's syndrome, Turner's syndrome) in about 3 per 1000 live births, whereas the second group is composed of a wide variety of conditions which are mostly much less common.

Clinical chemistry at present plays little part in the diagnosis and management of chromosomal abnormalities. On the other hand, chemical investigations (often very detailed in nature) are needed for the proper characterization of many of the genetic disorders. Their existence, even as rarities, has yielded a lot of information about the details of metabolic pathways in man.

This chapter will be concerned only with the second, less common, genetic group of inherited metabolic disorders. Brief mention is made elsewhere of the hormonal effects of Klinefelter's syndrome (p. 383).

Effects of inherited metabolic disorders

The primary defect in each of the Mendelian disorders is probably a change in the base sequence of a gene, which is composed of deoxyribonucleic acid (DNA). This change affects the synthesis of a protein, which may be a structural protein, or a transport protein, or an enzyme, etc. Many consequences may follow, depending on the functions of the protein affected by the primary alteration in the gene. These consequences can be illustrated by discussing an inherited defect that gives rise to a marked reduction (rarely a complete absence) of activity of an enzyme (E) involved in the metabolism of a compound A, at one stage in the following reaction sequence:

$$A \xrightarrow{E_1} B \xrightarrow{E_2} C \xrightarrow{E_3} D \xrightarrow{E_4 - E_n} R$$

We shall make the important assumption that the enzymic defect resulting from the disorder has made the reaction catalysed by the affected enzyme into the rate-limiting step in the metabolic pathway.

PHENYLKETONURIA

This is one of the disorders that is nowadays specifically sought in screening programmes carried out in the neonatal period. Unless recognized and treated from a very early age, phenylketonuria (PKU) causes progressively severe and irreversible mental deterioration.

The normal pathways of phenylalanine metabolism are shown in Fig. 21.1. The major pathway, involving conversion to tyrosine as its first step, depends on the hepatic enzyme phenylalanine hydroxylase. The pathways that give rise to the deaminated metabolites are normally only of minor importance. In the classical form of PKU,

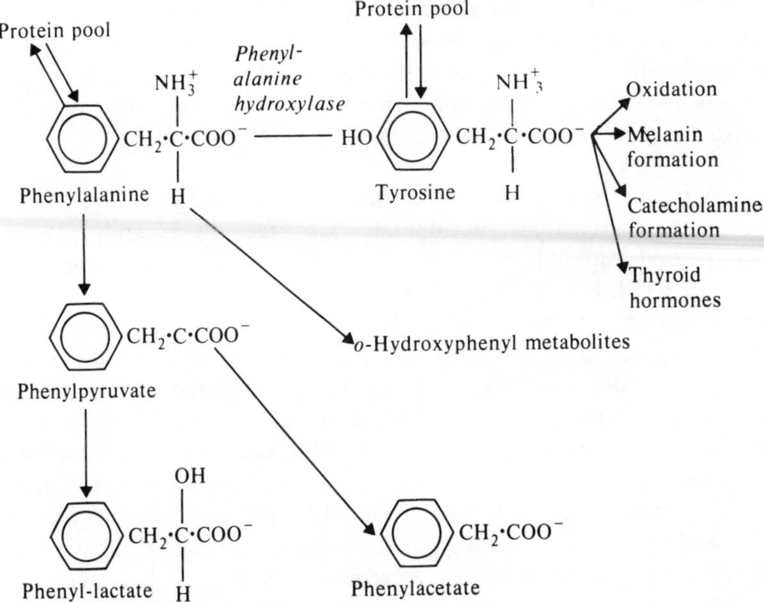

FIG. 21.1. *Metabolism of phenylalanine.* In the classical form (the commonest form) of phenylketonuria, the activity of phenylalanine hydroxylase is reduced and minor metabolites of phenylalanine are excreted in greater amount (e.g. phenylpyruvate, *o*-hydroxyphenylacetate).

phenylalanine hydroxylase activity is either undetectable or very much reduced.

Phenylalanine hydroxylase requires tetrahydrobiopterin (BH_4) as cofactor, BH_4 being oxidized to the quinonoid form of dihydrobiopterin (q-BH_2) at the same time as phenylalanine is converted to tyrosine. Dihydrobiopterin reductase (DHBR) normally reconverts most of the q-BH_2 that has been formed, but which is very unstable, back to BH_4. Another enzyme, dihydrobiopterin synthetase (DHBS), normally makes good any q-BH_2 that is lost due to its instability.

About 3% of patients with abnormalities of phenylalanine metabolism have a deficiency of DHBR or DHBS, or some other defect that is different from the deficiency of phenylalanine hydroxylase. All these metabolic errors carry a poor prognosis unless they are recognized, and the rarer forms need to be differentiated from the classical form of PKU (once the presence of PKU has been diagnosed) as they require different forms of treatment. We shall restrict further discussion of PKU to the classical form of the disease, and use this example to illustrate many of the principles that help to explain the diverse effects of inherited metabolic disorders. In terms of metabolic complexity, the classical form of PKU is one of the simplest examples to consider. It corresponds to a block at the first stage, A→B, in the general reaction sequence shown above. The following effects may be observed:

(1) *Accumulation of substrate.* Phenylalanine, the substrate of the blocked reaction, accumulates in the liver and plasma [phenylalanine] is much increased, unless dietary phenylalanine is restricted.
(2) *Reduced formation of product.* Tyrosine formation from phenylalanine is severely affected in these patients. However, tyrosine deficiency can be avoided if the normal dietary intake of tyrosine is adequately supplemented.
(3) *Alternative paths of metabolism.* Phenylalanine is metabolized along its alternative pathways, and there is increased urinary excretion of phenylpyruvate, phenyl-lactate and phenylacetate, as well as various *o*-hydroxyphenol metabolites (e.g. *o*-tyrosine and *o*-hydroxyphenylacetate). Restriction of dietary phenylalanine reduces the amounts of these various metabolites that are produced.
(4) *Effects on other reactions.* Accumulation of phenylalanine and its metabolites inhibits the transport of other amino acids into the

liver and brain, and their concentrations in these tissues fall. The activity of enzymes involved in the metabolism of other amino acids (e.g. tyrosine, tryptophan) may also be affected.

Other inherited metabolic disorders may exhibit additional features, discussed now in relation to a block at the stage C→ D, as an example of a block part-way along a metabolic sequence:

(1) *Reduced formation of an important product* (e.g. R) which is normally derived solely from the product (D) of the blocked reaction. In congenital adrenal hyperplasia (CAH) due to 21-hydroxylase deficiency (Fig. 18.6, p. 379), there is reduced formation of cortisol as a result of the impaired production of 11-deoxycortisol.

(2) *Interference with feedback control mechanisms* where the immediate product (D) or a more remote product (R) of the blocked reaction normally exerts a control function, directly or indirectly, on one of the steps preceding the site of the enzyme block. This effect is exemplified by CAH, and by blocks in the pathway leading to the formation of thyroid hormones (Fig. 17.1, p. 337). Failure to synthesize cortisol and thyroid hormones in normal amounts leads, respectively, to increased output of ACTH and TSH.

(3) *Accumulation of precursors* (e.g. A) of the substrate of the blocked reaction. This occurs if the preceding reactions are freely reversible, as in some of the glycogen storage diseases, e.g. Type I, in which glycogen accumulates as a result of glucose-6-phosphatase deficiency (Fig. 11.1, p. 237).

DIAGNOSIS OF INHERITED METABOLIC DISEASE

Problems of diagnosis are much influenced by the age of the patient at the time the illness presents, and by the nature of the onset, whether acute or insidious. Diagnosis may be greatly facilitated by there being a family history of inherited metabolic disease, especially if the precise defect has previously been established in a relative of the patient. However, in this section on *diagnosis,* we shall assume that the patient being investigated is the index case. We shall consider *screening* in a later section.

Diagnosis in the neonatal period

A wide variety of inherited metabolic diseases may present acutely in the first few days or weeks of life. Table 21.1 lists examples of some of the less uncommon conditions, but many more have been described. Some of the conditions listed have been identified with a single enzymic defect, but some are not yet fully understood.

THE NEED FOR OTHER INVESTIGATIONS

Inherited metabolic disorders represent only one group of many possible causes for a newborn child failing to thrive or being seriously ill. Apart from haematological, microbiological and radiological investigations, there are several chemical measurements that may be required, some for diagnostic and others for supportive reasons.

Side-room tests

It is worth testing a urine specimen in the side-room for reducing substances (e.g. Clinitest, Ames Co.) and phenylketones (Phenistix, Ames Co.) as long as this is not done at the expense of providing the laboratory with an adequate specimen for investigation.

Clinitest may detect fructose or galactose, in addition to glucose, and a specific test for glucose in urine can also be performed in the side-room (p. 220). Laboratory confirmation of a provisional diagnosis of galactosaemia or hereditary fructose intolerance should follow reasonably quickly, but meanwhile appropriate dietary treatment can be started without detriment to the patient.

Testing urine with Phenistix in the neonatal period has tended to fall into disrepute, because it is a much less reliable means of detecting PKU at this time than is the Guthrie test (see below). However, it may give an abnormal colour in patients with PKU and certain other conditions, including maple syrup urine disease, tyrosinaemia, and histidinaemia. An abnormal result with Phenistix, therefore, suggests the need for more detailed laboratory investigation, but a normal result does not exclude this possible requirement.

Other observations to record from side-room tests on urine should include the results of tests for the presence of bilirubin and ketone bodies.

General tests on blood

Depending on the infant's clinical condition, these may include blood gas studies and measurement of plasma concentrations of bilirubin, calcium, creatinine or urea, glucose, potassium and sodium.

TABLE 21.1. Inherited metabolic disorders.
Examples of disorders that may produce acute illness, especially in the neonatal period (After Haan and Danks, 1981)

Metabolic group	Description	Site of block
Amino acid disorders	Maple syrup urine disease	Oxidative decarboxylation of branched-chain ketoacids corresponding to valine, leucine and *iso*leucine
	Non-ketotic hyperglycinaemia	Glycine decarboxylase system
	Hereditary tyrosinaemia	*p*-Hydroxyphenylpyruvic acid oxidase
Carbohydrate disorders	Galactosaemia	Galactose-1-phosphate uridylyltransferase
	Hereditary fructose intolerance	Fructose-1-phosphate aldolase
	Glycogen storage disease, Type I	Glucose-6-phosphatase
	Glycogen storage disease, Type II	Lysosomal α-glucosidase
Organic acid disorders	*Iso*valeric acidaemia	*Iso*valeryl-CoA dehydrogenase
	Methylmalonic acidaemia	Metabolism of methylmalonyl CoA
	Propionic acidaemia	Propionyl-CoA carboxylase
Urea cycle disorders	Hyperammonaemia Type I	Ornithine carbamoyltransferase
	Citrullinaemia	Argininosuccinate synthetase
	Argininosuccinic aciduria	Argininosuccinate lyase
Steroid synthesis disorders	Congenital adrenal hyperplasia (salt-losing)	21-Hydroxylase (severe form)

TESTS FOR METABOLIC DISORDERS

Most of the inherited metabolic disorders that present in the neonatal period give rise to non-specific symptoms such as feeding difficulties, e.g. vomiting when protein or carbohydrate feeding is started, or failure to thrive. There may be respiratory difficulties, fits, jaundice, or hepatomegaly but rarely unusual signs such as the smell of maple syrup urine disease. In the salt-losing forms of congenital adrenal hyperplasia, the patient may present in adrenal crisis.

The aim must be to make a specific diagnosis, and the wide-ranging nature even of the conditions selected for inclusion in Table 21.1 indicates how difficult it can be to achieve this aim. The following investigations should help to narrow the differential diagnosis considerably:

(1) *Chromatography.* Serum or plasma, and urine studies can help to identify fairly precisely most of the conditions listed in the various groups contained in Table 21.1. Separate chromatographic examinations are performed for amino acids, monosaccharides, organic acids and steroid metabolites, but not all are necessarily required in every case.

(2) *Blood [NH_3].* This is probably the best initial investigation if a urea cycle defect is suspected.

Further tests to identify the site of a metabolic block

Final confirmation of the diagnosis of inherited metabolic disorder cannot usually be made on the basis of plasma and urine examinations, although identification of a precursor that has accumulated in plasma, or detection of abnormal amounts of the precursor or one of its metabolites in urine, may strongly indicate the site of the primary defect. Instead, it is usually necessary to obtain tissue preparations, e.g. leucocyte concentrates or erythrocyte haemolysates, or biopsy specimens, e.g. from liver, skin, thyroid or brain.

In an index case, intracellular enzyme studies may be essential if the precise diagnosis is to be established. They assume particular importance if the infant's condition deteriorates, and indeed the downward progression of patients with some inherited metabolic diseases can be very rapid in the neonatal period. The possible need for intracellular enzyme studies may not be appreciated until the infant is already very seriously ill. Under these circumstances, if death seems imminent, arrangements should be made for the collection and preservation of those specimens that a laboratory specializing in the identification of metabolic disorders will require, for later detailed studies.

Retrospective diagnosis

Once it is agreed that the infant will not survive, permission should be obtained for a needle biopsy of the liver to be performed while the patient is still alive. Other specimens to be obtained include serum and urine (i.e. in addition, if possible, to the residue of specimens previously collected and examined), blood for lymphocyte and erythrocyte studies, CSF and a specimen of skin for fibroblast culture. Specimens collected later, at post-mortem examination, are rarely as satisfactory for these metabolic studies, although if possible a post-mortem should be carried out also.

THE IMPORTANCE OF MAKING A DIAGNOSIS

The specific recognition of a metabolic disorder forms the basis for any subsequent search for affected relatives (p. 429). It may also be particularly relevant to planning the management and investigation of any subsequent pregnancy (see below).

Genetic counselling may be required for parents who have had a child affected by one of the inherited metabolic disorders. Such counselling is more likely to be acceptable to relatives if it is soundly based on the results of scientific investigation, and if it is not long delayed in being given.

Diagnosis later in childhood

Many other inherited metabolic disorders may present less acutely than the conditions listed in Table 21.1, but may nevertheless carry a very poor prognosis; several progress inexorably to death in infancy or childhood (Table 21.2). It is very important to make the diagnosis precisely in the index case, so as to be able to advise parents about possible future pregnancies. Knowledge of the exact metabolic defect in the index case is essential for guiding the choice of enzyme studies for the prenatal diagnosis of the disorder in subsequent pregnancies (p. 409).

The approach to the diagnosis of the specific metabolic defect in the index case with these patients is similar to the approach described in the neonatal period. However, with conditions such as those listed in Table 21.2, there is usually more time in which to reach the diagnosis because of the slower progression of the illness.

SCREENING FOR INHERITED METABOLIC DISEASE

In this section we shall describe screening programmes for the detection of treatable conditions in the general population of neonates. We shall also discuss screening for heterozygotes, an essential preliminary for informed genetic counselling. Finally we shall consider the investigation of the special high-risk group of inherited metabolic disorders in which amniocentesis with a view to prenatal diagnosis should be offered.

Screening in infancy

Screening programmes require considerable organization. Before embarking on them, the following questions need to be considered:

(1) What is the incidence of the disorder?
(2) Is the disorder life-threatening or liable to be severe?
(3) Is acceptable treatment available?
(4) Is a suitable screening test available?
(5) Can abnormal results be followed up?
(6) Is the cost acceptable?

For the neonatal period, some of these questions have become less

TABLE 21.2. Examples of inherited metabolic diseases that carry a very poor prognosis in the index case and for which termination of pregnancy may be advised in subsequent pregnancies.

Category and examples	Enzyme deficiency
Glycogen storage disease	
Pompe's disease	α-D-Glucosidase
Lipid storage diseases	
Fabry's disease*	α-D-Galactosidase
Gaucher's disease	Glucosylceramidase
Niemann-Pick disease	Sphingomyelin phosphodiesterase
Tay-Sachs disease (G_{M2} gangliosidosis)	β-N-Acetyl-D-hexosaminidase (Hexosaminidase A)
Mucopolysaccharidoses (MPS)	
Hurler syndrome (MPS IH)	α-L-Iduronidase
Sanfilippo syndrome A (MPS IIIA)	Sulphoglucosamine sulphamidase (Heparan-N-sulphatase)
Purine metabolism)	
Lesch-Nyhan syndrome*	Hypoxanthine phosphoribosyl-transferase

* Sex-linked recessive inheritance. The other conditions listed in this Table have autosomal recessive inheritance.

difficult to answer for diseases where screening is being considered as a possibility, as long as the new programme can be related to a pre-existing programme (e.g. PKU screening) without detriment to the arrangements already established.

Approximately 10% of deaths in infancy and childhood in Britain are due to genetic disorders; appreciable numbers of these give rise to disorders of metabolism. The incidence of individual disorders is low (Table 21.3), however, and this highlights the difficulties facing those responsible for organizing and supervising the operation of screening programmes.

Screening for phenylketonuria

Almost the whole population of infants in Britain is screened nowadays for PKU, within a few days of birth. The most widely employed initial screening procedure, the *Guthrie test,* is a microbiological test carried out on blood samples. The test uses a mutant strain of *Bacillus subtilis,* spores of which are incorporated into agar that includes β-thienylalanine, an analogue of phenylalanine that prevents bacterial growth unless there is phenylalanine available. Blood specimens from infants with PKU, and with hyperphenylalaninaemia due to other causes, enable *B.subtilis* growth to occur; the specimens consist of drops of blood placed on to filter paper supplied by the laboratory responsible for carrying out the test.

The sensitivity of the Guthrie test for PKU, with a detection level of

TABLE 21.3. Incidence of metabolic diseases detectable in the neonate.

Disorder	Approximate frequency (Number of cases/100 000 live births)
Cystic fibrosis	25–30
Hypothyroidism	20–30
Phenylketonuria (classical form)	10–20
Cystinuria	5–10
Hartnup disease	5–10
Histidinaemia	5–10
Mucopolysaccharidoses	5–10
Urea cycle defects	2–5
Galactosaemia	1–2
Homocystinuria	1
Maple syrup urine disease	1
Non-ketotic hyperglycinaemia	1

blood [phenylalanine] set at 240 μmol/L (4.0 mg/100 mL) approaches 100%. For example, over the 5-year period 1974–1978, out of 357 cases of PKU in the U.K., five were missed. In two of these five missed cases, the screening laboratory did not receive a specimen for examination. In the other three cases it seems likely that blood [phenylalanine] would have been raised in the specimen sent to the laboratory but that it failed to detect this abnormality. These U.K. results relate to a national screening programme in which blood specimens are collected from infants aged 7–14 days.

In some countries, blood specimens are taken from infants in the first three days of life, e.g. before the mother is discharged home. More cases of PKU tend to be missed under these circumstances, because the specimens are collected before infant feeding with protein has become established. False negatives in the Guthrie test may also occur if the mother or the infant is being treated with antibiotics; the laboratory needs to be advised of this, so as to be able to adopt a different analytical method (e.g. a fluorimetric method).

The specificity of the Guthrie test is also excellent, being over 99.5%. However, because of the rarity of PKU, the predictive value of a positive Guthrie test (as indeed of most screening tests) is low (p. 513). Over 90% of positive Guthrie tests are due to causes other than classical PKU, the commonest causes of false positives being transient tyrosinaemia and transient hyperphenylalaninaemia.

Positive results in the Guthrie test, suggesting the possible diagnosis of PKU, require further investigation to determine whether the infant does indeed have PKU, in which case appropriate dietary treatment will need to be instituted and monitored. Additional tests may also be needed to differentiate between classical PKU and the much rarer variants mentioned above (p. 420).

Screening for other metabolic disorders in the neonatal period

Several other inherited metabolic disorders can be detected in the neonatal period, by tests based on principles similar to the Guthrie technique. These include maple syrup urine disease, histidinaemia, homocystinuria and tyrosinaemia. In addition, other blood specimens, collected at the same time as those used to screen for these disorders of amino acid metabolism, can be used to screen for hypothyroidism and for galactosaemia.

The approximate frequency of some of the inherited metabolic diseases is given in Table 21.3. Suitable screening tests are only available for some of these conditions.

Screening for heterozygotes

This is at present mainly restricted to screening the relatives of patients found to have a metabolic disorder. The programme is usually restricted to the specific abnormality identified in the index patient.

The effect of the metabolic defect can sometimes be assessed directly and quantitatively, as in the haemoglobinopathies, in which normal and abnormal haemoglobins can both be measured. Usually, however, the effect of the mutant gene can only be detected by investigating the metabolic reaction, directly as an enzymic assay or indirectly by measuring the effects of the block or partial block when subjected to a loading test. For instance, a phenylalanine loading test can be used to screen for PKU heterozygotes.

Heterozygotes may show abnormal results in tests of the metabolic reaction. However, the results do not necessarily lie approximately half-way between results observed in patients (homozygotes) and in normal individuals; overlap between normals and heterozygotes is often observed. Such studies can provide potentially valuable information (e.g. plasma cholinesterase measurements, p. 153).

Prenatal diagnosis

The aim, with inherited metabolic diseases that will prove to be severely handicapping, untreatable and eventually fatal, is to be able to offer the mother, in a future pregnancy, the opportunity of a therapeutic abortion if the fetus is found to be affected in the same way as a previous child. Table 21.2 gives examples of some of the conditions that can be detected *in utero,* and for which therapeutic abortion is likely to be offered.

Diagnosis is made on the basis of enzyme studies carried out on fibroblasts cultured from amniotic fluid, usually obtained between fourteen and sixteen weeks' gestation. This should allow time for the studies to be completed by twenty weeks' gestation, regarded by many obstetricians as the last acceptable date for termination.

For prenatal diagnosis of metabolic disorders to be carried out reliably, it is essential that the specimen of amniotic fluid should be cultured and examined by one of the few laboratories that presently possess the requisite skills. It is also essential that the laboratory be informed about the *precise* nature of the metabolic disorder for which the cultured cells are to be examined. This diagnosis will, in most instances, have been established on the basis of enzyme studies carried out on leucocytes, or skin biopsy specimens from which fibroblasts were cultured, or on other suitable specimens *from the index case.*

If a pregnancy is terminated on the basis of advice derived from prenatal diagnostic investigations, the correctness of the diagnosis should be confirmed by further biochemical studies carried out on the aborted fetus.

Prenatal diagnosis could prove very valuable, in the future, for the recognition of those inherited metabolic disorders for which it might become possible to correct the specific enzymic deficiency by appropriate replacement therapy.

FURTHER READING

ELLIS, R., editor (1980). *Inborn Errors of Metabolism.* London: Croom Helm.

HAAN, E.A. and DANKS, D.M. (1981). Clinical investigation of suspected metabolic disease. In *Laboratory Investigation of Fetal Disease,* pp. 410–428. Ed. A.J. Barson. Bristol: Wright.

HOLTON, J.B. (1982). Diagnosis of inherited metabolic disease in severely ill children. *Annals of Clinical Biochemistry,* **19,** 389–395.

Leading article (1979). New varieties of P.K.U. *Lancet,* **1,** 304–305, and subsequent correspondence.

Leading article (1980). Population screening for carriers of recessively inherited disorders. *ibid.,* **2,** 679–680.

SINCLAIR, L. (1982). A new look at the inborn errors of metabolism. *Annals of Clinical Biochemistry,* **19,** 314–321.

STANBURY, J.B., WYNGAARDEN, J.B., FREDRICKSON, D.S., GOLDSTEIN, J.L. and BROWN, M.L., editors (1983). *The Metabolic Basis of Inherited Disease,* 5th Edition. New York: McGraw-Hill.

WALD, N., editor (1983). *Antenatal and Neonatal Screening.* Oxford: Oxford University Press.

WOLFF, O.H., SMITH, I., CARSON, N. *et al.* (1981). Routine neonatal screening for phenylketonuria in the United Kingdom 1964–78. *British Medical Journal,* **282,** 1680–1684.

Chapter 22

Clinical Chemistry in Paediatrics

Chemical investigations can be as valuable in the diagnosis and management of diseases affecting children as they are for adults, provided that (1) satisfactory specimens can be collected for analysis, and (2) the laboratory has available the appropriate range of investigations, capable of being performed reliably and quickly on small volumes of blood or other specimens.

Some hospitals make the collection of blood specimens from children the responsibility of a team of specimen-collecting staff, who then obtain on a co-ordinated basis the specimens required for the different laboratories, e.g. haematology and clinical chemistry. This minimizes the upset caused to patients, by reducing or eliminating the need to disturb them repeatedly. It does require co-ordination of requests for laboratory investigations, and an indication of the order of priority attaching to the different tests, to make sure that the clinically most important specimens are collected first.

Most paediatric clinical chemistry laboratories can perform a wide range of investigations. Reference values appropriate to the age-group and the diet should be used. For example:

(1) In the first few hours after birth, plasma [bilirubin] and [total T4] normally show wide variations in concentration.

(2) In the first few months, the nature of the diet significantly influences the reference values for plasma [calcium], [phosphate], [amino acids], etc.

(3) Some plasma constituents normally vary in their activity or concentration with different levels of osteoblastic activity or in relation to different stages of genital development (e.g. alkaline phosphatase activity, urate in boys, gonadotrophins and gonadal steroids).

Clayton and her colleagues have published one of the most comprehensive sets of reference values, for children in the U.K.

In this chapter a few diagnostic problems have been selected for consideration. Paediatric textbooks and journals contain much more information about paediatric clinical chemistry.

Biochemical problems in the neonatal period

CONVULSIONS

Anoxia, infections and trauma are all relatively common causes of convulsions in the neonatal period. These may cause metabolic disturbances that give rise to the convulsions, but sometimes metabolic disturbances are the primary cause. Chemical investigations may help in the diagnosis and management of these conditions.

Glucose

Plasma [glucose] tends to be lower in the newborn than in adults, with values in the range 0.5–2.0 mmol/L (9–35 mg/100 mL). According to the definition of hypoglycaemia in adults (plasma [glucose] less than 2.2 mmol/L or 40 mg/100 mL), hypoglycaemia occurs frequently in the neonatal period. The definition of what constitutes hypoglycaemia in infancy is arbitrary, but may be taken as a plasma [glucose] of 1.1 mmol/L (20 mg/100 mL) or less, on at least two occasions, when measured by an enzymic technique.

In the neonatal period, hypoglycaemia is most commonly due to decreased production of glucose; this often occurs in premature babies. Neonatal hypoglycaemia is also quite often due to hyperinsulinism, e.g. in infants of diabetic mothers.

Hypoglycaemia giving rise to convulsions only affects about one in every 500 live births, usually between 24 and 72 h after birth.

Blood [glucose] can be monitored on the ward (p. 34) and this practice is widely adopted nowadays, especially with premature and low birthweight infants. If a low reading is obtained, this may be taken as an indication for treatment, and for the blood [glucose] to be confirmed by measurement in the laboratory.

Hypoglycaemia in the neonatal period is often accompanied by ketosis, and urine tests for ketone bodies are then positive.

Rarely, in older infants, hypoglycaemia is secondary to galactosaemia, glycogen storage disease, hereditary fructose intolerance, leucine sensitivity, insulinoma or deficiency of glucocorticoids, growth hormone or thyroid hormones.

Calcium

Full-term infants, at birth, normally have a plasma [calcium] in the range 2.20–3.00 mmol/L and plasma [phosphate] in the range 1.0–2.3

mmol/L. These ranges differ considerably from the corresponding reference values for adults (p. 19).

After birth, plasma [calcium] tends to fall by about 10–20% in the first 2–3 days of life. It then returns over the course of the next 3–4 days towards the value that was present at birth, as the infant begins to secrete parathyroid hormone (PTH). Other mechanisms are also involved.

Neonatal hypocalcaemia giving rise to convulsions can be subdivided into two categories, occurring (1) within 48 h of birth and (2) between the fourth and tenth days of life.

Hypocalcaemia within the first 48 h, and giving rise to convulsions, occurs particularly in premature infants, in infants of diabetic mothers and infants that have been asphyxiated. It seems probable that the hypocalcaemia in these infants is due to a slow or inappropriate development of the normal physiological response to hypocalcaemia, e.g. the production of PTH. Usually, the hypocalcaemic tendency corrects itself spontaneously, although calcium gluconate may need to be given intravenously if the plasma [calcium] falls to less than 1.50 mmol/L (6.0 mg/100mL).

Late neonatal hypocalcaemia, occurring between the fourth and tenth days, is usually accompanied by signs of hyperexcitability of muscles. In the past, it was most often due to infants being fed on cow's milk, which has a high phosphate content. With appropriate changes to the composition of infants' feeds, this cause has largely disappeared. However, late neonatal hypocalcaemia is still liable to occur in infants whose mothers had a low intake of vitamin D during pregnancy; these infants may also have low plasma [magnesium]. Treatment with intravenous calcium, and often magnesium also, may be required.

Hypocalcaemia and defective bone mineralization, sometimes giving rise to rickets, may occur especially in premature infants. This is caused by the increased requirements of premature infants for calcium, phosphate and vitamin D, and may present at any time during the neonatal period, or the next 2–3 months. It is an important condition, but is itself a very rare cause of convulsions or muscular hyperexcitability.

Magnesium

Hypomagnesaemia is an occasional cause of neonatal convulsions. It tends to occur in association with decreased plasma [calcium] but, when the convulsions are due to low plasma [magnesium], administration of calcium fails to correct the clinical disturbance whereas treatment with magnesium does. The primary defect is probably in the intestinal absorption of magnesium. In untreated infants, plasma [magnesium] may be as low as 0.1 mmol/L (0.25 mg/100 mL).

Sodium

Abnormally high and, more rarely, abnormally low plasma [Na^+] have been reported in a small percentage of infants with convulsions.

BLOOD GAS ABNORMALITIES

Any cause of hypoxia or marked acid–base disturbance may give rise to convulsions in the neonatal period.

Respiratory difficulty in neonates is most often due to the *respiratory distress syndrome* (RDS). This occurs mainly in premature infants of less than thirty-eight weeks' gestation. There is atelectasis, probably due to deficiency of surfactant activity in the alveoli, a low arterial $P\text{O}_2$ and high $P\text{CO}_2$ and often an accompanying metabolic acidosis. Treatment by positive pressure ventilation and bicarbonate infusions should be monitored by determination of blood [H^+], $P\text{CO}_2$ and $P\text{O}_2$.

It is possible to predict the likelihood of RDS developing after birth by determining amniotic fluid [lecithin]. A few laboratories prefer to measure the lecithin : sphingomyelin (L:S) ratio in amniotic fluid (p. 413).

NEONATAL HYPERBILIRUBINAEMIA

The definition of neonatal hyperbilirubinaemia is necessarily somewhat arbitrary. Over 90% of normal babies have plasma [bilirubin] exceeding 34 μmol/L (2 mg/100 mL) at some time during the first week of life. In some babies, plasma [bilirubin] may rise much higher, up to 200 μmol/L, without any apparent pathological cause. However, babies with plasma [bilirubin] greater than 200 μmol/L, especially if this persists after the first week of life, must be regarded as having some additional cause for the hyperbilirubinaemia.

Once the plasma [unconjugated bilirubin] exceeds the capacity of albumin to bind bilirubin, unconjugated bilirubin can cross the blood-brain barrier and cause kernicterus in which there is brain damage, especially in the basal ganglia. The critical plasma [bilirubin] at which this occurs is about 340 μmol/L (20 mg/100 mL), but this level also depends on (1) plasma [albumin], (2) the presence of certain drugs (e.g. sulphonamides) which may occupy some of the available binding sites on albumin, and (3) acid–base disturbances, which may affect the equilibrium between albumin-bound and unbound bilirubin.

Neonatal hyperbilirubinaemia is most often associated with increased plasma [unconjugated bilirubin]. It is useful, diagnostically, to distinguish between unconjugated hyperbilirubinaemia (in which conditions plasma [conjugated bilirubin] is normal) and conjugated hyperbilirubinaemia. Patients in the latter group have increased plasma [conjugated bilirubin], but this may nevertheless constitute only 20–30% of the total bilirubin in plasma in these patients.

Physiological unconjugated hyperbilirubinaemia

This occurs very frequently. The factors that contribute to its development are not fully understood, but include:

(1) *Overproduction of bilirubin* from Hb, due to shortened red cell life-span and ineffective erythropoiesis.
(2) *Immaturity of the hepatic processes* of bilirubin uptake from plasma and conjugation in the hepatocyte.
(3) *Interference with hepatic transport functions* by drugs transferred from the mother across the placenta or present in human breast milk (e.g. progesterone and steroids with progesterone-like activity).
(4) *Reabsorption of unconjugated bilirubin* from the intestine. In the neonate β-glucuronidase is present in the small intestine, and this releases bilirubin from the conjugates. Since the bacterial flora that normally convert bilirubin to urobilinogen in the intestine are not fully developed at this age, some bilirubin can be reabsorbed and thus add to the load that the liver has to excrete.

Pathological causes of unconjugated hyperbilirubinaemia

In the neonate, these causes augment the tendency already present for physiological hyperbilirubinaemia to develop, especially in premature infants. The groupings of pathological causes are:

(1) *Increased haemolysis.* This may be due to Rhesus or ABO incompatibility, or to abnormalities within the red cell such as glucose-6-phosphate dehydrogenase deficiency or hereditary spherocytosis.

(2) *Defective hepatic uptake or conjugation.* This may occur in a large variety of disorders including prematurity, hypoglycaemia, hypothyroidism and inherited disorders of bilirubin metabolism, e.g. Gilbert's syndrome and the very much rarer Crigler-Najjar syndrome (p. 176).

Conjugated hyperbilirubinaemia

There are several causes of conjugated hyperbilirubinaemia in infancy. These may be grouped as follows:

(1) *Developmental abnormalities of the biliary tree.* The most important of these is extra-hepatic biliary atresia; intra-hepatic biliary atresia also occurs.
(2) *Neonatal hepatitis.* This is an ill-defined group which includes patients with infective causes of hepatitis (e.g. cytomegalovirus) or metabolic causes (e.g. α_1-antitrypsin deficiency, galactosaemia).

Chemical tests have so far been of little help in making the important clinical distinction between infants with jaundice due to extra-hepatic biliary atresia (in whom surgery is indicated) and infants with intra-hepatic lesions (in whom surgery is not indicated). The commonly performed 'liver function tests' usually show a predominantly cholestatic pattern, with large rises in plasma alkaline phosphatase activity and lipoprotein X detectable, but plasma ALT and AST activities are also increased.

It has been reported that measurement of plasma [lipoprotein X] before and after treatment with cholestyramine, which binds bile acids and other anions in the intestinal lumen, may help to distinguish biliary atresia from neonatal hepatitis. In infants with a patent bile duct system, plasma [lipoprotein X] falls with cholestyramine treatment, whereas in biliary atresia it does not fall and may even increase.

Bilirubin measurements

Examination of urine by side-room tests may not be practicable, but inspection of staining produced by urine and of the colour of any faecal material can be helpful. In jaundice due to increased plasma [unconjugated bilirubin], urine does not usually contain bilirubin unless plasma [bilirubin] is over 340 μmol/L (20 mg/100 mL); even then it only contains small amounts of bilirubin unless there is concomitant glomerular damage and proteinuria.

Measurements of plasma [bilirubin] are valuable in following the progress of hyperbilirubinaemia and in determining when therapeutic measures (e.g. exchange transfusion) are required. In this age-group, separate estimates of conjugated and unconjugated bilirubin in plasma are of much more value than in adults.

Failure to thrive

There are many possible metabolic causes of failure to thrive, including malnutrition, vitamin deficiency diseases and cystic fibrosis. Other inherited disorders of metabolism will not be further discussed here.

MALNUTRITION IN CHILDREN

Protein-energy malnutrition (PEM), in its severest forms, includes kwashiorkor and marasmus; there is a range of less severe clinical presentations. The pathology is complex. There may be other important factors, e.g. deficiency of essential fatty acids and the consequences of immune defence mechanisms impaired by malnutrition. Many chemical tests can be carried out, but are no substitute for serial measurement and recording of children's weights.

Measurement of plasma [albumin] is valuable in screening for children at risk of developing malnutrition but in whom the disease is still at a subclinical stage; the rate of albumin synthesis is diminished when protein intake is deficient. The most consistent abnormality in severe kwashiorkor is a low plasma [albumin]. In practice, plasma [albumin] is not likely to be determined unless clinical impressions suggest the need; results greater than 36 g/L (3.6 g/100 mL) are taken as normal. Plasma [albumin] below 30 g/L should be regarded as abnormally low, and below 25 g/L is associated with increasing degrees of oedema.

Concentrations of some amino acids in blood are also reduced in PEM, especially the three essential branched-chain amino acids (valine, leucine and *iso*leucine). The degree of abnormality of blood amino acid patterns is sometimes expressed as a ratio of the concentration of the non-essential (N) to essential (E) amino acids (the N:E ratio); the greater the value of this ratio, the more severe the degree of malnutrition.

Other early indices of PEM include changes in plasma [pre-albumin] and [transferrin], both of which fall, but the measurements do not add to the information gained from determining plasma [albumin].

If liver failure develops, many other tests become abnormal.

VITAMIN DEFICIENCY DISEASES

Nutritional vitamin deficiency is a potentially important cause of failure to thrive, since the growing child has relatively greater requirements than the mature adult.

Nutritional rickets continues to be important, even in developed countries. Plasma [calcium] and [phosphate] are usually reduced, but not always. Plasma alkaline phosphatase activity measurements may be difficult to interpret, because of the wide range of reference values (p. 148), but are generally increased. Measurement of plasma [25-HCC] may confirm the diagnosis of vitamin D deficiency (p. 267).

CYSTIC FIBROSIS

This is the commonest inherited metabolic disease in Caucasians; it occurs in about 1 in 3500 live births (Table 21.3), or possibly even more frequently. It is inherited as an autosomal recessive condition, normally affecting the exocrine glands. The basic biochemical defect in cystic fibrosis (CF) is unknown. Chemical investigations used for diagnostic purposes mostly depend on tests relating to sweat production or to pancreatic function. Tests for intestinal malabsorption such as faecal fat estimation and pancreatic function tests are described in Chapter 10. Discussion here will be confined to consideration of sweat tests and screening procedures.

Tests on sweat

The reference method for the diagnosis of CF by means of chemical tests is measurement of $[Na^+]$ and $[Cl^-]$ in sweat, obtained by iontophoresis from a small area of skin under standardized conditions; sweating is induced by the intradermal injection of pilocarpine. The test demands close attention to detail, if reliable results are to be obtained.

In healthy children and adults, the $[Na^+]$ and $[Cl^-]$ in pilocarpine-stimulated sweat are normally in the range 10–50 mmol/L; in patients with CF, they are nearly always in the range 70–140 mmol/L. However, some children with CF give equivocal results. With sweat $[K^+]$, there is some overlap between results from healthy children and CF patients.

Sweat tests in the neonatal period may be of no diagnostic value, as many very young infants fail to produce sweat fast enough, even in response to pilocarpine stimulation. Unless sweat flows freely, evaporation occurs and false positive results are obtained. The tests can usually be performed satisfactorily in infants from one month old onwards.

Sweat tests are too time-consuming, and demanding of attention to detail, to be used for screening purposes.

Screening tests for cystic fibrosis

At present, these are all based on the effects of CF on pancreatic function. Measurements can be made on meconium, blood or faeces.

Meconium [albumin] measurements have given rise to reports that are conflicting, large numbers of false positives being reported in some series and false negatives in others, when compared with the results of sweat tests. Infants with cystic fibrosis have meconium [albumin] that is higher than in healthy infants but positive results require to be confirmed with another type of test before instituting dietary treatment. Specimens of meconium deteriorate quickly unless cooled; this may account for unsatisfactory reports from some meconium albumin studies, in which an incidence of up to 30% false negatives has been reported. The incidence of false positives can be reduced by measuring the albumin : α_1-antitrypsin ratio in meconium.

Immunoreactive trypsin (IRT), measured in dried blood spots similar to those collected for the Guthrie test (p. 427), when screening for phenylketonuria, has been reported to be greatly increased in specimens collected from CF infants as compared with the IRT content in specimens from healthy infants. The test appears to be more specific than tests on meconium, but is at present much more expensive. It appears to be potentially the best CF screening method from among the tests so far evaluated.

Faecal enzyme activity tests have also been used to screen for pancreatic involvement in cystic fibrosis. Faecal specimens are examined for the presence of tryptic or chymotryptic activity, which may be reduced or undetectable in these patients. Earlier reports of faecal enzyme measurements tended to discount these tests as unreliable. However, the use of better substrates (e.g. benzoyl-arginine-*p*-nitroanilide) has led to more promising reports. The tests can be carried out on faecal specimens that have been smeared on a card and allowed to dry before being sent to the laboratory.

Associated abnormalities

Patients with cystic fibrosis may develop obstructive pulmonary disease with recurrent complications, due basically to the increased viscosity of mucous secretions. There may be acid–base disturbances, and effects on hepatic and renal function caused by the development of cor pulmonale and congestive cardiac failure.

Prenatal diagnosis of cystic fibrosis

The chance of heterozygous parents producing a child with cystic fibrosis is 1 in 4, and the relatively high incidence of the disease (Table 21.3) has focused attention on the possibility of prenatal diagnosis. Recently, gamma-glutamyl transferase (GGT) and aminopeptidase M activities have been reported to be abnormally low in amniotic fluid obtained from women who later gave birth to CF-affected infants. These tests may be used in future for prenatal diagnosis of cystic fibrosis.

Disorders of growth

Table 22.1 lists the principal causes of disordered growth. Some of these have a biochemical basis or produce marked metabolic effects, and are discussed elsewhere as indicated by the cross-references.

TABLE 22.1. Disordered growth.

Group	Examples	Cross-reference
Short stature		
Genetic	Familial short stature	
	Delayed development	
Intra-uterine	Low birth weight dwarfism	
Nutritional	Inadequate food supply	p. 437
	Malabsorption	p. 210
	Coeliac disease	p. 441
	Infections	
Systemic disease	Chronic renal disease	p. 299
	Congenital heart disease	
Endocrine	Growth hormone deficiency	p. 390
	Thyroid deficiency	p. 350
	Corticosteroid excess	p. 359
	Precocious puberty	
Syndromes	Turner's syndrome	
Social	Emotional deprivation	
Tall stature		
Genetic	Familial tall stature	
	Advanced development	
Endocrine	Growth hormone excess	p. 389
	Hyperthyroidism	p. 349
	Precocious puberty	
Syndromes	Klinefelter's syndrome (XXY)	p. 383
	XYY anomaly	

COELIAC DISEASE

This is a common cause of growth retardation. Its incidence in the U.K. has been reported to be 1 in 1850, but there are indications that the incidence may be higher.

The definitive method of diagnosis is small intestinal biopsy examination. Other diagnostic features include the improvement that is brought about by a gluten-free diet, and relapse that results from dietary relaxation. Chemical and haematological tests can, however, prove valuable as preliminary investigations, intestinal biopsy being reserved for those patients who show abnormal results in preliminary tests such as:

(1) *Faecal fat* (p. 206). Excretion may be greatly increased.
(2) *Xylose absorption* (p. 203). This may be considerably impaired.
(3) *Erythrocyte [folate]*. This may be markedly reduced.
(4) *Serum [vitamin B_{12}]*. This may be reduced if the terminal ileum becomes affected by the disease.

Patients with coeliac disease have abnormal secretin cells in the jejunal mucosa. They show a markedly reduced release of secretin and of GIP (p. 194) in response to test stimuli, e.g. a protein-containing meal or intra-duodenal acid, but a greatly increased release of a peptide with gut glucagon-like immunoreactivity from the ileum. This pattern of gut hormone release reflects the location of the intestinal lesion in coeliac disease; concentrations of hormones (e.g. gastrin) produced in areas unaffected by the disease remain normal.

Tumours in childhood

Deaths due to malignant disease make up an important category of childhood mortality. Neuroblastoma and related tumours (p. 453) account for approximately one-third of these patients. Congenital adrenal hyperplasia and adrenal tumours are discussed in Chapter 18.

FURTHER READING

AYNSLEY-GREEN, A. (1982). Hypoglycaemia in infants and children. *Clinics in Endocrinology and Metabolism*, **11,** 159–194.

CARBARNS, N.J.B., GOSDEN, C. and BROCK, D.J.H. (1983). Microvillar peptidase activity in amniotic fluid: possible use in the prenatal diagnosis of cystic fibrosis. *Lancet,* **1,** 329–331.

CLAYTON, B.E., JENKINS, P. and ROUND, J.M. (1980). *Paediatric Chemical Pathology: Clinical Tests and Reference Ranges.* Oxford: Blackwell Scientific Publications.

DODGE, J.A. and RYLEY, H.C. (1982). Screening for cystic fibrosis. *Archives of Disease in Childhood,* **57,** 774–780.

EVANS, S.E. and DURBIN, G.M. (1983). Aspects of the physiological and pathological background to neonatal clinical chemistry. *Annals of Clinical Biochemistry,* **20,** 193–207.

HILLMAN, L.S. and HADDAD, J.G. (1982). Hypocalcaemia and other abnormalities of mineral homeostasis during the neonatal period. In *Calcium Disorders,* pp. 248–276. Ed. D.A. Heath and S.J. Marx. London: Butterworth.

JAMES, W.P.T. (1977). Kwashiorkor and marasmus: old concepts and new developments. *Proceedings of the Royal Society of Medicine,* **70,** 611–615.

KELNAR, C.J.H. and HARVEY, D. (1977). Respiratory distress syndrome. *British Journal of Hospital Medicine,* **18,** 232–243.

LEBENTHAL, E. and BRANSKI, D. (1981). Childhood celiac disease — a reappraisal. *Journal of Pediatrics,* **98,** 681–690.

MEITES, S. and LEVITT, M.J. (1979). Skin-puncture and blood-collecting techniques for infants. *Clinical Chemistry,* **25,** 183–189.

WHITEHEAD, R.G. (1967). Biochemical tests in differential diagnosis of protein and calorie deficiencies. *Archives of Disease of Childhood,* **42,** 479–484.

Chapter 23

Clinical Chemistry in Geriatrics

The proportion of the U.K. population over 65 has increased considerably in the last thirty years, and similar trends have occurred or are to be anticipated in many other countries. Many elderly patients admitted to hospital, for assessment and treatment, are suffering from more than one disease. Problems of diagnosis are often complex. This chapter considers aspects of clinical chemistry selected for their particular relevance to the growing specialty of geriatrics. However, most of the other chapters in this book are also relevant to the investigation of elderly patients.

Reference values

Until recently, specimens from elderly patients sent to clinical chemistry laboratories for examination were reported without attention being given to possible special problems in interpretation. Results had to be assessed for significance in relation to reference values that most laboratories assumed were applicable to adult patients of all ages. It is now realized that this assumption is by no means always true.

It is important to recognize that there is no sharp distinction chemically between middle-age and old age, each with its separate set of reference values. Instead, the changes in reference values that are sometimes evident in those over 65 years old often merely represent the extension of changes that have been occurring gradually throughout adult life. Frequently, these changes continue in the same direction with advancing age, but this is not always the case. The ideal, however, would be to have available different sets of reference values for age-bands such as 55–65, 65–75, 75–85 and 85–95.

Several plasma constituents tend to increase in concentration with advancing age. Examples will be briefly discussed.

Plasma [urea] and [creatinine] both tend to increase, as an inevitable accompaniment of the progressive loss of nephrons, which starts in middle-age. The reduction in muscle mass and the smaller dietary intake of protein, both of which also tend to occur with older people, are insufficient to offset the effects of the loss of nephrons.

Plasma [cholesterol] increases progressively throughout adult life until, by the age of 65, mean values are 10–30% higher than those observed in the 20–30 age-group.

Plasma [glucose] tends to increase with age, due to progressive impairment of glucose tolerance in individuals who, however, do not go on to develop overt diabetes mellitus (discussed below).

Plasma alkaline phosphatase activity is probably higher in healthy people over 65 than in younger adults, but not all reports agree on this. Some reports suggest that there is no difference between reference values for adults in the 20–65 age-range and those for apparently healthy adults over 65, if care is taken to exclude people with Paget's disease, or unsuspected liver disease, or malignancy, and especially to exclude unsuspected malignant disease with secondary deposits in liver or bone. We consider that a plasma alkaline phosphatase activity more than 1.5 times the laboratory's upper reference value for adults should definitely be regarded as abnormal in an elderly person. High values are not necessarily serious, however, as they are often due to Paget's disease.

Some plasma constituents tend to decrease in concentration in the elderly. For instance plasma [total protein] and [albumin] both tend to fall with increasing age. However, the changes are unlikely to lead to difficulties in the interpretation of results, except possibly in the preliminary investigation of paraproteinaemia (p. 134).

Reference values in practice

It will be apparent, from this brief review, that it is very difficult to determine reference values for healthy individuals in the older age-groups, when the incidence of 'abnormalities' is so great.

For individual elderly patients, it is worth checking whether particular investigations have been carried out previously. Past records sometimes provide useful baseline data for comparison with findings observed in a later illness. Record linkage facilities in hospitals should allow rapid access to the data held on file, and these data should be able to be recalled quickly.

Diseases with chemical features requiring special consideration in geriatric patients

DIABETES MELLITUS

The presence of diabetes mellitus may not be suspected on side-room testing of urine for the presence of glucose or reducing substances, since the renal threshold tends to rise with age. Instead, measurement of plasma [glucose] 2 h after a 75 g load of glucose is the best screening test, and is to be preferred to measurements of fasting plasma [glucose]; the 2-h post-glucose value should not exceed 11.0 mmol/L (198 mg/100 ml).

Abnormal responses to a 75 g glucose load can be demonstrated in about 25% of the population over 75. However, impaired glucose tolerance (as defined on p. 225) should not necessarily be taken to imply that the patient will go on to develop diabetes mellitus.

Management of the elderly diabetic patient may need to depend on the help of relatives, who will require to be taught how to perform the tests at home. Because of the higher renal threshold for glucose, urine testing may not be as helpful in controlling the treatment of elderly diabetics as it can be in younger patients. Instead, home-monitoring of blood [glucose] will probably be needed (p. 34). Greater reliance may also have to be placed in some cases on haemoglobin A_{1c} determinations (p. 227); these may replace blood and urine glucose determinations as the best means of long-term control for some elderly diabetics.

THYROID DISEASE

Hypothyroidism is common among the elderly, some reports suggesting that it occurs in 2–6% of people over 70. Confirmation of the diagnosis by means of chemical investigations is less straightforward than in younger age-groups, because both plasma [total T4] and [free T4] tend to be low in elderly patients. The low plasma [total T4] is sometimes due to poor nutrition giving rise to reduced plasma [TBP], or to the ingestion of drugs (e.g. salicylates) that affect thyroid hormone-binding capacity. However, the commonest cause of misleadingly low results for plasma [total T4] and [free T4] in elderly patients is the presence of non-thyroidal illness. In the absence of thyroid disease, these patients have normal plasma [TSH], and the combination of low plasma T4 and normal TSH concentrations is thought probably to be due to a 'resetting' of the hypothalamic-pituitary-thyroid axis.

To overcome these difficulties in interpretation of T4 measurements, the diagnosis of hypothyroidism in the elderly should be based on the combination of (1) reduced plasma [total T4] or [free T4], and (2) increased plasma [TSH].

Hyperthyroidism has also been reported as a moderately common finding among the elderly, being detected in 1–2% of admissions to some geriatric assessment units. The clinical presentation in the elderly is often non-specific or atypical. Diagnostic tests, therefore, need to be particularly sensitive and specific.

Measurements of plasma [total T4] and [total T3], or of plasma [free T4] and [free T3], are the tests that should be performed initially. Thereafter, in doubtful cases (i.e. cases where the diagnosis has not been confirmed but is still considered possible), it is often necessary to perform a TRH test (p. 345); this may be the only test that is abnormal in these patients.

BONE DISEASE

The incidence of bone disease rises markedly in old age. Osteoporosis is the commonest, especially in women, but at present routinely available chemical investigations are not of value in its detection. Paget's disease is also very common, with a prevalence of 5–10% in people over 65, and the incidence of osteomalacia is also appreciable.

Paget's disease is one of the first diagnoses to be considered when increased plasma alkaline phosphatase activity, of bone origin, is found as an isolated abnormality in an elderly patient. Further, more specialized, chemical investigations are unlikely to be of value. Urinary hydroxyproline is usually found to be increased, but the conditions for performing this investigation require the patient to be put on a special diet and the test is little used.

TABLE 23.1. Admission screening of elderly patients.

Examination	Abnormalities commonly detected
Blood/plasma	
Calcium, phosphate and alkaline phosphatase	Hypocalcaemia – often associated with osteomalacia
Erythrocyte sedimentation rate	Non-specific indicator of disease – very high in paraproteinaemia
Glucose	Diabetes mellitus
Haemoglobin and blood film	Anaemia
Potassium	Hypokalaemia – often due to diuretic therapy
Total thyroxine	Hypothyroidism, hyperthyroidism
Total protein, albumin	Poor nutritional state
Urea, creatinine	Renal disease, post-renal uraemia

Note: Side-room testing of urine for glucose and protein, and faeces for occult blood, should form part of the initial investigation of all patients admitted to geriatric assessment units.

Osteomalacia is usually due to lack of exposure to sunlight combined with nutritional deficiency. Plasma [calcium] and [phosphate] may both be reduced and alkaline phosphatase activity increased in many cases. However, there can be difficulties in interpreting the results of these measurements, due to lack of precise data for their reference ranges in the elderly. Plasma [25–HCC] is often normal, but it may be reduced due to inadequate intake of vitamin D_3 and lack of endogenous synthesis (p. 264). The typical X-ray changes of osteomalacia may not be present. Definitive diagnosis may require trephine bone biopsy, but it is often more practicable to make the diagnosis on the basis of a therapeutic trial, using vitamin D or 1α-hydroxycholecalciferol. The response to the trial should primarily be assessed symptomatically but should also be monitored by serial measurements of plasma [calcium], [phosphate] and alkaline phosphatase activity.

Admission screening of elderly patients

Illnesses often present differently in the elderly, and a logical ordered approach to the selection of investigations, based on the patient's history and the findings on clinical examination, may not always be as appropriate in the elderly as in younger patients. Some diseases may present so insidiously, while nevertheless being amenable to treatment, that many geriatricians now make intensive use of laboratory investigations at an early stage of the patient's illness, simply as part of a wide-ranging standard pattern of assessment.

Much attention has been paid to the choice of tests to be incorporated into screening programmes of investigations for performance on elderly patients, for instance when first admitted to geriatric assessment units. Table 23.1 lists laboratory tests that have gained widespread acceptance together with an indication of relatively common explanations for the finding of abnormal results.

Measurements of plasma [T4] are nowadays included in most admission screening programmes for elderly patients. However, it must again be emphasized that abnormalities are often found in thyroid function tests when patients are malnourished, or affected by intercurrent illness, but without such abnormalities necessarily signifying the presence of thyroid disease. As far as possible, especially with geriatric patients, assessment of thyroid function by chemical tests should only be carried out on (1) patients in whom thyroid disease is suspected, (2) patients in whom malnutrition or any intercurrent infection has first been treated, and (3) patients who have recovered from the acute effects of a recent severe illness (e.g. myocardial infarction, major operation), or (4) as part of a routine screening of patients admitted to a geriatric assessment unit and in whom there are no obvious signs of the other non-thyroidal conditions already mentioned.

FURTHER READING

BAKER, M.R., PEACOCK, M. and NORDIN, B.E.C. (1980). The decline in vitamin D status with age. *Age and Ageing,* **9,** 249–252.

CAIRD, F.I. (1973). Problems of interpretation of laboratory findings in the old. *British Medical Journal,* **4,** 348–351.

CAIRD, F.I. (1975). Biochemical normality in old age. In *Eleventh Symposium on Advanced Medicine,* pp. 306–316. Ed. A.F. Lant. London: Pitman Medical.

CAMPBELL, A.J., REINKEN, J. and ALLAN, B.C. (1981). Thyroid disease in the elderly in the community. *Age and Ageing,* **10,** 47–52.

EXTON-SMITH, A.N. (1977). Malnutrition in the elderly. *Proceedings of the Royal Society of Medicine,* **70,** 615–619.

GARLAND, M.H. (1979). Problems in the elderly: Common endocrine disorders. *Hospital Update,* **5,** 825–832.

GREEN, M.F. (1974). Endocrine disorders in the elderly. *British Medical Journal,* **1,** 232–236.

HAVARD, C.W.H. (1981). The thyroid and ageing. *Clinics in Endocrinology and Metabolism,* **10,** 163–178.

HODKINSON, H.M. (1977). *Biochemical Diagnosis of the Elderly.* London: Chapman and Hall.

KAFETZ, K. (1983). Renal impairment in the elderly: a review. *Journal of the Royal Society of Medicine,* **76,** 398–401.

MORGAN, D.B. (1983). The impact of ageing—present and future. *Annals of Clinical Biochemistry,* **20,** 257–261.

Chapter 24
Neoplastic Disease

Neoplastic diseases account for about 20% of all deaths in Britain. Their effects are varied, and the diagnosis and management of several forms of neoplastic disease can be helped by chemical investigations.

ENDOCRINE TUMOURS

Several endocrine tumours have been discussed elsewhere, e.g. tumours of the adrenal cortex, gonads, pancreas, parathyroid and pituitary. Catecholamine-secreting tumours, 5-hydroxyindole-secreting tumours and medullary carcinoma of the thyroid will be considered here.

Many tumours that arise from organs other than the endocrine glands nevertheless produce endocrine and metabolic effects. These tumours, collectively referred to as 'non-endocrine' to denote their different origin, will be discussed in the section on tumour markers (p. 458).

Catecholamine-secreting tumours

The most important naturally occurring catecholamines are dopamine (3-hydroxytyramine), noradrenaline and adrenaline. Small amounts can be detected in blood, and small amounts of unmetabolized catecholamines are normally present in urine. Most of the catecholamines produced in the body, physiologically or as a result of disease, are inactivated by metabolism and the metabolites then excreted, mainly in the urine.

Catecholamines may be produced in large amounts by tumours of neural crest origin, and marked pharmacological effects may be produced when noradrenaline and adrenaline are released into the bloodstream. Dopamine has much less marked pharmacological effects.

PHAEOCHROMOCYTOMA

These tumours, which may secrete excessive amounts of

noradrenaline or adrenaline or both (usually both), mostly arise in the adrenal medulla. However, extra-adrenal phaeochromocytomas do occur, usually in the abdomen or pelvis. They give rise to excessive excretion of catecholamine metabolites in the urine (Fig. 24.1).

Phaeochromocytoma is a rare cause of hypertension, possibly being present in 0.5% of cases. The tumour is usually benign. Diagnosis is important since removal of a phaeochromocytoma, before complications develop, can cure the hypertension. The characteristic feature of the condition is episodic hypertension. Even when the hypertension is present all the time, episodic attacks of symptoms such as headache or

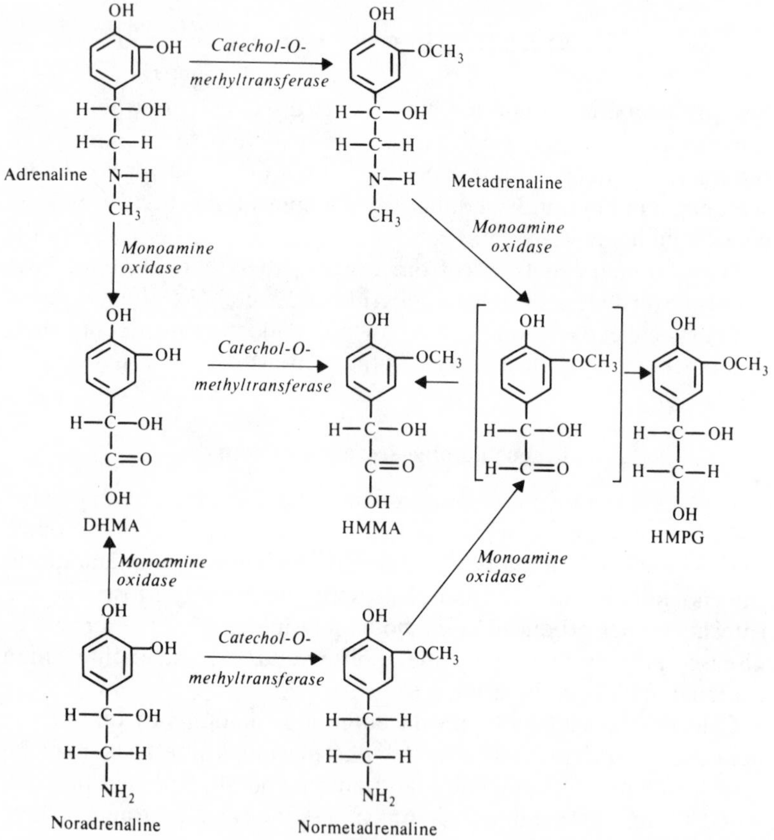

FIG. 24.1. *The main pathways in the metabolism of noradrenaline and adrenaline.* The principal product excreted in urine, in man, is HMMA. Noradrenaline and normetadrenaline, adrenaline and metadrenaline are all partly excreted as conjugated derivatives. Dihydroxymandelic acid (DHMA) is mostly converted to HMMA, and very little DHMA is normally present in urine.

sweating tend to occur. Rarely, predominantly adrenaline-secreting tumours may give rise to paroxysmal attacks of hypotension. These clinical features are important when selecting the time for collection of specimens for laboratory investigation.

Phaeochromocytoma sometimes occurs as a familial condition, in association with medullary carcinoma of the thyroid, and then quite often also with hyperparathyroidism (MEN IIa or Sipple syndrome, p. 457).

Urinary catecholamines and metabolites

Chemical tests usually involve measurement of 24-h urinary excretion of catecholamine metabolites. Of these, HMMA is normally excreted in the largest amount, the amount of 'total metadrenalines' in urine being about 10–20% of HMMA output; excretion of unmetabolized catecholamines is only a small percentage of the HMMA excretion. Most laboratories perform one or more of the following:

(1) *4-hydroxy, 3-methoxy mandelic acid* (HMMA), still sometimes incorrectly called vanillyl mandelic acid (VMA), is probably the most widely available chemical investigation. It is formed from both noradrenaline and adrenaline, and its measurement therefore cannot be used to estimate the separate output of the individual catecholamines.

(2) *Metadrenaline and normetadrenaline,* and their conjugated derivatives. Most laboratories that measure the methylated amines determine 'total metadrenalines' ('total metanephrines'), which include both of these metabolites and their conjugates. A few laboratories determine the individual compounds, thereby obtaining a measure of the separate output of the parent catecholamines. 'Total metadrenalines' is a widely performed test.

(3) *Noradrenaline and adrenaline,* and their conjugated derivatives. Some laboratories measure total urinary catecholamines, after hydrolysing the conjugates; others measure noradrenaline and adrenaline separately. Few laboratories perform these tests.

(4) *4-hydroxy, 3-methoxy phenylglycol* (HMPG). Few laboratories perform this measurement routinely.

Several points concerning the collection and timing of urine specimens, for the investigation of catecholamine metabolism, are worth noting:

(1) *Drugs.* Many drugs affect the release and metabolism of catecholamines (e.g. reserpine, monoamine oxidase inhibitors).

Others (e.g. methyldopa, labetalol) interfere with some methods of estimation. Patients should preferably be investigated for suspected phaeochromocytoma before treatment is started. However, if treatment has been started and cannot be discontinued even temporarily, details of drugs prescribed should be provided to the laboratory, and treatment be kept the same for at least two weeks before specimens are collected for analysis.

(2) *Diet.* Some analytical methods are liable to interference from dietary constituents. The laboratory's requirements for dietary restriction must therefore be observed.

(3) *Timing: Patients who have 'attacks'.* If the patient is normotensive between attacks, there is no point in sending specimens repeatedly for analysis when the blood pressure is normal. Instead, a single baseline set of values should be determined against which to compare results obtained following an attack, and the patient instructed to start the next 24-h urine collection when the next attack occurs.

(4) *Number of specimens: Patients with persistent hypertension.* Where investigations are indicated (see below), a *single* measurement of 24-h urinary excretion of HMMA or of 'total metadrenalines' is the most that can normally be justified, to exclude phaeochromocytoma.

Plasma catecholamines

It is now possible to measure plasma [adrenaline] and plasma [noradrenaline] reliably. However, the methods are complex and are not widely available. Also, the concentrations of catecholamines in plasma are liable to fluctuate markedly in response to stress, so urinary measurements seem likely to retain their value as diagnostic tests.

Extra-adrenal phaeochromocytoma can be localized by determining the concentration of plasma catecholamines in blood specimens obtained from veins draining various possible tumour sites. However, this is rarely needed.

Selection and interpretation of tests

In most patients in whom there is a clinical suspicion of phaeochromocytoma, the diagnosis can be made on the basis of an increased 24-h urinary excretion of HMMA or 'total metadrenalines'. However, it can be difficult to decide the significance of slightly increased outputs, since these may be observed in patients with essential hypertension and in otherwise healthy individuals subjected to stress, as well as in some patients with phaeochromocytoma.

Where the clinical index of suspicion is high (i.e. the patient has paroxysmal symptoms), it may be necessary (1) to repeat the measurement of urinary HMMA or 'total metadrenalines' on several occasions, and (2) to measure the urinary excretion of metabolites not normally measured in that laboratory (e.g. both 'total metadrenalines' and HMMA) or alternatively plasma [catecholamines] if these measurements are available.

Hypertension is common and phaeochromocytoma is rare. Where a patient presents with hypertension, without paroxysmal symptoms, i.e. *where the clinical index of suspicion is low,* there is much less justification for requesting chemical investigations of catecholamine metabolites. In general, only if the patient is a young hypertensive, or has other features that could suggest the presence of phaeochromocytoma, is screening of urine specimens for abnormal excretion of catecholamines or their metabolites worthwhile.

NEUROBLASTOMA AND GANGLIONEUROMA

This group of tumours is rare. However, it is one of the commonest among the tumours of childhood. Excessive formation of dopamine is a practically constant feature.

Dopamine overproduction is not associated with marked pharmacological effects, and patients with neuroblastoma or ganglioneuroma only occasionally have abnormalities of blood pressure despite the fact that the urinary output of HMMA and of 'total metadrenalines' is often markedly increased. This is because the catecholamines are metabolized by the tumour tissue before release of their pharmacologically inactive metabolites into the circulation.

Some patients with neuroblastoma and ganglioneuroblastoma excrete cystathionine. This amino acid is not normally present in urine. It has only been reported in these patients and, rarely, in hepatoblastoma or in an inherited metabolic disorder.

The following diagnostic tests may be performed on urine specimens, the measurements usually being related to urinary [creatinine] because of the difficulties of obtaining complete 24-h urine collections in children:

(1) *HMMA* or another quantitative measure of the output of noradrenaline and adrenaline. These are the tests most often performed initially. Abnormal excretion of HMMA, etc., in children, nearly always points to a diagnosis of neuroblastoma or ganglioneuroblastoma; very occasionally, it signifies the presence of phaeochromocytoma.

(2) *4-hydroxy, 3-methoxy phenylacetic acid* (homovanillic acid, HVA). This is the metabolite of dopamine excreted in largest amount; greatly increased excretion occurs in patients with neuroblastoma or ganglioneuroblastoma. Measurements of HVA can now be performed in many laboratories.

(3) *Cystathionine excretion.* This does not necessarily parallel the excretion of catecholamines and metabolites in these patients. Its detection in urine serves as a useful independent diagnostic pointer.

OTHER CATECHOLAMINE-SECRETING TUMOURS

Retinoblastoma, another rare condition, is sometimes the cause of increased urinary excretion of catecholamines and metabolites.

Carcinoid syndrome

Carcinoid tumours all produce 5-hydroxytryptamine (5-HT or serotonin). They can arise in any tissue derived from the endoderm, but much the commonest site is the terminal ileum. In some sites (e.g. the rectum), the tumours rarely metastasize, in which case they do not produce the symptoms of the carcinoid syndrome. Tumours arising in other sites may metastasize; in the case of ileal tumours, the liver is the main site of their metastases.

In general, since the liver can metabolize the pharmacologically active products of carcinoid tumours, it is only tumours that have given rise to secondary deposits in the liver, or tumours of which the venous drainage is not into the portal circulation (e.g. bronchial adenomas of the carcinoid type), that give rise to the carcinoid syndrome. The main presenting features include flushing attacks, abdominal colic and diarrhoea, and dyspnoea sometimes associated with asthmatic attacks. Valvular disease of the heart is often present.

The carcinoid syndrome is usually associated with a tumour of the terminal ileum and extensive secondary deposits in the liver. Most carcinoid tumours secrete excessive amounts of 5-HT and its metabolite 5-HIAA (Fig. 24.2). The 'atypical carcinoids' of foregut origin, however, contain excessive amounts of 5-hydroxytryptophan (5-HTP) and relatively little 5-HT, because the tumour tissue is deficient in 5-HTP decarboxylase; these atypical tumours may also secrete histamine. High concentrations of prostaglandin $F_{2\alpha}$ have also been reported in bronchial carcinoids.

Most carcinoid tumours show a positive argentaffin reaction, but 'atypical' carcinoid tumours do not.

Although 5-HT is an active pharmacological agent, the release of abnormal amounts of 5-HT into the circulation does not account for all the clinical features of the carcinoid syndrome. Many features can be attributed to the vasodilating kinin, bradykinin. Increased amounts of bradykinin have been demonstrated in plasma from patients with the carcinoid syndrome; it is formed after release into the circulation of kallikrein, a proteolytic enzyme present in carcinoid tissue.

Only about 1% of dietary tryptophan is normally metabolized along

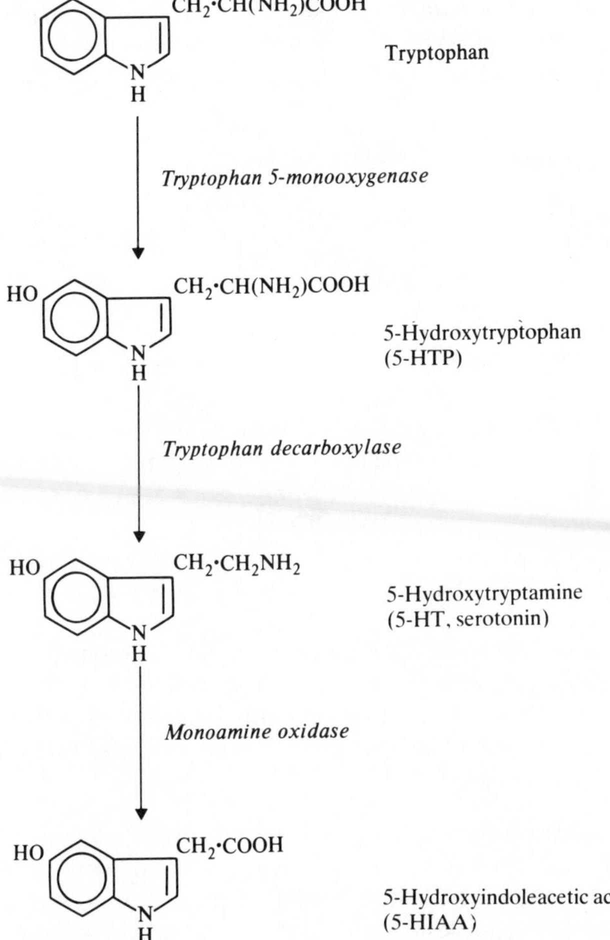

FIG. 24.2. The metabolism of tryptophan via the 5-hydroxyindole pathway. The other pathways of tryptophan metabolism are outlined in Fig. 13.2 (p. 270).

the 5-hydroxyindole pathway (Fig. 13.2, p. 270), but overproduction of 5-HT is a constant feature of the carcinoid syndrome. In these patients as much as 60% of dietary tryptophan may be metabolized to 5-HTP, 5-HT and finally 5-HIAA. There is sometimes a consequent shortage of tryptophan for incorporation into protein, and carcinoid tumours can give rise to severe hypoproteinaemia and oedema, even in the absence of cardiac complications. There may also be signs of nicotinic acid deficiency.

Some carcinoid tumours produce ACTH or ACTH-like peptides and may cause Cushing's syndrome (p. 359). Other metabolically active intestinal hormones (e.g. pancreatic polypeptide, gut glucagon-like immunoreactivity) have been detected in small amounts in some carcinoid tumours.

Chemical investigation of 5-HT metabolism

Measurement of 5-HIAA excretion in a 24-h urine specimen is the most widely performed investigation; the output is often greatly increased. The following points need to be observed before or during the collection of specimens:

(1) *Drugs.* Urinary 5-HIAA output should be determined before treatment is started, especially treatment with drugs such as reserpine which release 5-HT from body stores. However, if treatment has already started, the form of treatment should be stabilized for at least two weeks before 5-HIAA excretion is measured.

(2) *Diet.* Bananas and tomatoes contain large amounts of 5-HT, and should not be eaten the day before nor during the 24-h period of urine collection.

(3) *Timing of collection.* If the patient is having frequent attacks of symptoms, the time of starting the urine collection is unimportant. However, if attacks are only occasional (i.e. less often than daily), the patient should be instructed to begin the collection when the next attack takes place.

More information about the pattern of 5-hydroxyindole metabolites excreted may be required in some patients. It can be obtained from chromatographic examination of urine. If increased amounts of 5-HTP are found, this indicates the presence of an 'atypical carcinoid'. These are, in general, tumours in the stomach, pancreas or biliary tract (i.e. foregut tumours) and some bronchial tumours.

Other 5-HT-producing tumours

Various tumours have in some cases been found to be associated with abnormal metabolism of tryptophan. In these tumours, there are increased amounts of 5-HT, and urinary excretion of 5-HIAA may be increased. These include small cell carcinoma of the bronchus, carcinoma of the pancreas and islet cell tumour of the pancreas, hepatoma, and medullary carcinoma of the thyroid. Although overproduction of 5-HT by these tumours and excessive 5-HIAA excretion have been described, it is rare for them to be associated with the features of the carcinoid syndrome. These tumours show a negative argentaffin reaction, and do not resemble carcinoid tumours.

Medullary carcinoma of the thyroid

This is a tumour of the parafollicular, calcitonin-producing cells (C-cells) of the thyroid gland. The relatively common, non-familial form accounts for about 10% of cases of carcinoma of the thyroid.

Calcitonin, in pharmacological dosage, lowers the plasma [calcium] and decreases the tubular reabsorption of phosphate. It is not yet known for certain whether these effects are important for normal calcium homeostasis in man.

Patients with medullary carcinoma of the thyroid usually have a normal plasma [calcium] but increased fasting plasma calcitonin and katacalcin concentrations (p. 281). In those patients with normal plasma [calcitonin], stimuli such as an injection of pentagastrin or a drink of whisky (which stimulates the release of gastrin) provoke the release of calcitonin and katacalcin and their plasma concentrations rise markedly.

A few patients with medullary carcinoma of the thyroid have been described with intermittent or persistent hypocalcaemia, increased plasma [calcitonin] and sometimes a raised plasma histaminase activity.

Multiple endocrine neoplasia (MEN) syndromes

Medullary carcinoma of the thyroid occurs rarely as one component of two different forms of MEN II syndromes; the MEN I syndrome is discussed elsewhere (p. 291). The index cases of these familial syndromes usually present in childhood. The two forms of MEN II are:

(1) *MEN IIa* (MEN 2) or Sipple syndrome. In this form, the thyroid carcinoma occurs in association with phaeochromocytoma and with hyperplasia of the parathyroid glands; in about 20% of patients there is primary hyperparathyroidism.

(2) *MEN IIb* (MEN 3). In this form, the thyroid carcinoma again occurs in association with phaeochromocytoma. However, primary hyperparathyroidism is not a feature. MEN IIb is distinguishable from MEN IIa by the occurrence of multiple neuromatous lesions, and frequently by the presence of a myopathy and marfanoid features.

In MEN II, it is generally held to be advisable to measure fasting plasma [calcitonin] in relatives of the patient; if this is normal, the measurement should be repeated after stimulation by pentagastrin or whisky.

Other calcitonin-producing tumours

Increased plasma [calcitonin] sometimes occurs in association with carcinomas of several tissues other than the thyroid, but especially small cell carcinoma of the lung, and can be used as a marker for these. Plasma [calcium] is not reduced in these patients.

TUMOUR MARKERS

Many tumour markers have been described. So far, however, their measurement has been of little value in the detection of asymptomatic cancer.

The measurement of tumour markers may help with the early detection of local recurrence of malignant disease after treatment, or in detecting the development of metastases. The main groupings of tumour markers are hormones and hormone-like substances, enzymes or isoenzymes, and tumour antigens.

HORMONES AS TUMOUR MARKERS

The secretion of hormones by tumours of the endocrine glands is described as 'appropriate'. Recognition of various of these tumours by chemical tests has been discussed elsewhere. Hormones and hormone-like substances, mostly polypeptides, may also be produced by non-endocrine tumours. These 'inappropriate' or ectopic products together make up one important group of tumour markers. The principal examples are listed in Table 24.1, but many more have been described.

The production of hormones by non-endocrine tumours is, in general, not subject to any of the normal physiological feedback mechanisms of control. However, the products of these tumours may stimulate the target organs and have indirect negative feedback effects on the hypothalamus and pituitary, where hypothalamic or pituitary control normally applies.

Sometimes the clinical and chemical features of the 'inappropriate' hormone syndromes may differ significantly from the corresponding 'appropriate' syndromes, e.g. Cushing's syndrome (p. 359). At other times the distinction may be more difficult, in which case the existence of the non-endocrine tumour may not be suspected until the endocrine gland normally responsible for the production of the hormone has been removed.

Some tumours are particularly associated with the ectopic production of hormones. For example, small cell carcinoma of the lung may produce ectopic hormones in 25–30% of patients; ACTH and LPH, and ADH are the hormones most often released by these tumours.

The polypeptide hormones produced by tumours seem to be more variable in their primary amino acid structure than the corresponding hormones that are normally secreted by the 'appropriate' endocrine

TABLE 24.1. Tumour markers.

Class of marker	Examples	Tumour sources (examples)
Hormones	ACTH	Chromophobe adenoma of pituitary *Carcinoma of lung, colon, prostate, ovary*
	ADH	*Carcinoma of lung*
	Calcitonin	Medullary carcinoma of thyroid *Carcinoma of lung, pancreas, prostate, breast*
	HCG	Hydatidiform mole, choriocarcinoma *Carcinoma of liver, lung*
	Insulin	Insulinoma *Lymphosarcoma*
	Prolactin	*Carcinoma of lung*
	PTH	Parathyroid tumours *Carcinoma of lung, kidney, liver, breast*
Tumour antigens	AFP	Hepatoma, teratoma
	CEA	Carcinoma of colon and other gastrointestinal sites
Enzymes	Acid phosphatase	Carcinoma of prostate
	Alkaline phosphatase (Regan isoenzyme)	Carcinoma of gastrointestinal tract, liver

Note: Tissues giving rise to ectopic or inappropriate production of hormones are shown in italics.

gland. The tumour-generated polypeptides also seem to give rise to a greater number of hormone fragments. However, most of the ectopic material that has hormonal activity is probably identical in structure to the hormone produced normally. Confirmation of the diagnosis of a non-endocrine tumour depends, in most cases, on the same chemical investigations as have been described elsewhere for use in the diagnosis of the corresponding truly endocrine conditions.

ENZYMES AS TUMOUR MARKERS

Although many alterations in the activity of enzymes in plasma have been described in association with neoplastic disease, few specific associations have been identified. The best example is the association between increased plasma acid phosphatase activity and the presence of metastatic carcinoma of the prostate in bone (p. 150).

A small number of tumours, usually of gastrointestinal or liver origin, produce atypical isoenzymes of alkaline phosphatase. The commonest of these, the Regan isoenzyme (p. 148), strongly resembles the alkaline phosphatase that is present in placenta (heat-stable alkaline phosphatase, p. 409) but which is not normally found in non-pregnant adults.

Several other rare tumour-associated isoenzymes of alkaline phosphatase have been described. However, none of the atypical alkaline phosphatase isoenzymes has been shown to be of much diagnostic significance, although it has been suggested that very sensitive methods of analysis for heat-stable alkaline phosphatase may be able to detect early carcinoma.

Tumour antigens as markers

The two antigens that have been most fully evaluated are CEA and AFP. The extent of their usefulness and their limitations as diagnostic tests, and as aids to the management of patients who have been treated for cancer, are now reasonably well defined. Both CEA and AFP are normally detectable in small amounts in plasma (p. 121). Both may be increased in the presence of non-neoplastic conditions as well as in the presence of malignant disease, but the largest increases are nearly always due to malignancy.

CARCINOEMBRYONIC ANTIGEN

The National Institutes of Health (NIH) consensus statement (1981) recommended that measurements of plasma [CEA] should not be used in isolation to establish a diagnosis of cancer. However, in a patient with symptoms, if plasma [CEA] is increased to more than five times the upper reference value (i.e. to over 50 μg/L), this is considered to be strong evidence pointing to the presence of cancer.

The consensus statement also reviewed the place of plasma CEA measurements in reaching a pre-operative prognosis in patients with cancer. It concluded that plasma [CEA] should be measured pre-operatively in patients with colorectal or with bronchial carcinomas, the information so gained to be added to data for clinical and pathological staging. The statement stressed, however, that a low plasma [CEA] does not necessarily imply a good prognosis. About 20% of patients in whom malignant disease is later proved never have raised plasma [CEA]. Also, poorly differentiated colorectal carcinomas would seem, in general, not to synthesize CEA to such an extent, nor to release as much CEA, as better differentiated tumours.

The main place of plasma CEA measurements is in the postoperative monitoring of patients after surgical removal of colorectal carcinoma. Plasma [CEA] should be measured at follow-up assessments postoperatively. If the concentration was increased pre-operatively, it should return to normal within six weeks, following successful operation. Thereafter, if plasma [CEA] starts to rise again slowly, this may indicate local recurrence whereas rapidly increasing plasma [CEA] is usually associated with metastases.

α_1-FETOPROTEIN

At one time thought to be a specific tumour marker for primary hepatocellular carcinoma, it is now known that the concentration of AFP in plasma is increased in association with malignant tumours arising in many other organs besides the liver. Plasma [AFP] may also be increased in several non-neoplastic conditions; in these cases the increases may only be temporary, associated with the tissue's (usually the liver's) response to injury, and usually of lesser magnitude.

The main diagnostic value of plasma AFP measurements, in neoplastic disease, is in the recognition of hepatoma, where greatly increased amounts are detectable in 50–90% of patients, and malignant teratoma where 50–70% of patients have been reported to show marked increases. Increased plasma [AFP] is also found in about 20% of patients with gastric, pancreatic and biliary tract neoplasms.

Measurements of plasma [AFP] are used to monitor the course of malignant teratoma and its response to treatment and can be used for this purpose with hepatoma. They are of little value with those tumours in which plasma [AFP] is only occasionally increased.

Screening for neoplastic disease

Side-room tests for occult blood in faeces are widely used as screening tests for carcinoma of the gastrointestinal tract, especially colorectal cancer. Many other chemical measurements, mainly on blood specimens, have been proposed as means of detecting various forms of neoplastic disease at an early stage. However, the lack of any truly tumour-specific marker for the commonly occurring carcinomas means that the introduction of such screening tests would have to be backed by extensive facilities for follow-up investigation. These proposals, therefore, do not as yet begin to be either feasible or cost-effective.

At one time, CEA was thought to have potential as a plasma measurement, in screening for cancer. However, there is considerable overlap between plasma [CEA] observed in patients with inflammatory diseases and in benign and malignant tumours occurring in many sites besides the gastrointestinal tract. It is, therefore, impossible readily to interpret small departures of plasma [CEA] from normality. Also, plasma [CEA] is increased as a result of smoking. The NIH consensus statement concluded that plasma CEA has neither the sensitivity nor the specificity required for use as a screening test.

Miscellaneous chemical tests in neoplasia

Plasma calcium

Malignant disease is the commonest cause of hypercalcaemia in hospital patients (p. 283). It is not an early sign, and the presence of the tumour may already have been recognized.

Hypercalcaemia is usually caused by metastatic disease in bone. However, in these patients it may also result from immobilization, or from secretion of PTH or PTH-like peptides by the tumour.

Plasma proteins

Apart from the tumour markers already discussed, several non-specific changes in the concentrations of other proteins normally present in plasma have been described in patients with cancer. These changes mostly consist of increases in the acute phase reactants (p. 123).

Some tumours, notably carcinoma of the breast, produce various proteins normally only detectable in plasma in trace amounts. These include the milk proteins (casein, lactoferrin and lactalbumin) and pregnancy-associated α_2-glycoprotein.

Non-specific measures of response to treatment

Measurements of plasma acid phosphatase activity, or plasma [CEA], [AFP] or [HCG], can all be used to provide indices of response to treatment, or evidence of recurrence or the development of metastases, for appropriately selected tumours (Table 24.1). Other, less specific indices of general response by tumours to treatment have been investigated. None has become widely accepted as yet, but the following are two examples:

(1) *Polyamine excretion.* Putrescine, spermine and spermidine are excreted in increased amounts in many inflammatory and neoplastic conditions. Measurements of urinary polyamines do not provide a specific index of malignant disease, but have been used to assess response to treatment of tumours.
(2) *Cyclic GMP excretion.* Urinary cyclic GMP is increased in a large number of conditions, both inflammatory and neoplastic. The measurements have been used as a sensitive but non-specific index of response to treatment of tumours.

Conclusion

Knowledge of tumour markers is developing rapidly, and we have only touched on the subject in this chapter. Although some of the markers lack specificity, it seems possible that multiparametric combinations of tests will reduce this drawback. Also, new markers with much greater specificity for detecting particular tumours seem likely to be discovered; specific monoclonal antibodies may then be raised against these markers.

The sensitivity and specificity of chemical tests for neoplastic disease are likely to improve, and tests of established value will become more widely available. Consequently, the contributions of clinical chemistry and related laboratory investigations to (1) the early recognition of neoplastic disease, (2) the assessment of tumour size, rate of growth and prognosis, and (3) the monitoring of treatment and follow-up surveillance, can be expected to increase.

FURTHER READING

ALLISON, D.J., BROWN, M.J., JONES, J.H. and TIMMINS, J.B. (1983). Role of venous sampling in locating a phaeochromocytoma. *British Medical Journal,* **286,** 1122–1124.

ATKINSON, R.J. (1981). Tumour markers in the detection of gynaecological cancer. *British Journal of Hospital Medicine,* **26,** 381–386.

BUCKMAN, R. (1982). Tumour markers in clinical practice. *ibid.,* **27,** 9–20.

COOMBES, R.C. (1982). Metabolic manifestations of cancer. *ibid.,* **27,** 21–27.

DEPIERRE, D., JUNG, A., CULEBRAS, J. and ROTH, M. (1983). Polyamine excretion in the urine of cancer patients. *Journal of Clinical Chemistry and Clinical Biochemistry,* **21,** 35–37.

GOLDENBERG, D.M., NEVILLE, A.M., CARTER, A.C. *et al.* (1981). Carcinoembryonic antigen: its role as a marker in the management of cancer. **Summary of an NIH** consensus statement. *British Medical Journal,* **282,** 373–375.

GRAHAME-SMITH, D.G. (1968). The carcinoid syndrome. *American Journal of Cardiology,* **23,** 376–387.

HARDCASTLE, J.D., FARRANDS, P.A., BALFOUR, T.W., CHAMBERLAIN, J., AMAR, S.S. and SHELDON, M.G. (1983). Controlled trial of faecal occult blood testing in the detection of colorectal cancer. *Lancet,* **2,** 1–4.

Leading article (1972). The carcinoid syndrome — 20 years on. *Scottish Medical Journal,* **17,** 234–236.

Leading article (1983). Plasma calcitonin in medullary thyroid carcinoma. *Lancet,* **1,** 338–339.

NEVILLE A.M. and COOPER, E.H. (1976). Biochemical monitoring of cancer. *Annals of Clinical Biochemistry,* **13,** 283–305.

NEVILLE, A.M., GRIGOR, K. and HEYDERMAN, E. (1976). Clinicopathological role of tumour index substances in paediatric neoplasia. *Journal of Clinical Pathology,* **29,** 1026–1032.

PLOUIN, P.F., DUCLOS, J.M., MENARD, J., COMOY, E., BOHUON, C. and ALEXANDRE, J.M. (1981). Biochemical tests for diagnosis of phaeochromocytoma: urinary versus plasma determinations. *British Medical Journal,* **282,** 853–854.

RATCLIFFE, J.G. (1982). Ectopic production of hormones in malignant disease. In *Recent Advances in Endocrinology and Metabolism,* **2,** pp. 187–209. Ed. J.L.H. O'Riordan. Edinburgh: Churchill Livingstone.

ROBINSON, R. (1980). *Tumours that Secrete Catecholamines.* Chichester: John Wiley.

SCHWARTZ, M.K. (1973). Enzymes in cancer. *Clinical Chemistry,* **19,** 10–22.

SHERWOOD, L.M. (1980). The multiple causes of hypercalcaemia in malignant disease. *New England Journal of Medicine,* **303,** 1412–1413.

TURNER, G.A., ELLIS, R.D., GUTHRIE, D. *et al.* (1982). Urinary cyclic nucleotide concentrations in cancer and other conditions; cyclic GMP: a potential marker for cancer treatment. *Journal of Clinical Pathology,* **35,** 800–806.

Chapter 25

The CNS and CSF

Many neurological disorders would appear either to have a biochemical basis or to be associated with disturbances of metabolism, and much biochemically oriented research has been directed towards their better understanding. For instance, attempts have been made to explain various disorders of movement (e.g. epilepsy, Parkinson's disease, Huntington's chorea) and some of the major psychoses (e.g. schizophrenia) in terms of altered metabolism of neurotransmitters. So far, however, no routine diagnostic chemical tests have been introduced as a result of this work, although therapeutic control of some of the conditions has been improved on the basis of the theories, and the laboratory may then have some part to play, in therapeutic drug monitoring (p. 477).

Chemical investigations have a more positive role in the diagnosis and management of inherited metabolic disorders that involve the CNS. Also, chemical examination of the CSF is important in the investigation of many diseases that affect the CNS.

Inherited metabolic disorders

MENTAL RETARDATION

Many inherited metabolic disorders are associated with mental retardation. As a rule, the enzymic defect is general, affecting other organs besides the CNS, especially the liver. The defect can often be characterized by measuring the activity of the enzyme or by investigating the pattern of abnormal metabolites (Table 25.1). There is then a gap in understanding, since the way in which the defective enzyme or the accumulation of metabolites affects mental function is not clear.

The value of mass screening for neurometabolic disease has been established for only a limited number of conditions (e.g. PKU). For the rest, the main diagnostic effort is applied to family studies. There is a wide variety of clinical features associated with these inherited metabolic disorders that give rise to mental retardation, and specialized chemical determinations are needed to identify many of the conditions precisely.

TABLE 25.1. Inherited metabolic diseases that can cause mental retardation. (To illustrate the diverse nature of the metabolic abnormalities)

Groupings and examples	Enzyme defect or other abnormality	Diagnostic tests (examples)
Amino acid metabolism		
Phenylketonuria (p. 419)	Phenylalanine hydroxylase (in the classical form)	Guthrie test, urine chromatography
Hartnup disease (p. 98)	Epithelial transport defect	Urine chromatography
Carbohydrate metabolism		
Galactosaemia (p. 239)	Galactose-1-phosphate uridylyl-transferase, or galactokinase	Urine chromatography, red cell enzyme studies
Pompe's disease (p. 237)	α-D-Glucosidase	Tissue enzyme studies
Mucopolysaccharidoses		
Hurler's syndrome	α-L-Iduronidase	Urinary mucopolysaccharides
Lipid storage disorders		
Tay Sachs disease (p. 410)	β-N-Acetyl-D-hexosaminidase	Tissue enzyme and ganglioside studies
Gaucher's disease	Glucosylceramidase	Tissue enzyme and glycolipid studies
Purine metabolism		
Lesch-Nyhan syndrome (p. 334)	Hypoxanthine phosphoribosyl transferase	Red cell enzyme studies
Hormone synthesis		
Familial hypothyroidism	Iodide uptake defect, peroxidase defect, etc.	Blood [TSH] or [T4], tissue enzyme studies

Many of these disorders have an inevitably fatal outcome at present. However, there are some (e.g. PKU, p. 419; galactosaemia, p. 239) where early identification of the defect and institution of the appropriate treatment can prevent deterioration in mental function, or stop further deterioration from occurring.

DEGENERATIVE DISORDERS

Wilson's disease is an example of a treatable metabolic disorder which affects the CNS, and which can be diagnosed by chemical tests (p. 187).

Cerebrospinal fluid

The CSF approximates to an ultrafiltrate of plasma. There are, however, differences between the relative concentrations of low mol. mass substances in plasma and CSF, and between the relative concentrations of high mol. mass substances, for the following reasons:

(1) *Low molecular mass substances*

(a) *Differential rates of diffusion.* For instance, dissolved CO_2 diffuses into CSF more rapidly than HCO_3^-, so the $[H^+]$ of CSF (which depends on the HCO_3^-/H_2CO_3 ratio) may sometimes be significantly different from plasma $[H^+]$.

(b) *Effects of ultrafiltration.* For example, in plasma about 30–45% of the calcium and nearly all the bilirubin are normally bound to protein, but only the unbound components can cross the blood-brain barrier. There is an equilibrium between the free or unbound components in plasma and their amount in the CSF.

(2) *High molecular mass substances*

(a) *Differential rates of diffusion.* Smaller proteins (e.g. albumin) are present in CSF in relatively higher concentrations than larger proteins. Even so, because of the blood-brain barrier, CSF [total protein] is only about 0.5% of plasma [total protein].

(b) *Secretion of proteins* within the CNS, e.g. immunoglobulins (see below).

CHEMICAL EXAMINATION OF CSF

CSF is mostly collected by lumbar puncture; it should be clear and colourless. Abnormal appearances should be noted at the time of collection. Examinations for cells and micro-organisms are often more important than chemical investigations, so CSF specimens for these other tests should, in general, be collected first in case the flow proves inadequate for all the tests envisaged. Separate specimens should be sent to the various laboratories, to reduce the possibility of contamination occurring, and to ensure that samples are received by each laboratory as quickly as possible; some CSF constituents are labile.

At lumbar puncture it is normal practice to collect at least two samples of CSF for chemical analysis, (1) into a fluoride-oxalate tube for measurement of [glucose] and (2) into a plain tube (no preservative) for other analyses such as protein investigations. If a blood vessel is accidentally damaged, the CSF will initially contain blood but subsequent samples may not. Under these circumstances, only the later specimens should be sent for chemical examination.

Appearance

The fluid is normally clear and colourless. Turbidity is usually due to leucocytes, but may be due to micro-organisms. A clot may form in CSF samples when the [protein] is high.

Xanthochromia (yellow colour) may be apparent or, in the case of blood-stained specimens, it should be looked for after centrifuging. It is most often due to previous haemorrhage into the CSF, but may also occur when CSF [protein] is very high. Patients with jaundice may have yellow CSF due to the presence of bilirubin.

Blood in the CSF may indicate recent haemorrhage or the puncturing of a vein during lumbar puncture. The presence of even small amounts of fresh blood has marked effects on the results of CSF protein measurements.

Glucose

The CSF [glucose] in fluid from the ventricles or the cisterna magna does not normally differ significantly from plasma [glucose], whereas lumbar CSF [glucose] is usually about 0.5–1.0 mmol/L (10–20 mg/100 mL) lower than plasma [glucose].

CSF [glucose] may be very low in patients with hypoglycaemia; conversely, it is raised in hyperglycaemic states. It may also be low, or even undetectable, in patients with acute bacterial, cryptococcal or carcinomatous meningitis, probably due to consumption of glucose by polymorphonuclear leucocytes, other rapidly metabolizing cells or bacteria.

Measurements of CSF [glucose] are of most value in differential diagnosis when the CSF cell count is only moderately raised. In patients with tuberculous meningitis or cerebral abscess, it may be markedly reduced or undetectable, whereas in cases of meningitis or encephalitis due to viruses the CSF [glucose] is usually normal.

Total protein

Reference values for [protein] in lumbar CSF are 0.1–0.4 g/L (10–40 mg/100mL); this consists almost entirely of albumin. Lower concentrations are found in ventricular and cisternal CSF. Considerably higher values (up to 0.9 g/L) have been reported for lumbar CSF in the neonatal period.

CSF [protein] is increased in a large number of pathological conditions affecting the CNS. The main value of the measurement derives from the fact that, if there is an increased CSF [protein], this strongly indicates the presence of organic disease of the CNS. For example:

(1) *Infection of the CNS.* The increase may be slight, but is sometimes very marked. Capillary permeability is increased in acute and in many chronic inflammatory conditions of the CNS.
(2) *Demyelinating disorders.* The increase is often only moderate, with values usually in the range 0.5–1.0 g/L (50–100 mg/100 mL).
(3) *Primary and secondary neoplasms* involving the brain or the meninges. Very large increases in CSF [protein], to values in excess of 5 g/L, may be observed in patients who have a spinal block, usually due to a tumour which interferes with the circulation of CSF over the spinal cord. These specimens may be xanthochromic, and a clot may form on standing.

Blood in the CSF. The addition of 50 μL blood to a 2 mL sample of CSF in which the CSF [protein] is in fact normal will increase the [protein] approximately fourfold. It will also add high mol. mass proteins that are normally absent. This emphasizes the importance of obtaining CSF specimens atraumatically.

Immunoglobulins

CSF normally contains small amounts of IgG (8–64 mg/L), a trace of IgA but no IgM. Increases in CSF [immunoglobulins], particularly CSF [IgG], may be due to (1) an increased exudation of plasma proteins including immunoglobulins into the CSF, or (2) increased local synthesis of immunoglobulins. If there is increased local synthesis of IgG, the ratio CSF [IgG]:CSF [albumin] or CSF [IgG]:CSF [total protein] will be higher than normal.

Increased synthesis of IgG in the CNS occurs in a number of disorders including multiple sclerosis, neurosyphilis and subacute sclerosing panencephalitis. In all these conditions, a limited number of clones of B-cells produce immunoglobulins. These cause a number of discrete, oligoclonal bands demonstrable by electrophoresis of CSF.

It can be difficult to confirm a diagnosis of *multiple sclerosis*. The following tests can be performed:

(1) *CSF [IgG]*. This test lacks specificity. As the sole measurement, it suffers from the disadvantage that any cause of increased CSF [total protein] will also tend to cause an increase in CSF [IgG].
(2) *CSF [IgG]:CSF [albumin]*. This ratio renders the CSF [IgG] measurement more specific for CNS immunoglobulin synthesis, but is only abnormal in about 60–70% of cases.
(3) *CSF electrophoresis* for the presence of oligoclonal bands that are not present in serum. This test is abnormal in over 90% of cases of multiple sclerosis. However, the method is not widely available, and results require skill for their interpretation.

Measurement of CSF [IgM] can help to distinguish between patients with meningism, in whom IgM is undetectable, and patients with acute bacterial or viral meningitis. Increases in CSF [IgM] are particularly marked in patients with bacterial meningitis.

Enzymes

The activities of many enzymes in CSF have been measured, e.g. lactate dehydrogenase and its isoenzymes, aspartate aminotransferase, creatine kinase and its B–B isoenzyme, and adenylate kinase. Changes are usually confined to the CSF as the enzymes do not usually escape from the subarachnoid space. Plasma enzyme activities do not show any characteristic patterns of alteration as a consequence of CNS disorders.

None of these enzyme measurements on CSF specimens has an established role in diagnosis. With most of the enzymes so far investigated, increased activities have been detected in a wide variety of conditions. However, adenylate kinase may prove of value as a marker for the presence of cerebral tumours; adenylate kinase is not normally detectable in CSF.

Chloride

Measurement of CSF [Cl^-] is an obsolete investigation. It provides no information additional to that obtained from the history of the patient's illness, clinical examination, and assessment of the state of fluid and electrolyte balance.

FURTHER READING

BRADBURY, M. (1979). *The Concept of a Blood-Brain Barrier*. Chichester: John Wiley.

CROW, T.J. and JOHNSTONE, E.C. (1982). Biochemical bases of the psychoses. In *Biochemical Aspects of Human Disease,* pp. 481–518. Ed. R.S. Elkeles and A.S. Tavill. Oxford: Blackwell Scientific Publications.

DAVISON, A.N., editor (1976). *Biochemistry and Neurological Disease.* Oxford: Blackwell Scientific Publications.

EADIE, M.J. and TYRER, J.H. (1983). *Biochemical Neurology*. Lancaster: MTP Press.

GRANT, G.H. (1978). Examination of cerebrospinal fluid protein. *Association of Clinical Pathologists Broadsheet* **88.** London: British Medical Association.

LASCELLES, P.T., THOMPSON, E.J. and WARNER, D.S. (1983). Chemistry of the cerebrospinal fluid in health and disease. In *Biochemistry in Clinical Practice,* pp. 445–455. Ed. D.L. Williams and V. Marks. London: Heinemann.

MCKERAN, R.O. (1982). The biochemistry of disease of the central nervous system. In *Biochemical Aspects of Human Disease,* pp. 435–480. Ed. R.S. Elkeles and A.S. Tavill. Oxford: Blackwell Scientific Publications.

STANBURY, J.B., WYNGAARDEN, J.B., FREDRICKSON, D.S., GOLDSTEIN, J.L. and BROWN, M.S., editors (1983). *The Metabolic Basis of Inherited Disease*, 5th Edition. New York: McGraw-Hill.

STERN, J. (1983). Hereditary and acquired mental deficiency. In *Biochemistry in Clinical Practice*, pp. 489–523. Ed. D.L. Williams and V. Marks. London: Heinemann.

Chapter 26

Therapeutic Drug Monitoring and Chemical Toxicology

Measurements performed by clinical chemistry departments can nowadays play an important part in the investigation and management of patients, both in respect of the therapeutic uses of drugs and under circumstances where drugs and poisons have been taken in overdose or otherwise misused. The laboratory also has a part to play in monitoring the environment, and in the detection of some industrial and occupational hazards. This wide range of topics will be briefly reviewed in this chapter.

THERAPEUTIC DRUG MONITORING

In the last 10–15 years, it has become apparent that the control of several forms of drug treatment could be substantially improved if steady-state drug concentrations in blood specimens collected from individual patients were to be known. At the same time, it has become abundantly clear that therapeutic drug monitoring (TDM) is no substitute for careful clinical assessment and surveillance; also, that the dosage of many drugs can be controlled satisfactorily by other means, and that TDM has so far only been shown to be essential, or highly desirable, for a limited number of drugs.

Therapeutic drug monitoring has become an increasingly practical proposition, due partly to improvements in laboratory instrumentation, and partly to the development of quick, simple and yet at the same time specific methods, mostly based on immunoassays, for drug measurements at therapeutic levels. Plasma (or serum) drug concentrations can be used to guide the treatment of individual patients if the following are known:

(1) The therapeutic range of plasma concentrations for the drug.
(2) The clinical pharmacokinetics of the drug. For TDM to be of value, the criteria for establishing steady-state conditions should normally have been met.
(3) The timing of the drug's administration to the patient, in relation to the time of collecting the blood specimen.
(4) Details of other drug treatment, and relevant clinical information, e.g. about renal or hepatic function.

It is by no means always necessary to determine plasma drug concentrations in order to monitor and control their administration. For instance, the pharmacological actions of anticoagulants, antihypertensive drugs and monoamine oxidase inhibitors can be assessed and dosage adjusted on the basis of prothrombin time, blood pressure and platelet monoamine oxidase activity measurements, respectively. Nevertheless, there are several drugs where TDM has been found to be essential. In all instances, care must be exercised to collect specimens at the appropriate times (see below).

THERAPEUTIC RANGES

It can be difficult, in clinical pharmacology, to determine therapeutic ranges for drugs by relating laboratory observations to the clinical assessment of therapeutic effects. This is particularly true in the case of many psychiatric disorders, which are mostly heterogeneous both in terms of their aetiology and clinical manifestations, where patients' compliance may be imperfect, and where clinical assessment of therapeutic response is liable to introduce a considerable subjective element.

Published therapeutic ranges give an indication of inter-individual variation in the clinical response to drugs, usually under the therapeutically ideal conditions of single-drug therapy. These ranges can often do little more than serve as guidelines for TDM of individual patients, because of the many factors that can contribute to intra-individual variation such as age, diet and nutritional state, ingestion of other drugs, and the presence of hepatic or renal disease.

STEADY-STATE CONCENTRATIONS

Plasma and tissue concentrations of drugs are in equilibrium by the time drug distribution is complete and, for many drugs, the magnitude of the pharmacological response is determined by the tissue concentration. In the blood, most drugs circulate partly bound to erythrocytes or plasma proteins, the bound drug being in equilibrium with unbound drug, and it is the plasma [unbound drug] that is of particular interest in TDM studies. Unfortunately, the measurement of unbound, or free, drug concentrations presents difficulties and TDM mostly has to depend on measurements of plasma [total drug].

The importance of correct timing of blood specimen collection for TDM cannot be over-stressed. If the specimen is collected too soon after the start of drug therapy, i.e. before a steady-state concentration has been achieved, the result of the analysis will be misleadingly low. Increasing the dosage, on the basis of such a result, could produce toxic effects.

The elimination half-times of many drugs are now known, as is the time taken to establish steady-state concentrations in plasma, when the drug is given at regular intervals. In general, it takes about five times the elimination half-time of a drug for its plasma concentration to achieve a steady state with a particular dosage level. Thereafter, it is usual to adopt a standard time for collecting blood specimens for TDM, depending on the requirements appropriate for a particular drug. For example:

(1) *Lithium.* Blood should be collected 12–18 h after the last dose, as trough concentrations are particularly important for therapeutic control.
(2) *Digoxin.* Blood should be collected 6–10 h after the last dose, since plasma [digoxin] measured at this time has been shown to mirror most closely the concentration in the myocardium.

Indications for therapeutic drug monitoring

Provided that details are known about the relationship between plasma [drug] and its pharmacological effects, and in those cases where there is a satisfactory (sensitive, specific and reliable) method of assay available, TDM can now be of great value in the following circumstances:

(1) *Drugs with a low therapeutic ratio,* and whose effects cannot be gauged precisely by clinical observation or by other laboratory measurements. Lithium is a good example, and regular measurements of serum [lithium] (or plasma [lithium] using EDTA as anticoagulant) are essential if toxicity is to be avoided.
(2) *Checking for non-compliance,* as a possible explanation for therapeutic failure in a patient, especially if this occurs unexpectedly when drug treatment has previously been adequate. Some patients for whom treatment with psychotropic drugs has been prescribed may not comply with the treatment directions. Low plasma concentrations of the drug, e.g. 'tricyclic' antidepressants, may help to explain the lack of clinical effectiveness.
(3) *Signs of drug toxicity.* Some drugs (e.g. digoxin, phenytoin) may produce toxic manifestations (cardiac arrhythmias, fits) that can be difficult to distinguish from the symptoms and signs of the

diseases for which they have been prescribed. With most other anti-epileptic drugs, TDM is not essential for various reasons, but it is needed for the proper control of phenytoin dosage.

(4) *Renal function impaired,* especially when potentially toxic drugs are being given. Examples include the need for close monitoring of treatment with anti-arrhythmic drugs (e.g. procainamide) or with aminoglycoside antibiotics (e.g. gentamicin, tobramycin).

BINDING OF DRUGS

Many drugs circulate in plasma mainly bound to albumin, e.g. phenytoin, quinidine, salicylates and the 'tricyclic' antidepressants. Drug-binding may be considerably altered if there is hypoalbuminaemia (p. 116), which reduces the number of available drug-binding sites. It is also reduced in uraemia, which alters the drug-binding properties of albumin, and sometimes by the simultaneous administration of other drugs, e.g. valproic acid which may displace phenytoin from its binding to albumin.

Many basic (cationic) drugs bind to α_1-acid glycoprotein, one of the acute phase reactants (p. 123); its concentration is increased in many inflammatory conditions, the extent depending on the severity of the inflammation. Drugs that bind to α_1-acid glycoprotein include benzodiazepines, chlorpromazine, propranolol, quinidine and 'tricyclic' antidepressants. When plasma [α_1-acid glycoprotein] is increased, the percentage of unbound, free or pharmacologically active drug is decreased, and drug-clearance by the liver may be accelerated.

Depending upon the drug, and on its binding properties to plasma proteins, alterations in plasma [albumin] or [α_1-acid glycoprotein] may considerably influence the pharmacokinetics of the drug, and hence its activity. It is, however, impossible to predict the extent of these effects from measurements of the individual plasma proteins, because there is considerable inter-individual variation in the extent of drug-binding. This partly explains the growing interest in plasma [free drug] measurements.

Free drug levels

Several methods have been explored for measuring plasma [free drug], directly or indirectly, but none has yet been widely adopted for routine use. Equilibrium dialysis is slow and complex, but ultrafiltration techniques appear more promising. Another approach would be to measure drug concentrations in naturally occurring ultrafiltrates:

(1) *CSF.* Plasma [free drug] and CSF [drug] are closely similar or identical. However, the use of CSF measurements is clearly impracticable for routine TDM.

(2) *Saliva.* Technically, there are few difficulties in measuring salivary [drug], if satisfactory specimens can be obtained. Unfortunately, this can be difficult in seriously ill patients. Also, there are factors that complicate attempts to infer values for plasma [free drug] from measurements of salivary [drug]. The excretion of some drugs into saliva varies with salivary [H^+], and some free drugs are actively transported from the plasma space into saliva.

For these reasons, little is known as yet about therapeutic ranges of free drugs in plasma, and TDM must for the present continue to depend on measurements of plasma [total drug], despite their limitations. As previously emphasized, data from TDM must be interpreted in relation to the patient's clinical condition.

IMPORTANCE OF DRUG METABOLITES

Most drugs are metabolized to inactive products (by oxidation or reduction, hydroxylation, conjugation etc.), and then excreted. Some drugs are inactive when taken and are first rendered active, in the gastrointestinal tract or the liver or elsewhere, before exerting their actions and being subsequently rendered inactive by further metabolism and the metabolites excreted. This can mean that very specific analytical methods are required if, for instance, active drug and a structurally closely related but nevertheless inactive compound are to be differentiated.

In recent years, analytical methods have improved considerably and techniques have been introduced that can, where appropriate, measure active drug in the presence of an inactive precursor, or pro-drug, and inactive metabolites of the drug. Specific methods are required for TDM, since results from methods that fail to differentiate between active drug and related inactive compounds cannot readily be interpreted.

Amitriptyline and imipramine, and similar 'tricyclic' antidepressants, are examples of drugs that undergo *in vivo* metabolism to active demethylated metabolites (nortriptyline, desipramine, etc.). Some analytical methods provide results for the plasma [amitriptyline + nortriptyline] etc., depending on the drug that has been prescribed.

Primidone is an example of a drug that probably acts mostly by conversion to active metabolites, principally phenobarbitone. If TDM is to be used for primidone therapy, plasma [phenobarbitone] is the measurement required.

Therapeutic drug monitoring in clinical practice

There is now an extensive literature on therapeutic drug monitoring, and the indications for its use as well as the identification of less worthwhile applications is becoming clearer. The subject was reviewed by Richens and Marks (1981), who concluded that only a tiny fraction of the drugs then in use were worth monitoring. Table 26.1 lists examples of drugs where TDM is considered to be either essential or very helpful. The list is not intended to be exhaustive, and certainly there are differing views on whether TDM should be widely available now for lignocaine, phenobarbitone, quinidine, salicylates (when being prescribed in high dosage) and 'tricyclic' antidepressants.

The frequency with which TDM may be required for a particular drug is very variable. Some of the indications have already been given, namely:

(1) Failure to achieve therapeutic control, although dosage is apparently adequate.
(2) Loss of therapeutic control in a patient previously stabilized on treatment.
(3) To check for possible toxic effects, especially if several drugs are being given.
(4) To confirm that plasma [drug] remains at an acceptable level when either the dosage level of the drug itself is changed, or the accompanying therapy.

In addition, more frequent TDM may be needed in pregnancy or in children, and in patients with hepatic or renal disease, especially if drug metabolism or excretion is particularly dependent on either organ.

UNITS OF MEASUREMENT

Plasma drug concentrations, in TDM and in toxicological studies, may be expressed in mass (gravimetric) or in molar units of concentration, and Table 26.1 gives examples of both types of unit.

One of the advantages that it was claimed would follow the introduction and widespread adoption of the SI recommendations on units (p. 517), with their emphasis on the preferred use of molar units, was a better understanding of clinical pharmacology since there had hitherto been considerable diversity in the units used for reporting results of drug measurements in routine practice (e.g. mg/mL, mg/100 mL, mg/L). However, possibly because this multiplicity of units had been self-imposed, clinical pharmacologists and toxicologists had accepted them. Unfortunately, the introduction of molar units only added to the confusion, and serious concern was expressed by clinical pharmacologists, and indeed by the medical profession generally, about the danger to patients caused by yet another way of expressing results.

Whatever the units of measurement, the only guidance we can offer doctors is to take particular note of the units used for expressing the results of measurements of drug concentrations. Special care is needed when discussing results reported from a laboratory other than the one with which the doctor is accustomed to work. The practice of simply referring to the numerical values of drug measurements, without qualifying these by mention of the units, cannot be condoned since it may lead to serious, even fatal mistakes.

CHEMICAL TOXICOLOGY

Self-poisoning and accidental or deliberate poisoning with drugs normally prescribed for therapeutic purposes, or with other substances, has been one of the commonest causes of emergency admission to hospital for many years. The problems facing the clinician and the clinical chemistry department are very different from those connected with therapeutic drug monitoring.

On the assumption that the correct diagnosis is drug overdosage, in

TABLE 26.1. Therapeutic drug monitoring: The principal

Drug	Therapeutic Mass units
Carbamazepine	4–10 μg/mL
Digoxin	0.8–2.0 ng/mL
Gentamicin × (a) Peak	5–10 μg/mL
(b) Trough	<2 μg/mL
Lithium +	—
Phenytoin	10–20 μg/mL
Procainamide	4–10 μg/mL

× Gentamicin is listed as an example of those aminoglycoside antibiotics that should have their dosage monitored by TDM, because of their potential toxicity.

+ Serum, or EDTA as anticoagulant if plasma analysed.

the majority of cases clinical management is usually based on clearly defined criteria, e.g. those relating to the management of an unconscious patient. Frequently there is no specific treatment to help with the elimination of drug(s) taken in overdose, or to counteract their effects. However, there are some potentially lethal drugs for which active forms of treatment now exist. Because the treatments themselves are, in some instances, not without risk or free from unpleasant side-effects, the help of the laboratory is often sought urgently, and a rapid and reliable 24-h/day emergency service in chemical toxicology should be available in every major hospital.

Emergency toxicology

It is often impossible to rely upon the history given by the patient or relatives, friends or the police in cases of suspected drug overdose. Under these circumstances, the investigation of blood specimens for the possible presence of toxic agents, and determination of their concentration, may be required for the following reasons:

(1) *Specific treatment is available,* and the result for plasma [drug] will determine whether treatment should be used. Examples include paracetamol and iron, overdose of which may require treatment with N-acetylcysteine or desferrioxamine, respectively. If possible, the interval between the time of ingestion of the drug and the time of collection of the blood specimen should be established. Specimens collected within 4 h of ingestion of paracetamol may not be helpful in determining whether active treatment is needed; a low plasma [paracetamol] at this time may be due to the fact that the drug has not yet been fully absorbed.

drugs for which TDM has established value.

ranges in plasma Molar units	Time after last dose for collecting blood specimen
17–42 μmol/L	Just before next dose
1.0–2.6 nmol/L	6–10 h
—	30–60 min
—	Just before next dose
0.6–1.0 mmol/L	12–18 h
40–80 μmol/L	Not critical (long half-life)
17–42 μmol/L	18–24 h

(2) *The rate of elimination* is able to be increased by active treatment in some cases. For instance, forced alkaline diuresis increases the increases the rate of removal of salicylate from the body, and haemoperfusion can be used to accelerate the removal of some barbiturates, e.g. amylobarbitone. If poisoning with methanol or ethylene glycol is detected, the patient can be treated with ethanol to slow their conversion to toxic metabolites, and the poisons themselves removed by haemodialysis.

Most measurements for emergency chemical toxicology are carried out on blood specimens. This is because results for plasma [drug] not only help to determine the prognosis in some cases but also the requirement for active, specific treatment. The range of toxicological analyses widely available for these purposes is listed in Table 26.2. In addition, many of these patients require other chemical measurements of a non-toxicological nature; examples of those most often requested are also included in the Table.

Ethanol

The acute effects of over-indulgence in ethanol may result in admission to hospital. If the diagnosis is in doubt, blood [ethanol] can be measured by gas-liquid chromatography (GLC) or enzymically, with alcohol dehydrogenase. Some patients may have drunk methylated spirits rather than ethanol, and GLC can help with rapid identification of both methanol and ethanol in these patients.

Ethanol is frequently taken, together with other drugs, in self-poisoning attempts and may have an important potentiating effect on the pharmacological actions of other drugs. It is often worthwhile,

TABLE 26.2. Emergency toxicology: blood analyses.

*General chemical investigations***	
Glucose	Urea or creatinine
Total CO_2 or arterial blood gas analysis	
Drug measurements (available in a well-equipped laboratory)	
Barbiturates, quantitative and qualitative	Lithium
Carbon monoxide	Methanol
Ethanol	Paracetamol**
Ethylene glycol	Paraquat
Iron*	Salicylate**

** Very frequently performed
* Frequently performed

therefore, requesting measurements of blood [ethanol] in these patients.

Another reason for emergency requests for blood [ethanol] is to assist in the management of patients involved in road traffic accidents, especially if they have suffered head injuries.

NON-URGENT ANALYSES IN DIFFICULT CASES

Other specimens and materials, obtained at the time of the emergency admission, may prove helpful in the fuller investigation of patients who are seriously ill and in whom poisoning is thought to be a possible cause, but who fail to respond to conventional supportive medical treatment. Their examination may also help to exclude drug toxicity in seriously ill patients, when the diagnosis is uncertain. Many of the investigations that might be required may be too complex, or are needed too infrequently, for them to be included in the laboratory's 24-h/day repertoire of toxicology investigations. The following specimens and materials should be collected, whenever possible, labelled carefully and kept for later, non-urgent examinations, should these be required:

(1) *Any tablets or medicine* found near the patient, however small in amount. Pharmacy departments may be able to identify tablets and drug residues, but chemical analysis may also be necessary.

(2) *Vomit, stomach aspirate and the first portion of gastric lavage* (if performed) should be collected, each into a separate container, and stored in a refrigerator.

(3) *Urine, the first specimen passed.* Side-room tests for glucose and ketone bodies should be carried out on a small sample. The rest should be stored in a refrigerator.

(4) *Blood.* A specimen (10 mL, heparinized) should be collected, in addition to any specimens needed for emergency toxicology, and the plasma stored frozen.

Stomach contents sometimes contain large quantities of undigested and unmetabolized drug or poison, so can prove very helpful for drug-identification purposes in difficult cases. Urine, on the other hand, may contain metabolites and relatively little or no detectable quantity of the original drug or poison. Since examination of stomach contents can give no indication of the amount of material absorbed, and since the type of limited qualitative drug screen that can be carried out on urine specimens under emergency conditions does not significantly extend the range of information on which decisions about immediate treatment are taken, it should be apparent that urgent requests for toxicological examinations should be restricted to analyses on blood specimens (Table 26.2).

Other aspects of drug overdosage

There are some poisons for which specific treatment is available, including carbon monoxide and opium alkaloids, and for which it would not be appropriate to wait for laboratory investigations to be carried out before starting treatment. Either the nature of the poison may be self-evident or the urgency of the clinical condition so great as to demand rapid treatment. Treatment of CO poisoning, or morphine overdosage, should be started as soon as possible, and any specimens that might require to be analysed are collected for later examination. In cases of suspected cyanide poisoning, however, even though specific treatment is available, despite the urgency of the clinical condition the diagnosis should be confirmed before giving specific cyanide antidotes; this is because the antidotes themselves are potentially dangerous unless cyanide ions are present.

Cases of poisoning may develop complications that cannot always be anticipated at the time of admission to hospital. Even though it may not prove necessary to examine the specimens, it is advisable to obtain and to preserve for possible analysis those specimens (detailed above) most likely to provide helpful information, should difficulties arise later in diagnosis or management. It is essential to record the date and time and the conditions under which the specimens were obtained. If the patient's name is not known, other means of specifically identifying each specimen must be used. Unless these points are conscientiously observed, difficulties can occur later, especially if medicolegal issues arise.

DRUG SCREENING

The wide variety of compounds liable to be ingested in toxic quantities, and the complexity of the mixtures of drugs and metabolites that may be present in body fluids, some taken in overdose and others given in treatment, can present chemical toxicology and forensic science laboratories with very difficult analytical problems.

The range of powerful analytical techniques available to help solve these problems has increased markedly in recent years, and these are becoming more widely available in hospital laboratories. They include gas-liquid chromatography (GLC), high-performance liquid chromatography (HPLC) and gas chromatography-mass spectrometry (GC-MS), as well as a rapidly increasing number of immunoassay methods. Drug screening, which may involve long and painstaking analytical work and some or all of these techniques, may prove particularly helpful in the following circumstances:

(1) *Deeply unconscious patients,* where there is doubt about the nature and variety of drugs taken.
(2) *To exclude drug toxicity* in seriously ill patients, where poisoning is thought to be a possible cause.
(3) *Poisoning in children,* accidental or non-accidental.
(4) *Suspected brain death.*

The first two of these circumstances have already been discussed.

Poisoning in children

The diagnosis of poisoning in childhood may present difficulties not usually encountered in adults. The differential diagnosis may be wider, and the range of substances that may have been taken or administered is very large. Also, some compounds (e.g. ethanol) may produce much more marked toxic effects in young children than in adults. The position is further complicated by the inability of the patient, as a rule, to give useful information and the history obtained from the parent(s) may be vague or misleading, especially if responsible for unauthorized drug administration.

Non-accidental poisoning in children or child abuse, i.e. the deliberate administration of a potentially toxic compound with the intent of poisoning a child, is an increasingly common and very serious problem. Its recognition and management, sometimes including treatment directed specifically against the poison, require rapid and definitive methods of analysis, sometimes extending to include a full drug screen.

Suspected brain death

In cases of suspected brain death, where the cause is unknown, a drug screen is considered mandatory before life-support systems can be removed. With the widespread availability of life-support systems, and the growing demand for firm information on which to base difficult decisions, requests for drug screening in these patients are increasing. The analyses to be performed may require the best available modern techniques, including GC-MS, and considerable expertise.

THE INTERPRETATION OF TOXICOLOGICAL ANALYSES

Several important points must be borne in mind when interpreting data obtained for drug analyses.

Interference by drugs given for the purposes of treatment may sometimes lead to a failure by the laboratory to detect other drugs in samples from patients with symptoms thought to be due to poisoning. This effect is described as *masking*. Specimens for toxicological analysis should whenever possible be taken before any medication is given.

It is not widely appreciated that antibiotics can completely mask the presence of small amounts of other drugs that may be present, and which may require to be detected. Details of any treatment should therefore be provided, when submitting specimens for toxicological examination, specifying the names of all drugs (including antibiotics) and dosage. Where interference in the analytical method is to be anticipated, it may then be possible to modify the method of analysis so as to be able to detect the drug that has caused poisoning in the presence of other drugs given for therapeutic purposes.

With barbiturate determinations, the interpretation of quantitative data is markedly affected by a knowledge of the patient's previous history of barbiturate ingestion, since patients who have become habituated to barbiturates may be fully conscious with blood concentrations that would make a non-habituated patient comatose or even moribund.

Lack of specificity of some toxicological methods still in use can mean that they fail to differentiate between (1) pharmacologically active drug, (2) unaltered drug rendered pharmacologically inactive by binding (e.g. to plasma proteins), (3) metabolites closely related structurally to the parent drug but which are nevertheless pharmacologically inactive, and (4) other drugs, including drugs used in treatment.

OTHER INVESTIGATIONS IN DRUG OVERDOSAGE

Some drugs ingested in accidental or deliberate acts of poisoning are hepatotoxic, others are nephrotoxic, some have adverse effects on the haemopoietic system or the circulating erythrocytes and others on the lungs. While there is no pattern of abnormality specific for the hepatotoxic effects of drugs, some of the highest plasma aminotransferase activities recorded have been observed in patients who have ingested drugs such as paracetamol. This emphasizes the advisability of collecting specimens of blood to allow baseline sets of measurements (e.g. plasma [urea] and aminotransferase activities) to be obtained in patients who have been admitted because of acts of self-poisoning.

The effects of salicylate overdosage are very varied, especially in children. Dehydration is often severe, due to vomiting, and acid–base balance is often severely affected. If vomiting is marked, a metabolic alkalosis develops; otherwise, a mixed acid–base disturbance consisting of respiratory alkalosis and metabolic acidosis is commonly observed. In patients with salicylate overdosage, plasma [urea], [K^+] and [total CO_2] or full acid–base assessment are usually determined before active treatment (e.g. forced alkaline diuresis) is begun, and these measurements as well as further determinations of plasma [salicylate] are made to monitor the effects of treatment.

SPECIAL PROBLEMS

Solvent abuse and drug addiction are problems of great social importance. Confirmation of the diagnosis may require appropriate chemical investigations. Over-indulgence in ethanol is another serious social problem, and one that can have a wide variety of pathological effects, some of which will be briefly considered.

A wide range of volatile substances, many of them found in adhesive (glues) or aerosol sprays, have been reported as being involved in cases of volatile solvent abuse, 'glue sniffing' etc. Gas chromatographic analysis of blood specimens provides a rapid and specific means of detecting the volatile compounds, and HPLC of urine extracts enables identification of organic acids formed by the metabolism of toluene and xylene. As with other toxicological problems, the laboratory needs to be informed about details of any treatment that the patient has been given.

Detection of the illicit use of drugs is sometimes the responsibility of the clinical chemistry laboratory, but more often the task of a forensic scientist. Many drugs of addiction can now be specifically identified even when present in minute amounts, including morphine and its derivatives, cocaine, tetrahydrocannabinol, lysergic acid diethylamide (LSD) and amphetamine.

Chronic ethanol abuse

Ethanol is metabolized by alcohol dehydrogenase, by catalase and by a NADP-dependent microsomal enzyme oxidizing system (MEOS).

The relative importance of these systems varies, depending both on the blood [ethanol] itself and on the length of any period of chronic over-indulgence.

Blood [ethanol] is worth measuring in patients in whom the differential diagnosis includes the possibility of alcohol abuse. However, ethanol is not present normally in blood, except in traces, unless some has been drunk recently, so a negative result for this test does not necessarily exclude the diagnosis. For the diagnosis of alcoholism, the American National Council on Alcoholism has proposed as criteria the finding of blood [ethanol] greater than 65 mmol/L (300 mg/100 mL) at any time, or greater than 33 mmol/L in the absence of symptoms.

Plasma enzyme activity measurements have been widely used as pointers to the diagnosis of chronic alcohol abuse, notably plasma GGT measurements. However, by itself, this has proved to be a somewhat insensitive test, although its value as a diagnostic aid can be increased by also measuring the MCV. The combination of slight macrocytosis and increased plasma GGT activity is probably the best routinely available pair of measurements for the detection of alcohol abuse. However, neither test is specific for this purpose, so caution must be observed before acting upon their results. Plasma [urate] may also be increased in these patients, but again is neither a specific nor a sensitive pointer to the diagnosis.

Chronic ethanol abuse is one of the commonest causes of increased fasting plasma [triglycerides].

Although chronic over-indulgence in ethanol may produce these and other chemical abnormalities, the diagnosis is likely to depend greatly on the history and clinical examination unless blood [ethanol] is measured, and found to be abnormal.

INDUSTRIAL AND OCCUPATIONAL HAZARDS

There is a formidable list of potentially toxic compounds used in industry. Their use is controlled in Britain by the provisions of the Factories Acts and by the Health and Safety at Work, etc., Act. Concern arises when these provisions are not observed or when faulty equipment leads to a breakdown in the precautions; sometimes there is resulting environmental pollution. Hospital laboratories are sometimes called upon to help with the investigation of these problems, but more often they are the responsibility of the Health and Safety Executive.

METAL POISONS

Many metals are potentially toxic. Analytical methods for their detection in small amounts (e.g. in the air) and quantitation have greatly improved in recent years. Some of the techniques demand expensive and highly sophisticated equipment, available in only a few reference centres.

Mercury, cadmium and lead are heavy metals used in industry and elsewhere; they are all highly toxic to man. Their effects depend partly on the type of compound involved, whether inorganic or organometallic, and partly upon the route of absorption. In all cases the kidney is liable to be severely damaged, and other organs that may be affected include the liver and the nervous system.

Blood and urine measurements of the metal are important in the specific identification of the toxic agent. Since these potentially toxic metals are predominantly associated, in the blood, with erythrocytes it is essential that toxicological analyses be carried out on specimens of whole blood, not on plasma or serum. In some cases there is lack of agreement as to the concentration of the toxic agent to be regarded as potentially dangerous; this applies particularly to blood [lead].

Chemical investigations also provide valuable information about the severity of organ damage. Intoxication with mercury, lead or cadmium may give rise to an overflow amino aciduria (p. 97); lead poisoning also causes abnormalities in porphyrin metabolism. The early detection of lead toxicity, by its effects on porphyrin metabolism, is discussed elsewhere (p. 319).

Chronic renal failure

Although neither an industrial nor an occupational hazard, in the context of metal poisons it is important to note that patients who are being maintained on haemodialysis regimes are particularly at risk from poisoning by constituents of the dialysis fluid. Aluminium toxicity leading to dialysis dementia and to metabolic bone disease has been described. Prevention of aluminium toxicity requires periodic checks of aluminium content in the water supply and in the effluent from deionizers used with dialysers.

ORGANIC SOLVENTS

Many organic solvents used in industry are toxic, e.g. the chlorinated hydrocarbons and ethylene glycol. Toxicity may be due to accidental exposure, or sometimes to solvent abuse. The toxic agent can usually be identified specifically (p. 485). Chemical investigations may also be needed, to assess whether there has been hepatic or renal damage.

PESTICIDES AND HERBICIDES

Organophosphates (e.g. parathion, malathion and menazon) may cause poisoning among farm-workers by inhibiting acetylcholinesterase; plasma cholinesterase is also inhibited.

Measurements of plasma cholinesterase or erythrocyte acetylcholinesterase activity may be helpful in the recognition of exposure to organophosphates, especially if tests are repeated at intervals after exposure is thought to have occurred. Serial measurements provide comparative data and enhance the value of the tests; there is a wide variation between individuals in the baseline levels of cholinesterase activity.

Paraquat and diquat are extremely dangerous herbicides, if ingested. There are usually clinical features that suggest the diagnosis, before the onset of complications, and the diagnosis can be confirmed by chemical examination of blood or urine. Paraquat has severe effects, especially on the lungs and kidneys. Early diagnosis is essential if treatment is to have any chance of preventing a fatal outcome. Equally important is early exclusion of the diagnosis, as active treatment involves drastic measures which should be avoided, if possible. Paraquat estimations are among the tests that can be rapidly performed for diagnostic purposes (Table 26.2).

FURTHER READING

FLANAGAN, R.J., HUGGETT, A., SAYNOR, D.A., RAPER, S.M. and VOLANS, G.N. (1981). Value of toxicological investigation in the diagnosis of acute drug poisoning in children. *Lancet,* **2,** 682–685.

GEORGE, C.F. (1982). Interactions with digoxin: more problems. *British Medical Journal,* **284,** 291–292.

HELLIWELL, M., HAMPEL, G., SINCLAIR, E., HUGGETT, A. and FLANAGAN, R.J. (1979). Value of emergency toxicological investigations in differential diagnosis of coma. *ibid.,* **2,** 819–821.

KNOTT, C., HAMSHAW-THOMAS, A. and REYNOLDS, F. (1982). Phenytoin-valproate interaction: importance of saliva monitoring in epilepsy. *ibid.,* **284,** 13–16.

Leading article (1978). SI, moles and drugs. *ibid.,* **1,** 668–669 and related correspondence (pp. 716, 1277 and 1620).

Leading article (1979). Monitoring plasma concentrations of psychotropic drugs. *ibid.,* **2,** 513–514.

Leading article (1981). Carbon monoxide, an old enemy forgot. *Lancet,* **2,** 75–76.

LINDUP, W.E. and ORME M.C.L'E. (1981). Plasma protein binding of drugs. *British Medical Journal,* **282,** 212–214.

MARKS, V. (1983). Clinical pathology of alcohol. *Journal of Clinical Pathology,* **36,** 365–378.

RAMSAY, J.D. and FLANAGAN, R.J. (1982). The role of the laboratory in the investigation of solvent abuse. *Human Toxicology,* **1,** 299–311.

RICHENS, A. and MARKS, V., editors (1981). *Therapeutic Drug Monitoring*. Edinburgh: Churchill Livingstone.

VALE, J.A. and MEREDITH, T.J., editors (1981). *Poisoning Diagnosis and Treatment.* London: Update Books.

Chapter 27

Computers in Clinical Chemistry

Hospital laboratories were among the first areas of medical practice to experiment with the use of computers. Now, after approximately twenty years of progressively increasing experience, many clinical chemistry departments have come to depend on computers for assistance with practically every facet of their activities.

Several aspects of computer applications in hospital laboratories have implications for the practice of doctors who depend on the diagnostic services they have to offer. This applies particularly to the data-processing sector of the laboratory's activities, including the procedures for requesting investigations and the related subjects of record filing and the issue of reports. Computers are used for many other purposes by laboratories, e.g. for process-control of equipment operation and quality assessment of work performance, but this chapter will be mainly concerned with those aspects of laboratory computer applications that are particularly relevant to doctors as users of the diagnostic service.

Requesting procedures

Accurate information about the work to be carried out in a laboratory is the key to efficient computer-assisted operations. Whether the information about test requests is to be entered by data-preparation staff in the laboratory who have to key details of each request into the computer by transcription from request forms, or entered by doctors using terminals at a distance from the laboratory, certain procedures must be observed:

(1) *Patient identification data.* Every patient must be correctly and uniquely identified, so that the computer can either generate a new record entry in its files in the case of first requests, or add the latest request to a pre-existing record belonging to the right patient in the case of patients who have been previously investigated.

(2) *Test request information.* This needs to specify the investigations to be performed. It clearly also needs to include information about the nature of the specimen (e.g. venous blood, arterial blood, 24-h urine collection), the date and time of collection and an indication whether or not the request is urgent.

(3) *Report destination.* This may be a hospital ward or a general practitioner's surgery, or a doctor's private consulting room, etc.

These data represent the minimum requirements. On the first occasion that a request is entered for a patient, the basic identification data (unique number, surname and first names) are usually extended to include the date of birth, sex and other personal data, and the name of the doctor responsible for the care of the patient. Most laboratory request forms provide fields for assisting with the manual entry of these data; ideally, these should all be completed. Sometimes, machine-readable request forms are issued by medical records departments; these can considerably ease the task of completing request forms.

Other data that should be briefly noted on laboratory request forms include the clinical diagnosis and details of relevant treatment, including drug therapy. Many computer programs offer the facility of drawing attention to possible interference from particular drugs in either the analytical process or in the interpretation of results.

Clinical interest in the benefits to be derived from laboratory computers tends to focus on the range of reports, summarizing tables, graphical presentations, etc., and the rapidity with which they can be provided as output from the system. It is important to realize, however, that the full range of computer-generated benefits can only be derived if good quality data are entered into the computer in the first place. There is considerable room for improvement in the care taken by doctors in most British hospitals when initiating requests for laboratory investigations. Correct and consistent quoting of patient identification data on all request forms, especially the unique identification number, is unusual. As far as clinicians are concerned, the most noticeable effect of providing these data systematically is on the speed of response to telephoned enquiries about results. In the laboratory, many data-processing activities would be speeded up, and mistakes avoided, if request forms and specimen labels were always to be completed properly.

None of the details mentioned in this section is peculiar to computer-assisted laboratory operations. They were all discussed at greater length in Chapter 1; they contribute to better use of clinical chemistry laboratories, whether or not these are computer-assisted. However, there is a widely held belief that computers can somehow make good deficiencies in the information that clinicians provide to them. Their manifest inability to do so tends to undermine the confidence of doctors in computer-dependent laboratory records systems. The computer's requirements have, therefore, been emphasized.

Laboratory analytical work

Computers are regularly used to perform many operations within the laboratory. These include the organization and monitoring of work, by the automatic generation of work-sheets and the issue of periodical reminders about outstanding work that might otherwise be in danger of being overlooked. They also automatically re-allocate specimens for repeat analysis if the first analysis was unsuitable for reporting, for instance if the first result was off-scale, requiring the specimen to be re-analysed after dilution.

Many laboratory instruments now include their own microcomputers or microprocessors. These perform many functions such as on-line process-control (e.g. automatic fault recognition and trouble-shooting routines), data-acquisition and results calculation programs. Quality control data provided in real time has meant that standards of laboratory work, in well-equipped and computer-assisted laboratories, have greatly improved.

Laboratory records and reporting procedures

The preparation of laboratory records and the generation of even simple reports involve linkage operations between patient identification data, as entered on request forms, and the results of analytical work. Accurate linkage becomes more demanding if records of all work relating to individual patients (i.e. past and present work) are to be maintained, e.g. if cumulative summary reports are to be issued by the laboratory. The requirements for operating efficient laboratory records systems with the assistance of a computer reinforce the earlier comments about the need for reliable requesting procedures.

Two of the greatest benefits derived from computer-assisted operations in laboratories have been the elimination of transcription steps in the preparation of reports, and the issue of reports as printed documents. Transcription steps inevitably result in errors, and hand-written reports are often illegible.

From the clinician's point of view, the advantages of computer-assisted reporting procedures include the opportunity to see the latest sets of results on individual patients directly associated with previous

THE ROYAL INFIRMARY OF EDINBURGH Date Printed TUE, 2 AUG 1983 Page 1

DEPARTMENT OF CLINICAL CHEMISTRY

Patients name	Registration No.	Sex	Date of Birth/Age	Ward	Hospital	Consultant/General Practitioner
M	DCOO B	F	06.02.00	PE	DC	IMN

Clinical Details and Relevant Treatment	Duty Clinical Chemist
DEMENTIA	MISS L.

BLOOD Date of collection / Time of collection	2.06 1400	7.06	20.06 0900	1.07	26.07	28.07	2.08 * 0900		2.8 S.D.
Urea (2.5–6.6 mmol/L)	13.0		16.1	16.5	12.1	12.0	8.7		1.2
Sodium (132–144 mmol/L)	133		137	136	135	137	138		3.9
Potassium (3.3–4.7 mmol/L)	4.6		4.0	4.3	4.6	3.9	3.8		0.28
Total CO_2 (24–30 mmol/L)	21		23	21	19	19	20		2.2
Bilirubin (2–17 mol/L)	14	19	9	8		29	26		5.6
Alanine Aminotransferase (ALT) (10–40 units/L)	58	22	20	19		220	171		14
Alkaline Phosphatase (40–100 units/L)	242	155	210	162		392	321		17
Gamma-Glutamy Transferase (GGT) (M 10–55 Units/L, F 5–35 Units/L)	40		39	30		267	236		11
Total Protein (60–80 g/L)	76		80	72		77	70		2.8
Albumin (36–47 g/L)	36		38	34		36	33		2.2
Calcium (2.12–2.62 mmol/L)	2.17		2.29	2.17		2.25			0.11
Phosphate (Fasting 0.8–1.4 mmol/L)	1.08		1.16	1.10		1.25			0.22
Urate (M 0.12–0.42 mmol/L, F 0.12–0.36 mmol/L)	0.54			0.56		0.50	0.50		0.28
Creatinine (55–150 μmol/L)	117			148		169			31
Aspartate Aminotransferase (AST) (10–35 units/L)	89	27							5
Urea-Stable Lactate Dehydrodenase (USLD) (100–300 units/L)	467	265							56
Thyroxine (Under 65 70–150 nmol/L, Over 65 55–140 nmol/L)					113				25
Glucose (Fasting 3.6–5.8 mmol/L)	6.2								0.62
Iron (M 14–32 μmol/L, F 10–28 μmol/L)									
Iron-Binding Capacity (45–72 μmol/L)									

FIG. 27.1. Part of a computer-prepared cumulative summary report. The asterisk in the column headed '2.08' indicates that this is the column containing previously unreported results. The column headed '2.8 S.D.' indicates the extent by which the most recent result would need to alter for the difference to be significant, on analytical grounds, at the 5% level. For example, the change in plasma [urea] between 28th July 1983 (28.07) and 2nd August (2.08) was significant.

findings for each patient on legible, printed cumulative summary reports (e.g. Fig. 27.1). These reports draw attention to changes that have occurred since the patient was last investigated. Some reporting programs highlight marked changes by means of symbols (e.g. asterisks). In the example shown (Fig. 27.1), figures printed in the 2.8 SD column indicate the amount by which the most recent analysis would have to alter for the probability of the difference from the preceding or the subsequent analysis to be significant, on analytical grounds, at the 5% level (p. 24).

Cumulative summary reports help the laboratory, by drawing attention to the possibility that a serious error has occurred, but one which might not otherwise have been obvious. One example would be the unsuspected interchange of specimens or transposition of results, as can occur when keying a series of data into a computer. These reports also add to the interest of the work of a busy department, since the progress of seriously ill patients can be visualized on cumulative reports by laboratory staff.

Communication of results

Requests for emergency investigations and other forms of urgent work need to be processed and reported as quickly as possible. Laboratories make special arrangements to ensure that this category of requests receives the priority handling that is so clearly essential. However, many non-urgent requests for clinical chemistry investigations lose much of their value unless their reports also get back to the doctor who initiated the request as soon as possible, preferably on the day the specimen was collected.

Telephoned enquiries to laboratories, seeking information about work in progress, or about results contained in reports that have not been received or have since been mislaid, can be answered very quickly with the help of computer-held records systems. A patient's record can be called up almost instantaneously on a visual display unit, by entering the unique identification number. Where only the name of a patient is known, the computer can provide information on all patients of that name for whom it holds a record. The process of selecting the appropriate record can then proceed, but responding to the specific enquiry inevitably takes longer.

Developments in communications equipment mean that ward-based enquiry terminals and locally sited printers can now be used in association with laboratory-based computer systems. Reports can be printed on the wards, thereby obviating the delays inevitably associated with porter or postal delivery systems. Also, by using read-only programs, authorized clinical staff can access laboratory records relating to their patients, provided they can satisfy stipulated criteria for the maintenance of confidentiality and as long as they can quote the unique identification data needed to access the records they are seeking.

Records storage and analysis

Computer-held files of records provide a compact means of storage. Cumulative reports issued by laboratories are also potentially space-saving, but clinical staff need to cull patients' records and to discard earlier reports each time a new report is issued that includes the previous data.

For statistical and research purposes, records of laboratory work can be accessed retrospectively, and various types of enquiry or analysis carried out on the data that have accumulated. However, for such enquiries to achieve their full potential value, it is essential that the full range of data relating to each request be provided, by the proper completion of request forms; otherwise, some patients will be omitted from retrospective searches, even though they ought to have been included in the analysis of records chosen for inclusion in the group.

ASSISTANCE IN THE INTERPRETATION OF LABORATORY DATA

Computer-dependent reporting procedures can be programmed to help with the tasks of sifting through and interpreting masses of laboratory data, and to refine the interpretations themselves. These activities will be discussed under the following headings:

(1) Reference values.
(2) Data reduction.
(3) Effects of drugs.

Reference values

Some chemical measurements have reference values which vary with the sex of the individual, others vary with age, and some vary with both age and sex. It would be impracticable to associate the appropriate reference values with each result using standard manual reporting procedures. However, computer-assisted systems of reporting can perform these operations automatically and very quickly, provided the necessary reference data are on file.

A few laboratories have established separate sets of reference values for adults, for both sexes and for each decade. Given the relevant details for an individual patient, as part of the data entry from the request form, the computer can be programmed to print out the reference values appropriate for that patient, together with the set of results. This is a refinement of reporting practices that is not likely to be widely available for many years. In the meantime, with few exceptions, most clinical chemistry results on adult patients can be satisfactorily interpreted with the aid of single sets of reference values, and these are often preprinted on the report stationery (e.g. Fig. 27.1).

Data reduction

Where screening by chemical tests is practised (p. 501), this can mean that reports contain masses of data that some doctors find confusing (the 'data flood'). Confusion is even more likely to arise when a laboratory introduces multi-channel analytical equipment and thereafter issues reports containing results of all the analyses performed, whether or not they had all been requested in the first place.

Computer-dependent reporting techniques can help focus attention on results that appear to be important, for instance by the process of selective reporting. This involves printing out only the results of tests specifically requested *plus* those unrequested tests for which results differ by more than a stipulated amount from the reference values, the amount being specified in the computer's program of instructions.

For instance, if plasma [calcium] is measured automatically on all specimens for which one or more tests in a particular group of analyses is requested, the computer can be programmed to report all results in excess of 3.00 mmol/L (or other specified value), whether or not a plasma calcium measurement was requested. It is also possible to report results for unrequested tests performed on more than one occasion, if the latest set of results differs significantly (by a specified amount) from previous results.

Alternative ways of depicting clinical chemistry data, as a means of overcoming the adverse effects of the 'data flood', include the use of computer graphics. However, these techniques are still at an early stage of development.

Selective reporting cannot contribute much to data reduction, in the manners described, unless doctors make a habit of only requesting investigations on a discretionary basis (p. 6). Uncritical and repeated requesting of all the investigations that have been grouped by laboratories for technical reasons on to multi-channel analysers inevitably produces a flood of data.

Effects of drugs

Another refinement, only practicable where reporting procedures are computer-assisted, is to draw attention to the possibility that results of laboratory tests have been affected by the presence of drugs or metabolites in the specimen, or to the possible need to reconsider the significance of results in the light of the fact that the patient was receiving a particular form of drug treatment. In the U.S.A., records have been compiled listing about 10,000 effects that drugs can have on results of laboratory tests. If information about treatment is given on the laboratory request form, the computer's file can be searched to find out whether the drug(s) might have affected the technical performance of the laboratory analyses, or have an effect on the interpretation of the results. The report can then be annotated appropriately, by the computer.

COMPUTER-ASSISTED DIAGNOSIS

Computer-assisted interpretation of laboratory data can take the form of automatically generated diagnostic comments or suggestions. For instance, standard interpretative programs have been written for data derived from electrophoretic patterns of serum proteins and from lactate dehydrogenase isoenzyme separations, and for data from groups of 'liver function tests'. These are early examples of laboratory-based computer-assisted approaches to diagnosis. Other programs combine clinical data and the results of chemical laboratory tests, for instance in computer-assisted diagnosis of thyroid disease.

Comprehensive lists have been prepared, and can be held on computer file, detailing the effects of diseases on clinical laboratory tests. With most chemical investigations, it is still too early to produce manageable lists of possible diagnoses, starting from the results of an investigation or group of investigations. Similarly, listings of all the chemical abnormalities that have been described as occurring in particular conditions can only have limited value, until combined with data for the prevalence of specific conditions in particular populations. Thereafter, the ability of computers to store, to sort and to associate data on a large scale may prove to be of real help in diagnosis.

Programs requiring interaction between the doctor and the computer can assist in the diagnosis and management of disease, e.g.

patients with electrolyte and acid–base disorders. In one system, the doctor responds to specific questions posed by the computer program, in which the computer seeks information about physical signs and details of relevant previous investigations and treatment. These data are then associated with the results of chemical investigations after which the program prints out a list of diagnoses with an indication of the relative probabilities of each being correct. For hospitals with access to this kind of system, computer-assisted diagnosis may provide an alternative and acceptable way of managing patients with complex metabolic disorders. The successful general application of this type of approach to diagnosis and management of patients will require clinicians to accept the discipline inherent in complying with the rigidity imposed by computer programs.

Computer techniques have been used to investigate statistical relationships between the concentrations of various chemical constituents in the blood. The objective here is to identify patterns of laboratory data which might characterize a cluster of chemical findings that suggest a particular diagnosis, when considered as a group. It might also enable an illness to be recognized at an earlier stage in its development than would otherwise have been the case, if diagnosis had had to wait until symptoms and signs appeared.

Computer-assisted diagnosis is a feature of many screening programmes of investigation, the diagnostic assessments depending only in part on the results of multiple chemical analyses.

Computer-assisted instruction

Several types of computer-based simulation of diagnostic problems have been devised, and used for instructing medical students. It seems they may soon be widely adopted as teaching aids, and for self-assessment, in clinical chemistry and in many other branches of medicine.

Simulation models can be static or dynamic. In static models, the computer program stores information relating to a patient, and about pathological conditions that may enter into the differential diagnosis of the patient's condition, often based on data obtained from the case-notes of a particular individual. The student is shown how to interrogate the computer file, and the method thereafter adopted by the student to acquire information leading to a diagnosis can be monitored. Stresses can be applied by limiting access to sources of information, or by awarding penalty marks for ill-considered questions, or by applying time-constraints within which to reach a diagnosis.

Dynamic models can include some or all of the features already mentioned. In addition, the program can be extended so as to allow the condition of the patient, as stored in the program, to alter with time or in response to any treatment that the student may prescribe.

These simulations have been extended so as to provide quickly, on request, displays that can be used to inform the student why decisions made in proceeding towards a diagnosis, or why investigative methods selected or therapeutic actions adopted, were incorrect.

FURTHER READING

ABSON, J., PRALL, A. and WOOTTON, I.D.P. (1977). Data processing in pathology laboratories: the Phoenix system. *Annals of Clinical Biochemistry*, **14**, 307–329.

BARON, D.N. and FRASER, P.M. (1968). Medical applications of taxonomic methods. *British Medical Bulletin*, **24**, 236–240.

BLEICH, H.L. (1974). Computerised clinical diagnosis. *Federation Proceedings*, **33**, 2317–2319.

ELLIS, G., WORTHY, E. and GOLDBERG, D.M. (1980). Further experience with computer-assisted diagnosis of diseases of the liver and biliary tree. *Clinical Chemistry*, **26**, 1266–1271.

ENLANDER, D. (1981). Computerised graphic representation of clinical chemistry results. *Journal of Clinical Pathology*, **34**, 806–808.

FLYNN, F.V. and BALL, S.G. (1982). Comprehensive computerized data management in a chemical pathology laboratory with SOCRATES.*Medical Informatics*, **7**, 275–305.

FREIDMAN, R.B., ANDERSON, R.E., ENTINE, S.M. and HIRSHBERG, S.B. (1980). Effects of diseases on clinical laboratory tests. *Clinical Chemistry*, **26**, 1D–476D.

McPHERSON, K., HEALY, M.J.R., FLYNN, F.V., PIPER, K.A.J. and GARCIA-WEBB, P. (1978). The effect of age, sex and other factors on blood chemistry in health. *Clinica Chimica Acta*, **84**, 373–397.

MURRAY, T.S., CUPPLES, R.W., BARBER, J.H., HANNEY, D.R. and SCOTT, D.B. (1976). Computer-assisted learning in undergraduate medical teaching. *Lancet*, **1**, 474–476.

RAJ, P.P., KRICKA, L.J. and CLEWETT, A.J. (1982). Microcomputer simulations as aids in medical education: applications in clinical chemistry. *Medical Education*, **16**, 332–342.

ROBERTSON, E.A., STEIRTEGHEM, A.C. VAN., BYRKIT, J.E. and YOUNG, D.S. (1980). Biochemical individuality and the recognition of personal profiles with a computer. *Clinical Chemistry*, **26**, 30–36.

STERN, R.B., KNILL-JONES, R.P. and WILLIAMS, R. (1975). Use of computer program for diagnosing jaundice in district hospitals and specialized liver unit. *British Medical Journal*, **2**, 659–662.

YOUNG, D.S. (1976). Interpretation of clinical chemical data with the aid of automatic data processing. *Clinical Chemistry*, **22**, 1555–1561.

YOUNG, D.S., PESTANER, L.C. and GIBBERMAN, V. (1975). Effects of drugs on clinical laboratory tests. *ibid.*, **21**, 1D–432D.

Chapter 28

Screening for Disease Using Chemical Tests

Patterns of requesting chemical investigations fall into two broad categories, discretionary tests and screening investigations (pp. 1, 4). The main emphasis in this book has been on the discretionary approach because it is much easier to explain results of chemical tests when there has been a specific reason for requesting each test in the first place. It can be very difficult to explain abnormal findings for tests carried out without a specific clinical indication.

Developments in automatic analytical equipment, and in computer techniques of data-processing, have been followed by changes in the ways in which many laboratories organize and carry out their work. Frequently nowadays, in response to requests for single tests or for a limited group of tests, selected on the basis of clinical requirements, laboratories equipped with multi-channel analysers use these to perform both the test(s) requested on clinical grounds (i.e. the discretionary tests) and all the other tests that the analyser has been designed to perform. These laboratories may then report the results of:

(1) only the tests that were specifically requested (discretionary test reporting).
(2) all the tests performed on the multi-channel analyser.
(3) the tests that were specifically requested, *plus* the results of unrequested tests that were outside specified limits.

Various terms have been used to describe the reporting of unrequested tests, e.g. non-discretionary test reporting. It has, however, become more common practice to describe these non-discretionary applications of multi-channel analysis as examples of biochemical screening or profiling, and not to qualify the descriptions of reporting practices. Screening can be carried out on patients and on the apparently healthy population, and experience gained with the use of chemical investigations in these two different contexts will be discussed.

The problem of setting limits

One of the first decisions to be made for any screening programme is the setting of limits for the results of investigations. This decision carries with it the implication that, if the limits are exceeded, confirmatory or follow-up investigations will be performed. We discuss one example later, screening for PKU, to illustrate this problem. Whether screening tests are being carried out on patients admitted to hospital or on outpatients, or on the apparently healthy population, these tests have the following features in common:

(1) The doctor does not usually expect the result of the tests to be abnormal (if the doctor had expected any of the tests to be abnormal, he might well have already requested such tests on a discretionary basis).
(2) Large numbers of test results are being reported. In the absence of other criteria, these results are compared with reference values (e.g. Table 2.1).

Reference values are defined so as to include 95% of the healthy population, so the screening 'net' will tend to 'catch' many normal healthy individuals who happen to have a result just outside the reference range, for each different test performed. The problem is compounded when several tests are included in the screening programme (multiphasic screening). These features mean that, for some of the tests performed as screening investigations, conventional sets of reference values may no longer be appropriate.

Screening of hospital patients

Data from one of the first of the American assessments of admission profiles, i.e. screening patients by means of a large number of chemical tests on first admission to hospital, are shown in Table 28.1. Eleven chemical investigations were carried out on 623 consecutive admissions to hospital, and the results compared with the findings revealed by discretionary methods of requesting some or all of these same investigations, which applied to 477 of these 623 patients. Specimens for discretionary analysis were collected on the day of admission to hospital, and examined in the hospital's routine laboratory. Specimens for the profile studies were obtained on the following morning, and analysed in a separate laboratory specially equipped for the study with automatic analysers.

Approximately three times more investigations were carried out in the profile laboratory, as compared with the number of discretionary investigations performed in the routine laboratory, i.e. an extra 4647

TABLE 28.1. Comparison of routine clinical chemistry laboratory findings and admission profiles (data from Bryan *et al*., 1966).

	Routine laboratory	Profile laboratory
Total patients investigated	477	623
Total determinations on serum	2206	6853
Total abnormalities found	191	340
Patients with abnormalities	120	226

Distribution of 149 extra abnormalities revealed by the admission profiles:

Test	Extra abnormals	Medically significant	*Test*	Extra abnormals	Medically significant
Sugar	31	16	Calcium	6	1
Urea	7	4	Phosphate	10	2
Sodium	10	0	Protein	15	2
Potassium	25	1	Albumin	9	4
Chloride	2	0	Urate	28	23
Total CO_2	6	1			

measurements were made, resulting in the finding of an additional 149 abnormalities. This may seem a very small return for so much extra analytical work, but by carrying out eleven measurements on all patients abnormalities were detected in 226 patients, compared with only 120 patients investigated on a discretionary basis. The question nevertheless arises whether so many extra analyses were worth carrying out.

As far as the laboratory was concerned, less effort and fewer staff were required to carry out the 6853 profile investigations than to perform the 2206 discretionary tests, because the eleven profile measurements were being performed as a standard pattern of work, and were therefore much easier to organize and perform. If the capital cost of the automatic equipment is left out of account, for the present, the running costs per test were also considerably less for the profile laboratory than for the laboratory carrying out the discretionary work.

An attempt was made to assess the medical significance of the 149 additional abnormal findings revealed by the profile of tests (Table 28.1). Only some of the unrequested tests revealed abnormal findings that were classified as 'medically significant'. In this particular study, these additional investigations yielded data that were considered to have led to the earlier institution of an important treatment or change of diagnosis or treatment in *less than 2% of the patients.* This level of yield is fairly typical of hospital screening programmes, using presently available chemical methods of investigation in an unselective manner for adult patients.

In spite of this rather disappointing yield of important new diagnoses, screening of all patients when first admitted to hospital has been widely adopted, especially in the U.S.A. The range of tests included in the admission screening has been gradually extended. The following are examples of diagnoses that have been revealed, in small numbers, by admission screening of all hospital patients:

(1) *Diabetes mellitus.* Most of the patients with diabetes recognized on the basis of blood or plasma glucose measurements carried out as part of screening programmes on hospital patients have had a mild form of the disease, being in the elderly, obese group of diabetics (Type 2, p. 221).

(2) *Hyperparathyroidism.* In patients whose reason for attending hospital was apparently unrelated to a disturbance of calcium metabolism, the finding of hitherto unsuspected hypercalcaemia has been found to be due to primary hyperparathyroidism in over 80% of cases. Other causes are listed in Table 14.2 (p. 284).

(3) *Thyroid disease.* Many screening programmes now include measurements of plasma [total T4]. These have revealed small numbers of patients with hitherto unsuspected hypothyroidism, but have proved less satisfactory for the detection of unsuspected hyperthyroidism, partly because of interference by drugs, especially oral contraceptives (p. 397).

(4) *Anaemia.* Plasma iron measurements are not a sensitive index of iron deficiency (p. 309). However, their inclusion in screening programmes of chemical tests has helped reveal some cases of hitherto unsuspected iron deficiency; blood [Hb] is usually also determined as a screening test.

(5) *Renal tract disease.* The finding of a raised plasma [urea] should lead on to further investigations to determine the cause (Table 4.4). However, as a non-discretionary screening investigation, plasma [urea] reveals few new cases of renal tract disease, because the measurement would normally be frequently performed as a discretionary test.

The medical value of screening investigations performed on hospital patients might be expressed in terms of the number of important new diagnoses unexpectedly revealed, or revealed significantly earlier by screening than they would have been otherwise. This would require comparisons between the times when diagnoses were reached as a result of carrying out screening tests, and the times when the same diagnoses would have been made by discretionary methods of requesting. For instance, screening investigations might unexpectedly reveal hypercalcaemia in a patient and lead on to early diagnosis of primary hyperparathyroidism, whereas there might have been a delay of months or years before the same patient developed symptoms suggesting the need for a plasma calcium determination on a discretionary basis. It is very difficult to make this kind of comparison. Also, the range of new diagnoses revealed by screening tests is clearly influenced by the selection of the screening investigations, and this affects their medical value.

The cost of screening investigations needs to be compared with discretionary tests by a method that takes account of capital expenditure on laboratory equipment, and the cost in terms of medical resources of following up abnormalities revealed by tests that would not have been requested on a discretionary basis.

This is not the place for a discussion on the cost of laboratory equipment, except to state that many studies have demonstrated the surprisingly small *additional* cost to the laboratory of carrying out a range of screening tests on every patient rather than the much smaller number of tests that would have been requested on a discretionary basis, when all aspects of laboratory expenditure are taken into account. There is, however, little information on the costs of following up the unexpected abnormalities revealed by hospital screening.

There is little or no evidence that screening of patients by chemical tests has produced economies in general hospital costs; their use has not been shown to shorten the stay of patients in hospital, at least not in hospitals in the U.K.

MEDICAL PROBLEMS OF HOSPITAL SCREENING

Doctors may be presented with problems if they attempt to find explanations for all the abnormalities that are liable to be reported in respect of tests that they would *not* have requested on a discretionary basis, i.e. as part of a multi-channel chemical screening programme. Inevitably, they find themselves faced with having to review much larger quantities of data, much of it gratuitous, if, for instance, twenty sets of results are reported on all their patients whenever (on a discretionary basis) they might perhaps only have had clinical reasons to request about five of these tests.

Important information can all too easily be overlooked when masses of figures require to be considered. This is one effect of the so-called 'data flood'. Furthermore, the brain is not efficient at correlating several items of numerical information. For these reasons, interest has increasingly focused on methods of data reduction and exception reporting, using computer techniques, and on computer-assisted recognition of patterns of abnormality (p. 497).

A systematic attempt was made in Birmingham, England, as part of one of the earliest screening programmes in the U.K., to assess the medical value and some of the follow-up problems of screening. The range of screening tests included sixteen chemical investigations carried out on over 2000 patients when first admitted to hospital. Arrangements were made to find out, retrospectively, how much of the analytical work would normally have been requested on a discretionary basis. Clinicians were asked to answer questions on each and every one of the sixteen investigations reported, as follows:

(1) *Would you normally have requested this investigation?* Table 28.2 shows, for instance, that only 11.7% of all the plasma calcium determinations carried out would normally have been requested. *If an abnormal result was reported* for an investigation that would

TABLE 28.2. Admission screening studies.
All figures are expressed as a percentage of the total number of analyses carried out for each test.

	Tests normally requested		*Tests not normally requested*				
					Abnormal results revealed by screening tests		
Analysis	Total	Abnormal	Total	Abnormal	Expected	Diagnostic	Unexplained
Calcium	11.7	3.5	88.3	8.1	1.5	0.7	5.9
Glucose	12.5	2.8	87.5	8.8	1.2	1.6	6.0
Iron	6.6	3.8	93.4	21.6	9.3	3.5	8.8
Potassium	64.4	7.3	35.6	1.6	0.1	0.1	1.4
Urate	3.4	0.8	96.6	7.2	2.9	1.1	3.2
Urea	64.8	9.8	35.2	1.2	0.2	0.2	0.8

not normally have been requested, three further questions were asked:

(2) *Would you have expected the test to be abnormal,* on the basis of your knowledge of the patient's condition? Table 28.2 shows that 3.5% of the calcium results consisted of abnormal results that would have been revealed by discretionary methods of requesting and 1.5% were abnormal results that would have been expected to be abnormal if they had been requested.

(3) *Was this abnormal result unexpected and, if so, did it lead to a new or additional diagnosis, or a change in treatment?* These results are shown in the column headed 'diagnostic'. For calcium measurements, 0.7% of results were placed in this category.

(4) *Was this an unexpected result which could not be explained?* Table 28.2 shows that 5.9% of all the plasma calcium measurements fell into this category.

There are differences with each of the tests listed in Table 28.2 in the proportions of answers falling into each of the categories provided by these four questions, but there are also some constant features. All the tests show a considerable proportion of unexplained abnormalities among the results of those tests that would not normally have been requested. In some instances, it proved possible later (sometimes much later) to explain these previously unexplained results, but only after a variable amount of extra work, often requiring repeated clinical and laboratory assessments.

It is not so readily apparent from Table 28.2 just how much extra work was required by clinical staff in the careful scrutiny of results from the screening studies, when compared with the effort involved in considering reports that would have contained only the results of discretionary tests. In those hospitals where the results of multi-channel screening investigations have not been followed up by such active questioning as in the Birmingham studies, it seems as if all too often abnormalities that are not immediately explicable get overlooked or ignored.

Studies carried out in Adelaide have seriously questioned the medical value of multi-phasic screening of patients admitted to hospital. It is, nevertheless, probable that admission screening by means of chemical and haematological investigations will continue to be practised. Indeed, with chemical examinations, the practice seems likely to grow, even if only because it greatly simplifies laboratory organization. Fortunately, there is a reasonable prospect that developments in the range and nature of chemical tests available for inclusion in hospital screening programmes will lead to an increase in their medical value, especially if more selectivity is adopted in their application (e.g. to particular age-groups).

Screening the ambulant population

Two types of programme will be considered, case-finding surveys and well-population screening.

Case-finding depends on tests chosen for their ability to identify people with a particular disease, usually applied to a sector of the population known to represent a high risk group. It may also be applied to definable components of the general population where it is known that there is a relatively high incidence of the disease. Case-finding is designed to be a search for people who should benefit from being identified as having a specific abnormality, and thereafter treated as patients.

Well-population screening makes use of tests that are carried out to identify people who may possibly be sick, from among an apparently healthy population. At the same time, these tests are intended to give those people for whom results are normal good grounds for believing that they are indeed healthy.

GENERAL REQUIREMENTS

All screening programmes of the ambulant population should be planned on the basis of the following considerations:

(1) They should include measurements or observations which will contribute to the early detection of diseases that are important in terms of frequency of occurrence, or because they constitute a threat to life, or are important for some other reason.
(2) They should use tests or methods that are sensitive and specific, acceptable to the people to be screened, and able to be carried out in large numbers, unless the study is a very limited case-finding survey.
(3) The clinical, laboratory and other facilities necessary for following up abnormalities revealed by the programme, and defining more precisely the nature of any abnormalities detected, must exist.
(4) Acceptable forms of treatment must be available for any disease likely to be discovered as a result of the screening programme, and there should be the facilities (clinical, radiological, laboratory, etc.) needed to guide and control any treatment prescribed.

(5) The financial aspects of the screening programme should be defined in advance.

Sensitivity, specificity and disease prevalence

These are very important concepts, when choosing tests for inclusion in any screening programme of the ambulant population for the possible presence of disease, in case-finding surveys but even more so in more general well-population screening. The terms sensitivity and specificity are defined, for use in epidemiological studies, as follows:

(1) *Sensitivity:* The ability of a test to give a *positive* (abnormal) finding when the person tested truly has a disease, the presence of which is indicated by obtaining an abnormal result with the screening test.
(2) *Specificity:* The ability of a test to give a *negative* finding (i.e. a normal result) when the person tested is free from disease, the presence of which would have been indicated if a positive result had been observed.

Chemical tests used in screening programmes need not be diagnostic in themselves, and indeed they are very rarely diagnostic. Instead, the finding of an abnormal result can be expected to lead on to further tests, aimed at defining the cause of the abnormality. It is important, therefore, that the implications of describing the result of a screening test as abnormal be considered in advance, in relation to the potential seriousness of the medical conditions that are being sought, and the size of the workload that would be generated by following up all abnormalities revealed by screening. These considerations need to take account of the prevalence of diseases in the populations or groups that are being studied.

In practice, decisions have to be taken arbitrarily that affect the sensitivity and specificity of chemical tests used for population screening. For instance, Fig. 2.1 (p. 18) shows that there is no clear dividing line between a normal and an abnormal level of plasma alkaline phosphatase activity. Instead, the data can be thought of as representing at least two overlapping distributions. If alkaline phosphatase measurements are to be included in a set of screening investigations, depending on where the cut-off limit between normality and abnormality is set, the sensitivity and specificity of the test will inevitably be affected (Fig. 28.1).

Decisions about the setting of cut-off limits are influenced, on the one hand, by the economic and logistic consequences of having to

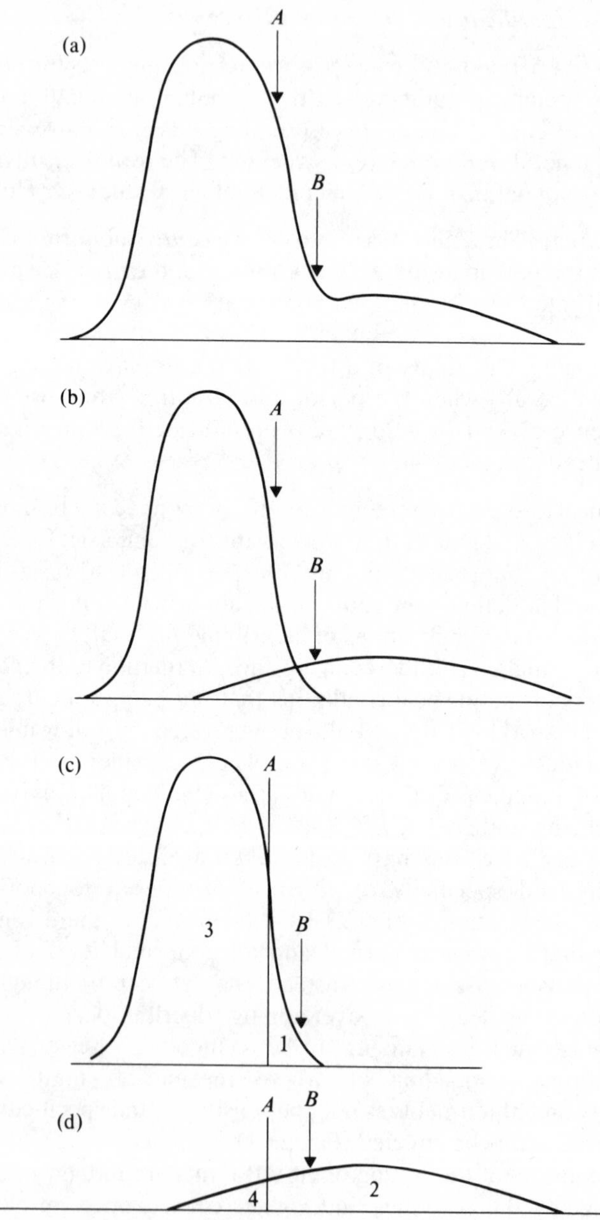
(a)
A
B
(b)
A
B
(c)
A
3
B
1
(d)
A
B
4
2

TABLE 28.3. Case-finding surveys. Examples of the application of chemical investigations.

Group	Condition being sought	Tests
Neonate	Phenylketonuria	Serum phenylalanine
Pregnant females	Anaemia	Haemoglobin, plasma iron
	Diabetes mellitus	Plasma and urine glucose
Industry, occupation	Lead poisoning	Blood lead, urine porphyrins
	Pesticide poisoning	Plasma cholinesterase
Old age	Scurvy	Leucocyte ascorbate

investigate large numbers of false positives if the limit is set too low, and on the other hand by the medical consequences of missing cases that the programme has set out to find, if the limit is set too high.

CASE-FINDING PROGRAMMES

These programmes aim to identify treatable disease, known to be important or to have an appreciable incidence in the sector of the population under study. Less often, they aim to identify several diseases in the group chosen for study. Table 28.3 gives examples of places where chemical investigations have been shown to play a definite part in case-finding programmes.

Screening for phenylketonuria

The comparative validity of various tests for PKU is shown in Fig. 28.2, which is based on a comparison between *children known to have PKU* and a population of normal children. Plasma [phenylalanine] shows the widest separation between the groups. When used for screening an

FIG. 28.1. Diagrammatic representation of the effects of altering the screening cut-off point from *A* to *B* on the incidence of false positive and false negative results for a variable that does not show a clear-cut division between 'health' and 'disease' in the population investigated.

The distribution of the results (a) has been resolved (b) into two overlapping Gaussian distributions, one for 'healthy' individuals and the other for 'diseased'. The subdivision of the 'normal' results into false positives (1) and true negatives (3), for a screening cut-off point set at *A*, as shown in (c), corresponds to a subdivision of the 'abnormal' results into true positives (2) and false negatives (4), as shown in (d). If the screening cut-off point had been set at *B*, the sensitivity would have been reduced (fewer true positives) but the specificity would have been increased (more true negatives). Modified from Wilson (1973).

unselected population of newborn *infants,* however, measurement of serum [phenylalanine] by the Guthrie test (p. 427) is not as perfect a test as Fig. 28.2 might suggest as there is an appreciable incidence of false positive results. This is because the prevalence of PKU is very low, and because (in view of the serious consequences of delays in diagnosis) false negative results from the screening programme cannot be tolerated.

Studies conducted on newborn infants (Table 28.4) can be used to illustrate the effect of setting the screening level at different values for serum [phenylalanine] on the incidence of false positive results. In these studies, 117 444 Guthrie tests were carried out on an unselected

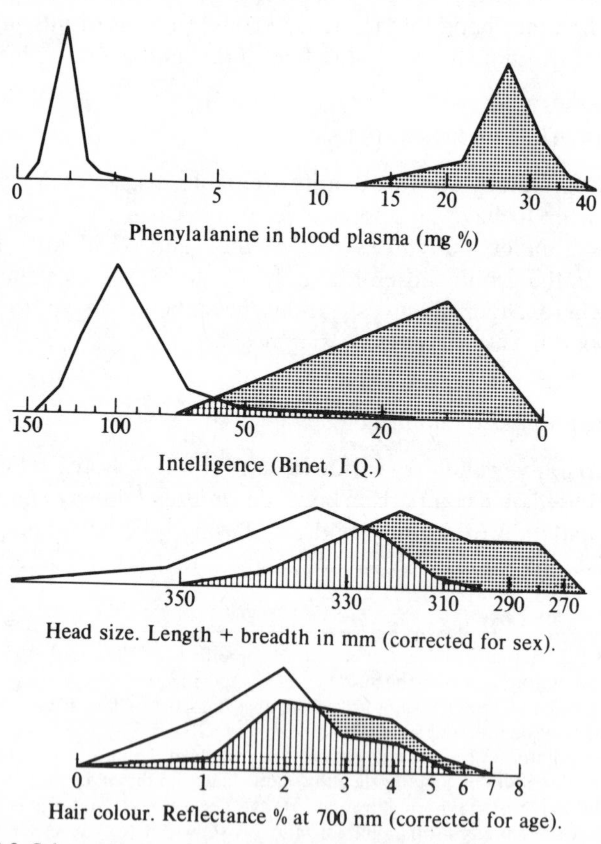

FIG. 28.2. Schematic representation of the discriminatory power of four indicators for PKU in a population of phenylketonuric and non-phenylketonuric children. Reproduced with permission from Penrose L.S. (1951), *Annals of Eugenics*, **16,** 134-141.

TABLE 28.4. Case-finding for phenylketonuria (data from Wilson, 1973)
Effect of altering the screening level of serum [phenylalanine] on the incidence of false positives among an unselected population of newborn infants. These data are based on 117 444 Guthrie tests, in a programme where the screening level was set at 240 μmol/L (4 mg/100mL), and which resulted in the identification of 8 cases of PKU.

Serum phenylalanine (μmol/L)	240	360	480	720	1200
False negatives	0	0	0	1	2
Sensitivity (%)	100	100	100	87.5	75
False positives	419	28	16	2	0
Specificity (%)	99.6	99.9	100	100	100
Predictive value (%)*	2	24	36	77	100

* Predictive value of a positive result is the percentage of truly positive results for phenylketonuria among all the positives revealed by the screening tests.

population of newborn infants. Eight cases of PKU were detected and follow-up showed that no cases had been missed, i.e. there were no false negatives. The Guthrie test, therefore, showed 100% sensitivity in this study, and its specificity was also very high. However, at the screening level used, i.e. serum [phenylalanine] of 240 μmol/L (4.0 mg/100 mL), 419 false positives were obtained, due to the test picking up other causes of hyperphenylalaninaemia.

The relative prevalence of PKU, i.e. phenylalanine hydroxylase deficiency, and of hyperphenylalaninaemia due to other causes in the population under investigation influences the predictive value for PKU of a positive result in the Guthrie test or in other tests (e.g. the Scriver test) used for measuring serum [phenylalanine] in the screening programme. The incidence of false positives in PKU screening programmes could be reduced, and the predictive value of a positive result increased, by setting the screening value for serum [phenylalanine] higher (Table 28.4). However, if the screening value were to be set too high, e.g. at 720 μmol/L (12 mg/100 ml) or above, false negatives would be obtained and the value of the screening programme called in question.

The data in Table 28.4 might appear to suggest that a serum [phenylalanine] of 480 μmol/L could have been safely used as the screening level, thereby greatly improving the predictive value of serum phenylalanine measurements. However, these data were obtained several years ago, and subsequent alterations in the nature of infant feeds have meant that a serum [phenylalanine] of 240 μmol/L is appropriate for use as the screening level.

With many chemical measurements, there is a much less clearly defined cut-off point between health and disease than in the case of phenylketonuria. This means that the dividing line between normality and abnormality for test results obtained in many case-finding surveys has to be a compromise between sensitivity and specificity. If the cut-off limit is set too low, the sensitivity of the test will be high but its specificity low, and there will be a large number of false positives. On the other hand, if the cut-off limit is set too high, the sensitivity will be low, there will be many false negatives, and the purpose of the screening programme will not have been achieved.

Screening programmes for the detection of neural tube defects, by measuring maternal serum [α_1-fetoprotein] at 16–18 weeks' gestation, provide examples of programmes where positive results must be checked since false positive results cannot be tolerated. Abnormal results for [MSAFP] in the screening programme must be confirmed and other follow-up examinations performed (ultrasound and, if necessary, amniocentesis, p. 410) before there can be any question of advising the termination of pregnancy.

MULTI-PHASIC SCREENING

This term means the 'application of two or more screening tests in combination' to the people being examined by the programme. The information collected may include anthropometric data, and results of hearing tests, respiratory function and visual acuity assessments, chest X-ray and ECG, and laboratory examinations of both a haematological and a chemical nature.

The chemical investigations performed as part of multi-phasic screening programmes are, in many instances, the same as those used when screening hospital patients. This means that the same type of abnormality has tended to be revealed, e.g. increased plasma [glucose] or low plasma [iron], but with a lower frequency than in the hospital population. Problems of handling masses of data, and questions relating to action to be taken when abnormalities are revealed but are unable to be explained, are greater with multi-phasic screening programmes carried out as part of a 'health-check' than with screening investigations carried out on patients admitted to hospital.

Regular screening assessments, for instance 'health-checks' conducted on an annual basis, may reveal patterns of abnormality which, in course of time and on the basis of retrospective assessment, can be seen to have been early pointers to the development of specific diseases. Multi-variate analysis of data obtained from such multi-phasic screening programmes is being used in this search for early pointers to the diagnosis of specific conditions.

Screening the apparently healthy, ambulant adult population by means of chemical tests has not yet been shown to be worthwhile, possibly because the tests used so far have been selected more on the basis of the capabilities of analytical equipment than for their diagnostic potential. They have tended to lack sensitivity or specificity, or both, and have had unacceptably low predictive value in a population where, by definition, disease prevalence can be expected to be low.

Conclusion

Screening methods of investigation carried out on hospital patients have proved their value in terms of their ability to streamline the flow of work through laboratories. They also have diagnostic value, revealing hitherto unsuspected abnormalities of medical importance. Although the diagnostic yield is low at present, it is likely to rise as more appropriate (i.e. more sensitive and specific) tests come to be included in the hospital screening group.

As far as screening the ambulant population is concerned, case-finding programmes with limited, defined objectives have also been shown to be valuable, and chemical investigations play a useful part in some of these (Table 28.3). However, well-population screening by means of chemical tests for early recognition of the presence of disease in the general population is still largely at the research stage. The contribution of each test to these screening programmes has to be assessed, and this requires prolonged and careful follow-up of the people screened.

No-one questions the basic tenets that prevention is better than cure, and that early detection of disease may offer the chance of arresting and even reversing the pathological process more readily than detection at a later stage (Fig. 1.1, p. 6). However, it still has to be shown, for example, that early detection of hyperlipidaemia improves the long-term prognosis for people with this chemical abnormality. Unresolved questions include the choice of tests to incorporate in well-population screening programmes. Problems in the definition of 'health' and 'disease' also arise, and are relevant in assessing the sensitivity and specificity of investigations for the detection of hitherto unsuspected disease.

FURTHER READING

BRADWELL, A.R., CARMALT, M.H.B. and WHITEHEAD, T.P. (1974). Explaining the unexpected abnormal results of biochemical profile investigations. *Lancet,* **2,** 1071–1074.

BRYAN, D.J., WEARNE, J.L., VIAU, A., MUSSER, A.W., SCHOONMAKER, E.W. and THIERS, R.E. (1966). Profile of admission chemical data by multi-channel automation: an evaluation experiment. *Clinical Chemistry,* **12,** 137–143.

CARMALT, M.H.B., FREEMAN, P., STEPHENS, A.J.H. and WHITEHEAD, T.P. (1970). Value of routine multiple blood tests in patients attending the general practitioner. *British Medical Journal,* **1,** 620–623.

COCHRANE, A.L. and HOLLAND, W.W. (1971). Validation of screening procedures. *British Medical Bulletin,* **27,** 3–8.

COLE, G.W. (1980). Maximizing the efficiency of an admissions test profile in a 680-bed hospital. *Clinical Chemistry,* **26,** 46–50.

DURBRIDGE, T.C. EDWARDS, F., EDWARDS, R.G. and ATKINSON, M. (1976). Evaluation of benefits of screening tests done immediately on admission to hospital. *ibid.,* **22,** 968–971.

GRINER, P.F. and GLASER, R.J. (1982). Misuse of laboratory tests and diagnostic procedures. *New England Journal of Medicine,* **307,** 1336–1339.

Lancet (1974). *Screening for Disease.* A series of articles available as a combined reprint. Lancet: London.

Leading article (1976). Admission multiphasic screening. *Lancet,* **2,** 1229–1230. (and letter from W.K.C. Morgan (1977) in *ibid,* **1,** 142).

Leading article (1978). Multiphasic screening in general practice. *ibid.,* **1,** 29.

MCKEOWN, T. (1976). An approach to screening policies. *Journal of the Royal College of Physicians of London,* **10,** 145–152.

MARKS, V. (1980). The choice between discretionary and profile testing. In *Centrifugal Analysers in Clinical Chemistry,* pp. 259–270. Ed. C.P. Price and K. Spencer. Eastbourne: Praeger.

RHODES, P. (1976). Institute of Medical and Veterinary Science and admission multiphasic screening. *British Medical Journal,* **2,** 804–805.

WILSON, J.M.G. (1973). Current trends and problems in health screening. *Journal of Clinical Pathology,* **26,** 555–563.

Appendix 1

Units Used in Reporting

Considerable variety in the units used by different clinical chemistry laboratories existed in the past, when reporting results obtained for the same type of measurement. This often caused confusion when results reported by one laboratory were compared with results obtained elsewhere for the same range of investigations, e.g. when a patient was transferred from one hospital to another. To overcome this confusion, about ten years ago the Système International d'Unités (SI units) were adopted by clinical chemistry laboratories in many countries. This system offers a coherent system of units that has proved applicable for reporting most aspects of laboratory work, apart from results of enzyme activity measurements (p.520) and toxicology (p. 477).

There are seven quantities on which SI units used in clinical chemistry are based, and from which most other units can be derived. All the units are based on the metric system and the whole system is coherent since the product or quotient of any two or more unit quantities is the unit of the resultant quantity. For instance, the metre is the basic unit of length, and unit volume = unit length × unit length × unit length. This means that the cubic metre (m^3) is the coherent SI derived unit of volume. In practice, when using the SI system in medicine, it is more convenient to use the litre (L) as the derived unit of volume rather than the cubic metre, and this is allowable.

Table A1.1 includes the basic SI units and examples of derived units particularly relevant to clinical chemistry. Many rules or guidelines governing the use of SI units are discussed in the references cited. Observance of the following rules is intended to lead to data being presented in a standardized form:

(1) Symbols are printed without a full stop, except when they come at the end of a sentence.
(2) Symbols do not have plurals.
(3) Multiples and submultiples are all decimal (Table A1.2).
(4) In a set of figures, the decimal point is a full stop (or comma) on the line; a raised point represents the multiplication sign.
(5) Whereas large numbers used to be subdivided by means of commas into groups of figures (e.g. 850,000 and 1,500,000), the

TABLE A1.1. SI units and examples of units derived from the basic SI units.

Quantity or measurement	Name of SI unit	Symbol	Expressed in terms of other SI units
Basic units			
Length	metre	m	—
Mass	kilogram	kg	—
Time	second	s	—
Electric current	ampere	A	—
Thermodynamic temperature	kelvin	K	—
Luminous intensity	candela	cd	—
Amount of substance	mole	mol	—
Derived units (examples)			
Frequency	hertz	Hz	l/s or s^{-1}
Force	newton	N	$kg.m/s^2$ or $kg.m.s^{-2}$
Work, energy, quantity of heat	joule	J	N.m or $kg.m^2.s^{-2}$
Pressure	pascal	Pa	N/m^2 or $kg.m^{-1}.s^{-2}$
Concentration			
(1) molar concentration	—	—	mol/L or $mol.L^{-1}$, etc.
(2) mass concentration	—	—	g/L or $g.L^{-1}$, etc.

Note The mole is defined as the amount of substance of a system which contains as many elementary units as there are carbon atoms in 0.012 kg of the pure nuclide carbon-12 (^{12}C). The elementary unit must be specified and may be an atom, a molecule, an ion etc., or a specified group of such entities.

same groupings may still be used but separated by spaces or by approved symbols (e.g. 850 000 and $1.5 \cdot 10^6$).

(6) With complex derived units (e.g. the pascal) the divider symbol (/, meaning per) may only be used once, or a combination of multiplication symbols and negative powers may be used (e.g. kg/$m.s^2$ or $kg.m^{-1}.s^{-2}$ for the pascal). The full stop in abbreviations for derived units indicates 'multiplied by'.

Concentrations are expressed on a molar basis (e.g. mmol/L or $mmol.L^{-1}$) in most instances where single substances of known atomic or molecular mass are being measured. Previously, these same substances had their concentrations expressed mostly as mass concentrations, and in a variety of units (e.g. mg/100 mL, mg/mL, meq/L). Mass concentrations are still used, in accordance with the rules of the system, for reporting some measurements, for instance results of protein determinations and of the examination of mixtures.

Table 2.1 (p. 18) gives examples of reference values for many commonly performed chemical determinations, with values expressed both in accordance with SI rules and in units that were largely replaced in Britain in 1974, following the changeover to molar units, where appropriate, as the preferred SI units. It should be apparent, from the examples contained in Table 2.1, just how important it can be to take note of the units in which reports are expressed when interpreting results. It can be very dangerous to confine attention to the numerical result, since serious mistakes in interpretation may then occur, sufficient to lead to fatal mistakes in treatment.

Apart from the litre, the other main departures of convenience from the SI recommendations are the continued usage of (1) the Celsius (or centigrade) scale of temperature, in preference to the Kelvin scale, and (2) other units of time (hour and day) as alternatives to the second.

The adoption of SI recommendations has helped to achieve uniformity in the expression of results. However, some sources of

TABLE A1.2. Prefixes for basic and derived SI units, with examples of decimal fractions and multiples widely used in medicine.

Fraction	Prefix	Symbol	Multiple	Prefix	Symbol
10^{-1}	deci	d			
10^{-3}	milli	m	10^3	kilo	k
10^{-6}	micro	μ	10^6	mega	M
10^{-9}	nano	n			
10^{-12}	pico	p			
10^{-15}	femto	f			

difficulty and confusion remain and doctors still need to be aware of other methods of expressing results, for at least three reasons:

(1) With some patients, medical records extend over many years, and it may still be necessary to consider results of tests carried out on these patients and reported in other units, e.g. before 1974.
(2) Molar units have not yet been widely adopted by several countries, notably the U.S.A. and Canada, and even parts of the Continent of Europe. Also, results in some parts of the scientific literature are still reported in other units. Sometimes the data are presented in two sets of units (molar and gravimetric), but editorial mistakes in making the conversions continue to be disturbingly frequent.
(3) Lack of agreement as to the most appropriate units to adopt for reporting measurements of drug concentrations (p. 477).

It is readily possible to convert data expressed in molar concentrations into the corresponding mass concentrations, and *vice versa,* by means of nomograms or by using the formula:

$$y(\text{mol/L}) = \frac{x(\text{g/L})}{\text{mol. mass}}$$

Some laboratories provide lists of factors to help convert mass concentration units expressed in grams per litre into the corresponding data expressed in moles per litre.

UNITS OF ENZYME ACTIVITY

For technical reasons, it is rarely practicable to measure enzyme concentrations directly. Instead, the results of laboratory measurements are usually expressed in units of enzyme activity. This practice tacitly assumes that enzyme activity is directly proportional to enzyme concentration, an assumption which may only be valid if the correct conditions are selected under which to perform the assay.

There are several different methods in routine use for measuring the activities of most of the enzymes currently regarded as being of diagnostic importance. In the past, when a method was first introduced for measuring a particular enzyme, arbitrary units (usually eponymous) tended to be used. For instance, King-Armstrong (K-A) units were widely adopted for expressing results of alkaline phosphatase activity; one K-A unit of activity liberates 1 mg of phenol from phenylphosphate after incubation for 15 min at pH 9.8 and 37°C. However, modifications and improvements to methods in some cases resulted in less enzyme being required to liberate the same amount of reaction product (phenol in the K-A unit example), and this led to changes in reference values. Prior to the introduction of International Units of enzyme activity, many laboratories had established their own sets of reference values, which depended on the details of the method adopted locally for these measurements of enzyme activity.

The International Unit (i.u.) of enzyme activity was introduced in an attempt to achieve a degree of standardization. One i.u. is defined as 'that amount of enzyme which, under given assay conditions, will catalyse the conversion of 1 μmol of substrate per minute'.

Unfortunately, the use of International Units does not greatly simplify the problems of lack of uniformity that were associated with the reporting of enzyme activity measurements in eponymous units, since the definition of one i.u. includes the words '*under given assay conditions*'. This means that the numerical value of enzyme activity measurements will depend on:

(1) The nature and concentration of the substrate.
(2) The direction in which the reaction proceeds.
(3) The nature, $[H^+]$ and molar concentration of the buffer.
(4) The presence of activators and inhibitors.
(5) The temperature of the reaction.

Although international recommendations specifying each of these five conditions have now been made for some commonly performed enzyme activity measurements, it will be some time before all laboratories adopt these recommendations and thereby achieve uniformity in the reporting of enzyme activity data.

A further complication is that the definition of an International Unit of enzyme activity refers to 'μmol/min', but the SI unit of time is the second. It has, therefore, been proposed that enzyme activity measurements should be expressed in the appropriate derived SI unit, 'mol/s'. This unit has been termed the *katal* and some laboratories now report enzyme activities in katals (1 nkatal = 16.7 i.u.), but the majority still use international units.

FURTHER READING

Baron, D.N., Broughton, P.M.G., Cohen, M., Lansley, T.S., Lewis, S.M. and Shinton, N.K. (1974). The use of SI units in reporting results obtained in hospital laboratories. *Journal of Clinical Pathology,* **27,** 590–597.

Doumas, B.T. (1979). IFCC documents and interpretation of SI units — A solution looking for a problem. *Clinical Chemistry,* **25,** 655–657.

Ellis, G., editor (1971). *Units, Symbols and Abbreviations.* London: Royal Society of Medicine.

McLauchlan, D.M., Raggatt, P.R. and Zilva, J.F. (1975). SI units: How systematic and how international are they? *Annals of Clinical Biochemistry,* **12,** 1–3.

World Health Organization (1977). *The SI for the Health Professions.* Geneva: W.H.O.

Appendix 2

Acid–Base Terminology

Laboratory Terms

ACID

An acid is a proton donor. It dissociates in solution to yield protons (H^+) and the corresponding base. For example:

$$H_2CO_3 \rightleftharpoons H^+ + HCO_3^-$$

BASE

A base is a proton acceptor. In the above equation the base (HCO_3^-) can accept a proton (H^+), thereby forming the corresponding undissociated acid (H_2CO_3).

BASE EXCESS (units: mmol/L)

This is the amount of strong acid that would be required to titrate one litre of fully oxygenated blood to a $[H^+]$ of 40 nmol/L (pH 7.40) at 37°C and under conditions where the $P\text{CO}_2$ is 5.3 kPa (40 mmHg). If alkali were to be required to perform this titration, the results would be expressed as negative quantities or else described as a *base deficit.*

BLOOD BICARBONATE (units: mmol/L)

The concentration of HCO_3^- in the plasma of a blood sample.

BUFFERS

Mixtures which, by their presence in solution, increase the amount of acid or base that must be added to cause unit change of pH. A buffer is made up of two components: (1) a weak acid (e.g. H_2CO_3) and (2) its corresponding base (e.g. HCO_3^-).

OXYGEN CONTENT (units: mmol/L, volumes/100mL)

The total O_2 content of blood.

P_{CO_2} (units: kPa, mmHg)

The partial pressure of CO_2 in the gas phase in equilibrium with CO_2 dissolved in a biological fluid.

P_{O_2} (units: kPa, mmHg)

The partial pressure of O_2 in the gas phase in equilibrium with O_2 dissolved in a biological fluid.

PER CENT OXYGEN SATURATION

The proportion of haemoglobin (Hb) present as oxyHb in a solution when some is present as Hb and some as oxyHb.

pH

The logarithm (to the base 10) of the reciprocal of the hydrogen ion concentration; $pH = -\log_{10}[H^+]$. pH meters measure the *activity* of H^+, not $[H^+]$. At physiological pH values, H^+ activity and $[H^+]$ are practically identical.

pKa

The logarithm (to the base 10) of the reciprocal of the ionization constant of each acid group in a buffer solution. The pKa of an acid is equal to the pH at which the [acid] and [base] are equal and is the pH at which the buffering capacity is greatest.

PLASMA BICARBONATE (units: mmol/L)

The concentration of HCO_3^- in plasma.

STANDARD BICARBONATE (units: mmol/L)

The concentration of bicarbonate in plasma of a blood specimen that has, following collection, been equilibrated with O_2–CO_2 mixtures at 37°C. These *in vitro* manipulations allow the bicarbonate concentration to be determined under precisely specified conditions, namely fully oxygenated blood, 37°C and P_{CO_2} of 5.3 kPa (40 mmHg).

TOTAL CARBON DIOXIDE CONCENTRATION (units: mmol/L)

The CO_2 released from a biological fluid by strong acid. The measurement, which includes dissolved CO_2, H_2CO_3, HCO_3^-, CO_3^{2-} and carbamino compounds, is referred to as plasma [totalCO_2] in the main text. The largest contribution comes from HCO_3^-. This is the most commonly performed acid–base measurement, carried out on venous blood (p. 74).

Physiological Terms

ACIDAEMIA

The [H^+] of the blood sample is above the upper reference value.

ACIDOSIS

An abnormal condition in which there is an accumulation of acid, or loss of base, and which would cause a significant rise in blood [H^+], or fall in pH, if there were to be no compensatory changes in response to the primary cause of the acid–base disturbance.

ALKALAEMIA

The [H^+] of the blood sample is below the lower reference value.

ALKALOSIS

An abnormal condition in which there is an accumulation of base, or loss of acid, and which would cause a significant fall in blood [H^+], or rise in pH, if there were to be no compensatory changes in response to the primary cause of the acid–base disturbance.

COMPENSATED

An adjective sometimes used to qualify the description of an acid–base disturbance when there is evidence to suggest that compensatory processes have occurred. If compensatory effects are discernible but the blood [H^+] is still outside the reference range, the disturbance is said to be partially compensated; if the blood [H^+] has been brought back within the range of reference values, the disturbance is said to be fully compensated.

COMPENSATORY (OR SECONDARY)

An adjective used (1) to qualify the description of a physiological process, renal or respiratory, occurring in response to a primary acid–base disturbance, or (2) to qualify a change in the composition of a blood sample (e.g. in the respiratory component) occurring in response to the primary disturbance.

METABOLIC (OR NON-RESPIRATORY)

Adjectives that help to describe the aetiology of an acid–base disturbance, used to qualify 'acidosis' when the *primary* disturbance involves either the retention of acids *other than* H_2CO_3 or the loss of base. The adjectives are also used to qualify 'alkalosis' when the primary acid–base disturbance involves either the loss of acids *other than* H_2CO_3 or the gain of base.

MIXED

An adjective used to qualify the description of an acid–base disturbance when it would appear, usually on clinical grounds, that two or more aetiological factors were primarily responsible for producing the acid–base disturbance.

RESPIRATORY

An adjective that helps to describe the aetiology of an acid–base disturbance by qualifying the nouns 'acidosis' and 'alkalosis' when the primary disturbance involves, respectively, the retention or the loss of H_2CO_3 together with a corresponding increase or decrease in $P\text{CO}_2$.

SIMPLE

An adjective sometimes used when it is known that a single aetiological factor is responsible for producing an acid–base disturbance.

UNCOMPENSATED

An adjective sometimes used to qualify the description of an acid–base disturbance when the blood $[H^+]$ is outside the range of reference values and there is no evidence to suggest the occurrence of an effective compensatory process.

FURTHER READING

ANDERSEN, O.S., ASTRUP, P., CAMPBELL, E.J.M., CHINARD, F.P., NAHAS, G.G. and WINTERS, R.W. (1966). Report of *ad hoc* committee on acid–base terminology. *Annals of the New York Academy of Sciences*, **133,** 251–258.

WINTERS, R.W. (1966). Terminology of acid–base disorders. *ibid,* **133,** 211–224.

Index

Chapter headings and some major sub-headings have been indicated in **bold type**; this means that the index entry usually relates to several pages. **Bold type** has also been used in some places, where a subject has two or more indexed entries, to denote the page which the authors suggest should be consulted first.

Entries that begin with a Greek letter or with a number, for instance, α_1-Fetoprotein and 1:25-Dihydroxycholecalciferol, have been indexed under the letters F and D, respectively.

Abbreviations viii
Abetalipoproteinaemia 260
Abnormal results **18,** 26
Abortion 385, 405, 408
Abscess 185, 469
Accuracy 23, 85
Acetoacetate
  blood 35, **217,** 229
  urine 32, 232
Acetone 32, **217,** 229
Acetylcholinesterase 153, 167, 412, 488
Acetyl CoA 229
N-Acetyl-cysteine 479
N-Acetyl-β-glucosaminidase 166
Achlorhydria 196, 198
Acid 522
Acidaemia 524
Acid-base balance **64**
  chronic renal failure 106
  definitions **522**
  diabetic coma 229, 231
  electrolyte metabolism **54,** 96, 282, 287
  fetal monitoring 413
  measurements 67
  mixed disturbances 72, 525
  respiratory distress syndrome 434
  results presentation 69
  salicylate overdosage 485
  simple disturbances 72, 76, 525
  specimen collection 68
  temperature effects 69
  treatment of disturbances 81
α_1-Acid glycoprotein 114, 123, 475
Acidification tests (urine) 93
Acidosis, definition 524, *see also* metabolic acidosis, respiratory acidosis
Acid phosphatase 12, **150,** 163, 460, 463
Acne vulgaris 383
Acrodermatitis enteropathica 275
Acromegaly 222, **389,** 394
ACTH stimulation test 367
Active chronic hepatitis 183
Acute cholecystitis 199
  hepatitis 97, 143, 144, **160,** 176, 179, 181, 184
  intermittent porphyria 315
  pancreatitis 144, 151, 166, **199,** 254
  phase reactants 123, 462
  renal disease 58, **103,** 143, 144, 146
Addiction (drugs) 485
Addison's disease 367, 394
Adenine 327
Adenylate kinase 470
Adenyl cyclase 216
Adipose tissue 216, 242
Admission screening 4, 447, 501
Adrenal cortex 354
  fetal hypoplasia 407
  hyperfunction **359,** 374, 391, 394
  hyperplasia, congenital 376, 421, 424
  hypofunction 58, 227, 304, **367,** 391, 394
  steroid therapy 372
  stimulation tests 362
  suppression tests 363
  tumours 364, 374, 391
Adrenaline 215, 449
Adrenal medulla 449
Adrenocorticotrophic hormone 354, 367, 374, 377, 380, **390,** 403, 456, 459
  plasma 357, 359, 365, 370, 391
Adrenogenital syndrome, *see* congenital adrenal hyperplasia
Afferent loop syndrome 199

Agammaglobulinaemia 128
Age (effect on reference values) 22, 147, 250, 399, 432, 443
ALA and related enzymes, *see* 5-aminolaevulinic acid
Alanine aminotransferase **144,** 158, 484
  liver disease 160, 181, **184,** 436
  oral contraceptives 401
  specimen collection 12
  vitamin B_6 deficiency 269
Albumin **115,** 132, 281, 325, 343, 435
  drug-binding 475
  :globulin ratio 181
  liver disease 179, 186
  meconium 439
  oral contraceptives 400
  plasma 47, 59, **115,** 444
  pregnancy 402
  protein-energy malnutrition 437
  transport functions **117,** 216, 338, 475
Alcohol 159, 186, 199, 256, 315, 320, 480, 485
Aldolase 165
Aldosterone 53, 354, **373,** 378
  renal tubule (effects on) 44, 96, 373
  trauma 59
Aldosteronism **374**
  secondary 46, 51, 96, 179, 376
Alkalaemia 524
Alkaline phosphatase **146,** 277, 520
  age-variation 147, 444
  bone disease 147, 267, 285, 295, 348, 446
  heat-stable 148, 403, 409, 460
  liver disease 147, 161, 181, 184
  malignant disease 148, 460, *see also* carcinoma
  multiple myeloma 136
  oral contraceptives 401
  pregnancy 147, 403, 409
  reference values 18, 148, 444
  Regan isoenzyme 148, 163, 460
  vitamin D deficiency 267
Alkalosis, definition 524
  see metabolic and respiratory alkalosis
Allopurinol 333
Aluminium toxicity 300, 487
Amenorrhoea 383, 385, 414
Amerlex Free T3 RIA kit 344
  Free T4 RIA kit 342
Amino acids 111, 170, 193, 423, 466
  liver disease 181
  malabsorption 205
  parenteral nutrition 61
  protein-energy malnutrition 437
  see also individual amino acids
Amino aciduria 97
  heavy metal poisoning 487
  liver disease 182, 187
5-Aminolaevulinic acid 313
  lead poisoning 319
  porphyrias 317
Aminopeptidase 149, 440
Ammonia 93, 182, 424
Ammonium chloride test 95
Amniocentesis 409, 411, 429
Amniotic fluid 166, 380, 412, 429, 434, 440
Amylase **151,** 188, 192, 199
  :creatinine ratio 200
  duodenal juice 201, 211
Amyloid 101, 136, 185
Anaemia 309, 310, 504
  see also folic acid, vitamin B_{12}
Analbuminaemia 116
Analgesic nephropathy 50, 90
Analytical precision 23
Androgens 340, 354, 356, 379, **381**
Anencephaly 407
Angioneurotic oedema 46
Angiotensin 44, 373, 376
  -converting enzyme (ACE) 373
Anion gap 75
  transport (hepatic) 173
Anorexia nervosa 371
Anterior pituitary 338, 354, **388**
  reserve 388, 394
Antibiotics 475, 484
Antibodies **124**
  adrenal 371
  islet cells 221
  liver disease 183
  thyroid 347
Anticoagulants 11, 258, 267
Anticonvulsants 222, 340, 474, 476
Antidiuretic hormone 395
  inappropriate secretion 48, 51, **395,** 459
  metabolic response to trauma 59
  renal concentration tests 91
  water, Na^+ depletion 41, 44, 91
  water, Na^+ excess 45
α_1-Antitrypsin 119, 436
Apolipoproteins **244,** 246, 260, 400
APRT deficiency 334
Arachidonic acid 243
Argentaffin reaction 454, 457

Arginine vasopressin, *see* antidiuretic hormone
Arterial disease 155, 257
Arylamidase 150
Ascending cholangitis 189
Ascites 179, 187, 375
Ascorbic acid **272,** 306, 324, 511
  saturation test 272
Aspartate aminotransferase **144,** 470
  drug overdose 484
  liver disease 160, 181, 184, 436
  muscle disease 158
  myocardial infarction 155
  oral contraceptives 401
  specimen collection 12
  vitamin B_6 269
Assisted ventilation 52
Asthma 81, 454
Atomic symbols xii
Atransferrinaemia 117
Atrial fibrillation 350
Atrophic gastritis 198
Atypical carcinoids 454
Augmented histamine test 196
Autoimmune disease 129, 347, 371
Automatic analysers 4, 35, 492, 500

Bacterial colonization (small intestine) 204, 208, 211
Balance (metabolic)
  acid-base 64
  calcium 277, 298
  electrolyte 39
  fluid **58**
  magnesium 301
  nitrogen 111
  water 39
Bantu (iron intoxication) 310
Barbiturates 315, 318, 476, 484
Bartter's syndrome 56
Basal metabolic rate 348
Base 67, 522
  deficit and excess 67, 522
BB isoenzyme (creatine kinase) 148, 163, 470
Bence-Jones myeloma 131
  protein 32, 100, 130
Benign paraproteinaemia 131
Benzodiazepines 475
Betamethasone, *see* dexamethasone
Bicarbonate 64, 522
  CSF 467
  plasma 67, 75, 229, 523
  renal excretion 93
  results presentation 70, 76
  therapy 81
Bile 41, 188, 207, 242, 339, 399
Bile acids (and salts) 171, 178, **207,** 242, 260
  'breath test' 208, 211
  plasma 178, 185, 401
  primary 171, 207
  secondary 172
Biliary tract disease 152, 161, 162, 176, 185, 188, 210, 436
Bilirubin **173,** 184, 311, 434
  amniotic fluid 413
  CSF 468
  gallstones 189
  jaundice 175, 434
  measurements 174, 436
  oral contraceptives 400
  pregnancy 404
  side-room tests 32, 184
  transport 171, 173, 434
Biological variation 24
Bisalbuminaemia 117
'Blind loop' syndrome, *see* 'stagnant gut' syndrome
Blood gas analysers 35, 69, 80
Blood specimens
  changes after collection 12, 69, 151, 219
  collection of specimens **10,** 57, 68, 182, 257, 275, 282, 431
  side-room tests **33,** 220, 232
  storage 11
  see also specimen collection and preservation
Blood volume 47
Body compartments 42, 47
Bohr effect 77
Bone **277,** 446, *see also* alkaline phosphatase, calcium, vitamin D
Bradykinin 455
Brain damage 145, 483
Breath tests 206, 208, 209
Bromocriptine 393, 415
Bromsulphthalein 171, 178, 401
BT PABA/^{14}C-PABA test 202, 211
Buffer 64, 522
Burns 166, 124

C-reactive protein 123

Cadmium poisoning 487
Caerulein 201
Calcidiol 265
Calciferol, *see* vitamin D
Calcitonin 279, 281, 457, 459
Calcium 265, **277,** 348, 457, 462, 505
  binding (to protein) 281
  :creatinine ratio (urine) 289
  hyperparathyroidism **284,** 290, 294, 297, 503
  hypoparathyroidism 295
  infusion test 290
  ionized 278, 282, 299
  malignant disease 291, 462
  metabolic bone disease 296
  multiple myeloma 136
  non-endocrine tumours 291, 458
  pancreatitis 200
  renal stones 107
  urinary 108, 289, 292
  vitamin D **267,** 284, 292, 433
Calculi 107, 188
Carbamino compounds 75, 78
Carbohydrate 169, **213,** 432, 444
  absorption 192, 203
  liver disease 182
  oral contraceptives 398
Carbon dioxide 12, 64
  total 74, 524
Carbonic acid 64, 467
Carbonic anhydrase 52, 65, 94
  inhibitors 73
Carbon monoxide 79, 325, 482
  tetrachloride 185, 487
Carboxyhaemoglobin 325
Carcinoembryonic antigen 121, 460, 463
Carcinoid syndrome 210, 270, **454**
Carcinoma 162, 458
  adrenal 364, 375
  bone 148, 163, 289
  breast 151, 388, 459
  bronchus 121, 164, 364, 457, 459, 461
  colorectal 121, 461
  gonads 121, 385, 408
  intestine 121, 462
  liver 148, 163, 453, 456, 461
  ovary 384, 385
  pancreas 202, 211, 457, 461
  prostate 150, 163, 460
  stomach 196, 209, 461
  thyroid 351, 457
Carrier proteins (plasma) 112, 114, 400, *see also* individual proteins
Case-finding 508, 511
Casein 463
Catecholamines 215, 449
Cell-mediated immunity 124, 128
Central nervous system 73, **465**
Cephalin 243
Cerebral abscess 469
Cerebrospinal fluid 232, **467,** 476
Ceruloplasmin 119, 180, 187, 400
Change (significant) 22
Chelating agents 274, 296, 309, 319, 479
Chemical toxicology **478**
Chenodeoxycholic acid 189
Chest injuries 81
Children 86, 89, 127, **431,** 483
Chloride 12, **39,** 46, 52, 192, 287
  anion gap 75
  CSF 471
  sweat 438
Chlorinated hydrocarbons 487
Cholecalciferol 264
Cholecystokinin (CCK) 195, 199
  secretin/CCK-PZ test 201
Cholera 46, 210
Cholestasis **185,** 319, 400, 404
  alkaline phosphatase 147, 161
  gamma-glutamyl transferase 149, 161
  lipoprotein X 183, 256
  prothrombin time 179
Cholesterol 170, 172, **242,** 348
  liver disease 183, 185, 188
  oral contraceptives 399
  reference values (plasma) 251, 259, 444
  transport 251
  see also HDL- and LDL-cholesterol
Cholestyramine 260, 436
Cholinesterase 153
  liver disease 160, 180
  organophosphorus poisoning 154, 488, 511
  see also acetylcholinesterase
Choriocarcinoma 408
Chorionic gonadotrophin (CG), *see* human CG
Chromosome abnormalities 418
Chronic bronchitis 81
  hepatitis 129, **160,** 179, 183, 186, 275, 340
  infection 309
  pancreatitis **200,** 211
  pyelonephritis 96
  renal disease 58, 96, **105,** 256, 293, 487

thyroiditis 129, 348
Chylomicrons 194, 244, **247,** 251
Circadian rhythm 393, *see also* nychthemeral rhythm
Cirrhosis 160, 176, 179, **186,** 310, 375
α_1-antitrypsin deficiency 119, 436
immunoglobulins 129, 180
serum electrophoresis 133, 135, 180
Clearance tests (renal) 84
Clofibrate 260
Clomiphene test 392, 415
Coagulation factors 122, 179, 212, 267, 400
Coeliac disease 195, 204, 441
Coenzyme A 217
Colloidal gold test 181
Coma (drug overdose) 478
Combined hyperlipidaemia (familial) 254
Complement-fixing antibodies 347
Computer-assisted diagnosis 189, 497
instruction 498
Computers (laboratory) **490**
Concentration tests (renal) **90**
Congenital adrenal hyperplasia **377,** 410, 421
erythropoietic porphyria 314
fibrinogen deficiency 12
hyperbilirubinaemia 177, 434
non-spherocytic haemolytic anaemia 322
Conjugation (hepatic) 172
Connecting peptide (insulin) 213, 235
Conn's syndrome 374
Continuous ambulatory peritoneal dialysis 300
Convulsions, neonatal 432
Copper 119, 187, 275
Coproporphyrinogens, coproporphyrins 312
Cori Types (glycogen storage diseases) 237
Coronary heart disease 155, 252
Corpus luteum 385, 404
Corticosteroids 222, 340, **354,** *see also* aldosterone, androgens, glucocorticoids, oestrogens and progesterone
Corticosterone 356, 378
Corticotrophin-releasing factor 354, 387, 390
Cortisol 53, 215, **354,** 377
-binding globulin 354, 357, 400, 403
:creatinine ratio (urine) 359
plasma **356,** 361, 369, 372, 387, 400, 403
saliva 357
urinary free 358, 361, 364, 373
see also glucocorticoids
Cortisone glucose tolerance test 223
Costs 4, 504
C-peptide 213, 235
Creatine kinase 140, 141, **145**
CSF 470
isoenzymes 145, 155, 163, 470
muscle disease 158
myocardial infarction 155
Creatinine
amniotic fluid 412
clearance 84
plasma 87, 89, 443
side-room tests 240
Crigler-Najjar syndrome 176, 177, 436
Crohn's disease 116, 273
Cryofibrinogen 123
Cryoglobulins 131
Cumulative reports 494
Cushing's syndrome 56, 222, 358, **359,** 384, 391, 394, 456
Cyanocobalamin, *see* vitamin B_{12}
Cyclic AMP 278, 289, 295
GMP 463
Cystathionine 453
Cystic fibrosis 186, 427, **438**
Cystinosis 98
Cystinuria 98, 108, 109, 427
Cytochromes 173, 305
Cytotoxic drugs 333

Data reduction 496, 505
DDAVP test 91, 395
Dehydration **41,** 53, 86, 132, 230
Dehydro*epi*androsterone 381, 405
7-Dehydrocholesterol 242, 264, 298
Demyelinating disorders 469
11-Deoxycorticosterone 378, 381
11-Deoxycortisol 358, 365, 378
Depressive illness 362
Desferrioxamine 479
Desmolase 377, 381
Detoxication 172
Dexamethasone 355, 370, 373
screening test 359
suppression test 363, 364
Diabetes insipidus 52, 90, 395
Diabetes mellitus **220,** 256, 310, 444

complications 101, 229
Cushing's syndrome 366
diagnostic criteria 224
glucose tolerance test 222, 445
haemoglobin A_1 227
potassium metabolism 58, 231
pregnancy 226, 403, 503
renal threshold for glucose 99, 445
screening investigations 220, 503, 511, 514
side-room tests 31, 34, 220, 232
Diabetic coma **229**
ketoacidosis 146, 217, 229
Diagnosis 2, 31, 189, 497
Dialysis dementia 300, 487
osteodystrophy 300
Diarrhoea 46, 51, 210
Diastase, *see* amylase
Dibromsulphthalein 178
Dibucaine number 154
Diet (effect on tests) 9, 20, 250, 328, 451, 456
Differential protein clearance 102
Digestive tract 41, **192**
Digoxin 474
Dihydrobiopterin 420
1:25-Dihydroxycholecalciferol 265, **278,** 292, 300
24:25-Dihydroxycholecalciferol 265, 278, 280
5α-Dihydrotestosterone 382
2:3-Diphosphoglycerate 78, 165
Diquat poisoning 488
Disaccharidase deficiency 204, 211
Discretionary tests **1,** 6, 500
Distal renal tubular acidosis 95
Distribution (of data) 16
Diuretic phase (renal failure) 104
Diuretic therapy 46, 50, 56, 73, 97, 284, 289, 302, 331, 334, 375, 480
Diurnal variation 10, 21, 307, *see also* nychthemeral rhythm
Dopamine 393, 449, 453
Down's syndrome 409, 418
Drugs
effects on tests 9, 21, 87, 91, 152, 356, 451, 456, 497
enzyme induction 149
overdosage **478**
screening 482
therapeutic monitoring 162, **472**
units of measurement 477
urine appearance 30
Dubin-Johnson syndrome 176, 178
Duchenne-type muscular dystrophy 146, 158
Duodenal atresia 411
intubation 200, 212
ulcer 196, 198
Dwarfism 390
Dysgammaglobulinaemia 128
Dyshormonogenesis (thyroid) 344

Ectopic ACTH syndrome 365
Edetate (EDTA) infusion test 278, 296
Elderly, *see* geriatric patients
'Electrolyte group' 4,59,75,366,371
Electrophoresis **113**
amniotic fluid 412
CSF protein 470
haemoglobin 322
immunoelectrophoresis 115, 135
isoenzyme 142, 145
lipoprotein 259
serum protein **113,** 132, 180
urine protein 134
Emergency work 13, 36, 479
Emphysema 120
Enterohepatic circulation **171,** 174, 207, 242, 260, 339
Enzymes **138**
cofactor saturation tests 263, 268
CSF 470
faecal 439
induction 147, 149, **181,** 186, 401
lipid transport 247
nomenclature 141
plasma **138,** 181
tissue 140, 167, 424
tumour markers 460
units of activity 140, **520**
urine 152, 166
Epilepsy 474, 476
Equipment (used by clinicians) 36, 413
Ergosterol 264
Errors 26, 492
Erythrocyte enzymes 164
acetylcholinesterase 488
cofactor saturation tests 263, 268
transketolase 268
Erythrocyte folate 271
magnesium 303
porphyrins 314, 319
Erythropoiesis 173, 175, 306

Erythropoietic porphyria, protoporphyria 314
Esbach's test 99
Essential fatty acids 243
fructosuria 241
pentosuria 240
Ethanol, *see* alcohol
Ethylene glycol 480, 487
Exception reporting 26
Exercise (effect on tests) 20, 146
Exfoliative dermatitis 117
Exomphalos 411, 412
Extracellular fluid **40,** 47, 53, 56, 82, 103, 132
Exudate 188

Facioscapulohumeral muscular dystrophy 158
FAD effect 269
Faeces
collection of specimens 13, 206
enzymes 166, 439
fat test 206, 211, 441
occult blood 33, 462
pH 205
porphyrins 315, 317
water 40, 210
Failure to thrive (children) 437
Familial combined hyperlipidaemia 254
hypercholesterolaemia 253
hypertriglyceridaemia 254
hypocalciuric hypercalcaemia 292
LCAT deficiency 247
TBG abnormalities 344
Fanconi syndrome 90, **98,** 298, 300, 335
Fasting hypoglycaemia 233
Fat absorption 193, 206
Fat-soluble vitamins 170, **264**
Fatty acids, *see* free fatty acids
Favism 165
Female sex hormones **384,** 397, 406, 414
Ferritin 306, 308
Ferrochelatase 313, 319
Fetal abnormalities 407, 409
maturity 406, 412
monitoring (labour) 35, 413
Fetoplacental function 402
unit 384, 405
α_1-Fetoprotein **121,** 180, 188, 410, 412, 461, 463, 514
Fetus **404**
Fibrin degradation products 122
Fibrinogen 122
Floating β-lipoproteins 255
Flocculation tests 181
Fluid balance 39, 58
depletion, *see* dehydration
deprivation test 92, 395
replacement 60, 232
Folic acid 271, 441
Follicle-stimulating hormone 372, 381, 388, **392,** 397, 404
subfertility 383, 414
Forced alkaline diuresis 480
Formiminoglutamate (FIGLU) 271
Fractional test meal 197
Franklin's disease 131
Fredrickson classification (lipoproteinaemias) 253
Free cortisol 247
Free fatty acids 183, 216, 247, 343
transport 229, 247
Free radicals 313
Free thyroxine 342, 350
index 343
Free tri-iodothyronine 344, 350
Fructose intolerance 241
Fructosuria 32, 240
Functional hyperinsulinism 236

Galactorrhoea 393, 415
Galactosaemia 32, 182, 186, 234, **239,** 422, 428, 432, 436
Galactose tolerance test 182
Gallstones 176, **188,** 199, 399
Gamma-glutamyl transferase 149, 440, 486
enzyme induction 181, 401
liver disease 161, **181,** 185, 189
Ganglioneuroma 453
Gastric function **195**
inhibitory peptide 214,441
juice 41, 166, 306
ulcer 152, 196, 198
Gastrin 195, 197
Gastrinoma 197, 291
Gastrointestinal peptides 194, 441
secretions 41, 210
tract disease 129, **192,** 273
Gaucher's disease 151, 466
Gaussian distribution 16, 510

Genetic counselling 425
  polymorphism 118
Genotypes (cholinesterase) 154
Geriatric patients 341, 350, 351, **443,** 551
Gestational diabetes 226, 403
Giantism 389
Gilbert's syndrome 176, **177,** 184, 436
Globulins (immuno-) 113, **124,** 133
  CSF 435
  hypergammaglobulinaemia 129
  hypogammaglobulinaemia 128
  liver disease 180, 185
Glomerular filtration rate 44,84
  function tests **83,**87
  proteinuria 101
Glomerulonephritis 100
Glucagon 195, 214, 235, 441
Glucocorticoids **354**
  adrenal hyperfunction 359
  adrenal hypofunction 359, 367
  metabolic response to trauma 58, 355
  oral contraceptives 357, 400
  plasma 356
  pregnancy 357, 403
  urinary 358
Glucose **213**
  CSF 468
  measurements 218
  plasma 215, 432, 444, 503
  preservation (in blood) 12, 219
  side-room tests 32, 34, 220, 232, 422, 481
  tolerance tests, *see below*
Glucose-dependent insulinotrophic peptide 195, 214, 456
Glucose oxidase 32, 218
Glucose-6-phosphatase 167, 216, 237
Glucose-6-phosphate dehydrogenase deficiency 164, 435
Glucosephosphate isomerase 164, 165
Glucose tolerance tests (GTT) 219, **222,** 348, 366, 390, 445
  chronic renal failure 106
  cortisone GTT 223
  diagnostic criteria (oral test) 224
  intravenous 223
  malabsorption 204, 227
  oral contraceptives 398
  renal threshold 99, 227, 445
Glue sniffing 485
Glutamic oxaloacetic transaminase, *see* aspartate aminotransferase
Glutamic pyruvic transaminase, *see* alanine aminotransferase
Glycerol 216
Glycocholic acid 208
Glycogen storage diseases 167, 218, 233, **236,** 334, 421, 432
Glycosuria **32,** 90, 98, 187, 220, **240,** 403, 422, *see also* glucose tolerance tests
Glycosylated haemoglobin 227, 445
Gn-RH stimulation test 388
Gonadal dysgenesis 383, 393
  failure 382, 414
  neoplasms 121, 385, 408
Gonadotrophin, *see* follicle-stimulating and luteinizing hormones
  -releasing hormone 382, **387,**392, 397,415
  subfertility 383, 414
  tumours, *see* human chorionic gonadotrophin
Gout 331
Granulosa cell tumour (ovary) 385
Graves' disease 345, 348
Growth 440
  hormone 215, 372, 389
Guanine 327
Gut glucagon-like immunoreactivity 195, 441
Guthrie test 422, **427,** 512
Gynaecology **397**

Habituation (to drugs) 484
Haem 305, 312
Haematin 325
Haematuria 32
Haemochromatosis 186, 222, 310
Haemoglobin 78, **321**
  A_1 228
  derivatives 323
  oxygen carriage 78
Haemoglobinopathy 321, 324, 429
Haemolysis 11, 35, 58, 118, 144, 156
Haemolytic anaemia 118, 164, **175,** 184, 189, 322
Haemosiderin 306, 316
Halothane 185
Haptoglobin 118, 180, 400
Hartnup disease 98, 270, 427
Hashimoto's disease 347
HDL-cholesterol 250, 257, 258, 399

Heart disease 155, 252, 254, 255
'Heart-specific' lactate dehydrogenase 143, 155
Heat-stable isoenzymes
  alkaline phosphatase **148,** 164, 403, 409
  lactate dehydrogenase 143
Heavy chains 125
  disease 131
  metal poisons 101, 319, 487
Henderson equation 65
Henderson-Hasselbalch equation 66
Hepatic anion transport 173
  disorders, *see* cholestasis, hepatitis, hepatocellular, and liver disease
  hyperbilirubinaemia 176
  lipase 247
  porphyria 315
Hepatitis
  acute 143, 145, 160, **184**
  chronic 144, 147, 149, **186,** 485
  neonatal 436
Hepatocellular carcinoma 121, 180, 453, 457, 461
  jaundice 176
  transport 171
Hereditary coproporphyria 316
  fructose intolerance 234, 241, 422, 432
  spherocytosis 435
  TBG excess 340
Hexachlorobenzene 316
Hexokinase 165, 218
High density lipoproteins 245, 250, 256
Hirsutism 383, 415
Histaminase 457
Histidinaemia 422, 427, 428
Histidine load test 271
Homocystinuria 427, 428
Homovanillic acid 454
Hormone-sensitive lipase 216, 247
Human chorionic gonadotrophin 383, 404, **407,** 459, 463
  leucocyte antigens 221, 311
  placental lactogen (HPL) 402, 404, **408**
Hurler's syndrome 466
Hydatidiform mole 408
Hydrochloric acid 192, 195
Hydrocortisone test 290, 292
Hydrogen ion
  blood 35, **64,** 413
  urine 32, 64, 93
Hydroxyapatite 277
3-Hydroxybutyrate 217, 229
Hydroxybutyrate dehydrogenase, *see* 'heart-specific' lactate dehydrogenase
1α-Hydroxycholecalciferol 106, 300
25-Hydroxycholecalciferol 106, 170, **265,** 278, 295, 438, 447
11-Hydroxycorticosteroids, *see* glucocorticoids
17-Hydroxycorticosteroids 358, 366
5-Hydroxyindole secreting tumours 454
5-Hydroxyindoleacetic acid 270, 454
3-Hydroxykynurenine 269
1α-Hydroxylase (vitamin D) 106, 265, 279
11β-Hydroxylase (cortisol) 365, 380 410
21-Hydroxylase (cortisol) 379, 410, 421
24-Hydroxylase (vitamin D) 265
25-Hydroxylase (vitamin D) 265
4-Hydroxy, 3-methoxy mandelic acid 450, 453
  phenylacetic acid 454
  phenylglycol 451
17α-Hydroxyprogesterone 358, 379 410
Hydroxyproline 272, 289, 446
5-Hydroxytryptamine, 5-hydroxytryptophan 270, 454
Hyperacidity (gastric) 196, 198
Hyperaldosteronism 51, 56, **374**
  secondary 46, 51, 96, 179, **376**
Hyperbilirubinaemia **175,** 184, 434
Hypercalcaemia 90, 136, 265, **283,** 462
Hypercalciuria 95, 108, **289**
Hypercapnia 80
Hyperchloraemic acidosis 52, 287
Hypercholesterolaemia 183, 185, 253, 256, 348
Hypergammaglobulinaemia 129, 132
Hyperglycaemia 214, 229, 468
Hyperkalaemia 56, 372
Hyperlipidaemia 48, 50, **253,** 259
  secondary 256
Hyperlipoproteinaemias 253
Hypermagnesaemia 106, 303
Hypernatraemia 51, 434
Hyperoxaluria 109
Hyperparathyroidism
  primary 108, 148, 151, **283,** 296, 451, 503
  secondary 106, **294,** 296, 299

tertiary 106, **290,** 296, 299
Hyperphenylalaninaemia 428, 513
Hyperprolactinaemia 393, 415
Hyperproteinaemia 50, 182
Hypertension 334, 374, 450
Hyperthyroidism 222, 289, **341, 349,** 402
management and follow-up 352
screening **350,** 445, 504
T3-thyrotoxicosis 342
TRH test **345,** 387, 394
Hypertriglyceridaemia 254, 256, 259, 486
Hyperuricaemia 136, 239, **330**
Hyperventilation 73, 229, 294
Hypoalbuminaemia **115,** 267, 343, 437
Hypobetalipoproteinaemia 261
Hypocalcaemia 265, **293,** 447, 457
acute pancreatitis 200
neonatal 432
renal failure 299
Hypochloridaemia 52
Hypocholesterolaemia 260, 348
Hypochromic anaemia 309, 322
Hypogammaglobulinaemia 128, 132, 212
Hypoglycaemia 214, 219, **233,** 237, 468
insulin-induced 197, 362, 370, 391, 394
neonatal 235, 432
Hypogonadism
female 393
male 382, 393
Hypokalaemia **55,** 90, 95, 210, 232, 294, 366, 375, 376
Hypolipidaemia 260
Hypolipoproteinaemia 260
Hypomagnesaemia 294, **303,** 433
Hyponatraemia 50, 372, 395, 434
Hypoparathyroidism 294
Hypophosphataemia 99, 267, 285, 447
Hypophosphatasia 148
Hypopituitarism 227, 388
Hypoproteinaemia 46, 51, 101, 116, **132,** 179, 402, 437, 456
Hypotension 86, 451
Hypothalamic-pituitary-adrenal axis 354, 362, 371, 390, 394
-gonadal axis 382, 392, 414
-thyroid axis 338, 344, 387, 394, 445
Hypothalamus 90, **387**
Hypothyroidism 146, 256, **341,** 350, 370, 445, 466
management and follow-up 352
pituitary disease 345, 346, 392
screening 351, 427, 445, 504
thyroid antibodies 348
TRH test 345, 387, 394
Hypotriglyceridaemia 261
Hypouricaemia 335
Hypoxaemia 80
Hypoxanthine phosphoribosyl transferase 328, 332, 334
Hysterical overbreathing 73, 294

^{131}I, ^{132}I 346, 358
Identification **8,** 481, 490
Idiopathic haemochromatosis 310
hypercalciuria 108
Idiosyncratic jaundice 401, 404
IgA **125,** 130, 180, 186
IgD 125
IgE 125
IgG **125,** 130, 180, 186, 469
IgM **125,** 130, 180, 185, 469
Ileal disease 116, 207, 209, 211
Immobilization 108
Immunoglobulins (CSF) 469
Immunoglobulins (plasma) **124,** 400
deficiency 128
fragments 131
increase 129
liver disease 180, 186
plasma (changes with age) 127
Immunological methods 114, 135
Immunoreactive trypsin 202, 439
Inappropriate secretion of hormones 395, 458
Inborn errors of metabolism, *see* inherited metabolic disorders
Indican, urinary 209
Individual variation 20, 473
Indocyanine green 171, 178
Indomethacin 333
Induction of enzymes 149, **181,** 186, 315, 401
Industrial hazards 319, 486
Ineffective erythropoiesis 143, 173, 175, 435
Infancy 127, 431
Infections 129, 309
Infectious hepatitis 176
Infiltrations (liver) 185
Inherited metabolic disorders 409, **418,** 465
carbohydrate metabolism 236

congenital adrenal hyperplasia 377
incidence 427
phenylketonuria 419
screening 426, 465, 511
see also individual disease entries
Injury, *see* trauma
Insulin 53, 213, 229, 235
Insulin-dependent (Type 1) diabetes 221
Insulin-hypoglycaemia tests
adrenal disease 355, 362, 370
gastric function 197
pituitary disease 390, 394
Insulinoma 234, 291, 432
Intact nephron hypothesis 83
Intermediate density lipoproteins (IDL) 245, 248
International unit (enzyme activity) 140, 521
Interpretation of results **15,** 26, 49, 71, 484, 495
Interstitial cell stimulating hormone, *see* luteinizing hormone
Intestinal absorption 193, **203,** 205, 209
function tests 195, 203, 206
malabsorption 210, 227
obstruction 152
secretions 41
Intracellular fluid 40, 53, 54, 56, 77
Intrahepatic cholestasis 176, *see also* cholestasis
Intramuscular injection 146
Intravenous hyperalimentation 61
infusion 11, 182
Intrinsic factor 192, 209
Inulin clearance 84
Iodide, iodine 336, 347
Iodotyrosines 336, 344
Ion difference (anion gap) 75
Ion-selective electrodes 35, 48
Iron **305**
absorption tests 212, 305
-binding capacity 308, 401, 403
deficiency 307, 309, 504
overload 310
plasma 10, 63, **307,** 401, 403
poisoning 479
screening 309, 504
stores 305, 308
transport 306
Islet cell tumour 234, 291, 457
Isoenzymes 141
acid phosphatase 150, 163, 460, 463
alkaline phosphatase 146, 163, 403, 409, 460
creatine kinase 145, 155, 163, 470
lactate dehydrogenase 141, 156
see also individual enzyme entries

Jaundice 175, 186, 256
enzyme tests 161, 181
neonatal 120, 434
oral contraceptives 400
pregnancy 404
side-room tests 29, 183
Juxtaglomerular apparatus 373

Kallikrein 455
Katacalcin 279, 281, 457
Katal 521
Kernicterus 178, 434
Ketoacidosis 32, 73, **229,** 231
Ketone bodies
plasma 35, 217, 231
urine 32, 232
17-Ketosteroids 358, 382
Kidney, *see* renal disease
Klinefelter's syndrome 383, 393
Kwashiorkor 236, 261, 264, 302, **437**
Kynurenine 270

Lactalbumin 463
Lactase deficiency 204, 211
Lactate 77, 217
diabetic coma 230
hyperuricaemia 334
Lactate dehydrogenase 12, **141,** 167, 217, 497
haemolytic disease 164
liver disease 161, 181
malignant disease 163
muscle disease 158
myocardial infarction 155
Lactic acidosis 77, **218,** 230, 239, 334
Lactoferrin 202, 212, 463
Lactose 32, 192, 204, 240, 404
LDL-cholesterol 257, 260, 348, 399
Lead poisoning 319, 487, 511
LE cell test 183
Lecithin 188, 413, 434

Lecithin-cholesterol acyl transferase 246, 251
  familial deficiency 247
Lesch-Nyhan syndrome 334, 410, 466
L-Leucine 235, 432, 437
Leucine aminopeptidase 150
Leucocyte ascorbate 272, 511
  potassium 54
Leukaemia 58, 101, 128, 129, 284, 309
Life-support systems 483
Ligandin 174
Light chain disease 131
  chains 125, 131
Lignac-Fanconi disease 98
Limb girdle muscular dystrophy 158
Linoleic acid 243
Lipase
  hepatic 247
  lipoprotein 247, 256
  mobilizing 216, 247
  pancreatic 152, 193, 199
Lipids 61, 170, 216, **242**
  absorption 193, 206
  liver disease 183, 239
  oral contraceptives 399
α-Lipoprotein deficiency 260
Lipoprotein lipase deficiency 256
Lipoprotein X 183, 185, 247, 436
Lipoproteins **244,** 247, 253
  electrophoresis 259
Lipotrophin 357, 390, 459
Liquorice 375
Lithium 90, 289, 474
Liver disease 97, **169,** 310
  albumin 116, **179,** 186, 190
  alkaline phosphatase 147, **161,** 181, 183, 186, 190
  amino acids 97, 182, 187
  aminotransferases 144, 145, **160,** 181, 183, 186, 190
  α_1-antitrypsin 119, 436
  bile acids 178, 207
  bilirubin 173, **183,** 401, 404, 434
  ceruloplasmin 119, 180, 187
  α_1-fetoprotein 121, 180, 188, 461
  lactate dehydrogenase 143, 161, 181
  protein electrophoresis 132, 134, 180
  prothrombin time 179
  serological tests 183
  see also cholestasis, jaundice, hepatitis
'Liver function tests' 4, 189, 436
  oral contraceptives 400
Low density lipoproteins (LDL) 245, 249, 254
Lumbar puncture 468
Lundh test 202
Lung disease 80, 160, 434
  acid-base status **70**
  α_1-antitrypsin 119, 436
Lupus erythematosus 101, 129
Luteinizing hormone 372, 381, 388, **392,** 397, 404
  subfertility 383, 414
Lymphoma 101, 129
Lysozyme 102

'M' (monoclonal) components 129, 136
Macro-amylasaemia 152
α_2-Macroglobulin 102, 114, **121,** 400
Macroglobulinaemia 131, 135
Magnesium 106, **301,** 433
Malabsorption 116, 132, 193, **210,** 227, 261, 298
Male sex hormones 381
Malignant disease, *see* neoplastic disease, carcinoma
Malignant hyperpyrexia 159
Malnutrition (children) 116, 233, 371, **437**
Maltose 192, 205
Maple syrup urine disease 236, 422, 428
Marasmus 302, 437
Masking (of drugs) 484
Maternal serum α_1-fetoprotein 121, **410,** 514
Maturity onset (Type 2) diabetes 221
Maximum tubular reabsorption 288, 289
MB isoenzyme (creatine kinase) 145, 155
MCV 486
Meals (effect on blood composition) 147, 213, 247, 263, 275, 285
Mean 16
Meconium albumin 439
Medullary carcinoma (thyroid) 451, 457
Megaloblastic anaemia 143, 209, 271
Melanocyte-stimulating hormone 394
Membrane-associated enzymes 161, *see also* individual enzymes
Mendelian disorders 418
Meningitis 469, 470
Menke's disease 275
MEN syndromes 198, **291,** 451, 457
Menstrual cycle 20, 307, 382, 398

Mental retardation 465
Mercury poisoning 101, 487
Mesenteric infarction 199
Metabolic acidosis **71,** 75, 80, 525
potassium depletion 53, 56, 58, 231
renal failure 73, **104,** 106, 299
renal tubular acidosis 52, 53, 73, **95,** 300
respiratory distress syndrome 434
see also hyperchloraemic acidosis, ketoacidosis, lactic acidosis
Metabolic alkalosis **71,** 76, 80, 525
Cushing's syndrome 366
potassium depletion 53, 56, 366, 376
vomiting 52
Metabolic balance, *see* balance (metabolic)
Metabolic bone disease 267, 285, **296**
component (acid-base balance) 66, 525
response to trauma 58, 123, 355
Metadrenaline 451
Metal poisons 101, 310, 319, 487
Metanephrine 451
Methaemalbuminaemia 200, 322, **325**
Methaemoglobin 323
Methaemoglobinaemia 79, 324
Methanol 480
Method (of analysis) 22, 112, 218, 356, 520
Methylmalonic acid 271
Methyltestosterone 176
Metyrapone test 365
Micelles 194
β_2-Microglobulin 102
Milk alkali syndrome 73, 284
Middle molecules 106
Mineralocorticoids **373**
Misuse of laboratories 27, 37
Mobilizing lipase 216, 247
Mole (definition) 518
Monoamine oxidase inhibitors 451
Monoclonal band 114, **133,** 134
Monoglycerides 243
Morphine 152
Motor neurone disease 160
Mucopolysaccharidoses 466
Multi-channel analysers 5, 496, 500, 505
Multi-phasic screening 5, 514
Multiple endocrine neoplasia, *see* MEN syndromes
Multiple myeloma 129, 134, 284
pregnancy 411
sclerosis 470
Mumps 152
Muscular dystrophy 143, 144, 158
exercise (effect on blood constituents) 20, 146
Myasthenia gravis 81
Myeloproliferative disorders 284, 333
Myocardial infarction 124, 257, 259, 275, 342
enzyme tests 143, 149, **155**
Myoglobin 30, 32, 173, 305
Myopathies 159
Myotonias 159
Myxoedema, *see* hypothyroidism

NADH-methaemoglobin reductase 324
Natriuretic hormone 44
Negative feedback control 59, 279, 338, 354, 365, 377, 387, 392, 459
Nelson's syndrome 394
Neonatal hypocalcaemia 293, 433
hypoglycaemia 235, 432
period 273, 421, **432**
jaundice 140, 434
Neoplastic disease **449**
hypercalcaemia 283, 462
enzyme tests 162, 167, **460**
tumour markers 164, **458**
see also carcinoma
Nephrocalcinosis 107, 284
Nephrogenic diabetes insipidus 93, 395
Nephrotic syndrome **99,** 116, 128, 133, 256, 282, 340, 357, 375
Nesidioblastosis 233, 236
Neural tube defects 121, **410,** 514
Neuroblastoma 441, 453
Neurogenic muscle disease 160
Neurohypophysis 394
Nicotinamide 98, 270
Nicotinic acid 260, 270, 456
Nitrogen metabolism 104, 107, 111, 171, 182
'Non-endocrine' tumours 364, 366, 391, 394, 395, **458**
Non-insulin-dependent (Type 2) diabetes 221
Non-respiratory component (acid-base balance) 66, 525
Non-toxic multinodular goitre 345
Noradrenaline 449
Normal range 16, 25

Normetadrenaline, normetanephrine 451
Novobiocin 176
Nucleic acids 327
5′-Nucleotidase 150
Nucleotides 270, 278, 287, 295, 327
Nychthemeral rhythm 21, 355, 357, 361, 365, 391

Obesity 81, 328, 360
Objectives xiii
Obligatory losses (water) 43
Obstetrics **397**
Obstructive jaundice, *see* cholestasis
Occult blood (faeces) 33, 462
Occupational hazards 319, 486
Oedema 46, 101, 132, 437, 456
Oesophageal atresia 411
Oestradiol-17β, oestriol 384, 397, 406, 414
Oestrogens 256, 354, 356, **384,** 393, 397
  :creatinine ratio (urine) 407
  plasma 385, 407, 408
  therapy 150, 340, 341, 357
  urinary 406, 408, 414
Oestrogen-secreting tumours 385
Oestrone 384, 397, 406
Old age 86, 350, 351, **443**
Oligomenorrhoea 383
Oliguric phase (acute renal failure) 103
Oral contraceptives 222, 319, 393, **397**
  plasma corticosteroids 357, 368
  plasma iron 307, 400
  plasma lipids 250, 256, 399
  thyroxine-binding proteins 341, 342
Organic solvents (poisoning) 185, 480, 487
Organophosphorus poisoning 154, 488
Orthostatic proteinuria 101
Osmolality
  plasma 42, **47,** 92, 229, 231, 395
  urine **29,** 47, 90, 104
Osteitis fibrosa 106, 299
Osteodystrophy, uraemic 106, 299
Osteomalacia 267, 296, 298, 446
  alkaline phosphatase 147
  chronic renal failure 106
  intestinal malabsorption 212
Osteoporosis 296, 297, 446
Osteosclerosis 106, 299
Ovarian function 397, 414
Overbreathing 73, 294
Overflow amino aciduria 97
  proteinuria 100
Overuse of laboratories 27
Ovulation, detection of 415
Oxalate stones 108
Oximetry, transcutaneous 80
17-Oxogenic steroids 358
17-Oxosteroids 358, 382
Oxygen content 522
  dissociation curve 78, 323
  saturation, percentage 79, 523
  transport **77,** 309
Oxyhaemoglobin 77, 309

Paediatrics **431,** 483
Paget's disease 148, 151, 289, **301,** 444, 446
Pancreatic disease **198,** 210, 222
  juice 41, 166, 199, 202
  polypeptide 195, 456
  stimulation tests 201, 212
Pancreatitis
  acute 151, 199
  chronic 200, 206
Paracentesis 188
Paracetamol 90, 185, 479
Paralytic ileus 55
Paraproteinaemia 129, 134, *see also* multiple myeloma
Paraquat poisoning 488
Parathyroid hormone (PTH) 265, **278,** 459, 462
  plasma PTH 287, 294, 296
Parenteral fluids 60
Pascal, definition 518
Patients, identification 8, 490
PBG deaminase 312, 315
$P\text{CO}_2$ 65, **70,** 80, 231, 434, 523
Pellagra 98, 271
Pentagastrin test 196
Pentosuria 240
Pepsinogen 192
Peptic ulcer 152, 196, 198
Peptides, gastrointestinal **194,** 210, 393, 441, 456
Per cent oxygen saturation 79, 523
Peripheral neuritis 160
Peritoneal fluid 167
Pernicious anaemia 175, 196, 198, 209, 271

Peroxidase 173
P_{50} 77
pH 64, 523
 CSF 467
 intracellular 77
 plasma 66
 temperature effects (blood) 69
 urine 32, 93
Phaeochromocytoma 222, **449,** 457
Phenobarbitone 476
Phenothiazines 176, 185, 393
Phenylalanine 419, 513
Phenylketonuria 97, 236, 270, **419,** 427, 511
Phenylpyruvic acid 419
Phenytoin 222, 474
Phosphate
 excretion (urine) 98, 187, 280, 286, 288, 298
 plasma 12, 106, 267, 285, 295
Phosphoenolpyruvate 165
Phosphofructokinase 165
3-Phosphoglycerate 78
Phospholipids 183, 185, 243
5-Phosphoribosyl-1-pyrophosphate 327
Phosphorylase kinase 215, 238
Phosphoryl-ethanolamine 149
Photosensitivity 313
Pituitary, anterior 345, 346, 372, **388,** 414
 functional reserve 388, 394
 posterior 90, **394**
Placental function **405**
 hormones 404, *see also* individual hormones
 lactogen 402, 404, 408
 sulphatase 407
Plasma bicarbonate 67, 523
 cells 131
 proteins 111, *see also* individual proteins
 renin activity 376
 volume 47, 402
Plasmin, plasminogen 123
Pleural fluid 167, 188
Pluriglandular syndrome, *see* MEN syndromes
Poisoning 311, 319, **478,** 483
Poliomyelitis 160
Polyamines 463
Polycystic ovaries 384, 415
Polycythaemia 322, 333
Polydipsia 90, 220, 395
Polymyositis 159
Polyneuritis 81, 268, 314, 486
Polyuria **90,** 93, 96, 98, 220, 284, 395
Porphobilinogen 30, 32, **311,** 318
Porphyria cutanea tarda 316
Porphyrin metabolism **311**
Porphyrinuria 318, 511
Portal-systemic encephalopathy 182
Posterior pituitary 90, **394**
Post-hepatic hyperbilirubinaemia, *see* cholestasis
Postoperative management 40, **58,** 157
Posture (effect on tests) 10, 20, 101, 132, 224, 282
Potassium 35,41,**53,**96,105,210
 acid-base disturbances 54,97,231
 aldosteronism 374
 Cushing's syndrome 366
 diabetic ketoacidosis 58, 229
 renal failure 103, 105
Po_2 77, 80, 231, 434, 523
PP effect 269
Pre-albumin 117, 437
Precision 23
Prednisone, prednisolone 255, 356, 373
Pre-eclampsia 404
Pregnancy **401**
 alkaline phosphatase 147, 403, 409
 -associated proteins 463
 corticosteroids 357
 fibrinogen deficiency 122
 gestational diabetes 226
 hormone-binding proteins 340, 357, 402
 oestrogens 365, 406
 plasma iron and TIBC 307, 308
 screening for disease 409, 511
 side-room tests 315
 -specific β_1-glycoprotein 402, 409
 thyroxine-binding proteins 340, 342, 402
 toxaemia 404
Pregnanediol 384, 405, 414, 415
Pregnanetriol 379, 385
Pre-hepatic hyperbilirubinaemia 175, 177, 184, 435
Prenatal diagnosis **409,** 429, 440
Prevalence (of disease) 26, 509
Primary aldosteronism 46, **374**
 biliary cirrhosis 129, 176, 180, 185
 hyperlipoproteinaemias 253
 hyperoxaluria 109
 hyperparathyroidism 108, **284**

Probability 18, 24
Profile investigations 3, 500, *see also* screening
Progesterone 384, 397, 404, 414, **416**
Pro-insulin 213, 235
Prolactin 387, 389, **393,** 414, 459
Pro-opiocortin 390, 394
Prostaglandins 243, 291, 454
Prostatic carcinoma 145, 460
Protein-bound iodine 341
Protein electrophoresis, *see* electrophoresis
Protein-energy malnutrition 112, 236, 261, 273, **437**
Protein-losing gastroenteropathy 116, 129, 210
Proteins **111,** 170
  absorption 193, 205
  CSF 469
  oral contraceptives 400
  plasma 47, **111,** 132, 179, 462
  pregnancy 402, 409, 463
  tumour markers 121, **458**
  urine 32, **99,** 130, 232, 403
  see also individual proteins and lipoproteins, hepatitis, renal disease
Proteinuria 90, **99,** 116
  Bence-Jones 32, 101, **130,** 134
  differential clearance (selectivity) 102
  side-room tests **32,** 99, 232, 403
Prothrombin 179, 267
  time 179, 186, 212, 267
Protoporphyria 314
Protoporphyrin, protoporphyrinogen 305, 311, 313, 315, 319
Provitamin D_3 264, 298
Proximal renal tubular acidosis 95
Pseudocholinesterase 153
Pseudogout 332
Pseudohypoparathyroidism 290, 295
Psoriasis 333
Psychogenic diabetes insipidus 90,93, 395
Pulmonary embolism 160
  infarction 143
  oedema 81
Purine metabolism **327**
Pyloric stenosis 52, *see also* vomiting
Pyridoxal-5-phosphate, pyridoxine 269
Pyrimidines 327
Pyrophosphate arthropathy 332
Pyruvate kinase deficiency 165
  tolerance test 268

Quality control 23, 37, 492

Race (effect on blood constituents) 22, 250
Radioactive uptake tests (thyroid) 346
Radioimmunoassay 114, 150, 287, 342, 356
Raynaud's phenomenon 131
Reactive hypoglycaemia 224, 234
Records 492
Rectal examination 151
Red cell, *see* erythrocyte
Reducing substances 32, 218, 219, 240
Reference values **15,** 495
  blood 18, 19, 259, 328, 355, 399, 432
  effects of age 22, 147, 153, 251, 399, 431, 443
  effects of sex 148, 259
  urine 19
Regan isoenzyme 148, 163, 460
Remnant hyperlipoproteinaemia 255, 259, 260
Renal artery stenosis 86
Renal disease **83,** 103, 504
  acid-base disturbances 73, **95,** 104, 106
  amino aciduria 97, 187
  failure **103,** 487
  glycosuria 99, 227
  hyperparathyroidism 106, 294, 299
  stones 98, **107,** 284, 332, 333, 335
  tubular acidosis 73, **95,** 97, 300
  uraemic osteodystrophy 299
Renal excretion (normal)
  acid 64, 94
  calcium 284, 289
  nitrogen 84, 97, 111, 171
  potassium 53, 56, 96
  proteins 100, 152
  sodium 42, 44, 50, 96
  water 41, 47
Renal function tests
  glomerular 44, **83,** 443
  side-room **32,** 99, 232, 403, 511
  tubular 29, **89,** 395
Renin 373, 376
  -angiotensin system 44, 373
Reports of investigations **15,** 492, **517**
Requests for investigations **1,** 490
  discretionary tests **1, 6,** 500, 504
  forms 8, 490

screening investigations **3,** 6, **500,** 515
urgent 13, 36, 479
Reserpine 451, 456
Residual individual variation 21
Resin uptake test 343
Respiratory acidosis 71,80,525
alkalosis 52, 72, 294, 525
component 66
distress syndrome 434
insufficiency 80
Results of investigations
abnormal **18,** 500, 509
interpretation **15,** 69, 497, 505
significant changes 22
see also reference values
Retinoblastoma 454
Retinol 264
transport proteins 114, 117
Reverse T3 337, 339, 348
Rhesus incompatibility 175, 413, 435
Rheumatoid arthritis 275
Riboflavin 269
Rickets 267, 296, **298,** 433, 438
alkaline phosphatase 147, 148, 267, 296
malabsorption 212
renal 95, 296, 298
vitamin D-resistant 98, 106, 148, 298, 300
Rotor syndrome 176, 178

Salicylates
overdosage 73, 77, 480, 485
protein-binding 340, 475
therapeutic monitoring 477
urate excretion 331
Saliva 41, 192
cortisol 357
progesterone 385, 405, 414, 415
therapeutic drug monitoring 476
Salvage pathway (nucleotides) 328, 330
Sarcoidosis 108, 185, 292
Saturation tests (vitamins) 263,272
Schilling test 209, 271
Scoline apnoea 153
Screening tests **3,** 6, **500**
adrenal dysfunction 359, 367
colorectal cancer 121, 462
cystic fibrosis 439
hypercalcaemia 283
inherited metabolic disease **426,** 465
neural tube defect 121, 410
phenylketonuria 427, 511
thyroid disease **350,** 351, 445
Scurvy 272, 511
Secondary aldosteronism 46, 51, 96, 179, **375**
diabetes 222
hyperlipidaemia 256
hyperparathyroidism 106, 294
hypogammaglobulinaemia 128
Secretin 199, 441
Secretin/CCK-PZ test 201, 212
Secretions, digestive tract 41
Secretory IgA 126
Selectivity (proteinuria) 102
Self-poisoning 478
Sensitivity (of a test) 139, 158, 509
Septicaemic shock 218
Serological tests 183, 347
Serotonin 454
Serum protein electrophoresis, *see* electrophoresis
uptake tests (thyroid) 343
Sex, effect on reference values (blood) 22, 148, 250, 328
Sex hormone-binding globulin 382, 384, 400
Sex hormones
female **384,** 397, 406, 414
male 381
Short stature 390
Sickle-cell disease 101, 321
Side-room tests **28**
on blood 34, 220, 231, 511
commercial products 32, 220, 422
on faeces 33, 462
on urine 29, 99, 183, 220, 232, 240, 403, 422, 511
Significant difference 22
Sipple syndrome 451
SI units 19, 477, 517
Sjögren's syndrome 129
Sodium 18, 35, **39,** 96, 210, 402, 434
adrenal dysfunction 366, 372, 379
cystic fibrosis 438
diabetic ketoacidosis 230
inappropriate secretion of ADH 395
renal failure 103, 105
Solvent abuse 485
Somatostatin 389
Specific gravity (urine) 29, 47, 90

Specimen collection **9,** 20
  blood 9, 20, 57, 68, 151, 219, 257, 282, 473
  CSF 468
  drug overdosage 479, 481
  exercise 20, 146
  faeces 13, 206
  haemolysis 11, 35, 58, 144
  meals (blood) 147, 213, 275
  posture 10, 20, 101, 132
  technique 10, 282
  time of day (blood) 21, 257, 307, 356, 357
  urine 12, 29, 85, 451, 456
  see also venous stasis
Specimen identification **8,** 481, 490
Specimen preservation **9,** 219, 468
  storage 11, 29, 151, 357, 481
  transport 11, 68
Sphingomyelin 243, 413, 434
Spinal block 469
Split fat 206
'Stagnant gut' syndrome 204, 207, 211
Standard bicarbonate 67, 523
  deviation 16, 23
Standards of performance 22
Starvation 233, 334, *see also* protein-energy malnutrition
Status asthmaticus 81
Steatorrhoea 206, 210
Stein-Leventhal syndrome 384, 415
Stercobilinogen, *see* urobilinogen
Steroid hormones **354,** 373, 381, 384
Stimulative hypoglycaemia 235
Stomach 41, 192
Stones
  gallstones 188
  renal stones 98, **107,** 284, 332, 333, 335
Stress 355
Subfertility **383,** 393, **414,** 416
Sucrose 192, 204
Sugar
  blood/plasma 34, 215, **218,** 511
  urinary 32, 220, 232, **240,** 403, 511
Sulphaemoglobin 324
Sweat 40, 41, 43
  test 438
Synacthen tests 367, 369
Syncytiotrophoblast 407, 408
Syndrome of inappropriate secretion of ADH 395
Système International (SI) 19, 477, **517**
Tangier disease (260)
Tay-Sachs disease 410, 466
Technetium (^{99m}Tc) 346
Teratoma 180, 408, 461
Tertiary hyperparathyroidism 106, **290,** 296, 300
Testicular feminization syndrome 382, 385
Testosterone 381, 415
Tetany 293, 299, 302
Tetracosactrin tests 367, 369, 373
Tetrahydrobiopterin 420
Thalassaemia 306, 321, 323
Therapeutic monitoring 2, 34, 81, **472,** 492
  ranges 473
Thiamin 268
Thiazide diuretics 45, 56, 222, 331, 334, 375
Thirst 42
Threatened abortion 407
Thrombocytopenia 151
Thymol turbidity 181
Thyrocalcitonin, *see* calcitonin
Thyroglobulin 336, 351
Thyroid **336**
  auto-antibodies 347
  enterohepatic circulation 172, 337
  function tests (selective use) **349**
  growth immunoglobulins 348
  hormone-binding capacity 340, 343, 445
  hormones in plasma 340, 349
  hormones in urine 339
Thyroid-stimulating
  hormone 336, 338, 344, **350,** 352, 372, 387, **391,** 445
  immunoglobulins 348
Thyroiditis 129, 348
Thyrotoxicosis, *see* hyperthyroidism
Thyrotrophin-releasing hormone 338, 345, 350, **387,** 392, 394
Thyroxine 336, **340,** 350, 387, 400, 445, 504
  binding to proteins 338, 343
  free (in plasma) 338, 340, **342,** 350, 400, 402, 445
  metabolism 338, 341, 348, 431, 445
  oral contraceptives 341, 342, 400
  pregnancy 340, 342, 402
  screening 350, 351, 445, 504
Thyroxine-binding proteins **338,** 340, 342, 344, 350, 351, 400

Time of day (specimen collection) 21, 257, 307, 356, 357
Tissue enzymes 167, 424
Titratable acidity 93
TmP/GFR 286, 288
Tolbutamide 235
Tophi 332
Total body water 47
  CO_2 74, 230, 287, 524
  iron-binding capacity 308, 401
  parenteral nutrition 61, 273
  protein (plasma) 47, 132, 444
  T4:TBG ratio 344
Toxaemia of pregnancy 404
Toxicology **478**
TPP effect 268
Trace elements 61, **273**
Transaminases, *see* alanine and aspartate aminotransferases
Transcortin 354, 357, 400, 403
Transferrin 102, **117,** 306, 400, 401, 403, 437
Transketolase 268
Transport proteins 112, 114, 117, *see also* individual proteins
Transport systems
  intestine 98, 205, 211
  liver 171
  renal tubule 97
Transudate 188
Trauma
  acute phase reactants 123
  creatine kinase 146, 157, 158
  metabolic response **58**
TRH test **345,** 350, 388, 392
Tricarboxylic acid cycle 299
Tricyclic antidepressants 475, 476
Triglycerides 193, 216, 243
  breath test 206
  plasma 183, 185, **252,** 258, 399, 402, 486
  transport 247
Tri-iodothyronine (T3) 336, 338, **341,** 350, 400, 402
  binding to proteins 338, 343
  free (in plasma) 338, **342,** 350, 400, 402
  resin test 343
  reverse 337, 339, 348
  -thyrotoxicosis 342
Trophoblast 404, 407
  tumours 408
Trypsin 152, 199, 439
Tryptophan 98, 269, 270, 455
  load test 269
'Tubeless' gastric analysis 197
Tuberculosis 185, 188, 395
Tubular function (renal) 89
  proteinuria 101
Tumour markers 148, **457**
Turbidity tests 181
Tyrosinaemia 422, 428
Tyrosine 336, 420

'Ubiquitous enzymes' 164
Ulcerative colitis 46, 56, 116
Ultracentrifuge (lipoproteins) 244, 255, 259
Ultrasound 411
Unconjugated bilirubin 173, 184, 436
Units
  enzyme activity 140, **520**
  SI 19, 477, **517**
Unselective proteinuria 103
Unsplit fat 206
Uraemia 87, **103,** 218, 296, 475
  osteodystrophy 106, 299
  toxins 106
Urate **328**
  plasma 136, **328,** 404, 486
  stones 108, 136, 330
Urea 111, 171, 182
  clearance 87
  plasma 47, 58, 87, 443, 504
  pregnancy 402, 404
  renal failure 104, 294, 296, 299
  side-room tests 34
  water and electrolyte imbalance 45, 47, 88, 210, 231
Urea-stable lactate dehydrogenase 143
Ureteric transplant (into colon) 52
Urgent investigations **13,** 36, 479
Uricosuric drugs 331, 333, 335
Urinary tract obstruction 87
Urine 12, 29, 83, 111
  acidification tests 93
  catecholamines 451
  concentration tests 29, **90,** 395
  cortisol 358, 360, 400, 403
  electrolytes 50, 55, 96, 104
  enzymes 152, 166
  5-hydroxyindoles 456
  oestrogens 406, 408
  osmolality 29, 47, 90, 104, 395

Index

linogen 32, 317
ins 30, 317
nediol 414
oom tests 32, 99, 183, 220, 232
n 29, 174
rphyrinogens and uroporphyrins
11
of data 27

aginal fluid enzymes 166
Vagotomy 197
Vanillyl mandelic acid 451
Variation (in results) 20, 24
Variegate porphyria 315
Vasoactive intestinal peptide 195, 210, 291, 393
Vasopressin test 91, 398
Venous stasis 11, 282, 357, 376
Verner-Morrison syndrome 194, 210, 211
Very low density lipoproteins 194, 216, **245,** 248, 254, 399
Vipoma 291
Viral hepatitis, *see* hepatitis
Virilism 415
Vitamins 60, 209, **262**
  deficiency 165, 209, 262, 438
  overdosage 108, 284
Vitamin A 114, 117, 264
Vitamin B_1, *see* thiamin
Vitamin B_2, *see* riboflavin
Vitamin B_{12} 208, 209, 211, 271, 441, *see also* pernicious anaemia
Vitamin C, *see* ascorbic acid
Vitamin D 170, **264,** 280
  deficiency 267, 298, 438
  overdosage 108, 284
  -resistant rickets 98, 106, 149, 300
Vitamin K 267
  liver disease 179, 186
  malabsorption 179, 186, 212
Voluntary overbreathing 73, 294
Vomiting 45, 50, 52, 56, 73
Von Gierke's disease 238

Waldenstrom's macroglobulinaemia 131, 136
Ward-based equipment **35,** 413
Water **39,** 366
  depletion 40, 43, 45, 50, 210, 231
  deprivation tests **91,** 395
  excess 46, 50, 132, 395
  renal failure 104
  retention 59, 395
Water-soluble vitamins 61, **268**
Well-population screening 5, 508
WHO classification (primary lipoproteinaemias) 253
Wilson's disease 119, 187
Wrong and Davies test (urine acidification) 95

Xanthine
  oxidase 330, 333, 334, 335
  stones 108, 335
Xanthoma 254, 256
Xanthochromia (of CSF) 468
Xanthurenate 269, 270
Xeroderma, xerophthalmia 264
Xylose absorption tests 203, 209, 211, 441
L-Xylulose 240

Zinc 63,274
  protoporphyrin 319
Zollinger-Ellison syndrome 194, 197, **198**